INTRODUCTION TO SEQUENTIAL SMOOTHING AND PREDICTION

INTRODUCTION TO SEQUENTIAL SMOOTHING AND PREDICTION

NORMAN MORRISON, Ph. D.

Chief Scientist, Harris-Intertype Corporation
Printing Equipment Research Center

McGRAW-HILL BOOK COMPANY New York St. Louis
San Francisco London Sydney Toronto Mexico Panama

INTRODUCTION TO SEQUENTIAL SMOOTHING AND PREDICTION

 Library of Congress Catalog Card Number 69-17187

43182

1234567890 MAMM 754321069

TO THE MEMORY OF MY FATHER
BENZION MORRISON
TO WHOM I OWE SO MUCH

PREFACE

Our objective has been to provide, in a systematic tutorial way, an introduction to the field of sequential smoothing and prediction of numerical data. The areas of application lie almost anywhere that digital computers are in operation – data analysis, process control, operations research, missile guidance, satellite work and astronautics being but a few examples.

Our approach has been to assume that the quantities we are attempting to estimate are *deterministic* functions which have been observed or measured in the presence of additive random errors. The estimation of *random* variables is beyond the scope of this book although there is, of course, a very strong connection between the two areas. In our final chapter we touch very briefly on this problem. Nor are we concerned with the theory of numerical filtering in the frequency-domain.†

The book commences with a review of necessary background material. This extends over the first six chapters and covers the essentials of numerical analysis, differential equations, perturbation methods, probability and statistics and some basic ideas from estimation theory. Thereafter, the main contents are presented in three parts, entitled Fixed-Memory Filtering, Expanding-Memory Filtering and Fading-Memory Filtering. This division is a very natural one and provides an extremely systematic way in which to develop the ideas and techniques involved.

We do not claim completeness in any of the areas discussed, nor do we claim rigor. The author is an engineer, trained as one by engineers and by

†See e.g. Kuo, F. and Kaiser, J. F., "System Analysis by Digital Computer," John Wiley & Sons, New York, 1966, Chapter 7.

applied mathematicians. Correctness and ease of understanding is what have been sought throughout and, above all, the needs of the reader entering this field for the first time, have been constantly kept in mind.

There are actually two, essentially separable books, entwined in this volume. As a first reading, one could omit Chapters 8, 10, 11, 12, 14 and 15. The material in Chapters 1 through 7, 9 and 13 forms a consistent introduction and covers the very basic area of *polynomial estimators.* The second reading could then include the slightly more advanced material in the remaining chapters, in which the ideas of polynomial estimation are generalized and extended to cover the area of *state estimation of deterministic processes governed by arbitrary differential equations.*

Concerning the background expected of the reader, we assume that he is familiar with the material presented in a first course on linear algebra, and in a first course on probability or statistics. A review is given, in Chapter 5, of the necessary techniques in the latter areas, but we do not formally review matrix theory in any one place, although from time to time helpful comments are inserted. Heavy use is made of matrix notation throughout the book, since any other approach would be utterly hopeless. However, the depth of proficiency called for in the area of matrix theory is very slight, knowledge being assumed in only the following areas: solvability of sets of linear equations, rank of a set of vectors and of a matrix, eigenvalues of a real symmetric matrix, quadratic forms, positive-definite and positive semidefinite matrices and congruence transformations.

Every attempt has been made to keep the symbol usage as simple as possible. This is perhaps not always accomplished, but it is hoped, at least, that consistency is maintained. A small mental investment on the part of the reader will be called for in this regard.

We have numbered every equation, whether it is to be referenced or not. The symbol (3.2.5) refers to Chapter 3, Section 2, Equation 5. Likewise, the symbol [9.5] refers to the fifth reference cited at the end of the ninth chapter.

The ideas presented in this book should perhaps best be thought of as the primary colors on an artist's palette – the onus is on the user to blend them correctly, in order to achieve the desired end-result. If we succeed in taking the reader along the first few steps of what is already a very old field – Gauss would have found very little new in this volume – then we will have accomplished our purpose.

N. Morrison

ACKNOWLEDGMENTS

The author is extremely grateful to his colleagues at Bell Telephone Laboratories for the assistance which they gave him while this book was in preparation. He was always made to feel welcome by them, and they gave freely both of their time and their ideas. He is also grateful for the opportunity to have taught this material in the out-of-hours course series which were offered at Bell Laboratories. Direct contact with the participants in those courses provided an invaluable amount of criticism which, it is hoped, fell on fertile ground.

It is difficult to name individuals who contributed to this work, for there were so many. An attempt to enumerate them would look like sections from the Bell Laboratories internal telephone directory. However, the author wishes to thank Drs. S. A. Burr, M. P. Epstein, J. C. Hsu, L. L. MacDougall, R. Sherman and P. P. Wang for their efforts in reading the manuscript and making many suggestions.

There are two people to whom the author is particularly grateful for the very substantial contributions which they made, namely Dr. Paul J. Buxbaum and Dr. Alfons J. Claus. They introduced the author to many new ideas and to original techniques developed by themselves, and then generously provided copies of their memoranda and notes to facilitate his plunder further. It is fair to say that without the stimulus and assistance provided by these two friends this book would never have come about.

Without typists and clerical assistance nothing would have been accomplished at all, and for the sort of typing assistance which he received, almost as if the work were their own complete responsibility, the author is deeply indebted to Mrs. N. G. Campbell and Mrs. B. W. Oravits.

N. Morrison

GREEK ALPHABET

Greek Letter	Greek Name	Greek Letter	Greek Name
α	alpha	Ξ	capital xi
β	beta	ξ	xi
Γ	capital gamma	Π	capital pi
γ	gamma	π	pi
Δ	capital delta	P	capital rho
δ	delta	ρ	rho
ϵ	epsilon	σ	sigma
ζ	zeta	τ	tau
η	eta	Φ	capital phi
Θ	capital theta	φ	phi
θ	theta	χ	chi
Λ	capital lambda	Ψ	capital psi
λ	lambda	ψ	psi
μ	mu	Ω	capital omega
ν	nu	ω	omega

CONTENTS

PART 1

BACKGROUND MATERIAL

This book is divided into four parts. The first of these extends over Chapters 1 through 6, and is devoted to a presentation of background material. It constitutes the basis on which all of our subsequent discussion on smoothing and prediction will depend.

Great care has been taken, wherever possible, to ensure that redundant material be left out, and only what was felt to be absolutely essential has been included. As a result, almost every equation that we develop in Part 1 will be applied in the later parts of the book.

1 INTRODUCTION

Throughout this book we are primarily concerned with the estimation of *deterministic* functions of the real variable t, which is essentially the *time* variable. Of course t could be regarded in any other way and the resulting estimation procedures would still be valid in that context. However, since time is what we have in mind, we deliberately refrain from assuming that the entire t-axis is available to us at every stage. Rather, we shall assume that there exists a point called *the present*, which moves along the t-axis in a positively increasing direction, and only points to the left of the present, namely *the past*, are accessible to us. Points to the right of the present are *the future*, and become available only at the real-time rate at which the present moves along the t-axis and uncovers them.

The functions, whose estimation we shall concern ourselves with, are supposed to be generated by what we term a *process*. Examples of processes are chemical reactions, bodies moving under the action of a force – *in fact any system which is governed by a differential equation, whether linear or nonlinear*. By assumption, the functions of interest will be continuous for all t, but we do not assume that their form is known. Instead we will suppose that we have some prior knowledge about the process, which enables us to construct a *model* of it, defined by a *differential equation*.

Under ideal conditions the model will be identical to the true process, but in practice it is usually an approximation to it. The approximations may be either deliberate, in order to reduce the complexity of the equations, or else forced on us, by virtue of our inability to understand the true process completely. Estimation errors will arise because of discrepancies between the model and the process, and we will consider these errors in greater detail in our discussion.

Three successively more general types of model equations will be considered. Let $X(t)$ be a vector of functions,† i.e.

$$X(t) \equiv \begin{pmatrix} x_0(t) \\ x_1(t) \\ \vdots \\ x_m(t) \end{pmatrix} \tag{1.1}$$

and define its time derivative as

$$\frac{d}{dt}X(t) \equiv \begin{pmatrix} \frac{d}{dt}x_0(t) \\ \frac{d}{dt}x_1(t) \\ \vdots \\ \frac{d}{dt}x_m(t) \end{pmatrix} \tag{1.2}$$

Then we write the three classes of differential equations as

$$\frac{d}{dt}X(t) = AX(t) \tag{1.3}$$

$$\frac{d}{dt}X(t) = A(t)X(t) \tag{1.4}$$

$$\frac{d}{dt}X(t) = F[X(t),t] \tag{1.5}$$

The vector $X(t)$ is called the *state-vector* of the model and its components are known as the *state-variables*. In (1.3) $\dot{X}(t)$‡ is seen to be related to $X(t)$

†Our matrix and vector subscripts start from 0 and not from 1. This slight deviation from customary usage was motivated by the fact that derivatives start from the zeroth, degrees of polynomials start from zero and factorials start naturally from zero. These items appear frequently in the material to come, and forced on us the decision to start our matrix and vector subscripts from zero as well.

‡We shall use the notation $\dot{X}(t)$ and $(d/dt)\,X(t)$ interchangeably.

by a linear transformation, defined by the constant matrix A. Such a system is accordingly called a *constant-coefficient linear differential equation.*

In (1.4) we have the general form of the *time-varying linear differential equation* in which $\dot{X}(t)$ is a linear transformation on $X(t)$, with the transformation matrix $A(t)$ having time-varying components.

In (1.5) $\dot{X}(t)$ is shown related to $X(t)$ by a vector of functions of the form

$$\begin{aligned} \dot{x}_0(t) &= f_0[x_0(t), x_1(t), \ldots, x_m(t), t] \\ \dot{x}_1(t) &= f_1[x_0(t), x_1(t), \ldots, x_m(t), t] \\ &\vdots \\ \dot{x}_m(t) &= f_m[x_0(t), x_1(t), \ldots, x_m(t), t] \end{aligned} \tag{1.6}$$

where each of the functions $f_0, f_1, \ldots, f_m$ is a possibly *nonlinear* function of the vector $X(t)$ and possibly of t as well. Equation (1.5) is thus the general form of a *nonlinear differential equation.*

If we assume that initial conditions are specified by a vector of deterministic numbers, then it is clear that in all of the above three cases $X(t)$ will be a deterministic vector-function of time, and it is essentially with such models that we are concerned. However, in the final chapter we will consider very briefly the situation where $X(t)$ is governed by a differential equation of the form

$$\dot{X}(t) = A(t)X(t) + U(t) \tag{1.7}$$

in which $U(t)$ is a vector of *white random variables.* This clearly makes $X(t)$ a vector of random variables and so, strictly speaking, the estimation of such a vector is beyond our scope. However the brief treatment given in Chapter 15 is included because it throws further light on the algorithms developed in the earlier chapters.

Given that a process is modelled by one of the three differential equations in (1.3), (1.4) or (1.5), there still exists an infinity of possible trajectories along which the state-vector $X(t)$ might be evolving. Without a set of initial conditions, and we assume that this is not explicitly available, *it is not known which trajectory is the one currently being generated.* We accordingly require some further information in order to enable us to narrow down our choice and to select a trajectory as being the one along which we believe the process to be evolving.

This further information is provided us by one or more measuring instruments, assumed to be at our disposal. These enable us to observe the process, and they provide us with *vectors of observations* which, under ideal conditions, are linear or possibly nonlinear transformations on some or all of the state-variables. In practice however, the measuring instruments introduce *errors* which are assumed to be *additive.*

Thus, letting $X(t_n)$ be the vector of model state-variables at $t = t_n$ and letting Y_n be the vector of observations obtained at that time,† *we consider three successively more general observation schemes*:

$$Y_n = MX(t_n) + N_n \tag{1.8}$$

$$Y_n = M_n X(t_n) + N_n \tag{1.9}$$

$$Y_n = G[X(t_n), t_n] + N_n \tag{1.10}$$

In (1.8) we show Y_n as being equal to a constant linear transformation on $X(t_n)$, as defined by the matrix M, plus a vector of errors symbolized by N_n. In (1.9) Y_n is related to $X(t_n)$ by the time-varying linear transformation M_n, plus an error vector N_n. Finally in (1.10) the vector Y_n is related to $X(t_n)$ by a system of nonlinear equations in which both the elements of $X(t_n)$ and possibly t_n as well constitute the independent variables.

An example would be the following. Let the state-variables of the model be the position and velocity, in Cartesian coordinates, of a body in straight line motion. Then its differential equation will be

$$\frac{d}{dt}\begin{pmatrix} x_0(t) \\ x_1(t) \\ x_2(t) \\ \dot{x}_0(t) \\ \dot{x}_1(t) \\ \dot{x}_2(t) \end{pmatrix} = \begin{pmatrix} 0 & 0 & 0 & 1 & 0 & 0 \\ 0 & 0 & 0 & 0 & 1 & 0 \\ 0 & 0 & 0 & 0 & 0 & 1 \\ 0 & 0 & 0 & 0 & 0 & 0 \\ 0 & 0 & 0 & 0 & 0 & 0 \\ 0 & 0 & 0 & 0 & 0 & 0 \end{pmatrix} \begin{pmatrix} x_0(t) \\ x_1(t) \\ x_2(t) \\ \dot{x}_0(t) \\ \dot{x}_1(t) \\ \dot{x}_2(t) \end{pmatrix} \tag{1.11}$$

which is of the form of (1.3). Also, let Y_n be the vector of observations made directly on the position coordinates of the body. Then the observation relation would be

†The vectors Y and X need not have the same numbers of elements.

$$\begin{pmatrix} y_0 \\ y_1 \\ y_2 \end{pmatrix}_n = \begin{pmatrix} 1 & 0 & 0 & 0 & 0 & 0 \\ 0 & 1 & 0 & 0 & 0 & 0 \\ 0 & 0 & 1 & 0 & 0 & 0 \end{pmatrix} \begin{pmatrix} x_0(t_n) \\ x_1(t_n) \\ x_2(t_n) \\ \dot{x}_0(t_n) \\ \dot{x}_1(t_n) \\ \dot{x}_2(t_n) \end{pmatrix} + \begin{pmatrix} \nu_0 \\ \nu_1 \\ \nu_2 \end{pmatrix}_n \tag{1.12}$$

which is of the form of (1.8).

Alternatively, let Y_n be the observations obtained by a radar, located at the Cartesian origin, of the *range*, *azimuth* and *elevation* of the body. Then the observation relation would be

$$(y_0)_n = \left(x_0^2 + x_1^2 + x_2^2\right)^{1/2}\Big|_{t=t_n} + (\nu_0)_n \tag{1.13}$$

$$(y_1)_n = \tan^{-1}\left(\frac{x_1}{x_0}\right)\Bigg|_{t=t_n} + (\nu_1)_n \tag{1.14}$$

$$(y_2)_n = \tan^{-1}\left[\frac{x_2}{\left(x_0^2 + x_1^2\right)^{1/2}}\right]\Bigg|_{t=t_n} + (\nu_2)_n \tag{1.15}$$

which is of the form of (1.10). ♦♦†

The primary purpose of this book is to show how the theory of differential equations (introduced through the model) can be combined with techniques from estimation theory (introduced through the observations), so that one trajectory of the possible infinity, referred to previously, can be selected. Criteria will be set up so that the selection process is best in some clearly defined sense. It will be the basic task, in each of the chapters from Chapter 7 on, to derive the algorithm which, based on a model and a set of observations, will select the trajectory which best satisfies the chosen criterion.

In addition to setting up the algorithms, we will also study the errors in the estimate, i.e., in the selected trajectory. The errors which we consider will be seen to arise from two sources – errors in the model and errors in the observations. The former give rise, in the estimate, to what are termed the *systematic* or *bias* errors, and the latter to what we call the *random* errors.

†The symbol ♦♦ will be used to signify the end of an example or the completion of a proof (theorems, lemmas, etc.).

The observations are assumed to be streaming in to a computer, whose function it is to accept and store them and to operate on them in order to arrive at a vector of numbers which constitutes an estimate of the process state-vector. These numerical values then stream out of the computer and serve as the most up-to-date estimates of the trajectory currently under way. As successive observations arrive, the algorithm will possibly modify the selected trajectory.

We will examine in detail three classes of smoothing procedures. The first will be the *Fixed-Memory Filters* in which the trajectory is always chosen on the basis of observations taken over a fixed time-interval into the past. The *Expanding-Memory Filters*, on the other hand, base their current estimates on all observations made up to the present, and as time moves forward, this is naturally an expanding set. Finally the *Fading-Memory Filters* perform their trajectory selection on the basis of all observations made up to the present, but a stress-factor is applied so that the older or staler an observation becomes, the less influence it exerts on the current estimate.

In addition to setting up algorithms and studying their error properties, we will also concern ourselves with some of the important practical problems, such as computer memory-space requirements, the amount of computation needed to execute the algorithms, numerical difficulties that can be expected, etc. It will emerge that each of the three classes of filters to be discussed has its own set of properties in respect to errors, memory-space and computational details. Depending upon the specific situation the user must then select the best compromise. There is no perfect filtering algorithm for use in all cases.

The output of a filter is a vector which we shall always show with a star, e.g. X^*, and the star should be immediately associated with the word "estimate." It is also necessary that we append *two subscripts* to X^*, the first to designate the time-instant at which this vector estimate applies, and the second to show the time-instant of the most recent observation on which the estimate is based. Thus, suppose X^* is an estimate of $X(t_p)$ and let it be based on observations up to and including those made at t_q. Then we show it as $X^*_{p,q}$. We will, on occasions, refer to t_p as the *validity instant*, meaning of course the instant for which this estimate of the true state-vector is valid. The instant t_q should be thought of as the "most recent observation" instant.

We distinguish three cases. If $t_p > t_q$ then $X^*_{p,q}$ is called a *prediction* and the algorithm producing it is called a *predictor*. If $t_p = t_q$, then $X^*_{p,q}$ is said to be an *updated estimate*, and finally if $t_p < t_q$, then we say that we are *retrodicting* and $X^*_{p,q}$ is called a *retrodiction*.

Many of the algorithms to be developed put out predictions of the state-vector which are valid for the time at which the next observation vector is to be made. Such an estimate is written $X^*_{n+1,n}$ and is called a *1-step prediction.* We shall frequently use the term *1-step predictor*, by which we mean an algorithm which is putting out, on successive cycles, the sequence of vectors

$$\ldots, X^*_{n-1,n-2}, X^*_{n,n-1}, X^*_{n+1,n}, \ldots$$

Other words which we shall use are *filter* and *estimator*, in addition to or in place of the word algorithm. The word *filter* is intended, as in electrical engineering, to convey the ideal of a *wanted* component (the true process) being allowed to pass, with an *unwanted* component (the observation errors) being retarded and diminished. An *estimator* accepts information in the presence of errors and estimates the values of certain quantities or parameters. The word *smoother*, in use in the literature, essentially coincides with the word filter.†

Our algorithms will accomplish various combinations of the above ideas. Thus we might develop an algorithm which puts out smoothed 1-step predictions of position and velocity together with estimates of a parameter in the process, based on observations of position. It will be an estimator, a predictor and a smoother.

In addition to obtaining $X^*_{p,q}$, it will frequently be the case that our algorithm also puts out estimates of the statistics of $X^*_{p,q}$. These will be in the form of its *covariance matrix*, a topic which we shall discuss in greater detail in Chapter 5. The covariance matrix of $X^*_{p,q}$ will be written $S^*_{p,q}$.

Along with the two subscripts discussed above, it will also frequently be necessary to show the matrix subscripts. Thus

$$X^*_{n,n} \equiv \begin{pmatrix} x^*_0 \\ x^*_1 \end{pmatrix}_{n,n} \tag{1.16}$$

shows that the updated estimate $X^*_{n,n}$ has two components $(x^*_0)_{n,n}$ and $(x^*_1)_{n,n}$. Likewise its covariance matrix would be

$$S^*_{n,n} \equiv \begin{pmatrix} s^*_{0,0} & s^*_{0,1} \\ s^*_{1,0} & s^*_{1,1} \end{pmatrix}_{n,n} \tag{1.17}$$

†One convention, sometimes adopted, is as follows. Let $X^*_{p,q}$ be the output of an algorithm. Then if $p < q$, we are *smoothing*, if $p = q$, we are *filtering* and if $p > q$ we are *predicting*. We shall however, use the looser meanings discussed above.

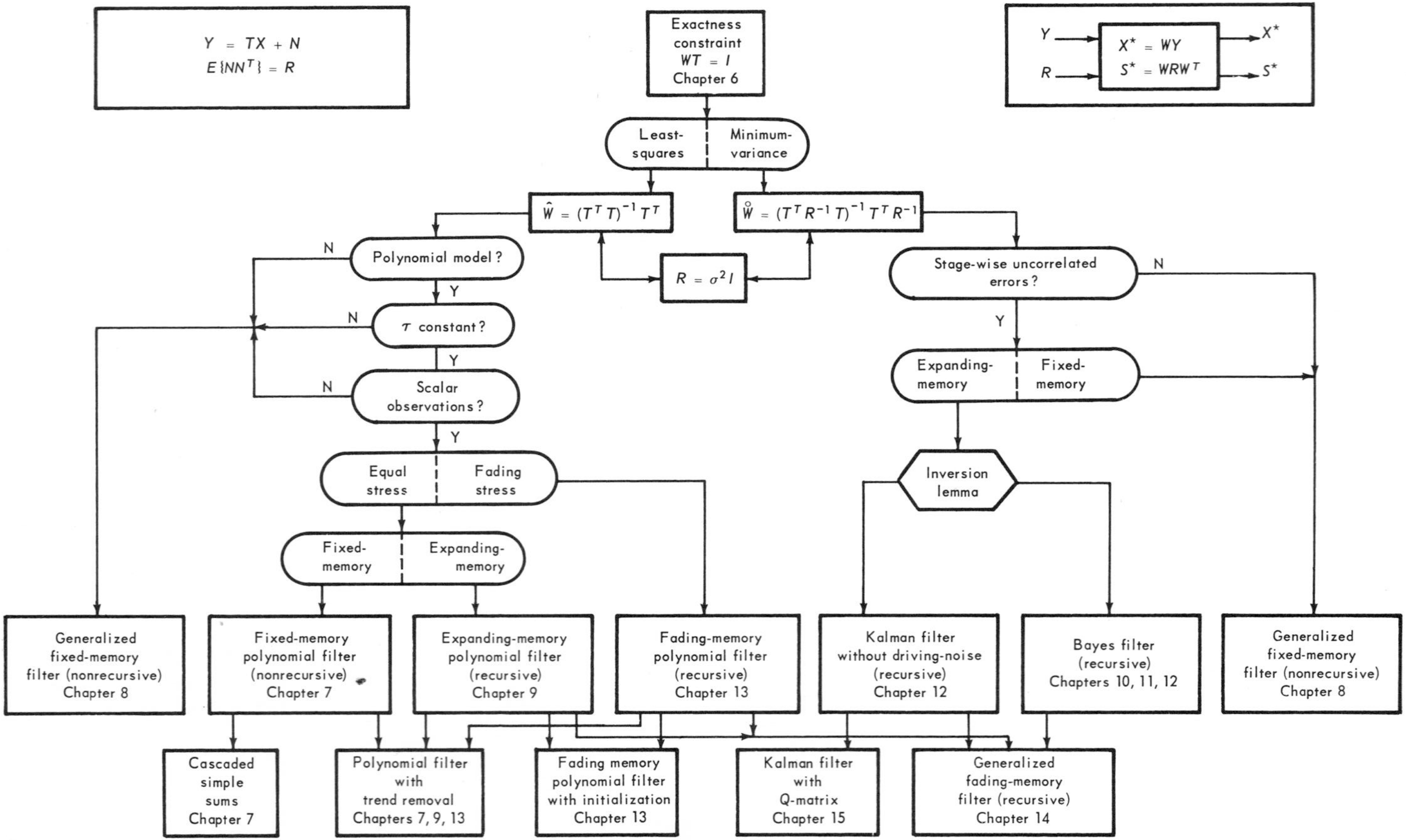

Fig. 1.1 *Genealogy of the principal filtering schemes.*

showing the four components $(s^*_{0,0})_{n,n}$, $(s^*_{0,1})_{n,n}$, $(s^*_{1,0})_{n,n}$ and $(s^*_{1,1})_{n,n}$. Observe that the matrix subscripts are *inside* the matrix brackets and the time subscripts are *outside.*

The letter n will be reserved exclusively for use as the general form of the *time-ordering* subscript. Thus the sequence of observations

$$y_1, y_2, \ldots, y_{n-1}, y_n, \ldots$$

has as its general term y_n whose predecessor is y_{n-1} and whose successor will be y_{n+1}. Likewise the general term of a one-step predictor will be $X^*_{n+1,n}$ whose predecessor was $X^*_{n,n-1}$. The general term of an updated estimator is $X^*_{n,n}$ with successor $X^*_{n+1,n+1}$.

On occasions we shall be forced to modify the double subscript notation very slightly and to write $X_{t_n+\zeta,\, t_n}$, by which we of course mean a ζ-second prediction.

The above notation will be used repeatedly and will be frequently redefined in the earlier chapters as we add to it. The reader should have no trouble becoming familiar with it and, like any system of notation, it is not perfect, having both advantages and drawbacks. However, we have striven for consistency, and with a little patience the symbolism should soon become easily understood.

In Figure 1.1 we display, in flow-chart form, the essential contents of the book, as contained in Chapters 6 through 15. As the development progresses, the reader should refer back to the chart, where he will be able to identify the titles and symbols appearing in each of the blocks. The chart is useful in defining the assumptions which lead to each of the algorithms and in establishing the interrelationships between them.

We now embark on a review of some background material needed for the development of the smoothing and prediction techniques to follow.

2

DISCRETE FUNCTION ANALYSIS

2.1 INTRODUCTION

In the course of developing various numerical filters, we will frequently be dealing with *sampled* or *discretized* functions of the form

$$f_n \equiv f(n\tau) \equiv f(t)\Big|_{t=n\tau} \tag{2.1.1}$$

While $f(t)$ may be amenable to analysis using the differential and integral calculus, the same is not true for f_n. However a corresponding calculus does exist for such discretized functions, and it is the purpose of this chapter to outline as much of what we accordingly call *discrete function analysis* as will be required for subsequent use.

Note from (2.1.1) that we are considering equispaced sampling, the separation being τ seconds. This is true for the entire present chapter.

2.2 THE SHIFTING OPERATORS

Assume that $f(t)$ is defined for all t. Sampling it at $t = n\tau$, where $n = 0, 1, 2, \ldots$, gives the sequence

$$f_0 \equiv f(0),\ f_1 \equiv f(\tau),\ \ldots,\ f_n \equiv f(n\tau)$$

We define the *backward-shifting operator* q by the equation

$$q^k f_n \equiv f_{n-k} \tag{2.2.1}$$

where k is any integer.

We sometimes prefer to use the forward-shifting operator, E, defined by

$$E^k f_n \equiv f_{n+k} \tag{2.2.2}$$

Then

$$\begin{aligned} q f_n &= f_{n-1} \\ E f_n &= f_{n+1} \end{aligned} \tag{2.2.3}$$

If we define the product $q^m q^k$ by

$$q^m q^k f_n = q^m \left(q^k f_n\right) \tag{2.2.4}$$

then

$$q^m q^k = q^{m+k} = q^k q^m \tag{2.2.5}$$

Moreover

$$q^m E^k = q^{m-k} = E^k q^m \tag{2.2.6}$$

Thus the q and E operators commute with themselves and with each other. For $m = 0$ the operator q^m is defined by (2.2.1) to give

$$q^0 f_n = f_n \tag{2.2.7}$$

The shifting operators are *linear* since

$$q^m (c_0 f_n + c_1 g_n) = c_0 q^m f_n + c_1 q^m g_n \tag{2.2.8}$$

Also since

$$\left(\alpha_0 q^m + \alpha_1 q^k\right) f_n = \alpha_0 q^m f_n + \alpha_1 q^k f_n \tag{2.2.9}$$

we see that polynomials in q (or E) can be defined. Thus,

$$\left(\sum_{j=0}^{i} \alpha_j q^j\right) f_n \equiv \sum_{j=0}^{i} \alpha_j f_{n-j} \tag{2.2.10}$$

By (2.2.3)

$$qEf_n = f_n \tag{2.2.11}$$

and so q and E are each other's inverse. The inverse of q^m is q^{-m} or E^m. Every shifting operator thus has an inverse.

We now ask whether the operator $\sum_{j=0}^{i} \alpha_j q^j$ defined in (2.2.10) has an inverse? Consider first the simpler case

$$(1 - q) g_n = f_n \tag{2.2.12}$$

This is simply the operational form of the *linear recursion relation*

$$g_n = g_{n-1} + f_n \tag{2.2.13}$$

in which g_n is computed as a linear combination of its predecessor and the input f_n. By iterating (2.2.13) starting with $n = 0$, we obtain

$$g_n = g_{-1} + \sum_{k=0}^{n} f_{n-k} = g_{-1} + \left(\sum_{k=0}^{n} q^k\right) f_n \tag{2.2.14}$$

We shall assume, henceforth, that any of the systems to which we shall apply these operators is initially in a *completely relaxed state,* i.e., both g_n and f_n are *identically zero* prior to some time. Without loss of generality, we let that time be $t = 0$. Then g_{-1} in (2.2.14) is zero. Moreover, we can also then write

$$\left(\sum_{k=0}^{n} q^k\right) f_n = \left(\sum_{k=0}^{\infty} q^k\right) f_n \tag{2.2.15}$$

and so (2.2.14) gives, under this assumption

$$g_n = \left(\sum_{k=0}^{\infty} q^k\right) f_n \tag{2.2.16}$$

If we now compare (2.2.16) with (2.2.12) we see that the inverse of $1 - q$ is $\sum_{k=0}^{\infty} q^k$. But $1/(1 - q)$ if treated algebraically gives (by direct division)

$$(1 - q)^{-1} = 1 + q + q^2 + \cdots = \sum_{k=0}^{\infty} q^k \tag{2.2.17}$$

We see then that we can find the inverse of $1 - q$ by *algebraic reciprocation,* i.e.

$$\text{Inv}(1 - q) = \frac{1}{1 - q} = (1 - q)^{-1} \tag{2.2.18}$$

In a like manner, we can show that for any number β

$$\text{Inv}(q - \beta) = \frac{1}{q - \beta} = (q - \beta)^{-1} \tag{2.2.19}$$

Moreover, if

$$(q - \beta_1)(q - \beta_2)\, g_n = f_n \tag{2.2.20}$$

then, by transposing the operators one at a time, it follows that

$$g_n = \frac{1}{(q - \beta_2)} \frac{1}{(q - \beta_1)} f_n \tag{2.2.21}$$

Thus the inverse of $(q - \beta_1)(q - \beta_2)$ is its algebraic reciprocal.

Now, any polynomial in q (or E) can be factored, i.e., for any set of α's there exist β's so that

$$\sum_{j=0}^{i} \alpha_j q^j = \alpha_i \prod_{j=1}^{i} (q - \beta_j) \tag{2.2.22}$$

the order of the factors being immaterial, since it is easily seen that they commute with each other. Then

$$\text{Inv}\left(\sum_{j=0}^{i} \alpha_j q^j\right) = \text{Inv}\left[\alpha_i \prod_{j=1}^{i} (q - \beta_j)\right] = \frac{1}{\alpha_i} \prod_{j=1}^{i} \left(\frac{1}{q - \beta_j}\right) \tag{2.2.23}$$

and so

$$\operatorname{Inv}\left(\sum_{j=0}^{i} \alpha_j q^j\right) = \frac{1}{\sum_{j=0}^{i} \alpha_j q^j} \tag{2.2.24}$$

showing that $\sum_{j=0}^{i} \alpha_j q^j$ does in fact have an inverse – its algebraic reciprocal. This means that any *rational function* of q such as

$$R(q) \equiv \frac{\sum_{j=0}^{k} \gamma_j q^j}{\sum_{j=0}^{i} \alpha_j q^j} \tag{2.2.25}$$

is a well defined operator. Moreover $R(q)$ can be combined with other such operators using $+$, $-$, $\times$ and $\div$ in their ordinary algebraic sense to form further rational operators involving q or F.

Any linear recursion relation is now seen to have an *equivalent operational form* involving the shifting operators. Thus if g_n is related to f_n by the linear recursion

$$g_n = -\sum_{j=1}^{i} \alpha_j g_{n-j} + \sum_{j=0}^{k} \gamma_j f_{n-j} \tag{2.2.26}$$

then using the q - operator we can write it as:

$$\left(\sum_{j=0}^{i} \alpha_j q^j\right) g_n = \left(\sum_{j=0}^{k} \gamma_j q^j\right) f_n \tag{2.2.27}$$

(where $\alpha_0 = 1$) and so

$$g_n = \frac{\sum_{j=0}^{k} \gamma_j q^j}{\sum_{j=0}^{i} \alpha_j q^j} f_n \tag{2.2.28}$$

We are thus able to express any linear recursion relation of the form (2.2.26) by the equivalent statement

$$g_n = R(q) f_n \tag{2.2.29}$$

where $R(q)$ is a rational function in q. We shall return at the end of this chapter to the solution of (2.2.29) for g_n given f_n.

A word of caution is in order. *All* of the above development was based on the assumption that the coefficients of the operators were *constants*, i.e., *independent of* n. When this is not the case, careful attention must be given to the rules formulated for q and E. (See Ex. 2.3.) Such situations will arise in later chapters and we will accordingly handle them with the care they deserve.

2.3 THE DIFFERENCE OPERATORS

Define the backward-difference operator ∇ by†

$$\nabla f_n \equiv f_n - f_{n-1} \tag{2.3.1}$$

and the forward-difference operator Δ by†

$$\Delta f_n \equiv f_{n+1} - f_n \tag{2.3.2}$$

It then follows that

$$\begin{aligned} \nabla &= 1 - q \\ \Delta &= E - 1 = q^{-1}(1 - q) \end{aligned} \tag{2.3.3}$$

Hence their inverses are, respectively

$$\nabla^{-1} = \frac{1}{1 - q} = 1 + q + q^2 + \cdots \tag{2.3.4}$$

and

$$\Delta^{-1} = q(1 + q + q^2 + \cdots) \tag{2.3.5}$$

†For ∇ read "Nabla" and for Δ read "Delta." Sometimes we shall subscript ∇ or Δ to show which variable they are operating on e.g., $\nabla_x (x - i)^j \equiv (x - i)^j - (x - 1 - i)^j$

By (2.3.3)

$$\nabla^m = (1 - q)^m \tag{2.3.6}$$

for any integer m. Thus for r, m, and k integers

$$\begin{aligned} \nabla^m \nabla^k &= \nabla^{m+k} = \nabla^k \nabla^m \\ \nabla^m \nabla^{-m} &= 1 \\ \nabla^m (\nabla^k \nabla^r) &= \nabla^{m+k+r} = (\nabla^m \nabla^k) \nabla^r \end{aligned} \tag{2.3.7}$$

The integral powers of ∇ thus form a set of operators which commute, have inverses and associate. Moreover they are all linear operators. Writing out (2.3.6) for $m = 2$ gives

$$\nabla^2 y_n = (1 - 2q + q^2) y_n = y_n - 2y_{n-1} + y_{n-2} \tag{2.3.8}$$

and so on for ∇^3 etc.

Since, by (2.3.3), Δ^m (for m an integer) is defined as a rational function in q, it too forms a set of linear operators which commute, have inverses and associate. Moreover any rational combinations of ∇^m, Δ^r, and q^k will thus also commute and associate with one another, and will constitute linear operators.

Given the values of f_n for various n, we can set up a table of *backward differences* as follows:

n	f_n	∇f_n	$\nabla^2 f_n$	$\nabla^3 f_n$	$\nabla^4 f_n$
-2	-4				
-1	1	5			
0	0	-1	-6		
1	-1	-1	0	6	
2	4	5	6	6	0
3	21	17	12	6	0

In this instance f_n is the cubic $n^3 - 2n$ although any function f_n can be treated in this way.

In a like manner, for the above f_n we can set up the table of *forward differences* of successively higher powers:

n	f_n	Δf_n	$\Delta^2 f_n$	$\Delta^3 f_n$	$\Delta^4 f_n$
-2	-4	5	-6	6	0
-1	1	-1	0	6	0
0	0	-1	6	6	
1	-1	5	12		
2	4	17			
3	21				

The operators ∇ and Δ come closest to the idea of differentiation. Using a Taylor series expansion, we have

$$f(t - \tau) = f(t) - \tau\dot{f}(t) + \frac{\tau^2}{2!}\ddot{f}(t - \zeta\tau) \qquad 0 < \zeta < 1 \tag{2.3.9}$$

and so we see that

$$f_n - f_{n-1} = \tau\dot{f}_n - \frac{\tau^2}{2!}\ddot{f}_{n-\zeta} \tag{2.3.10}$$

Thus to within an error term $(\tau^2/2)\ddot{f}_{n-\zeta}$

$$\nabla f_n = \tau\dot{f}_n \tag{2.3.11}$$

Likewise

$$f_{n+1} - f_n = \tau\dot{f}_n + \frac{\tau^2}{2!}\ddot{f}_{n+\xi} \qquad 0 < \xi < 1 \tag{2.3.12}$$

giving

$$\Delta f_n = \tau\dot{f}_n \tag{2.3.13}$$

with an error $(\tau^2/2)\ddot{f}_{n+\xi}$.

Neglecting the error terms, (2.3.10) and (2.3.12) give us the *square-law* (or Euler's law) numerical integration rules

$$f_n = f_{n-1} + \tau\dot{f}_n \tag{2.3.14}$$

$$f_n = f_{n-1} + \tau\dot{f}_{n-1} \tag{2.3.15}$$

and their average gives the *trapezoidal rule*

$$f_n = f_{n-1} + \tau\left(\frac{\dot{f}_n + \dot{f}_{n-1}}{2}\right) \tag{2.3.16}$$

whose error term

$$\epsilon = \frac{\tau^2}{2}\left(\ddot{f}_{n-1+\xi} - \ddot{f}_{n-\zeta}\right) \tag{2.3.17}$$

is easily shown to be of the form

$$\epsilon = \frac{\tau^3}{2}(\zeta + \xi - 1)\dddot{f}_{n-\mu} \qquad 0 < \mu < 1 \qquad \text{(see Ex. 2.5)} \tag{2.3.18}$$

By averaging (2.3.11) and (2.3.13) we obtain the differentiation rule

$$\dot{f}_n = \frac{\Delta f_n + \nabla f_n}{2\tau} \tag{2.3.19}$$

whose error is a term in τ^2. It is readily verified that the above is equivalent to

$$\dot{f}_n = \frac{f_{n+1} - f_{n-1}}{2\tau} \tag{2.3.20}$$

2.4 THE FACTORIAL FUNCTIONS

In the preceding section we discussed the difference operators ∇ and Δ which correspond, in some sense, to differentiation. There is no counterpart to q or E in the analysis of functions of a continuous variable. In this section we consider the concept in discrete function analysis which is analogous to powers such as x^2, x^3, etc.

Define the *backward factorial function* of order k by

$$x^{(k)} \equiv x(x-1)\ldots(x-k+1) \tag{2.4.1}$$

where k is a positive integer. Thus $x^{(k)}$ is the product of k terms starting from x and counting down by unity.

Likewise, the *forward factorial function* of order k is defined as

$$x^{[k]} \equiv x(x+1)\ldots(x+k-1) \tag{2.4.2}$$

again the product of k terms, but counted *up* by unity.

For $k = 0$, we define

$$\begin{aligned} x^{(0)} &\equiv 1 \\ x^{[0]} &\equiv 1 \end{aligned} \tag{2.4.3}$$

In what follows we confine our remarks chiefly to $x^{(k)}$, the backward factorial function. However, the reader should bear in mind that to each statement involving $x^{(k)}$, there is an analogous statement involving the forward factorial function $x^{[k]}$.

When x is a positive integer, but less than the integer k, then

$$x^{(k)} = x(x-\ )\ldots 1\,.\,0\,.\,(-1)\ldots(x-k+1) \tag{2.4.4}$$

and so for this case, $x^{(k)}$ is zero. For the case where x is an integer equal to k, we obtain

$$k^{(k)} = k(k-1)\ldots 3\cdot 2\cdot 1 \tag{2.4.5}$$

which is known as k - factorial, and written $k!$

When x is a variable, $x^{(k)}$ is a polynomial in x of degree k. We expand $x^{(k)}$ for $k = 0, 1, 2, \ldots$ to give

$$\begin{aligned} x^{(0)} &= 1 \\ x^{(1)} &= x \\ x^{(2)} &= x^2 - x \\ x^{(3)} &= x^3 - 3x^2 + 2x, \text{ etc.} \end{aligned} \tag{2.4.6}$$

which can be written as the matrix equation

$$\begin{pmatrix} x^{(0)} \\ x^{(1)} \\ x^{(2)} \\ x^{(3)} \\ \vdots \end{pmatrix} = \begin{pmatrix} 1 & 0 & 0 & 0 & 0 & \\ 0 & 1 & 0 & 0 & 0 & \\ 0 & -1 & 1 & 0 & 0 & \\ 0 & 2 & -3 & 1 & 0 & \ldots \\ \vdots & \vdots & \vdots & \vdots & \vdots & \end{pmatrix} \begin{pmatrix} 1 \\ x \\ x^2 \\ x^3 \\ \vdots \end{pmatrix} \tag{2.4.7}$$

The matrix of this equation is called the *Stirling matrix of the first kind,* named for its discoverer. We symbolize it as S, and it is seen to give $x^{(i)}$ as a polynomial in x, i.e.,

$$x^{(i)} = \sum_{j=0}^{i} [S]_{ij} x^j \qquad \text{(see Note)} \qquad (2.4.8)$$

The matrix S can be obtained by the recursion (see e.g. [2.1])

$$[S]_{i,j} = [S]_{i-1,j-1} - (i-1)[S]_{i-1,j} \qquad i,j \geq 1 \qquad (2.4.9)$$

To start the recursion we use

$$\begin{aligned} [S]_{0,0} &= 1 \\ [S]_{0,j} &= 0 = [S]_{i,0} \qquad i,j \geq 1 \end{aligned} \qquad (2.4.10)$$

Using this process we generate the Stirling first-kind matrix up to $i,j = 10$, displayed in Table 2.1.

A series of the form $\sum_{i=0}^{k} \alpha_i x^{(i)}$ is called a *Newton* series, in distinction to the *power* series $\sum_{i=0}^{k} \beta_i x^i$. Any Newton series can now be written as a power series as follows:

$$\sum_{i=0}^{k} \alpha_i x^{(i)} = (\alpha_0, \alpha_1, \ldots, \alpha_k) \begin{pmatrix} x^{(0)} \\ x^{(1)} \\ \vdots \\ x^{(k)} \end{pmatrix}$$

$$= (\alpha_0, \alpha_1, \ldots, \alpha_k) S \begin{pmatrix} 1 \\ x \\ \vdots \\ x^k \end{pmatrix}$$

$$= \sum_{i=0}^{k} \beta_i x^i \qquad (2.4.11)$$

Note: The symbol $[S]_{ij}$ means the i, j^{th} element of the matrix S.

Table 2.1 S (The Stirling Matrix of the First Kind)

$i \backslash j$	0	1	2	3	4	5	6	7	8	9	10
0	1	0	0	0	0	0	0	0	0	0	0
1	0	1	0	0	0	0	0	0	0	0	0
2	0	−1	1	0	0	0	0	0	0	0	0
3	0	2	−3	1	0	0	0	0	0	0	0
4	0	−6	11	−6	1	0	0	0	0	0	0
5	0	24	−50	35	−10	1	0	0	0	0	0
6	0	−120	274	−225	85	−15	1	0	0	0	0
7	0	720	−1764	1624	−735	175	−21	1	0	0	0
8	0	−5040	13068	−13132	6769	−1960	322	−28	1	0	0
9	0	40320	−109584	118124	−67284	22449	−4536	546	−36	1	0
10	0	−362880	1026576	−1172700	723680	−269325	63273	−9450	870	−45	1

where

$$(\beta_0, \beta_1, \ldots, \beta_k) \equiv (\alpha_0, \alpha_1, \ldots, \alpha_k) S \tag{2.4.12}$$

the S in this case being the Stirling first-kind matrix of order $k + 1$.

The inverse relation to (2.4.7), namely

$$\begin{pmatrix} 1 \\ x \\ x^2 \\ x^3 \\ \vdots \end{pmatrix} = \begin{pmatrix} 1 & 0 & 0 & 0 & 0 & \ldots \\ 0 & 1 & 0 & 0 & 0 & \ldots \\ 0 & 1 & 1 & 0 & 0 & \ldots \\ 0 & 1 & 3 & 1 & 0 & \ldots \\ \vdots & \vdots & \vdots & \vdots & \vdots & \end{pmatrix} \begin{pmatrix} x^{(0)} \\ x^{(1)} \\ x^{(2)} \\ x^{(3)} \\ \vdots \end{pmatrix} \tag{2.4.13}$$

defines the *Stirling matrix of the second kind.* We symbolize it as S^{-1} and it can be obtained either by inverting S, or else from the recursion (see [2.1]),

$$[S^{-1}]_{i,j} = [S^{-1}]_{i-1,j-1} + j[S^{-1}]_{i-1,j} \qquad i, j \geq 1 \tag{2.4.14}$$

with boundary conditions

$$\begin{aligned} [S^{-1}]_{0,0} &= 1 \\ [S^{-1}]_{0,j} &= 0 = [S^{-1}]_{i,0} \qquad i, j \geq 1 \end{aligned} \tag{2.4.15}$$

Table 2.2 gives S^{-1} up to $i, j = 10$.

We shall encounter two matrices in Chapter 4 which are closely related to S and S^{-1}. They are the *associate* Stirling first- and second-kind matrices, S and S^{-1} which serve to relate the backward and forward differences of a polynomial to its derivatives.

Resuming our discussion on the factorial function $x^{(k)}$, we have, as an analog to the equation

$$\frac{d^m}{dx^m} x^k = k^{(m)} x^{k-m} \tag{2.4.16}$$

the relations

$$\nabla^m x^{(k)} = k^{(m)} (x - m)^{(k-m)} \tag{2.4.17}$$

Table 2.2 S^{-1} **(The Stirling Matrix of the Second Kind)**

$i \backslash j$	0	1	2	3	4	5	6	7	8	9	10
0	1	0	0	0	0	0	0	0	0	0	0
1	0	1	0	0	0	0	0	0	0	0	0
2	0	1	1	0	0	0	0	0	0	0	0
3	0	1	3	1	0	0	0	0	0	0	0
4	0	1	7	6	1	0	0	0	0	0	0
5	0	1	15	25	10	1	0	0	0	0	0
6	0	1	31	90	65	15	1	0	0	0	0
7	0	1	63	301	350	140	21	1	0	0	0
8	0	1	127	966	1701	1050	266	28	1	0	0
9	0	1	255	3025	7770	6951	2646	462	36	1	0
10	0	1	511	9330	34105	42525	22827	5880	750	45	1

and

$$\Delta^m x^{(k)} = k^{(m)} x^{(k-m)} \tag{2.4.18}$$

where k and m are positive integers. Both are easily proved by induction. (See Ex. 2.8.)

The factorial function can often be conveniently broken up. Thus for example

$$\begin{aligned} x^{(m)} &= x^{(r)}(x-r)^{(m-r)} \\ &= x^{(r)}(x-r)(x-r-1)^{(m-r-1)} \end{aligned} \tag{2.4.19}$$

giving two possible forms. Note that in the latter, as r goes from zero to $m-1$, the factor $x-r$ scans through all of the factors of $x^{(m)}$.

It is also sometimes useful to factor -1 from each term in $x^{(m)}$. Thus,

$$\begin{aligned} x^{(m)} &= x(x-1)(x-2)\ldots(x-m+1) \\ &= (-1)^m(-x)(-x+1)\ldots(-x+m-1) \\ &= (-1)^m(-x+m-1)^{(m)} \\ &= (-1)^m(-x)^{[m]} \end{aligned} \tag{2.4.20}$$

showing how $x^{(m)}$ and $x^{[m]}$are related.

Although we have defined $x^{(m)}$ only for $m > 0$, it is possible to give a consistent definition for $m < 0$ or when m is not an integer. However, these cases will not be used, and we mention them only for completeness.

2.5 THE BINOMIAL COEFFICIENTS

Let k be any integer, x any number. Then the *binomial coefficient* $\binom{x}{k}$ is defined as

$$\binom{x}{k} \equiv \begin{cases} \dfrac{x^{(k)}}{k!}, & k > 0 \\ 1, & k = 0 \\ 0, & k < 0 \end{cases} \tag{2.5.1}$$

If, in particular, x is a non-negative integer, say m, then we see that

$$\binom{m}{k} = \frac{m!}{k!(m-k)!} \tag{2.5.2}$$

and so

$$\binom{m}{k} = \binom{m}{m-k} \tag{2.5.3}$$

If $k > m$, then $m - k < 0$ and so by (2.5.1) and (2.5.3)

$$\binom{m}{k} = 0 \qquad k > m \text{ an integer} \tag{2.5.4}$$

The binomial coefficients $\binom{m}{k}$ can be conveniently displayed using the well-known *Pascal's triangle,*

$$\begin{array}{ccccccccccccccccccccc}
&&&&&&&&&&1&&&&&&&&&& \\
&&&&&&&&&1&&1&&&&&&&&& \\
&&&&&&&&1&&2&&1&&&&&&&& \\
&&&&&&&1&&3&&3&&1&&&&&&& \\
&&&&&&1&&4&&6&&4&&1&&&&&& \\
&&&&&1&&5&&10&&10&&5&&1&&&&& \\
&&&&1&&6&&15&&20&&15&&6&&1&&&& \\
&&&1&&7&&21&&35&&35&&21&&7&&1&&& \\
&&1&&8&&28&&56&&70&&56&&28&&8&&1&& \\
&1&&9&&36&&84&&126&&126&&84&&36&&9&&1& \\
1&&10&&45&&120&&210&&252&&210&&120&&45&&10&&1 \\
-&-
\end{array}$$

in which each number is the sum of the two above it, and numbers not shown to the right and left are zeros. Then, starting with $m = 0$, the mth horizontal row is $\binom{m}{k}$.

Using the Gamma function, we can extend $\binom{x}{k}$ to the case where k is not an integer. While this generalization will not be needed anywhere in this book, we mention it in passing for completeness.

Perhaps one of the most useful relations involving the binomial coefficients is the recursion

$$\binom{x+1}{k} = \binom{x}{k} + \binom{x}{k-1} \tag{2.5.5}$$

Its proof follows easily from a direct expansion using (2.5.1) (see Ex. 2.9).

As a first application of (2.2.5) we are able to write

$$\nabla_x \binom{x}{k} = \binom{x-1}{k-1} \tag{2.5.6}$$

and

$$\Delta_x \binom{x}{k} = \binom{x}{k-1} \qquad \text{(See Ex. 2.9)} \tag{2.5.7}$$

Equation (2.5.5) is the cornerstone of the proof for the well-known binomial expansion theorem:

$$(1+y)^m = \sum_{k=0}^{\infty} \binom{m}{k} y^k \qquad |y| < 1 \tag{2.5.8}$$

The proof for m a positive integer is quite simple, and follows readily from (2.5.5) when the method of induction is applied. It is also a simple matter to extend the proof to cover negative integral exponents. Thus

$$(1+y)^{-\mu} = \sum_{k=0}^{\infty} \binom{-\mu}{k} y^k \qquad \mu > 0 \tag{2.5.9}$$

which means that for m any integer, positive or negative,

$$(1+y)^m = \sum_{k=0}^{\infty} \binom{m}{k} y^k \tag{2.5.10}$$

When m is a positive integer, $\binom{m}{k}$ is zero for $k > m$ and so the series terminates to give

$$(1+y)^m = \sum_{k=0}^{m} \binom{m}{k} y^k \tag{2.5.11}$$

From (2.5.10) we deduce that

$$(1-y)^m = \sum_{k=0}^{\infty} (-1)^k \binom{m}{k} y^k \tag{2.5.12}$$

Using the binomial expansion we are able to express ∇ and Δ in terms of q and E as follows. By (2.3.3), for m a positive integer

$$\begin{aligned}\nabla^m &= (1 - q)^m \\ &= \sum_{k=0}^{m} (-1)^k \binom{m}{k} q^k\end{aligned} \tag{2.5.13}$$

and so

$$\nabla^m f_n = \sum_{k=0}^{m} (-1)^k \binom{m}{k} f_{n-k} \tag{2.5.14}$$

Likewise

$$\begin{aligned}\Delta^m &= (E - 1)^m \\ &= E^m (1 - E^{-1})^m \\ &= \sum_{k=0}^{m} (-1)^k \binom{m}{k} E^{m-k}\end{aligned} \tag{2.5.15}$$

giving

$$\Delta^m f_n = \sum_{k=0}^{m} (-1)^k \binom{m}{k} f_{n+m-k} \tag{2.5.16}$$

Returning to (2.4.17) we see that ∇^m reduces the degree of the polynomial $x^{(k)}$, from k to $k - m$. In particular, when $m > k$ we see that

$$\nabla^m x^{(k)} = 0 \qquad m > k \tag{2.5.17}$$

We have seen in Section 2.4, that any polynomial $\sum_{i=0}^{r} \beta_i x^i$ can be written as $\sum_{i=0}^{r} \alpha_i x^{(i)}$ using the Stirling matrix, and so for $m > r$ (2.5.17) implies that

$$\nabla^m \left(\sum_{i=0}^{r} \beta_i x^i \right) = 0 \tag{2.5.18}$$

The backward-difference operator ∇^m thus *annihilates all polynomials of degree* $m - 1$ *or less.* Thus for example, if

$$f_n = \sum_{i=0}^{m-1} \beta_i n^i \tag{2.5.19}$$

then it follows that

$$\nabla^m f_n = 0 \tag{2.5.20}$$

and by an analogous argument,

$$\Delta^m f_n = 0 \tag{2.5.21}$$

Using (2.5.14) and (2.4.17) we can now write, (for $m > 0$ an integer)

$$\nabla^m x^{(r)} = r^{(m)}(x - m)^{(r-m)} = \sum_{k=0}^{m} (-1)^k \binom{m}{k} (x - k)^{(r)} \tag{2.5.22}$$

giving very useful alternate expressions for $\nabla^m x^{(r)}$.

In the above paragraphs, powers of ∇ were expanded in terms of q. We now reverse the procedure and obtain powers of q in terms of ∇. Thus

$$\begin{aligned} f_{n-h} &= q^h f_n \\ &= (1 - \nabla)^h f_n \\ &= \sum_{k=0}^{\infty} (-1)^k \binom{h}{k} \nabla^k f_n \end{aligned} \tag{2.5.23}$$

This gives a method of *interpolation.*

As an example, let f_n be given by a set of values for $n = 0,1,2,3,4$. We can then set up the table of successive backward differences of f_n:

n	f_n	∇f_n	$\nabla^2 f_n$	$\nabla^3 f_n$	$\nabla^4 f_n$
0	1				
1	1.5	0.5			
2	2.25	0.75	0.25		
3	3.375	1.125	0.375	0.125	
4	5.0625	1.6875	0.5625	0.1875	0.0625
.	.	.	.	.	.
.	.	.	.	.	.
.	.	.	.	.	.

Referring to the final line of the table, (2.5.23) gives

$$f_{4-h} = 5.0625 - 1.6875h + 0.5625\left[\frac{h(h-1)}{2}\right] - 0.1875\left[\frac{h(h-1)(h-2)}{6}\right] + 0.0625\left[\frac{h(h-1)(h-2)(h-3)}{24}\right] - \cdots \tag{2.5.24}$$

which is seen to be the Newton series counterpart to the power series expansion about the point 4.

By truncating this equation at various points, we obtain polynomials in h of varying degrees, which pass through successively more values of f_n. Thus,

$$f_{4-h} = 5.0625 - 1.6875h \tag{2.5.25}$$

gives the first-degree polynomial interpolator passing through f_4 $(h = 0)$ and f_3 $(h = 1)$. The formula

$$f_{4-h} = 5.0625 - 1.6875h + 0.5625\left[\frac{h(h-1)}{2}\right] \tag{2.5.26}$$

is the quadratic passing through f_4, f_3, and f_2, and so on. ◆◆

Although the above derivation was based on h being an integer, we extend it (without proof) to nonintegral h, giving for example, for $h = 1/2$ and a quadratic fit,

$$f_{3\frac{1}{2}} = 5.0625 - 1.6875(1/2) + 0.5625\left[\frac{1/2(-1/2)}{2}\right] = 4.1484 \qquad (2.5.27)$$

The function is actually

$$f_n = \left(\frac{3}{2}\right)^n \qquad (2.5.28)$$

and so

$$f_{3\frac{1}{2}} = 4.1335 \qquad (2.5.29)$$

showing that a quadratic interpolator in this case gives an error of about 1/3 percent. ◆◆

By setting $h = -1$, say, we can use (2.5.24) as an *extrapolator* to predict what f_5 will be. The prediction of course will be subject to error, depending on the goodness of fit.

It might seem, as we take more and more terms in (2.5.24), that the approximation improves. While this is often true it is not *always* so, and it is possible for the errors to begin to diverge as the number of terms is increased beyond a certain value. This anomaly is known as the *Runge phenomenon* and the interested reader is referred to [2.3].

An analogous interpolation formula can be developed using Δ rather than ∇. (See Ex. 2.14.)

We have touched only briefly on the concepts of numerical differentiation, numerical integration and interpolation. For an extremely good treatment of these topics the reader is referred to [2.2].

2.6 USEFUL IDENTITIES†

We have shown that if f_n is a polynomial in n of degree $m - 1$ then

$$\nabla^m f_n = 0 \qquad (2.6.1)$$

This means that by (2.5.14), if d is an integer less than m,

$$\sum_{k=0}^{m} (-1)^k \binom{m}{k} (n - k)^d = 0 \qquad (2.6.2)$$

†The author is indebted to J. Riordan of Bell Telephone Laboratories for much of the material in this section.

Since this is true for any n, we see that the operator

$$\sum_{k=0}^{m} (-1)^k \binom{m}{k} q^k$$

annihilates all polynomial sequences obtained from

$$f_n = n^d \tag{2.6.3}$$

when d is less than m.

As an example let $d = 2$, and consider the sequence

$$\ldots, 49, 64, 81, 100, \ldots$$

Then, for $m = 3$, (2.6.2) gives

$$\binom{3}{0} 49 - \binom{3}{1} 64 + \binom{3}{2} 81 - \binom{3}{3} 100 = 0 \tag{2.6.4}$$

By the symmetry of the binomial coefficients, we can reverse the order of the sequence. Moreover, *all linear combinations of sequences of inadequate degree* will also be annihilated. Thus, combining the first-degree sequence

$$\ldots, 1, 0, -1, -2, \ldots$$

with the quadratic sequence

$$\ldots, 4, 9, 16, 25, \ldots$$

gives

$$\ldots, 5, 9, 15, 23, \ldots$$

and again (2.6.2) gives

$$\binom{3}{0} 5 - \binom{3}{1} 9 + \binom{3}{2} 15 - \binom{3}{3} 23 = 0 \tag{2.6.5}$$

◆◆

We have defined $\binom{m}{k}$ in such a way that

$$\binom{m}{k} = \binom{m}{m-k} \tag{2.6.6}$$

We call this equivalence an *A-transformation.* We also have the equivalence

$$\binom{m}{k}\binom{k}{p} = \binom{m}{p}\binom{m-p}{k-p} \tag{2.6.7}$$

Proof is by direct expansion of both sides (see Ex. 2.1.6). We call (2.6.7) a *B-transformation.*

By alternately applying B and A-transformations we obtain the following chain of equivalences:

$$\begin{aligned}\binom{m}{k}\binom{k}{p} &= \binom{m}{p}\binom{m-p}{k-p}\\ &= \binom{m}{m-p}\binom{m-p}{k-p}\\ &= \binom{m}{k-p}\binom{m-k+p}{m-k}\\ &= \binom{m}{m-k+p}\binom{m-k+p}{m-k}\\ &= \binom{m}{m-k}\binom{k}{p}\end{aligned} \tag{2.6.8}$$

Suppose on the other hand that we start with an A-transformation rather than a B, and again alternate thereafter. We obtain:

$$\begin{aligned}\binom{m}{k}\binom{k}{p} &= \binom{m}{k}\binom{k}{k-p}\\ &= \binom{m}{k-p}\binom{m-k+p}{p}\\ &= \binom{m}{m-k+p}\binom{m-k+p}{p}\\ &= \binom{m}{p}\binom{m-p}{m-k}\\ &= \binom{m}{m-p}\binom{m-p}{m-k}\\ &= \binom{m}{m-k}\binom{k}{k-p}\end{aligned} \tag{2.6.9}$$

giving 6 *further* equivalences to $\binom{m}{k}\binom{k}{p}$.

We consider next the results of successively iterating (2.5.5) in various ways. As a start,

$$\begin{aligned}\binom{m}{k} &= \binom{m-1}{k} + \binom{m-1}{k-1}\\ &= \binom{m-1}{k} + \left[\binom{m-2}{k-1} + \binom{m-2}{k-2}\right]\\ &= \binom{m-1}{k} + \binom{m-2}{k-1} + \left[\binom{m-3}{k-2} + \binom{m-3}{k-3}\right]\end{aligned} \tag{2.6.10}$$

This can be terminated at any stage or else it can be continued until zeros set in, giving

$$\binom{m}{k} = \sum_{\nu=0}^{k} \binom{m-1-\nu}{k-\nu} \tag{2.6.11}$$

Alternatively, we may iterate (2.5.5) as follows

$$\begin{aligned}\binom{m}{k} &= \binom{m-1}{k} + \binom{m-1}{k-1}\\ &= \left[\binom{m-2}{k} + \binom{m-2}{k-1}\right] + \left[\binom{m-2}{k-1} + \binom{m-2}{k-2}\right]\\ &= \binom{m-2}{k} + 2\binom{m-2}{k-1} + \binom{m-2}{k-2}\end{aligned} \tag{2.6.12}$$

Again

$$\begin{aligned}\binom{m}{k} &= \left[\binom{m-3}{k} + \binom{m-3}{k-1}\right] + 2\left[\binom{m-3}{k-1} + \binom{m-3}{k-2}\right]\\ &+ \left[\binom{m-3}{k-2} + \binom{m-3}{k-3}\right] = \binom{m-3}{k} + 3\binom{m-3}{k-1} + 3\binom{m-3}{k-2} + \binom{m-3}{k-3}\end{aligned} \tag{2.6.13}$$

This is seen to be

$$\binom{m}{k} = \sum_{\nu=0}^{3} \binom{3}{\nu}\binom{m-3}{k-\nu} \tag{2.6.14}$$

and in general, it is easy to see that the iteration can be continued to give

$$\binom{m}{k} = \sum_{\nu=0}^{k} \binom{j}{\nu}\binom{m-j}{k-\nu} \tag{2.6.15}$$

This is known as the *Vandermonde convolution formula.* Note that it is of the form

$$\binom{a+c}{b+d} = \sum \binom{a}{b}\binom{c}{d} \tag{2.6.16}$$

Note also that ν, the variable of summation, is present *in only the lower positions* of the right of (2.6.15).

By various transformations, many variants are possible. Thus setting $m - j = \mu$ in (2.6.15), we obtain

$$\binom{\mu+j}{k} = \sum_{\nu=0}^{k} \binom{j}{\nu}\binom{\mu}{k-\nu} \tag{2.6.17}$$

and so on.

The above identities were essentially based on (2.5.5). We have not even begun to exhaust the powers of that equation, but for the present purpose our collection will be adequate. (See Ex. 2.19.)

2.7 SUMMATION FORMULAE

We are now able to develop a number of summation formulae which will be of frequent use. As a rather trivial start, we infer from (2.3.1) that

$$\sum_{n=a}^{b} \nabla f_n = f_n \Big|_{a-1}^{b} \qquad b \geq a \tag{2.7.1}$$

and (2.3.2) gives

$$\sum_{n=a}^{b} \Delta f_n = f_n \Big|_{a}^{b+1} \qquad b \geq a \tag{2.7.2}$$

These, of course, are analogous to

$$\int_a^b d[f(x)] = f(x)\Big|_a^b \tag{2.7.3}$$

However, note the important differences in the evaluation points $(a - 1, b + 1)$ on the right of (2.7.1) and (2.7.2) respectively.

Consider next the relation

$$\sum_{n=a}^{b} n^{(m)} = \frac{n^{(m+1)}}{m+1}\Bigg|_a^{b+1} \qquad b \geq a \tag{2.7.4}$$

Proof is either by induction or by a combination of (2.7.2) with (2.4.18) (see Ex. 2.20). We observe that this corresponds to

$$\int_a^b x^m\,dx = \frac{x^{m+1}}{m+1}\Bigg|_a^b \tag{2.7.5}$$

Again note the appearance of $b + 1$ rather than b on the right of (2.7.4).

We consider the analog to integration by parts, namely

$$\int_a^b u\,dv = (uv)\Big|_a^b - \int_a^b v\,du \tag{2.7.6}$$

Consider

$$\begin{aligned}\nabla(u_n v_n) &= u_n v_n - u_{n-1} v_{n-1}\\ &= u_n v_n - u_{n-1} v_n + u_{n-1} v_n - u_{n-1} v_{n-1}\\ &= v_n \nabla u_n + u_{n-1} \nabla v_n\end{aligned} \tag{2.7.7}$$

Hence, using (2.7.1) we obtain two equivalent formulae for *backward summation by parts:*

$$\begin{aligned}\sum_{n=a}^{b} u_{n-1} \nabla v_n &= (u_n v_n)\Big|_{a-1}^{b} - \sum_{n=a}^{b} v_n \nabla u_n\\ \sum_{n=a}^{b} v_n \nabla u_n &= (u_n v_n)\Big|_{a-1}^{b} - \sum_{n=a}^{b} u_{n-1} \nabla v_n\end{aligned} \tag{2.7.8}$$

In a like manner, we are able to obtain the rules for *forward summation by parts:*

$$\sum_{n=a}^{b} u_{n+1}\Delta v_n = (u_n v_n)\Big|_a^{b+1} - \sum_{n=a}^{b} v_n \Delta u_n$$

$$\sum_{n=a}^{b} v_n\Delta u_n = (u_n v_n)\Big|_a^{b+1} - \sum_{n=a}^{b} u_{n+1} \Delta v_n \tag{2.7.9}$$

(See Ex. 2.21.) The above four equations are sometimes known as Abel's transformations.

We conclude this section with a brief discussion on the summation of $\sum_{n=0}^{L} n^j$. Consider first, by (2.7.4)

$$\sum_{n=0}^{L} n^{(j)} = \frac{n^{(j+1)}}{j+1}\Bigg|_0^{L+1} = \frac{(L+1)^{(j+1)}}{j+1} \tag{2.7.10}$$

which we call C_j. We now recall that, by the use of the Stirling matrix of the second kind, we can write

$$n^k = \sum_{j=0}^{k} [S^{-1}]_{kj}\, n^{(j)} \tag{2.7.11}$$

and so

$$\sum_{n=0}^{L} n^k = \sum_{n=0}^{L}\left(\sum_{j=0}^{k} [S^{-1}]_{kj}\, n^{(j)}\right)$$

$$= \sum_{j=0}^{k} [S^{-1}]_{kj}\left(\sum_{n=0}^{L} n^{(j)}\right) \tag{2.7.12}$$

$$= \sum_{j=0}^{k} (S^{-1})_{kj}\, C_j \qquad \text{(by (2.7.10))}.$$

But C_j can be expressed as a polynomial in L using the Stirling matrix of the first kind, i.e.,

$$
\begin{aligned}
C_j &\equiv \frac{(L+1)^{(j+1)}}{j+1} \\
&= \left(\frac{L+1}{j+1}\right) L^{(j)} \qquad (2.7.13) \\
&= \left(\frac{L+1}{j+1}\right) \sum_{\nu=0}^{j} [S]_{j\nu} L^{\nu}
\end{aligned}
$$

Thus, by combining the preceding two equations, we obtain

$$
\sum_{n=0}^{L} n^k = (L+1) \sum_{j=0}^{k} [S^{-1}]_{kj} \frac{1}{j+1} \sum_{\nu=0}^{j} [S]_{j\nu} L^{\nu} \qquad (2.7.14)
$$

As an example, let $k = 3$. Then from Table 2.2 on p. 25, we see that the numbers $[S^{-1}]_{3j}$ are 0, 1, 3, 1. Using these together with $[S]_{ij}$ from Table 2.1, (2.7.14) gives

$$
\begin{aligned}
\sum_{n=0}^{L} n^3 &= (L+1)\left[0 + \frac{1}{2}(L) + \frac{3}{3}(-L + L^2) + \frac{1}{4}(2L - 3L^2 + L^3)\right] \\
&= \frac{L^2 (L+1)^2}{4} \qquad (2.7.15)
\end{aligned}
$$

◆◆

Proceeding in this manner we obtain the following. Define

$$
S_j \equiv \sum_{n=0}^{L} n^j \qquad (2.7.16)
$$

Then

$$
\begin{aligned}
S_0 &= L+1 \\
S_1 &= \frac{1}{2}L(L+1) \\
S_2 &= \frac{1}{6}L(L+1)(2L+1) \\
S_3 &= \frac{1}{4}L^2(L+1)^2 \\
S_4 &= \frac{1}{30}L(L+1)(2L+1)(3L^2+3L-1) \\
S_5 &= \frac{1}{12}L^2(L+1)^2(2L^2+2L-1) \\
S_6 &= \frac{1}{42}L(L+1)(2L+1)(3L^4+6L^3-3L+1) \qquad \text{(No } L^2 \text{ term)} \\
S_7 &= \frac{1}{24}L^2(L+1)^2(3L^4+6L^3-L^2-4L+2) \\
S_8 &= \frac{1}{90}L(L+1)(2L+1)(5L^6+15L^5+5L^4-15L^3-L^2+9L-3) \\
S_9 &= \frac{1}{20}L^2(L+1)^2(2L^6+6L^5+L^4-8L^3+L^2+6L-3) \\
S_{10} &= \frac{1}{66}L(L+1)(2L+1)(3L^8+12L^7+8L^6-18L^5-10L^4+24L^3 \\
&\quad +2L^2-15L+5)
\end{aligned}
\tag{2.7.17}
$$

The method used in the derivation of the above formulae should be compared to the approach in [2.4].

2.8 DIFFERENCE EQUATIONS

We close this introductory chapter with a brief discussion on the theory of difference equations. As an example of the sort of equations we wish to solve, consider the linear recursion

$$g_n = \theta g_{n-1} + n \tag{2.8.1}$$

By direct substitution we can easily verify that this is satisfied by

$$g_n = \alpha\theta^n - \frac{\theta}{(1-\theta)^2} + \frac{n}{1-\theta} \tag{2.8.2}$$

for any α.

Operationally the above recursion has the equivalent representation

$$(1 - q\theta)\, g_n = n \tag{2.8.3}$$

and is an example of a *linear difference equation with constant coefficients.* In general, such an equation has the operational form

$$D(q)\, g_n = N(q)\, f_n \tag{2.8.4}$$

or

$$g_n = \frac{N(q)}{D(q)} f_n \tag{2.8.5}$$

where D and N are polynomials in q (or E) and where, by assumption, those polynomials have constant coefficients. f_n is known as the *forcing* or *driving* function, and the unknown g_n is called the *response,* the *output* or the *solution.*

The equation (2.8.5) is conveniently represented by Figure 2.1

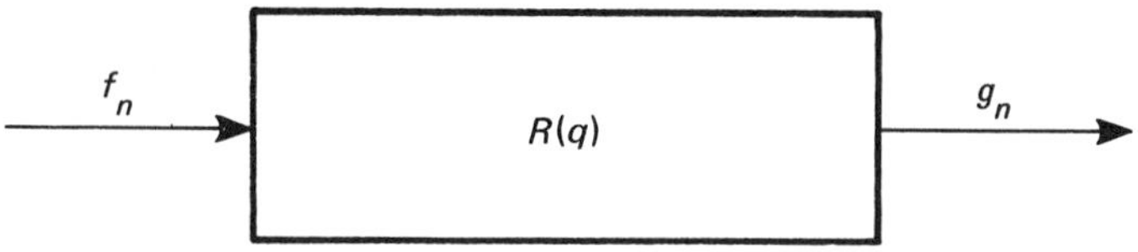

Fig. 2.1 *Difference-equation block diagram.*

in which the operator

$$R(q) \equiv \frac{N(q)}{D(q)} \tag{2.8.6}$$

acts on f_n to produce g_n.

Since $R(q)$ is a linear operator (see Section 2.2) we see that if f_n is the sum of two functions, $(f_1)_n$ and $(f_2)_n$ say, then g_n will be given by

$$g_n = (g_1)_n + (g_2)_n \tag{2.8.7}$$

where

$$\begin{aligned} (g_1)_n &= R(q)\,(f_1)_n \\ (g_2)_n &= R(q)\,(f_2)_n \end{aligned} \tag{2.8.8}$$

The solution can thus be obtained by studying each of the forcing functions individually and then adding the results.

This usage of the linearity property of $R(q)$ will prove useful in subsequent discussion when $(f_1)_n$ will be a wanted function and $(f_2)_n$ an unwanted one. $R(q)$ will be the filter whose purpose is to permit the former to pass through while at the same time reducing the latter.

The solution to the general case of (2.8.4) is based, almost entirely, on the results of the special case

$$(1 - q\lambda)^m g_n = 0 \tag{2.8.9}$$

We solve this as follows.

Suppose that p_n is any function of n, and consider by (2.5.12),

$$(1 - q\lambda)^m (\lambda^n p_n) = \sum_{k=0}^{m} (-1)^k \binom{m}{k} \lambda^k q^k (\lambda^n p_n)$$

$$= \lambda^n \sum_{k=0}^{m} (-1)^k \binom{m}{k} p_{n-k} \tag{2.8.10}$$

Thus by (2.5.14)

$$(1 - q\lambda)^m (\lambda^n p_n) = \lambda^n \nabla^m p_n \tag{2.8.11}$$

But, as we recall from (2.5.19) and (2.5.20), if p_n is *any polynomial in* n *of degree* $m - 1$ *then* $\nabla^m p_n = 0$. We have thus shown that the difference equation

$$(1 - q\lambda)^m g_n = 0 \tag{2.8.12}$$

has as its general solution

$$g_n = \lambda^n p_n \tag{2.8.13}$$

where p_n is any polynomial in n, of degree *one less than the multiplicity* of the factor $1 - q\lambda$ in the operator of (2.8.12).

The solution (2.8.13) contains m arbitrary constants. Thus, if $m = 2$, say, the difference equation

$$(1 - q\lambda)^2 g_n = 0 \tag{2.8.14}$$

has as its *general* solution

$$g_n = (\alpha_0 + n\alpha_1)\lambda^n \tag{2.8.15}$$

which contains two unspecified constants. Given two independent initial conditions we can solve for α_0 and α_1 and obtain the required *particular* solution.

Consider next the case

$$(1 - q\theta)^3(1 - q\zeta)^2 g_n = 0 \tag{2.8.16}$$

We know from Section 2.2, that the factors of the q - operator can be commuted. Let the general solution to the difference equation

$$(1 - q\zeta)^2 g_n = 0 \tag{2.8.17}$$

be the function a_n, and let the general solution to

$$(1 - q\theta)^3 g_n = 0 \tag{2.8.18}$$

be called b_n. Then applying the operator of (2.8.16):

$$\begin{aligned}\left[(1 - q\theta)^3(1 - q\zeta)^2\right](a_n + b_n) &= (1 - q\theta)^3\left[(1 - q\zeta)^2 a_n\right] \\ &\quad + (1 - q\zeta)^2\left[(1 - q\theta)^3 b_n\right] \\ &= 0\end{aligned} \tag{2.8.19}$$

showing that the general solution for (2.8.16) is the sum of the solutions for (2.8.17) and (2.8.18).

In general then, since any operator of the form

$$D(q) = 1 + \gamma_1 q + \gamma_2 q^2 + \cdots + \gamma_r q^r \tag{2.8.20}$$

can be factored into the product

$$D(q) = \prod_{i=1}^{\mu} (1 - q\lambda_i)^{m_i} \qquad \text{(See Note)} \tag{2.8.21}$$

Note: We assume that (2.8.20) has μ distinct roots $\lambda_1, \lambda_2, \ldots, \lambda_\mu$, and that these have multiplicities $m_1, m_2, \ldots, m_\mu$ respectively.

we see that the equation

$$D(q)\, g_n = 0 \tag{2.8.22}$$

has as its solution

$$g_n = a_n + b_n + c_n + \cdots \tag{2.8.23}$$

where

$$\begin{aligned} a_n &= \lambda_1{}^n p_1(n) \\ b_n &= \lambda_2{}^n p_2(n) \\ &\;\vdots \end{aligned} \tag{2.8.24}$$

and where the $p_i(n)$ are polynomials in n of degree one less than m_i, the multiplicity of the factor $(1 - q\lambda_i)$ in (2.8.21). There will be $m_1 + m_2 + \cdots + m_\mu = r$ arbitrary constants in (2.8.20) and so r initial conditions are needed to generate a particular solution.

The equation (2.8.22) is called a *homogeneous*, constant coefficient, linear difference equation, and is seen to arise from the general case

$$g_n = R(q)\, f_n \tag{2.8.25}$$

if at some instant f_n is set to, and thereafter maintained at, zero. The resultant function g_n is known as the *solution to the homogeneous part.* The degree of the denominator of $R(q)$, namely $D(q)$, is called the *order* of the difference equation.†

Consider, as an example, the second order equation

$$(1 + aq + bq^2)\, g_n = N(q)\, f_n \tag{2.8.26}$$

i.e.,

$$g_n = -ag_{n-1} - bg_{n-2} + N(q)\, f_n \tag{2.8.27}$$

†We exclude the case where $D(q)$ is of the form $q^2(1 + 2q + 3q^2)$ say, and regard this as a *quadratic*, and *not* as a quartic in q. However, the reader can verify that the term q^2 leads to a polynomial of degree unity multiplied by *zero* to the nth, i.e., to zero.

and suppose that at some time (say $n = 0$) f_n is removed, leaving the homogeneous portion

$$g_n = -ag_{n-1} - bg_{n-2} \qquad n \geq 0 \tag{2.8.28}$$

If g_{-1} and g_{-2} are *both zero,* then g_n will be zero for $n \geq 0$. The system is said to be *completely relaxed* and remains that way until stimulated. On the other hand, if g_{-1} and g_{-2} are *not both zero,* then g_n will follow the form given by the particular solution of (2.8.28) with specified initial values on g_{-1} and g_{-2}. The ways in which the system behaves when the forcing function is removed are known as its *natural modes.* They are all solutions of (2.8.28) for various initial conditions.

Setting

$$g_n = \lambda^n \qquad \lambda \neq 0 \tag{2.8.29}$$

in (2.8.28) gives (after permissible concellations)

$$\lambda^2 + a\lambda + b = 0 \tag{2.8.30}$$

which is known as the *characteristic equation* of the system. Its solutions are the values of λ needed to set up the form (2.8.21). In general we note that the characteristic equation is obtained from the denominator of (2.8.6) by setting $q = \lambda^{-1}$, i.e.,

$$D(q)\Big|_{q=\lambda^{-1}} = 0 \tag{2.8.31}$$

and clearing powers to leave a polynomial in λ. The values of λ which solve the characteristic equation are called the *eigenvalues* of the system. The functions $\lambda^n p_n$ of (2.8.24) are called the *eigenfunctions,* and we see that the natural modes are simply sums of the eigenfunctions.

Consider, next, the forcing function

$$f_n = \begin{cases} 1, & n = r \\ 0, & \text{otherwise} \end{cases} \tag{2.8.32}$$

This function is called a *unit impulse* and can be written as the Kronecker delta

$$f_n = \delta_{n,r} \tag{2.8.33}$$

In particular, $\delta_{n,0}$ is unity at $n = 0$ and zero otherwise.

One natural mode of particular interest is called the *impulse response* and symbolized I_n. It arises when the system is initially completely relaxed ($g_n = 0$ for $n < 0$) and a forcing function consisting of a unit impulse is applied at $n = 0$.

As an example, we consider the impulse response of the system

$$g_n = \frac{1 - q}{1 - \frac{1}{2}q} f_n \tag{2.8.34}$$

which is the operation form of the system

$$g_n = \frac{1}{2}g_{n-1} + f_n - f_{n-1} \tag{2.8.35}$$

The term f_n is taken to be $\delta_{n,0}$ and so f_{n-1} is $\delta_{n-1,0}$. But, by its definition, $\delta_{n-1,0}$ is unity when $n - 1 = 0$ and zero otherwise, and so

$$\delta_{n-1,0} = \delta_{n,1} \tag{2.8.36}$$

Hence (2.8.35) can be written as

$$g_n = \frac{1}{2}g_{n-1} + \delta_{n,0} - \delta_{n,1} \tag{2.8.37}$$

By virtue of the linearity of this system, one can analyze the effects of the two impulses separately (c/f (2.8.8)). Thus we consider first the forced system

$$(g_1)_n = \begin{cases} \frac{1}{2}(g_1)_{n-1} + \delta_{n,0} & n \geq 0 \\ 0 & n < 0 \end{cases} \tag{2.8.38}$$

But this is now seen to be precisely equivalent to the *homogeneous* system

$$(g_1)_n - \frac{1}{2}(g_1)_{n-1} = 0 \qquad n > 0 \tag{2.8.39}$$

with initial conditions

$$(g_1)_0 = 1 \tag{2.8.40}$$

and so the solution to (2.8.38) is just a particular solution of this homogeneous equation. In fact, by (2.8.39) and (2.8.40)

$$(g_1)_n = \begin{cases} 0 & n < 0 \\ \left(\frac{1}{2}\right)^n & n \geq 0 \end{cases} \tag{2.8.41}$$

Repeating the argument for the second impulse in (2.8.37), we obtain as the other component of g_n,

$$(g_2)_n = \begin{cases} 0 & n < 1 \\ -\left(\frac{1}{2}\right)^{n-1} & n \geq 1 \end{cases} \tag{2.8.42}$$

The solution of (2.8.37) is then the sum of $(g_1)_n$ and $(g_2)_n$, giving finally, as the impulse response,

$$I_n = \begin{cases} 0 & n < 0 \\ 1 & n = 0 \\ -\left(\frac{1}{2}\right)^n & n \geq 1 \end{cases} \tag{2.8.43}$$

◆◆

We have thus verified *that the impulse response is simply a certain sum of the natural modes,* each with appropriate initial conditions. This result is easily seen to be true in general.

The impulse response can now be used to give the solution for *any* forcing function whatever. First we note that if the response of a system to the impulse $\delta_{n,0}$ is I_n, then the response to $\delta_{n-r,0}$ will be I_{n-r}. But $\delta_{n-r,0}$ is the same as $\delta_{n,r}$. Thus $\delta_{n,r}$ will produce I_{n-r}.

Next, we observe that any forcing function f_n can be written as the sum

$$f_n = f_0\delta_{n,0} + f_1\delta_{n,1} + \cdots + f_r\delta_{n,r} + \cdots \tag{2.8.44}$$

and so the response to f_n will be the sum of the responses to its individual components. This, we now know, will be simply

$$g_n = f_0 I_n + f_1 I_{n-1} + \cdots + f_r I_{n-r} + \cdots \tag{2.8.45}$$

i.e.,

$$g_n = \sum_{r=0}^{\infty} f_r I_{n-r} \tag{2.8.46}$$

Finally, since I_{n-r} is identically zero for $n - r$ negative, we see that the above sum terminates when $r = n$, and so we obtain

$$g_n = \sum_{r=0}^{n} f_r I_{n-r} \tag{2.8.47}$$

Equation (2.8.47) is a very fundamental result, and is called a *convolution product* in which the forcing function and the impulse response have been *convolved* to give the response g_n.

The impulse response I_n is governed solely by the operator $R(q)$ in (2.8.25). It can thus be determined *once and for all* for a given system. Then, when a forcing function is given, the response is a convolution of that forcing function with the impulse response. In the final analysis every forced response is thus directly obtainable from the solution to the homogeneous part. (See Ex. 2.25.)

The convolution product (2.8.47) represents a particular solution to (2.8.25) for the case where $g_n = 0$ for $n < 0$. If to (2.8.47) we add the general solution of the homogeneous part, we see that we will have thereby obtained the *general forced solution,* capable of satisfying any set of initial conditions on g_n with any given forcing function f_n.

In some cases of practical interest we can obtain a closed form solution for g_n rather than the series given by the convolution sum. One method commonly used is known as the method of *undetermined coefficients* and an even more versatile method is known as the method of *variation of parameters.* The interested reader is referred to [2.5] for a discussion of both of the above methods.

EXERCISES

2.1 Consider the system

$$g_n - g_{n-1} = f_n \quad \text{with} \quad f_n = 0 \quad \text{for} \quad n < 0 \tag{I}$$

a) Assuming that $g_{-1} = 0$, verify by iteration that

$$g_n = \sum_{k=0}^{n} f_{n-k} \tag{II}$$

b) Now write (I) as $(1 - q)g_n = f_n$, and using (2.2.18) verify again that (II) is true.

2.2 Consider the system

$$g_n - \frac{1}{2} g_{n-1} = f_n \quad \text{with} \quad f_n = 0 \quad \text{for} \quad n < 0 \tag{I}$$

a) Assuming that $g_{-1} = 0$, verify by iteration that

$$g_n = \sum_{k=0}^{n} \left(\frac{1}{2}\right)^k f_{n-k} \tag{II}$$

b) Now write (I) as $\left(1 - \frac{1}{2}q\right) g_n = f_n$, and using (2.2.18) verify again that (II) is true.

2.3 Consider the system of equations

$$u^*_{n+1} = u^*_n + \alpha\delta_n \tag{I}$$

$$y^*_{n+1} = y^*_n + u^*_{n+1} + \beta\delta_n \tag{II}$$

$$y^*_{n+1} = 2y^*_n - y^*_{n-1} + \frac{4}{n+1}\delta_n - \frac{2(2n-1)}{(n+1)n}\delta_{n-1} \tag{III}$$

where we wish to solve for α and β. Write the first two equations as

$$\begin{pmatrix} 1-q & 0 \\ -1 & 1-q \end{pmatrix} \begin{pmatrix} u^*_{n+1} \\ y^*_{n+1} \end{pmatrix} = \begin{pmatrix} \alpha \\ \beta \end{pmatrix} \delta_n$$

and verify that this gives

$$y^*_{n+1} = \frac{\alpha\delta_n + (1-q)\beta\delta_n}{1 - 2q + q^2} \tag{IV}$$

Now equate (IV) to (III) obtaining

$$\alpha = \frac{2}{(n+1)n} \qquad \beta = \frac{2(2n-1)}{(n+1)n}$$

However, the fact that α and β emerge as *functions of n* now means that our procedure is in error. Repeat the entire process but using α_n and β_n rather than α and β as the unknowns, thereby obtaining the *correct* answers

$$\alpha_n = \frac{6}{(n+2)(n+1)} \qquad \beta_n = \frac{2(2n+1)}{(n+2)(n+1)}$$

2.4 Starting from (2.3.1) verify that

$$\nabla^3 f_n \equiv \nabla\left[\nabla(\nabla f_n)\right] = f_n - 3f_{n-1} + 3f_{n-2} - f_{n-3}$$

Apply this to $f_n \equiv n^2$, $f_n \equiv n^3$. Form the backward and forward-difference tables (see pp. 18 and 19) for $f_n \equiv n^2$, $f_n \equiv n^3$ in the range $-3 \le n \le 3$.

2.5 Using Taylor's expansion show that if $0 < \zeta < 1$, and $0 < \xi < 1$, then

$$g[(n - 1 + \xi)\tau] = g[(n - \zeta)\tau] + (\zeta + \xi - 1)\tau\dot{g}[(n - \mu)\tau]$$

where $0 < \mu < 1$ and so infer that (2.3.18) follows from (2.3.17).

2.6 Using the function

$$\dot{f}_n \equiv \sin\left(\frac{n}{8}\frac{\pi}{2}\right)$$

obtain (from tables) the values of $\dot{f}_n$ for $0 \le n \le 8$. Now use the three integration methods (2.3.14), (2.3.15), (2.3.16) to approximate

$$\int_0^{\pi/2} \sin x \, dx$$

Compare the three results obtained with the true answer in the light of the error terms in (2.3.10) and (2.3.18).

2.7 a) Verify that $x^{(k)}$ is a polynomial in x of degree k whose zeros are at $x = 0, 1, 2, \ldots, k - 1$. Write out $(-1)^{(k)}$ and verify that it equals $(-1)^k k!$

b) Express $x^{(5)}$ as a power series in x by direct multiplication. Compare the resultant coefficients to the 5th row of Table 2.1.

c) Verify that the matrices of (2.4.7) and (2.4.13) are inverses of one another.

2.8 Prove, by induction on m, that

$$\nabla^m x^{(k)} = k^{(m)}(x - m)^{(k-m)}$$

and

$$\Delta^m x^{(k)} = k^{(m)} x^{(k-m)}$$

2.9 By direct expansion on both sides prove that

$$\binom{x+1}{k} = \binom{x}{k} + \binom{x}{k-1}$$

Using this result, verify that

a) $\nabla_x \binom{x}{k} = \binom{x-1}{k-1}$ b) $\Delta_x \binom{x}{k} = \binom{x}{k-1}$

2.10 Verify that

a) $x^{[m]} = (-1)^m (-x)^{(m)}$, b) $x^{(m)} = (-1)^m (-x)^{[m]}$.

2.11 a) Verify that $\binom{5}{3} = \binom{5}{2}$, $\binom{5}{4} = \binom{5}{1}$.

b) Using the form $\binom{x}{k} = \dfrac{x^{(k)}}{k!}$, verify that $\binom{5}{6} = 0$,

$\binom{-1}{j} = (-1)^j$

2.12 Verify that $(1 - x)^{-1} = 1 + x + x^2 + \cdots$

a) by applying the binomial theorem

b) by direct division.

What happens if we set $x = 2$?

2.13 Verify (2.5.22) using $m = 1$, $r = 2$, and $m = 2$, $r = 4$.

2.14 Starting from the form $f_{n+h} = E^h f_n$, use (2.3.3) to express E in terms of Δ and hence verify that

$$f_{n+h} = \sum_{k=0}^{\infty} \binom{h}{k} \Delta^k f_n$$

Compare this to (2.5.23) and interpolate the function

$$f_n \equiv \left(\frac{3}{2}\right)^n$$

at the point $n = 3\frac{1}{2}$ by *forward* differences as was done on p. 31 using backward differences.

2.15 Form the sequence of numbers $n^2 - 2n + 4$ for $0 \le n \le 5$ and verify, by direct use of those numbers (c/f (2.6.4)) that (2.6.2) is true (use $m = 1, 2, 3, 4$). What is the counterpart of (2.6.2) is we start from $\Delta^m f_n = 0$?

2.16 Verify (2.6.7) by direct expansion of both sides.

2.17 Using various values for m and k, (e.g., $m = 5$, $k = 3$) verify (2.6.11).

2.18 a) Using $m = 6$, $k = 3$, $j = 3$ verify (2.6.15). Set $\mu = 3$, $j = 3$, $k = 3$ and verify (2.6.17).

b) Show that

$$\sum_{\nu=0}^{j} \binom{i}{\nu}\binom{j}{\nu} = \binom{i+j}{i} = \binom{i+j}{j}$$

Hint: Use Vandermonde convolution.

2.19 Starting from (2.6.11) verify that

$$\binom{m}{k} = \binom{m-1}{k} + \binom{m-2}{k-1} + \binom{m-3}{k-2} + \cdots = \sum_{\nu=0}^{k} \binom{\nu+0}{\nu}\binom{m-\nu-1}{k-\nu}$$

Then by (2.5.5), verify that this now gives

$$\binom{m}{k} = \binom{m-2}{k} + 2\binom{m-3}{k-1} + 3\binom{m-4}{k-2} + \cdots$$

$$= \sum_{\nu=0}^{k} \binom{\nu+1}{\nu}\binom{m-\nu-2}{k-\nu}$$

and in general, using (2.5.5) repeatedly, verify that

$$\binom{m}{k} = \binom{m-j}{k} + \binom{1+j-1}{1}\binom{m-1-j}{k-1}$$

$$+ \binom{2+j-1}{2}\binom{m-2-j}{k-1} + \cdots$$

i.e., that

$$\binom{m}{k} = \sum_{\nu=0}^{k} \binom{\nu+j-1}{\nu}\binom{m-\nu-j}{k-\nu} \qquad 0 \le j \le m$$

This form is similar but *not* identical to the Vandermonde convolution.

2.20 Prove (2.7.4) by induction on b. Now prove it by combining (2.7.2) with (2.4.18).

2.21 Prove (2.7.9).

2.22 Using (2.7.17) find

$$\sum_{n=0}^{10} n \qquad \sum_{n=0}^{4} n^2 \qquad \sum_{n=0}^{3} n^3$$

Verify each by direct expansion.

2.23 Solve the following

a) $g_n - g_{n-1} = 0 \qquad (g_0 = 1)$

b) $g_n - 2g_{n-1} + g_{n-2} = 0 \qquad (g_0 = 1, g_1 = 2)$

c) $g_n - \frac{1}{2}g_{n-1} = 0 \qquad (g_0 = 1)$

d) $g_n + \frac{1}{4}g_{n-2} = 0 \qquad (g_0 = g_1 = 1)$

e) $g_{n+1} - 3g_n + 2g_{n-1} = 0 \qquad (g_0 = 1, g_1 = 3/2)$

f) $g_{n+1} + 4g_{n-1} = 0 \qquad (g_0 = g_1 = 1)$

Which of these systems dies out as $n \to \infty$? Which builds up at an exponential rate as $n \to \infty$? Plot the eigenvalues of each equation on the complex plane. What can we infer about the location of the eigenvalues and the behavior of the corresponding system as $n \to \infty$? Does the position of the eigenvalues in relation to the unit circle have any significance?

2.24 Verify (2.8.42) and (2.8.43).

2.25 Find the impulse response of the system

$$g_n - \frac{1}{2}g_{n-1} = f_n$$

Hence, use (2.8.47) to solve

a) $g_n - \frac{1}{2}g_{n-1} = 1 \qquad n \geq 0$

$g_n = 0 \qquad n < 0$

b) $g_n - \frac{1}{2}g_{n-1} = n \qquad n \geq 0$

$g_n = 0 \qquad n < 0$

REFERENCES

1. Hamming, R. W., "Numerical Methods for Scientists and Engineers," McGraw-Hill Book Company, 1962, p. 18 et. seq.
2. Hildebrand, F. B., "Introduction to Numerical Analysis," McGraw-Hill Book Company, 1956.

3. Lanczos, C., "Applied Analysis," Prentice-Hall, 1956, p. 348.
4. Blackman, R. B., "Data Smoothing and Prediction," Addison-Wesley, 1965, p. 176.
5. Hildebrand, F. B., "Methods of Applied Mathematics," Prentice-Hall, 1952.

3

THE DISCRETE ORTHOGONAL POLYNOMIALS

3.1 INTRODUCTION

Consider the set of polynomials $\varphi(x;0)$, $\varphi(x;1)$, $\varphi(x;2)$, ..., $\varphi(x;j)$, ... where x is the polynomial argument and j is the degree, and let $w(x)$ be a function of x which is nonnegative in $a \leq x \leq b$. If these polynomials satisfy the condition

$$\int_a^b \varphi(x;i)\varphi(x;j)\,w(x)\,dx = 0 \qquad i \neq j \tag{3.1.1}$$

then we say that *these are the polynomials which are orthogonal over the range* $[a, b]$ *with respect to the weight-function* $w(x)$.

The reader is no doubt well aware of the orthogonal polynomials of Legendre and Laguerre, which satisfy

$$\int_{-1}^{1} \varphi(x;i)\varphi(x;j)\,dx = 0 \qquad i \neq j \tag{3.1.2}$$

and

$$\int_0^{\infty} \varphi(x;i)\,\varphi(x;j)\,e^{-\alpha x}\,dx = 0 \qquad \begin{matrix} i \neq j \\ \alpha > 0 \end{matrix} \tag{3.1.3}$$

respectively. These polynomials play a prominent role in the solution of certain differential equations, in electrical filter theory, and in general approximation theory. (See e.g. [3.1].)

Of much greater interest to us here are the corresponding polynomials which are orthogonal over a *discrete* set of equidistant points, defined by

$$\sum_{x=0}^{L} p(x;i,L)\,p(x;j,L) = 0 \qquad i \neq j \tag{3.1.4}$$

and

$$\sum_{x=0}^{\infty} p(x;i,\theta)\,p(x;j,\theta)\,\theta^x = 0 \qquad \begin{matrix} i \neq j \\ |\theta| < 1 \end{matrix} \tag{3.1.5}$$

These are termed respectively, the *discrete* Legendre and Laguerre polynomials. Note the triple arguments in each case. The first set, $\{p(x;j,L)\}$, are polynomials in x, of degree j, orthogonal over the discrete range $0 \leq x \leq L$ with respect to the weight-function unity. On the other hand $\{p(x;j,\theta)\}$ is the set of polynomials in x, of degree j, orthogonal over the discrete range $0 \leq x \leq \infty$ with respect to the weight-function θ^x. It is important to bear in mind that while the orthogonality condition involves the argument x *only at the integral values* 0,1,2,. . ., the polynomials are themselves nevertheless *continuous functions* of x – polynomials in the normal sense.

The discrete Legendre polynomials can be seen from (3.1.4) to have a weight-function equivalent to

$$w(x) = \begin{cases} 1, & 0 \leq x \leq L \\ 0, & \text{otherwise} \end{cases} \tag{3.1.6}$$

where L is a positive *fixed* integer. Thus $w(x)$ is unity over a fixed interval and zero elsewhere. In Chapter 7, these polynomials will form the basis for the Fixed-Memory Polynomial Filters.

In Chapter 9, on the other hand, we will show that the cycle counting number n can be used in place of L in (3.1.4). Then, as n increases through the integers, we obtain successive sets of polynomials which are orthogonal over an *expanding* interval. From these we will construct the Expanding-Memory Polynomial Filters.

Finally, we see in (3.1.5) that the weight-function of the discrete Laguerre polynomials is an exponential function which *fades out* in a well behaved manner as x increases. These polynomials will form the basis for the Fading-Memory Polynomial Filters of Chapter 13.

We now develop the expressions for their general forms, for use in the later chapters mentioned above.

3.2 THE DISCRETE LEGENDRE POLYNOMIALS†

We derive first, the general form of the discrete Legendre polynomial $p(x;j,L)$. In order to do this, it will be necessary to draw on some of the results obtained in the previous chapter dealing with discrete function analysis.

Let $p(x;j,L)$ be a polynomial in x of degree j, with parameter L, having the property that

$$\sum_{x=0}^{L} p(x;i,L)\,p(x;j,L) = 0 \qquad i \neq j \tag{3.2.1}$$

One approach to obtaining the form of the polynomials so defined would be to use a Schmidt orthogonalization procedure. Thus, we would take

$$p(x;0,L) = 1 \tag{3.2.2}$$

and assume that

$$p(x;1,L) = \alpha + \beta x \tag{3.2.3}$$

†In [3.1], Hildebrand considers briefly the question of orthogonality over a discretized interval and derives the Gram polynomials, sometimes also called the Chebyshev polynomials (see e.g. [3.2, p. 788]). These polynomials are related to ours by a shift of the origin. We have extended Hildebrand's approach to obtain the discrete Legendre and Laguerre polynomials.

For an alternate, and extremely elegant derivation of the discrete Legendre polynomials, the reader is referred to [3.3].

where α and β are constants to be determined. We now use these in (3.2.1) which then enables us to solve for $p(x;1,L)$ to within a single constant. Proceeding in this way we can derive the form of as many of the polynomials as we have energy for, each to within an unspecified constant.

The drawback of the above method is that the *general* form of the polynomials is not then obtained. A much more systematic approach is to proceed as follows.

Suppose that $f(x;k)$ is *any* polynomial of degree k. Then it is evident that there exist constants β_i such that $f(x;k)$ can be synthesized by the linear combination

$$f(x;k) = \sum_{i=0}^{k} \beta_i p(x;i,L) \tag{3.2.4}$$

That being the case, it follows immediately by (3.2.1) that

$$\sum_{x=0}^{L} f(x;k)\,p(x;j,L) = 0 \qquad 0 \leq k < j \tag{3.2.5}$$

This will be used in place of (3.2.1) to obtain the required form of $p(x;j,L)$. Since k cannot be negative and since k is less than j, we are by implication considering only $j \geq 1$. The case $j = 0$ is obviously solved by (3.2.2).

We apply summation by parts (see p. 37) to (3.2.5), and to facilitate this we define a function $g(x;j,L)$ by the relation

$$p(x;j,L) = \nabla^j g(x;j,L) \tag{3.2.6}$$

where the ∇ acts on the variable x according to the definition

$$\nabla g(x;j,L) \equiv g(x;j,L) - g(x-1;j,L) \tag{3.2.7}$$

(Note that $g(x;j,L)$ is *not* envisaged as a "polynomial in x of degree j with parameter L" as is the case with $p(x;j,L)$. We have merely adopted an analogous symbolism in both cases for simplicity. In fact, as is evident from (3.2.6), $g(x;j,L)$ must be a polynomial in x of degree $2j$.)

Since any polynomial of degree less than j is annihilated by ∇^j, we see that (3.2.6) only defines $g(x;j,L)$ to within an *arbitrary additive polynomial* of degree $j-1$. This will be of significance later.

Combining (3.2.6) and (3.2.5) we now have

$$\sum_{x=0}^{L} f(x;k)\,\nabla^j g(x;j,L) = 0 \qquad j \geq 1 \tag{3.2.8}$$

and, making the associations $f(x;k) \sim u_{x-1}$ and $\nabla^{j-1} g(x;j,L) \sim v_x$, (2.7.8) then gives us

$$0 = f(x+1;k)\,\nabla^{j-1} g(x;j,L)\Big|_{-1}^{L} - \sum_{x=0}^{L} \nabla f(x+1,k)\,\nabla^{j-1} g(x;j,L) \tag{3.2.9}$$

By iterating this procedure we obtain

$$\begin{aligned} 0 = f(x+1;k)\,\nabla^{j-1} g(x;j,L)\Big|_{-1}^{L} - \nabla f(x+2;k)\,\nabla^{j-2} g(x;j,L)\Big|_{-1}^{L} + \cdots \\ + (-1)^{j-1}\Big[\nabla^{j-1} f(x+j;k)\Big]\, g(x;j,L)\Big|_{-1}^{L} \qquad \text{(see Ex. 3.1)}, \end{aligned} \tag{3.2.10}$$

where the iteration terminates since, by assumption,

$$\nabla^j f(x;k) \equiv 0 \tag{3.2.11}$$

Let $j = 1$. Then we must have $k = 0$, and so (3.2.10) gives

$$f(x+1;0)\, g(x;1,L)\Big|_{-1}^{L} = 0 \tag{3.2.12}$$

Since this must hold for *any* polynomial f, of degree zero, we have

$$g(L;1,L) = g(-1;1,L) \tag{3.2.13}$$

But by (3.2.6), $g(x;1,L)$ is defined in terms of $p(x;1,L)$ *to within an arbitrary additive constant.* Hence we can take (3.2.13) to mean

$$g(L;1,L) = g(-1;1,L) = 0 \tag{3.2.14}$$

Repeating this argument for $j = 2, 3, \ldots$, and remembering that in each case (3.2.10) must hold for any polynomial $f(x;k)$ of degree $j-1$ or less, we obtain as one possible set of boundary conditions on $g(x;j,L)$ that

$$g(x;j,L)\Big|_{x=L} = \nabla g(x;j,L)\Big|_{x=L} = \cdots = \nabla^{j-1}g(x;j,L)\Big|_{x=L} = 0$$
$$g(x;j,L)\Big|_{x=-1} = \nabla g(x;j,L)\Big|_{x=-1} = \cdots = \nabla^{j-1}g(x;j,L)\Big|_{x=-1} = 0 \tag{3.2.15}$$

Moreover, by (3.2.6)

$$\nabla^{2j+1}g(x;j,L) = 0 \tag{3.2.16}$$

This difference equation and the $2j$ chosen boundary conditions of (3.2.15) can be solved to give $g(x;j,L)$ to within a *constant multiplier.* By trial and error, or otherwise (see Ex. 3.2), we obtain the following polynomial in x of degree $2j$:

$$g(x;j,L) = a(j,L)\sum_{\nu=0}^{j}(-1)^{\nu}\binom{j}{\nu}\frac{(x+j)^{(j+\nu)}}{(L+j)^{(j+\nu)}} \tag{3.2.17}$$

where the mutliplier $a(j,L)$ is as yet unspecified.

It then follows from (3.2.6) that

$$p(x;j,L) = \nabla^{j}g(x;j,L)$$
$$= a(j,L)\frac{j^{(j)}}{(L+j)^{(j)}}\sum_{\nu=0}^{j}(-1)^{\nu}\binom{j}{\nu}\binom{j+\nu}{\nu}\frac{x^{(\nu)}}{L^{(\nu)}} \tag{3.2.18}$$

(see Ex. 3.3).

The constant $a(j,L)$ is arbitrary. By setting

$$a(j,L) = \binom{L+j}{j} \tag{3.2.19}$$

we obtain

$$p(x;j,L) = \sum_{\nu=0}^{j}(-1)^{\nu}\binom{j}{\nu}\binom{j+\nu}{\nu}\frac{x^{(\nu)}}{L^{(\nu)}} \tag{3.2.20}$$

This is the discrete Legendre polynomial of degree j. The first five polynomials ($j = 0, 1, \ldots, 4$) are:

$$
\begin{aligned}
p(x;0,L) &= 1 \\
p(x;1,L) &= 1 - 2\frac{x}{L} \\
p(x;2,L) &= 1 - 6\frac{x}{L} + 6\frac{x(x-1)}{L(L-1)} \\
p(x;3,L) &= 1 - 12\frac{x}{L} + 30\frac{x(x-1)}{L(L-1)} - 20\frac{x(x-1)(x-2)}{L(L-1)(L-2)} \\
p(x;4,L) &= 1 - 20\frac{x}{L} + 90\frac{x(x-1)}{L(L-1)} - 140\frac{x(x-1)(x-2)}{L(L-1)(L-2)} \\
&\quad + 70\frac{x(x-1)(x-2)(x-3)}{L(L-1)(L-2)(L-3)}
\end{aligned}
\tag{3.2.21}
$$

Note that $p(x;j,L)$ is only defined for $j \leq L$, and so the set $\{p(x;j,L)\}$ has precisely $L+1$ members, namely $p(x;0,L) \ldots p(x;L,L)$. This is in marked contrast with the "continuous" Legendre polynomials defined by (3.1.2) which form a set having an *infinite* number of members.

The reader should investigate the location of the zeros of $g(x;j,L)$ by examining (2.2.15). It will then emerge that (3.2.6) is a perfect analogy to Rodriguez' theorem [3.2, p. 785] as related to the *continuous* Legendre polynomials which satisfy (3.1.2). This duality between the discrete and continuous polynomials is very evident throughout, and is a consequence of the many dualities, established in Chapter 2, between the calculi of continuous and discrete functions.

We require one further basic result, namely $c(j,L)$ defined by

$$
[c(j,L)]^2 \equiv \sum_{x=0}^{L} [p(x;j,L)]^2 \tag{3.2.22}
$$

In order to obtain this we proceed as follows. Let (3.2.20) be written as the Newton series

$$
p(x;j,L) = \sum_{\nu=0}^{j} \alpha(\nu,L)\, x^{(\nu)} \tag{3.2.23}
$$

Then (3.2.22) becomes

$$
\begin{aligned}
[c(j,L)]^2 &= \sum_{x=0}^{L} p(x;j,L) \sum_{\nu=0}^{j} \alpha(\nu,L)\, x^{(\nu)} \\
&= \alpha(j,L) \sum_{x=0}^{L} x^{(j)} p(x;j,L) \qquad \text{(by (3.2.5))} \qquad (3.2.24)\\
&= \alpha(j,L) \sum_{x=0}^{L} x^{(j)} \nabla^j g(x;j,L) \qquad \text{(by (3.2.6))}
\end{aligned}
$$

which we sum by parts to give

$$
[c(j,L)]^2 = (-1)^j j!\, \alpha(j,L) \sum_{x=0}^{L} g(x;j,L) \qquad (3.2.25)
$$

(see Ex. 3.4).

Now by (3.2.20), the coefficient of $x^{(j)}$ is

$$
\alpha(j,L) = (-1)^j \binom{2j}{j} \frac{1}{L^{(j)}} \qquad (3.2.26)
$$

and from (3.2.17) and (3.2.19),

$$
\begin{aligned}
\sum_{x=0}^{L} g(x;j,L) &= \sum_{x=0}^{L} \binom{L+j}{j} \sum_{\nu=0}^{j} (-1)^{\nu} \binom{j}{\nu} \frac{(x+j)^{(j+\nu)}}{(L+j)^{(j+\nu)}} \qquad (3.2.27)\\
&= \binom{L+j}{j} \sum_{\nu=0}^{j} (-1)^{\nu} \binom{j}{\nu} \frac{L+j+1}{j+\nu+1}
\end{aligned}
$$

(see Ex. 3.5).

It is shown in Ex. 3.6 that

$$
\sum_{\nu=0}^{j} (-1)^{\nu} \binom{j}{\nu} \frac{1}{j+\nu+1} = \left[(2j+1)\binom{2j}{j} \right]^{-1} \qquad (3.2.28)
$$

Hence

$$\sum_{x=0}^{L} g(x;j,L) = \frac{(L+j+1)^{(j+1)}}{j!\,(2j+1)\binom{2j}{j}} \tag{3.2.29}$$

Thus using (3.2.26) and (3.2.29) in (3.2.25), we obtain finally

$$[c(j,L)]^2 = \frac{(L+j+1)^{(j+1)}}{(2j+1)L^{(j)}} \tag{3.2.30}$$

For future reference, the first five values of $[c(j,L)]^2$ are:

$$\begin{aligned}
[c(0,L)]^2 &= L+1 \\
[c(1,L)]^2 &= \frac{(L+2)(L+1)}{3L} \\
[c(2,L)]^2 &= \frac{(L+3)(L+2)(L+1)}{5L(L-1)} \\
[c(3,L)]^2 &= \frac{(L+4)(L+3)(L+2)(L+1)}{7L(L-1)(L-2)} \\
[c(4,L)]^2 &= \frac{(L+5)(L+4)(L+3)(L+2)(L+1)}{9L(L-1)(L-2)(L-3)}
\end{aligned} \tag{3.2.31}$$

Equations (3.2.20) and (3.2.30) are the two fundamental equations for the discrete Legendre polynomials. We now turn our attention to the derivation of the general form of the discrete Laguerre polynomials.

3.3 THE DISCRETE LAGUERRE POLYNOMIALS†

Let $p(x;j,\theta)$ be a polynomial in x of degree j, with parameter θ, such that

$$\sum_{x=0}^{\infty} p(x;i,\theta)\,p(x;j,\theta)\,\theta^x = 0 \qquad j \neq i \tag{3.3.1}$$

where $|\theta| < 1$. Summation is over the positive integers although x itself is a

†The interested reader is referred to [3.4] for a more complete discussion on these polynomials.

continuous variable. We term these polynomials the discrete Laguerre polynomials and their form and properties will be seen to be very similar to the continuous Laguerre polynomials satisfying (3.1.3). These polynomials will constitute the basis on which we build the Fading-Memory Polynomial Filters in Chapter 13, and it is the purpose of this section to derive the general expression for $p(x;j,\theta)$ as well as for the quantity $\sum_{x=0}^{\infty} [p(x;j,\theta)]^2\theta^x$.

Following the same arguments as those given in the previous section, we see quite readily that (3.3.1) is equivalent to the statement

$$\sum_{x=0}^{\infty} f(x;k)\,p(x;j,\theta)\,\theta^x = 0 \tag{3.3.2}$$

where $f(x;k)$ is any polynomial of degree $k < j$.† We now obtain the form of $p(x;j,\theta)$ from (3.3.2) by the use of summation by parts.

To facilitate this, we first define the function $g(x;j,\theta)$ by the equation

$$\theta^x p(x;j,\theta) = \Delta^j g(x;j,\theta) \tag{3.3.3}$$

where Δ is the forward-difference operator which operates on the variable x by a definition which is analogous to (3.2.7) where the backward-difference operator ∇ was defined. Setting (3.3.3) into (3.3.2), we obtain

$$\sum_{x=0}^{\infty} f(x;k)\,\Delta^j g(x;j,\theta) = 0 \qquad j > 0 \tag{3.3.4}$$

We now make the associations $f(x;k) \sim u_{x+1}$ and $\Delta^{j-1}g(x;j,\theta) \sim v_x$, and by applying summation by parts‡ repeatedly to (3.3.4), we obtain (see Ex. 3.7)

$$0 = f(x-1;k)\,\Delta^{j-1}g(x;j,\theta)\Big|_0^{\infty} - \Delta f(x-2;k)\,\Delta^{j-2}g(x;j,\theta)\Big|_0^{\infty} + \cdots$$
$$+ (-1)^{j-1}[\Delta^{j-1}f(x-j;k)]\,g(x;j,\theta)\Big|_0^{\infty} \tag{3.3.5}$$

Since (3.3.5) must hold for *any* polynomial $f(x;k)$ where $0 \le k < j$, we use

†We are thus considering only the cases $j \ge 1$.
‡See (2.7.9).

the same arguments as we made to justify (3.2.15) in Section 3.2 to give, in this case,

$$g(x;j,\theta)\Big|_{x=\infty} = \Delta g(x;j,\theta)\Big|_{x=\infty} = \cdots = \Delta^{j-1} g(x;j,\theta)\Big|_{x=\infty} = 0$$

$$g(x;j,\theta)\Big|_{x=0} = \Delta g(x;j,\theta)\Big|_{x=0} = \cdots = \Delta^{j-1} g(x;j,\theta)\Big|_{x=0} = 0 \tag{3.3.6}$$

Moreover by (3.3.3), $\theta^{-x}\Delta^j g(x;j,\theta)$ is a polynomial of degree j, and so

$$\Delta^{j+1}[\theta^{-x}\Delta^j g(x;j,\theta)] = 0 \tag{3.3.7}$$

This boundary value problem defines one of the possible generators for the discrete Laguerre polynomials. The difference equation (3.3.7) plus the choice of the boundary conditions of (3.3.6) gives us, either formally or by inspection (see Ex. 3.8),

$$g(x;j,\theta) = \theta^x \binom{x}{j} \tag{3.3.8}$$

If we now apply (3.3.3) to (3.3.8) we obtain

$$p(x;j,\theta) = \theta^{-x} \Delta^j \theta^x \binom{x}{j}\Bigg]$$

$$= \sum_{\nu=0}^{j} (-1)^{j-\nu} \binom{j}{\nu} \theta^{\nu} \binom{x+\nu}{j} \tag{3.3.9}$$

(see 2.5.16).

Using the method outlined in Ex. 3.9, (3.3.9) reduces to the more convenient form

$$p(x;j,\theta) = \theta^j \sum_{\nu=0}^{j} (-1)^{\nu} \binom{j}{\nu} \left(\frac{1-\theta}{\theta}\right)^{\nu} \binom{x}{\nu} \tag{3.3.10}$$

and this is the required expression for the discrete Laguerre polynomial of degree j.

We list the first five polynomials:

$$p(x;0,\theta) = 1$$

$$p(x;1,\theta) = \theta\left[1 - \left(\frac{1-\theta}{\theta}\right)x\right]$$

$$p(x;2,\theta) = \theta^2\left[1 - 2\left(\frac{1-\theta}{\theta}\right)x + \left(\frac{1-\theta}{\theta}\right)^2\frac{x(x-1)}{2!}\right]$$

$$p(x;3,\theta) = \theta^3\left[1 - 3\left(\frac{1-\theta}{\theta}\right)x + 3\left(\frac{1-\theta}{\theta}\right)^2\frac{x(x-1)}{2!} - \left(\frac{1-\theta}{\theta}\right)^3\frac{x(x-1)(x-2)}{3!}\right]$$

$$p(x;4,\theta) = \theta^4\left[1 - 4\left(\frac{1-\theta}{\theta}\right)x + 6\left(\frac{1-\theta}{\theta}\right)^2\frac{x(x-1)}{2!} - 4\left(\frac{1-\theta}{\theta}\right)^3\frac{x(x-1)(x-2)}{3!} + \left(\frac{1-\theta}{\theta}\right)^4\frac{x(x-1)(x-2)(x-3)}{4!}\right] \tag{3.3.11}$$

This set of polynomials is seen to have an infinite number of elements in contrast to those of the previous section, which form a finite set for any given value of L.

We close this chapter with the derivation of the important quantity

$$[c(j,\theta)]^2 \equiv \sum_{x=0}^{\infty}[p(x;j,\theta)]^2\theta^x \tag{3.3.12}$$

Following the approach used in the derivation of the corresponding quantity for the discrete Legendre polynomials in the preceding section, we arrive at

$$[c(j,\theta)]^2 = (-1)^j j!\,\alpha(j,\theta)\sum_{x=0}^{\infty} g(x;j,\theta) \tag{3.3.13}$$

(see Ex. 3.10)

where $\alpha(j,\theta)$ is the coefficient of $x^{(j)}$ in (3.3.10). This is seen to be

$$\alpha(j,\theta) = \frac{(-1)^j(1-\theta)^j}{j!} \tag{3.3.14}$$

Following the method outlined in Ex. 3.11, we also obtain

$$\sum_{x=0}^{\infty} g(x; j, \theta) = \frac{\theta^j}{(1 - \theta)^{j+1}} \tag{3.3.15}$$

and so finally by (3.3.13), (3.3.14) and (3.3.15)

$$[c(j, \theta)]^2 = \frac{\theta^j}{1 - \theta} \tag{3.3.16}$$

EXERCISES

3.1 Verify (3.2.9) and (3.2.10).

3.2 Given that (c/f (3.2.17) and (3.2.19))

$$g(x; j, L) = \binom{L + j}{j} \sum_{\nu=0}^{j} (-1)^{\nu} \binom{j}{\nu} \frac{(x + j)^{(j+\nu)}}{(L + j)^{(j+\nu)}}$$

verify that

$$\nabla^i g(x; j, L) = \binom{L + j}{j} \sum_{\nu=0}^{j} (-1)^{\nu} \binom{j}{\nu} \frac{(j + \nu)^{(i)} (x + j - i)^{(j+\nu-i)}}{(L + j)^{(j+\nu)}}$$

Now verify that

$$\nabla^i g(x; j, L) = 0 \qquad \text{for } x = L, -1$$
$$\text{and } i = 0, 1, \ldots, j - 1$$

Hint: $(L + j)^{(j+\nu)} = (L + j)^{(i)}(L + j - i)^{(j+\nu-i)}$.

Also verify that

$$\nabla^{2j+1} g(x; j, L) \equiv 0$$

3.3 From the results of Ex. 3.2, verify that (c/f (3.2.18))

$$\nabla^j g(x; j, L) = \sum_{\nu=0}^{j} (-1)^{\nu} \binom{j}{\nu} \binom{j + \nu}{\nu} \frac{x^{(\nu)}}{L^{(\nu)}}$$

3.4 Show, using summation by parts, that (c/f (3.2.25))

$$\sum_{x=0}^{L} x^{(j)} \nabla^j g(x; j, L) = (-1)^j j! \sum_{x=0}^{L} g(x; j, L)$$

3.5 Show that

$$\sum_{x=0}^{L} \frac{(x + j)^{(j+\nu)}}{(L + j)^{(j+\nu)}} = \frac{L + j + 1}{j + \nu + 1}$$

Hence verify that (3.2.27) is correct.

3.6 Let

$$f_j(m) = \sum_{\nu=0}^{j} (-1)^{\nu} \binom{j}{\nu} \left(\frac{m}{m + \nu} \right)$$

Then show that

$$f_j(m) = f_{j-1}(m) - \frac{m}{m+1} f_{j-1}(m + 1)$$

Hint: $\binom{j}{\nu} = \binom{j-1}{\nu} + \binom{j-1}{\nu-1}$

Now verify that

$$f_0(m) = 1 = \frac{1}{\binom{m+0}{0}}$$

and so, using the above recursion, infer that

$$f_j(m) = \frac{1}{\binom{m+j}{j}}$$

Now deduce that

$$\sum_{\nu=0}^{j} (-1)^{\nu} \binom{j}{\nu} \frac{1}{j + \nu + 1} = \left[(2j + 1) \binom{2j}{j} \right]^{-1}$$

3.7 Verify (3.3.5).

3.8 Given that (c/f (3.3.8))

$$g(x;j,\theta) = \theta^x \binom{x}{j}$$

use (2.5.16) to verify that

$$\Delta^i g(x;j,\theta) = \sum_{\nu=0}^{i} (-1)^{i-\nu} \binom{i}{\nu} \theta^{x+\nu} \binom{x+\nu}{j}$$

Now deduce that for $|\theta| < 1$, $i \leq j - 1$,

$$\left.\begin{aligned} \lim_{x\to\infty} \Delta^i g(x;j,\theta) &= 0 \\ \Delta^i g(0;j,\theta) &= 0 \end{aligned}\right\} i = 0, 1, \ldots, j-1 \qquad \text{(c/f (3.3.6))}$$

Also verify that (c/f (3.3.7))

$$\Delta^{j+1}[\theta^{-x} \Delta^j g(x;j,\theta)] = 0$$

3.9 Given that (c/f (3.3.9))

$$p(x;j,\theta) = \sum_{\zeta=0}^{j} (-1)^{j-\zeta} \binom{j}{\zeta} \theta^{\zeta} \binom{x+\zeta}{j}$$

apply the Vandermonde convolution formula (see (2.6.17))

$$\binom{x+\zeta}{j} = \sum_{\mu=0}^{j} \binom{x}{j-\mu} \binom{\zeta}{\mu}$$

to obtain

$$p(x;j,\theta) = \sum_{\mu=0}^{j} \binom{x}{j-\mu} (-1)^{j-\mu} \sum_{\zeta=0}^{j} (-1)^{\zeta-\mu} \binom{j}{\zeta} \binom{\zeta}{\mu} \theta^{\zeta}$$

Now manipulate the term (see (2.6.8) or (2.6.9))

$$\binom{j}{\zeta}\binom{\zeta}{\mu}$$

to obtain

$$p(x;j,\theta) = \sum_{\mu=0}^{j} (-1)^{j-\mu}\binom{x}{j-\mu}\binom{j}{j-\mu}\theta^{\mu} \sum_{\zeta=\mu}^{j} (-1)^{\zeta-\mu}\binom{j-\mu}{\zeta-\mu}\theta^{\zeta-\mu}$$

and so finally, make use of the binomial expansion theorem to obtain

$$p(x;j,\theta) = \theta^{j} \sum_{\nu=0}^{j} (-1)^{\nu}\binom{j}{\nu}\left(\frac{1-\theta}{\theta}\right)^{\nu}\binom{x}{\nu}$$

3.10 Verify (3.3.13) from (3.3.12).

3.11 Define

$$S_j \equiv \sum_{x=0}^{\infty}\binom{x}{j}\theta^{x}$$

Then show, using summation by parts,† that

$$S_j = \frac{\theta}{1-\theta} S_{j-1} \qquad j \geq 1$$

Verify that

$$S_0 = \frac{1}{1-\theta}$$

and hence that

$$S_j = \frac{\theta^{j}}{(1-\theta)^{j+1}}$$

†See (2.7.9) and (2.5.7).

Finally infer that (c/f (3.3.15))

$$\sum_{x=0}^{\infty} g(x;j,\theta) = \frac{\theta^j}{(1-\theta)^{j+1}}$$

where $g(x;j,\theta)$ is the discrete Laguerre polynomial generator of (3.3.8).

REFERENCES

1. Hildebrand, F. B., "Introduction to Numerical Analysis," McGraw-Hill Book Company, 1956, Chapter 7.
2. Abramowitz, M., and Stegun, I. A., "Handbook of Mathematical Functions," National Bureau of Standards, *Applied Math. Series No. 55*, June, 1964.
3. Milne, W. E., "Numerical Calculus," Princeton University Press, 1949, Chapter 9.
4. Gottlieb, M. J., "Concerning Some Polynomials Orthogonal on a Finite or Enumerable Set of Points," *Am. J. Math.*, 60, 453-458, 1938.

4

STATE-VECTORS AND TRANSITION MATRICES

4.1 INTRODUCTION

The first step in implementing a smoothing algorithm is to select a model which we believe adequately describes the process which we are observing. This model is defined in the form of a differential equation.

As a simple example it may be decided that the process can be adequately described by a second degree polynomial, and so the model is taken to be

$$\frac{d^3x(t)}{dt^3} = 0$$

Alternatively, as a result of investigating the physics of the true process, some extremely complicated nonlinear differential equation may be arrived at, to be used as the model.

It is the purpose of this chapter to investigate the methods whereby a chosen model is actually implemented for use in the smoothing algorithms which we propose to develop in the later chapters.

4.2 THE DERIVATIVE POLYNOMIAL MODEL

Perhaps the simplest way in which to model a process is to assume that it can be adequately described, at least locally, by a polynomial of appropriate degree. The efficacy of polynomials as models is based on their well-known ability to approximate any continuous function over a finite interval to any degree of precision. (We refer to the celebrated Weierstrass approximation theorem, for a statement and proof of which see e.g. [4.1].)

As will be seen later, by virtue of their extreme simplicity, polynomials give rise to extremely compact smoothing algorithms. *Moreover they can generally be used for smoothing over short enough intervals with very little actual knowledge of the true process.* It is thus not surprising then, that they are in widespread use in smoothing and prediction work. We accordingly commence our study of models with the polynomial.

Assume that we have a process under observation, which we have decided to characterize as a polynomial. Let $x(t)$ be a polynomial of degree m and let D symbolize differentiation. Thus we abbreviate as follows:

$$D^j x_n \equiv \frac{d^j}{dt^j} x(t)\Big|_{t = t_n} \tag{4.2.1}$$

We consider first the situation where the sampling instants $t_0, t_1, \ldots, t_n$ are *unequally* spaced.

As j goes from 0 to m and n progresses through its successive values, (4.2.1) gives rise to a sequence of vectors of the form

$$X(t_n) \equiv \begin{pmatrix} x \\ \dot{x} \\ \cdot \\ \cdot \\ \cdot \\ D^m x \end{pmatrix}_n \tag{4.2.2}$$

each of which completely defines the polynomial $x(t)$. Thus, given the vector of numbers $X(t_n)$, we can form

$$x(t_n + \zeta) = x_n + \zeta \dot{x}_n + \frac{\zeta^2}{2!} \ddot{x}_n + \cdots + \frac{\zeta^m}{m!} D^m x_n \tag{4.2.3}$$

which is seen simply to be the polynomial $x(t)$, expanded about the point $t = t_n$. Thus for any n, the vector X_n of (4.2.2) provides us with all we need to know about the *state* of the assumed form of the process. It is accordingly referred to as a *state-vector* of the chosen model.†

By the use of expansions similar to (4.2.3) we can write

$$D^i x(t_n + \zeta) = \sum_{j=i}^{m} \frac{\zeta^{j-i}}{(j-i)!} D^j x_n \qquad 0 \le i \le m \tag{4.2.4}$$

Define the matrix $\Phi(\zeta)$ whose i, j^{th} element is

$$[\Phi(\zeta)]_{ij} \equiv \frac{\zeta^{j-i}}{(j-i)!} \qquad 0 \le i, j \le m \tag{4.2.5}$$

where by definition $1/(j - i)!$ is zero when $j < i$. Then (4.2.4) is equivalent to

$$X(t_n + \zeta) = \Phi(\zeta) X(t_n) \tag{4.2.6}$$

As an example, let $m = 2$. Then (4.2.6) becomes‡

$$\begin{pmatrix} x \\ \dot{x} \\ \ddot{x} \end{pmatrix}_{t_n + \zeta} = \begin{pmatrix} 1 & \zeta & \frac{\zeta^2}{2!} \\ & 1 & \zeta \\ & & 1 \end{pmatrix} \begin{pmatrix} x \\ \dot{x} \\ \ddot{x} \end{pmatrix}_{t_n} \tag{4.2.7}$$

which is simply a matrix Taylor's expansion of $x(t)$ and its derivatives about $t = t_n$. ♦♦

Equation (4.2.6) is of great value in that it provides us with a method for computing the value of the state-vector at any instant, given its value at any other instant. It is known as a *transition equation* for the chosen model, and $\Phi(\zeta)$ as the *transition matrix.*

When the sampling instants t_n are *equally* spaced, a simplification is possible. Let the separation between sampling instants be τ seconds. Then $t_n = n\tau$, and if we set $\zeta = h\tau$ say,

$$x(t_n + \zeta) = x(n\tau + h\tau) \equiv x_{n+h} \tag{4.2.8}$$

†As will be seen, more than one choice of state-vector for a given model is possible.

‡Whenever matrices appear as in (4.2.7) with missing elements, then those elements are zeros.

In this case (4.2.4) becomes

$$D^i x_{n+h} = \sum_{j=i}^{m} \frac{(h\tau)^{j-i}}{(j-i)!} D^j x_n \qquad 0 \le i \le m \tag{4.2.9}$$

and the definition for $\Phi(\zeta)$ is replaced by

$$[\Phi(h\tau)]_{ij} \equiv \frac{(h\tau)^{j-i}}{(j-i)!} \qquad 0 \le i, j \le m \tag{4.2.10}$$

giving the transition equation

$$X_{n+h} = \Phi(h\tau) X_n \tag{4.2.11}$$

It is now possible, and frequently convenient, to move τ out of the transition matrix and into the state-vector. We do this by writing (4.2.9) as

$$D^i x_{n+h} = \frac{i!}{\tau^i} \sum_{j=i}^{m} \binom{j}{i} h^{j-i} \frac{\tau^j}{j!} D^j x_n \tag{4.2.12}$$

giving

$$\frac{\tau^i}{i!} D^i x_{n+h} = \sum_{j=i}^{m} \binom{j}{i} h^{j-i} \frac{\tau^j}{j!} D^j x_n \tag{4.2.13}$$

Defining the state-vector

$$Z_n \equiv \begin{pmatrix} x \\ \tau \dot{x} \\ \frac{\tau^2}{2!} \ddot{x} \\ \vdots \\ \frac{\tau^m}{m!} D^m x \end{pmatrix}_n \tag{4.2.14}$$

and the transition matrix $\Phi(h)$ by

$$[\Phi(h)]_{ij} \equiv \binom{j}{i} h^{j-i} \qquad 0 \le i, j \le m \tag{4.2.15}$$

enables us to write (4.2.13) as

$$Z_{n+h} = \Phi(h)\, Z_n \tag{4.2.16}$$

where we see from (4.2.15) *that* $\Phi(h)$ *is now independent of* τ.

As an example, for $h = 1$, $m = 2$, (4.2.16) gives

$$\begin{pmatrix} x \\ \tau\dot{x} \\ \dfrac{\tau^2}{2!}\ddot{x} \end{pmatrix}_{n+1} = \begin{pmatrix} 1 & 1 & 1 \\ & 1 & 2 \\ & & 1 \end{pmatrix} \begin{pmatrix} x \\ \tau\dot{x} \\ \dfrac{\tau^2}{2!}\ddot{x} \end{pmatrix}_{n} \tag{4.2.17}$$

Closer examination of $\Phi(h)$ of (4.2.15) for $h = 1$, shows that it is simply Pascal's triangle (p. 27) arranged so as to form an upper-triangular matrix.

◆◆

The elements of the state-vector Z_n defined in (4.2.14) are seen to be equal to the elements of X_n of (4.2.2) to within scalar multipliers which depend on τ. In fact

$$X_n = D(\tau)\, Z_n \tag{4.2.18}$$

where $D(\tau)$ is a diagonal matrix defined by

$$[D(\tau)]_{ij} \equiv \frac{j!}{\tau^j}\delta_{ij} \qquad 0 \le i, j \le m \tag{4.2.19}$$

δ_{ij} being the Kronecker delta. Thus given Z_n and τ we can easily compute the values of x_n and its derivatives by the use of (4.2.18).

Although we have used the symbol Φ on three separate occasions, namely (4.2.5), (4.2.10) and (4.2.15), the matrices are of course not identical. However, they do have the common property of precisely stating that the model in question is a polynomial. We accordingly refer to them (and certain others to follow) as *polynomial transition matrices.*

They share a number of properties in common. In the first place they all satisfy

$$\Phi(0) = I \tag{4.2.20}$$

which follows immediately, either from their definitions or else from their associated transition relations.

From (4.2.16) we see that

$$Z_n = [\Phi(h)]^{-1} Z_{n+h} \tag{4.2.21}$$

where the inverse is obviously defined, since by inspection $\Phi(h)$ has a determinant equal to unity and is thus nonsingular.† Using (4.2.16) again, we set n to $n - h$ and then h to $-h$ giving

$$Z_n = \Phi(-h) Z_{n+h} \tag{4.2.22}$$

which, when compared with the preceding equation, shows that

$$[\Phi(h)]^{-1} = \Phi(-h) \tag{4.2.23}$$

This relation clearly holds for *any* polynomial transition matrix since it is based entirely on (4.2.16) in which Φ itself was not specified.

As we shall see presently, both (4.2.23) as well as (4.2.20) are special cases of the more general relation

$$[\Phi(h)]^k = \Phi(kh) \tag{4.2.24}$$

satisfied by the transition matrices of any system which satisfies a *constant-coefficient linear differential equation.* Since polynomials are certainly within this class, they too have transition matrices satisfying (4.2.24). We shall return to the proof of (4.2.24) later.

We pause for an example. Setting $h = 1$ in (4.2.15), we see by (4.2.23) that for $m = 3$ say

$$\begin{pmatrix} 1 & 1 & 1 & 1 \\ & 1 & 2 & 3 \\ & & 1 & 3 \\ & & & 1 \end{pmatrix}^{-1} = \begin{pmatrix} 1 & -1 & 1 & -1 \\ & 1 & -2 & 3 \\ & & 1 & -3 \\ & & & 1 \end{pmatrix} \tag{4.2.25}$$

†The fact that these matrices are nonsingular is a property shared by transition matrices in general.

showing that Pascal's triangle does indeed possess some delightful properties. ◆◆

The transition matrices $\Phi(h)$ are used in filtering algorithms *as manipulators of the validity instant.*† Thus, suppose by some means we arrive at an algorithm of the form

$$Z^*_{n,n} = WY_{(n)} \tag{4.2.26}$$

where W is a matrix of weights, $Y_{(n)}$ is a vector of observations defined by

$$Y_{(n)} \equiv \begin{pmatrix} y_n \\ y_{n-1} \\ \vdots \\ y_{n-L} \end{pmatrix} \tag{4.2.27}$$

and where $Z^*_{n,n}$ is an estimate of Z_n in (4.2.14), which is valid at time t_n, based on observations up to t_n, i.e., an *updated estimate* (see p. 8).

Assume that we now need a 1-*step prediction* of Z_n, and let the separation between observations be τ seconds. Then the required prediction can be obtained from (4.2.26) by operating on its output with the transition matrix as follows:

$$Z^*_{n+1,n} = \Phi(1)Z^*_{n,n} \tag{4.2.28}$$

The matrix Φ is defined in (4.2.15).

Alternatively we can obtain the 1-step prediction directly by the use of

$$Z^*_{n+1,n} = \Phi(1)WY_{(n)} \tag{4.2.29}$$

where W and $Y_{(n)}$ are defined as before. This can now be rewritten as

$$Z^*_{n+1,n} = W'Y_{(n)} \tag{4.2.30}$$

where

$$W' \equiv \Phi(1)W \tag{4.2.31}$$

showing how the transition matrix for the chosen model can be absorbed in the weight matrix of the algorithm.

†i.e., the first of the dual subscripts of an estimate vector, e.g., $X^*_{p,q}$.

This use of the transition matrix to move us along a trajectory will occur very frequently, and it can be generalized quite readily to give us predictions and retrodictions to any desired point. It is the explicit way in which we implement the model which we have chosen. Whether the observations are equally spaced or not, if the model chosen is a polynomial, the developments of this section should provide us with an appropriate state-vector/transition matrix pair which mathematically realizes that choice.

The reader is referred to Examples 4.1 through 4.5 which pertain to this section.

4.3 THE DIFFERENCE VECTORS

In the previous section we showed how the choice of a polynomial model could be mathematically realized by state-vector/transition matrix pairs of various kinds. However, in all three cases considered, the state-vectors shared the common property of being based on *derivatives* of the polynomial. In this section, we again assume a polynomial model but specified instead by a state-vector of *backward-differences* rather than derivatives. This approach will prove to be extremely useful when, at a later stage, we apply the orthogonal polynomials of the previous chapter to the synthesis of filtering algorithms.

Accordingly, let $x(t)$ be a polynomial of degree m, and assume *equally spaced* sampling instants. Letting q be the backward-shifting operator defined in Section 2.2, we have

$$x_{n+h} = q^{-h} x_n \tag{4.3.1}$$

Now, $q = 1 - \nabla$ and so

$$\begin{aligned} \nabla^i x_{n+h} &= \nabla^i q^{-h} x_n \\ &= \nabla^i (1 - \nabla)^{-h} x_n \\ &= \sum_{\nu=0}^{m-i} (-1)^\nu \binom{-h}{\nu} \nabla^{\nu+i} x_n \end{aligned} \tag{4.3.2}$$

where the sum terminates at $\nu = m - i$ since, by assumption, $x(t)$ is a polynomial of degree m. Setting $\nu = j - i$, this becomes (see Ex. 4.6)

$$\nabla^i x_{n+h} = \sum_{j=i}^{m} \binom{h + j - i - 1}{j - i} \nabla^j x_n \tag{4.3.3}$$

which is clearly a transition relation of the form

$$U_{n+h} = \Phi(h)\, U_n \tag{4.3.4}$$

with state-vector

$$U_n \equiv \begin{pmatrix} x \\ \nabla x \\ \vdots \\ \nabla^m x \end{pmatrix}_n \tag{4.3.5}$$

and transition matrix defined by

$$[\Phi(h)]_{ij} \equiv \binom{h + j - i - 1}{j - i} \qquad 0 \le i, j \le m \tag{4.3.6}$$

Note that the intersample time τ does not appear anywhere in (4.3.4). (See Ex. 4.7.)

It follows from (4.3.4) that for this matrix

$$\begin{aligned} \Phi(0) &= I \\ [\Phi(h)]^{-1} &= \Phi(-h) \end{aligned} \tag{4.3.7}$$

showing, of course, that $\Phi(h)$ is nonsingular for any h. We shall prove later that

$$[\Phi(h)]^k = \Phi(kh) \tag{4.3.8}$$

which subsumes the two preceding equations.

As an example, we form $\Phi(1)$ from (4.3.6) to obtain (for $m = 3$),

$$\begin{pmatrix} x \\ \nabla x \\ \nabla^2 x \\ \nabla^3 x \end{pmatrix}_{n+1} = \begin{pmatrix} 1 & 1 & 1 & 1 \\ & 1 & 1 & 1 \\ & & 1 & 1 \\ & & & 1 \end{pmatrix} \begin{pmatrix} x \\ \nabla x \\ \nabla^2 x \\ \nabla^3 x \end{pmatrix}_n \tag{4.3.9}$$

Upon closer examination, all of the four equations implicit above are seen to be simply reorganizations of

$$x_{n+1} = 4x_n - 6x_{n-1} + 4x_{n-2} - x_{n-3} \tag{4.3.10}$$

or equivalently

$$\nabla^4 x_n = 0 \tag{4.3.11}$$

which, of course, defines x_n as a cubic. ◆◆

As a second example we use (4.3.8) with $k = -1$ and $h = 1$ to give, (for $m = 3$),

$$\begin{pmatrix} 1 & 1 & 1 & 1 \\ & 1 & 1 & 1 \\ & & 1 & 1 \\ & & & 1 \end{pmatrix}^{-1} = \begin{pmatrix} 1 & -1 & 0 & 0 \\ & 1 & -1 & 0 \\ & & 1 & -1 \\ & & & 1 \end{pmatrix} \tag{4.3.12}$$

◆◆

4.4 THE LINK MATRICES

While the state-vectors consisting of backward or forward differences of x_n, as discussed in the preceding section, will prove to be extremely useful in the derivation of filtering algorithms, they are not too convenient for expressing final answers. One seldom requires smoothed differences and we almost always prefer smoothed derivatives as the final filter outputs. However, as we now show, the *derivatives* of a polynomial can be obtained by *forming appropriate linear combinations of its differences* using the Stirling matrices discussed in Chapter 2.

As before, let x_n be a sampled polynomial of degree m. Then

$$\begin{aligned} qx_n &= x_{n-1} \\ &= x_n - \tau\dot{x}_n + \frac{\tau^2}{2!}\ddot{x}_n - \cdots + (-1)^m \frac{(\tau D)^m}{m!} x_n \\ &= \left[1 - \tau D + \frac{(\tau D)^2}{2!} - \cdots\right] x_n \\ &= e^{-\tau D} x_n \end{aligned} \tag{4.4.1}$$

This gives the operator relationship (valid for polynomials),

$$q = e^{-\tau D} \tag{4.4.2}$$

We make use of this as follows

$$\begin{aligned}
\nabla^i x_n &= (1 - q)^i x_n \\
&= (1 - e^{-\tau D})^i x_n \\
&= \sum_{\nu=0}^{i} (-1)^\nu \binom{i}{\nu} e^{-\nu\tau D} x_n \\
&= \sum_{j=0}^{m} \left[\sum_{\nu=0}^{i} (-1)^\nu \binom{i}{\nu} (-\nu)^j \right] \frac{(\tau D)^j}{j!} x_n
\end{aligned} \tag{4.4.3}$$

Dividing through by $i!$ gives

$$\frac{1}{i!} \nabla^i x_n = \sum_{j=0}^{m} [S^{-1}]_{ij} \frac{(\tau D)^j}{j!} x_n \tag{4.4.4}$$

where the matrix S^{-1} is defined by

$$[S^{-1}]_{ij} \equiv \frac{1}{i!} \sum_{\nu=0}^{i} (-1)^\nu \binom{i}{\nu} (-\nu)^j \tag{4.4.5}$$

We now define the state-vectors

$$V_n \equiv \begin{pmatrix} x \\ \nabla x \\ \frac{1}{2!} \nabla^2 x \\ \vdots \\ \frac{1}{m!} \nabla^m x \end{pmatrix}_n \qquad Z_n \equiv \begin{pmatrix} x \\ \tau \dot{x} \\ \frac{\tau^2}{2!} \ddot{x} \\ \vdots \\ \frac{\tau^m}{m!} D^m x \end{pmatrix}_n \tag{4.4.6}$$

and we then see that (4.4.4) can be written

$$V_n = S^{-1} Z_n \tag{4.4.7}$$

Assuming for the moment that S^{-1} is nonsingular, we thus obtain

$$Z_n = SV_n \tag{4.4.8}$$

which gives the vector of derivatives (to within the scale factor $\tau^j/j!$) in terms of the backward differences (scaled by $1/i!$).

By direct computation on (4.4.5) (see Ex. 4.8) we obtain the first few elements of S^{-1} as

$$S^{-1} = \begin{pmatrix} 1 & 0 & 0 & 0 \\ 0 & 1 & -1 & 1 \\ 0 & 0 & 1 & -3 \\ 0 & 0 & 0 & 1 \end{pmatrix} \tag{4.4.9}$$

which is clearly nonsingular. Starting with the bottom row, we can easily verify that its inverse is

$$S = \begin{pmatrix} 1 & 0 & 0 & 0 \\ 0 & 1 & 1 & 2 \\ 0 & 0 & 1 & 3 \\ 0 & 0 & 0 & 1 \end{pmatrix} \tag{4.4.10}$$

and so by (4.4.8), for $m = 3$ (i.e., a cubic),

$$\begin{pmatrix} x \\ \tau \dot{x} \\ \dfrac{\tau^2}{2!}\ddot{x} \\ \dfrac{\tau^3}{3!}\dddot{x} \end{pmatrix}_n = \begin{pmatrix} 1 & 0 & 0 & 0 \\ 0 & 1 & 1 & 2 \\ 0 & 0 & 1 & 3 \\ 0 & 0 & 0 & 1 \end{pmatrix} \begin{pmatrix} x \\ \nabla x \\ \dfrac{1}{2!}\nabla^2 x \\ \dfrac{1}{3!}\nabla^3 x \end{pmatrix}_n \tag{4.4.11}$$

Equation (4.4.8) thus provides the link between the derivative and the backward-difference state-vectors of a polynomial. We accordingly call S a *link matrix*. (See Ex. 4.9.)

Comparing S of the above example to S, the Stirling matrix of the first kind (Table 2.1 on p. 23), we see that the elements of S can be obtained by *transposing* the matrix S and *removing all minus signs.* For this reason we have called S the *associate* Stirling matrix of the first kind and we display it in Table 4.1 up to $i, j = 10$.†

The matrix S^{-1} can be obtained analogously from S^{-1}, the Stirling matrix of the second kind (Table 2.2 on p. 25), by transposing that table and negating the *odd* superdiagonals. This gives S^{-1}, the associate Stirling matrix of the second kind, as displayed in Table 4.2.‡ By using the matrix S together with appropriate scale factors, we are thus able to obtain derivative state-vectors from the backward-difference state-vectors. We can, of course, also use the transition matrices of the previous sections in combination with the link matrices, as demonstrated in the following example.

In subsequent work we shall obtain an algorithm which gives us $V^*_{n,n}$, the updated estimate (0-step prediction) of the *backward-difference* state-vector of (4.4.6). In order to then obtain a 1-step prediction of the *derivative* state-vector $Z^*_{n+1,n}$ of that equation, we will use

$$Z^*_{n+1,n} = \Phi(1)\, S V^*_{n,n} \tag{4.4.12}$$

where S is the associate Stirling matrix of the first kind and $\Phi(1)$ is defined in (4.2.15). ♦♦

The reader can follow the derivation which led to (4.4.8) and satisfy himself that the *forward-difference* and derivative state-vectors, namely

$$V_n \equiv \begin{pmatrix} x \\ \Delta x \\ \dfrac{1}{2!}\Delta^2 x \\ \vdots \\ \dfrac{1}{m!}\Delta^m x \end{pmatrix}_n \qquad Z_n \equiv \begin{pmatrix} x \\ \tau \dot{x} \\ \dfrac{\tau^2}{2!}\ddot{x} \\ \vdots \\ \dfrac{\tau^m}{m!} D^m x \end{pmatrix}_n \tag{4.4.13}$$

†We give a recursion formula for generating S and S^{-1} in the notes at the end of this chapter.

‡Observe from Tables 4.1, 4.2 that both S and S^{-1} are nonsingular, regardless of their order, by virtue of the fact that they are upper-triangular with 1's on the diagonal.

Table 4.1 S (The Associate Stirling Matrix of the First Kind)

i \ j	0	1	2	3	4	5	6	7	8	9	10
0	1	0	0	0	0	0	0	0	0	0	0
1	0	1	1	2	6	24	120	720	5040	40320	362880
2	0	0	1	3	11	50	274	1764	13068	109584	1026576
3	0	0	0	1	6	35	225	1624	13132	118124	1172700
4	0	0	0	0	1	10	85	735	6769	67284	723680
5	0	0	0	0	0	1	15	175	1960	22449	269325
6	0	0	0	0	0	0	1	21	322	4536	63273
7	0	0	0	0	0	0	0	1	28	546	9450
8	0	0	0	0	0	0	0	0	1	36	870
9	0	0	0	0	0	0	0	0	0	1	45
10	0	0	0	0	0	0	0	0	0	0	1

Table 4.2 S^{-1} **(The Associate Stirling Matrix of the Second Kind)**

i \ j	0	1	2	3	4	5	6	7	8	9	10
0	1	0	0	0	0	0	0	0	0	0	0
1	0	1	−1	1	−1	1	−1	1	−1	1	−1
2	0	0	1	−3	7	−15	31	−63	127	−255	511
3	0	0	0	1	−6	25	−90	301	−966	3025	−9330
4	0	0	0	0	1	−10	65	−350	1701	−7770	34105
5	0	0	0	0	0	1	−15	140	−1050	6951	−42525
6	0	0	0	0	0	0	1	−21	266	−2646	22827
7	0	0	0	0	0	0	0	1	−28	462	−5880
8	0	0	0	0	0	0	0	0	1	−36	750
9	0	0	0	0	0	0	0	0	0	1	−45
10	0	0	0	0	0	0	0	0	0	0	1

are likewise related by

$$Z_n = S^T V_n \tag{4.4.14}$$

where S^T is simply the *transpose* of the Stirling matrix of the first kind (see p. 23).

4.5 THE LAGRANGE ESTIMATOR

One of the classical problems in numerical analysis is the construction of the polynomial which passes through a set of equally spaced ordinates. One method in wide use is that due to Lagrange and results in expressions for the polynomial and its derivatives as linear combinations of the ordinates through which the polynomial is to pass. We present here a method for performing Lagrange interpolation which involves considerably less effort than the classical approach (see e.g. [4.2]), while at the same time serving as a vehicle for the ideas being developed in this chapter. It also leads directly to a programmable algorithm.

Suppose we have a process $x(t)$, *not necessarily a polynomial,* which is sampled without errors at equispaced instants τ seconds apart. This gives us a train of observations

$$y_0 = x(0),\ y_1 = x(\tau),\ \ldots,\ y_n = x(n\tau)$$

Assume that we are located in time just after $t = n\tau$ but before $(n+1)\tau$, and that we wish to obtain predictions of what $x(t)$ and its derivatives will be at $(n+1)\tau$.

We decide to do this by passing a polynomial of degree L through the most recent $L+1$ observations

$$y_{n-L},\ y_{n-L+1},\ \ldots,\ y_n$$

and then to obtain estimates of the required 1-step predictions by the evaluation of this polynomial and its derivatives at $(n+1)\tau$.

Assume that n is fixed, i.e., *we freeze the further passage of time* so that, for the moment, no additional observations are made. Then the polynomial passing through the 3 points y_{n-2}, y_{n-1}, y_n is shown in Figure 4.1 where for definiteness we are considering the quadratic case $(L = 2)$.

We symbolize the required polynomial by $[p^*(r)]_n$, where the star symbolizes estimation and the continuous variable r is the *polynomial argument,* with the r-origin, as shown in the figure, being taken at the beginning of the interval containing the observations of interest. The subscript n in $[p^*(r)]_n$ is

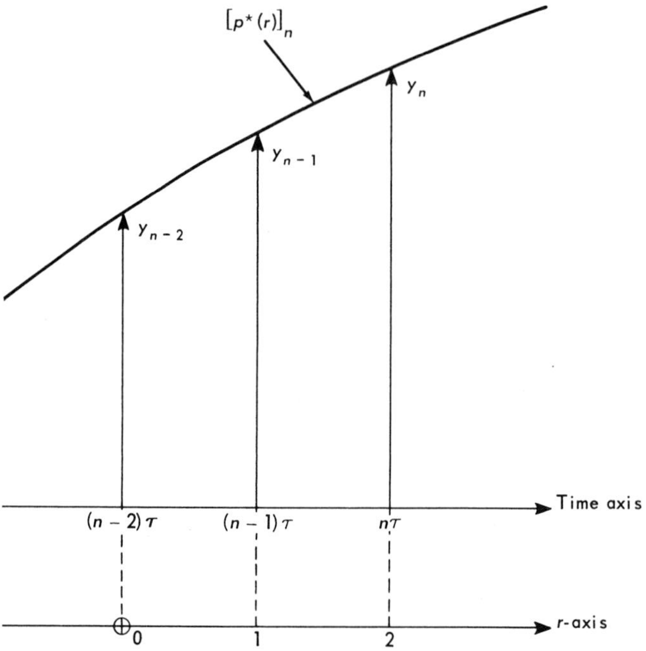

Fig. 4.1 *The interpolation polynomial* $[p^*(r)]_n$.

to remind us that this polynomial is based on the set of observations whose most recent member is y_n. It thereby serves to differentiate between this interpolating polynomial and its predecessors and successors, which may or may not be the same polynomial as $[p^*(r)]_n$.

Since, by assumption, $[p^*(r)]_n$ is to pass through y_{n-2}, y_{n-1}, y_n, we must have

$$
\begin{aligned}
[p^*(2)]_n &= y_n \\
[p^*(1)]_n &= y_{n-1} \\
[p^*(0)]_n &= y_{n-2}
\end{aligned}
\tag{4.5.1}
$$

We now evaluate successive backward differences† of $[p^*(r)]_n$ at $r = 2$, by the use of (4.5.1), i.e.,

$$
\begin{pmatrix} \nabla^0 p^*(2) \\ \nabla^1 p^*(2) \\ \nabla^2 p^*(2) \end{pmatrix}_n = \begin{pmatrix} y_n \\ y_n - y_{n-1} \\ y_n - 2y_{n-1} + y_{n-2} \end{pmatrix}
\tag{4.5.2}
$$

†With respect to r.

and defining the vector $V^*_{n,n}$ by

$$V^*_{n,n} = \begin{pmatrix} \frac{1}{0!}\nabla^0 p^*(2) \\ \frac{1}{1!}\nabla^1 p^*(2) \\ \frac{1}{2!}\nabla^2 p^*(2) \end{pmatrix}_n \tag{4.5.3}$$

we write (4.5.2) as

$$V^*_{n,n} = \begin{pmatrix} \frac{1}{0!} & 0 & 0 \\ 0 & \frac{1}{1!} & 0 \\ 0 & 0 & \frac{1}{2!} \end{pmatrix} \begin{pmatrix} 1 & 0 & 0 \\ 1 & -1 & 0 \\ 1 & -2 & 1 \end{pmatrix} \begin{pmatrix} y_n \\ y_{n-1} \\ y_{n-2} \end{pmatrix} \tag{4.5.4}$$

We call the vector of observations in the above equation $Y_{(n)}$, i.e., we define

$$Y_{(n)} \equiv \begin{pmatrix} y_n \\ y_{n-1} \\ y_{n-2} \end{pmatrix} \tag{4.5.5}$$

(Note that the subscript of $Y_{(n)}$ is *parenthesized* to show that it is not a sampling instant but rather the ordered set of observations whose *most recent* is y_n.) Then (4.5.4) can be written

$$V^*_{n,n} = F^{-1} B Y_{(n)} \tag{4.5.6}$$

where F^{-1} is the inverse of the matrix

$$F \equiv \begin{pmatrix} 0! & 0 & 0 \\ 0 & 1! & 0 \\ 0 & 0 & 2! \end{pmatrix} \tag{4.5.7}$$

i.e., F is the diagonal matrix of factorials, and where

$$B \equiv \begin{pmatrix} 1 & 0 & 0 \\ 1 & -1 & 0 \\ 1 & -2 & 1 \end{pmatrix} \tag{4.5.8}$$

is a weight matrix easily constructed using the binomial coefficients.

We set out to obtain the 1-step prediction of the derivative state-vector, and this is now easily derived as follows. Let

$$Z^*_{n,n} \equiv \begin{pmatrix} x^* \\ \tau \dot{x}^* \\ \frac{\tau^2}{2!} \ddot{x}^* \end{pmatrix}_{n,n} \tag{4.5.9}$$

Then by (4.4.12)

$$Z^*_{n+1,n} = \Phi(1)\, SV^*_{n,n} \tag{4.5.10}$$

and, taking into account (4.5.6), we thus have

$$Z^*_{n+1,n} = \Phi(1)\, SF^{-1}BY_{(n)} \tag{4.5.11}$$

Finally the derivatives themselves, namely

$$X^*_{n+1,n} \equiv \begin{pmatrix} x^* \\ \dot{x}^* \\ \ddot{x}^* \end{pmatrix}_{n+1,n} \tag{4.5.12}$$

may be obtained using (4.2.18), i.e.

$$X^*_{n+1,n} = D(\tau)\, \Phi(1)\, SF^{-1}BY_{(n)} \tag{4.5.13}$$

This expression, while perhaps appearing to be a bit formidable, is actually easily reduced to

$$X^*_{n+1,n} = WY_{(n)} \tag{4.5.14}$$

in which form it is readily implemented in a computer. For the quadratic case the reader can easily verify that (4.5.13) gives us, (assuming $\tau = 1$)

$$W = \begin{pmatrix} 1 & 0 & 0 \\ & 1 & 0 \\ & & 2 \end{pmatrix}\begin{pmatrix} 1 & 1 & 1 \\ & 1 & 2 \\ & & 1 \end{pmatrix}\begin{pmatrix} 1 & 0 & 0 \\ & 1 & 1 \\ & & 1 \end{pmatrix}\begin{pmatrix} 1 & 0 & 0 \\ & 1 & 0 \\ & & \frac{1}{2} \end{pmatrix}\begin{pmatrix} 1 & & \\ 1 & -1 & \\ 1 & -2 & 1 \end{pmatrix}$$

$$= \begin{pmatrix} 3 & -3 & 1 \\ \frac{5}{2} & -4 & \frac{3}{2} \\ 1 & -2 & 1 \end{pmatrix} \qquad (4.5.15)$$

W is called the *weight matrix* since it provides the numbers by which the observations $Y_{(n)}$ are weighted in (4.5.14) to give $X^*_{n+1,n}$.

We can easily generalize as follows. Let

$$\begin{aligned} [B]_{ij} &\equiv (-1)^j \binom{i}{j} \qquad 0 \le i, j \le L \\ [F]_{ij} &\equiv j!\,\delta_{ij} \qquad 0 \le i, j \le L \end{aligned} \qquad (4.5.16)$$

(Note that F^{-1} and not F is used in (4.5.13).) The matrix S can be obtained for any L up to 10 from Table 4.1 on p. 85. $\Phi(h)$ is defined in (4.2.15) and $D(\tau)$ in (4.2.19). Then

$$X^*_{n+h,n} = D(\tau)\,\Phi(h)\,SF^{-1}BY_{(n)} \qquad (4.5.17)$$

gives the h-step prediction of the derivative state-vector based on estimation by polynomial interpolation.

The weight matrix of (4.5.17) namely

$$W(h, \tau) \equiv D(\tau)\,\Phi(h)\,SF^{-1}B \qquad (4.5.18)$$

can be evaluated once and for all when τ, h and L are fixed, thereby yielding the $L + 1$ algorithms contained in

$$X^*_{n+h,n} = W(h, \tau)\,Y_{(n)} \qquad (4.5.19)$$

in their final computational form.

Of course we now unfreeze n and permit it to cycle through successive values, thereby obtaining successive estimates of the h-step prediction of the derivatives of the process producing the observations. As the numbers which form $Y_{(n)}$ are obtained and fed to the algorithm, so the vectors $X^*_{n+h,n}$ are produced and constitute the output of the algorithm. The scheme is readily implemented on a digital computer and can be seen to be a simple, real-time way of obtaining estimates based on the interpolating polynomial. (See Ex. 4.10.)

4.6 CONSTANT-COEFFICIENT LINEAR DIFFERENTIAL EQUATIONS

Until now, we have considered only one form of model, namely the polynomial. Instead of viewing a polynomial in its functional form

$$x(t) = \sum_{j=0}^{m} \alpha_j t^j \tag{4.6.1}$$

we could also view it in the equivalent form

$$D^{m+1}x(t) = 0 \tag{4.6.2}$$

This differential equation generates, as its solutions, the *entire class* of polynomials of degree m as given by (4.6.1).

Suppose, instead of (4.6.2) we consider the somewhat more general *linear, constant-coefficient*, differential equation

$$(D^{m+1} + \gamma_m D^m + \cdots + \gamma_1 D + \gamma_0)x(t) = 0 \tag{4.6.3}$$

where the γ_i are constants. Thus (4.6.2) is a special case of (4.6.3). For definiteness we let $m = 2$, and consider

$$(D^3 + \gamma_2 D^2 + \gamma_1 D + \gamma_0)x(t) = 0 \tag{4.6.4}$$

Define as the state-vector

$$X(t) \equiv \begin{pmatrix} x(t) \\ \dot{x}(t) \\ \ddot{x}(t) \end{pmatrix} \tag{4.6.5}$$

Then using (4.6.4) to obtain $\dddot{x}$ in terms of x, $\dot{x}$ and $\ddot{x}$, we see that

$$\frac{d}{dt} X(t) \equiv \begin{pmatrix} \dot{x} \\ \ddot{x} \\ \dddot{x} \end{pmatrix}_t = \begin{pmatrix} 0 & 1 & 0 \\ 0 & 0 & 1 \\ -\gamma_0 & -\gamma_1 & -\gamma_2 \end{pmatrix} \begin{pmatrix} x \\ \dot{x} \\ \ddot{x} \end{pmatrix}_t \tag{4.6.6}$$

i.e., (4.6.4) can be stated as

$$\frac{d}{dt} X(t) = C\,X(t) \tag{4.6.7}$$

where

$$C \equiv \begin{pmatrix} 0 & 1 & 0 \\ 0 & 0 & 1 \\ -\gamma_0 & -\gamma_1 & -\gamma_2 \end{pmatrix} \tag{4.6.8}$$

As a final generalization, we can consider the transformation

$$\hat{X}(t) = G\,X(t) \tag{4.6.9}$$

where G is a 3 x 3 nonsingular constant matrix. Then (4.6.7) becomes

$$\frac{d}{dt} \hat{X}(t) = A\hat{X}(t) \tag{4.6.10}$$

where

$$A \equiv GCG^{-1} \tag{4.6.11}$$

and since G is arbitrary, A is *any* 3 x 3 matrix of constants.

Now, it is well known from the elementary theory of differential equations, that the solutions of (4.6.4) are of the form

$$x(t) = \sum_{j=1}^{k} p_j(t)\, e^{\lambda_j t} \tag{4.6.12}$$

where $p_j(t)$ is a polynomial in t and λ_j is a zero of the characteristic equation of (4.6.4) namely

$$\lambda^3 + \gamma_2 \lambda^2 + \gamma_1 \lambda + \gamma_0 = 0 \tag{4.6.13}$$

and where k equals the number of *distinct* roots of this equation. The degree of the polynomial $p_j(t)$ is one less than the multiplicity with which λ_j occurs as a root of (4.6.13).

Since (4.6.10) is simply a restatement of (4.6.4), it follows that the solution to (4.6.10) is likewise of the form

$$x(t) = \sum_{j=0}^{k} p_j(t) e^{\lambda_j t} \tag{4.6.14}$$

where the coefficients of $p_j(t)$ depend on initial conditions.

When $\gamma_2 = \gamma_1 = \gamma_0 = 0$, (4.6.13) becomes

$$\lambda^3 = 0 \tag{4.6.15}$$

which means that we have a single root, namely zero, of triple multiplicity. In this case (4.6.14) takes on the form

$$x(t) = \alpha_0 + \alpha_1 t + \alpha_2 t^2 \tag{4.6.16}$$

which is, of course, the general polynomial of degree 2.

We see then that in general, the following three equations are equivalent

$$(D^{m+1} + \gamma_m D^m + \cdots + \gamma_1 D + \gamma_0) x(t) = 0 \tag{4.6.17}$$

$$\frac{d}{dt} X(t) = AX(t) \tag{4.6.18}$$

$$x(t) = \sum_{j=0}^{k} p_j(t) e^{\lambda_j t} \tag{4.6.19}$$

and that the polynomial model considered previously is simply just a special case of this more general model, *comprised of products of polynomials and exponentials.*

In practice, the polynomials (not multiplied by exponentials) are probably the only really useful subset of these functions. This accounts for the detailed treatment given to the polynomial transition matrices in the earlier sections of the chapter. However, the subsequent developments of the present section are a necessary preamble to the extensions which we intend to make later, and for that reason we continue our analysis further.

We now examine the solutions of (4.6.18) in more detail. First we write the vector Taylor's series

$$X(t+\zeta) = X(t) + \zeta DX(t) + \frac{\zeta^2}{2!} D^2X(t) + \cdots$$
$$= \sum_{\nu=0}^{\infty} \frac{\zeta^\nu}{\nu!} D^\nu X(t) \tag{4.6.20}$$

But by (4.6.18)

$$D^\nu X(t) = A^\nu X(t) \tag{4.6.21}$$

and so (4.6.20) can be written

$$X(t+\zeta) = \left[\sum_{\nu=0}^{\infty} \frac{(\zeta A)^\nu}{\nu!}\right] X(t) \tag{4.6.22}$$

It is shown in texts on matrix theory (see e.g. [4.3, p. 335 et seq]) that the infinite sum of matrices above converges to a well-defined matrix $G(\zeta A)$ for any A and ζ, with the property that

$$G(\zeta_1 A)\,G(\zeta_2 A) = G[(\zeta_1+\zeta_2)A]$$
$$[G(\zeta_1 A)]^k = G(k\zeta_1 A) \tag{4.6.23}$$
$$\frac{d}{d\zeta} G(\zeta A) = G(\zeta A)\,A$$

Since this is reminiscent of the *scalar exponential function* we accordingly write

$$G(\zeta A) \equiv \exp(\zeta A) \equiv \sum_{\nu=0}^{\infty} \frac{(\zeta A)^\nu}{\nu!} \tag{4.6.24}$$
(See Note),

and (4.6.22) becomes

$$X(t+\zeta) = \exp(\zeta A)\,X(t) \tag{4.6.25}$$

Note: The matrix exponential $\exp(\zeta A)$ does *not* share all of the properties of the scalar $\exp(\zeta a)$. Thus, e.g., $\exp(A)\exp(B) \neq \exp(A+B)$ unless A and B commute.

Thus, given the state-vector valid at time t, (4.6.25) gives the value of the state-vector at any other time. This equation is in fact a transition relation, and so we write it as

$$X(t_n + \zeta) = \Phi(\zeta) X(t_n) \tag{4.6.26}$$

where

$$\Phi(\zeta) \equiv \exp(\zeta A) \tag{4.6.27}$$

For now the function $\exp(\zeta A)$ can be thought of by its power-series, i.e.

$$\exp(\zeta A) = I + \zeta A + \frac{\zeta^2}{2!} A^2 + \frac{\zeta^3}{3!} A^3 + \cdots \tag{4.6.28}$$

although there are more powerful ways possible to actually evaluate it, as we shall presently show.

For the case of the polynomial model we now see that we can obtain each of the transition matrices of the preceding sections by the use of (4.6.27). (See Ex. 4.11 and 4.12.) Thus for any of the polynomial transition matrices already derived, there exists an associated matrix A such that (4.6.27) is satisfied.

It follows immediately from (4.6.27) and (4.6.23) that

$$[\Phi(\zeta)]^k = \Phi(k\zeta) \tag{4.6.29}$$

thereby proving (4.2.24) and (4.3.8). We can also deduce from (4.6.28) that $\Phi(0) = I$. Moreover, setting $k = -1$ in (4.6.29) shows that $\Phi(\zeta)$ is nonsingular for any ζ (and any A).

Returning to (4.6.26) we have

$$X(\zeta) = \Phi(\zeta) X(0) \tag{4.6.30}$$

and so, by differentiation of both sides,

$$\left[\frac{d}{d\zeta} \Phi(\zeta)\right] X(0) = \frac{d}{d\zeta} X(\zeta) \tag{4.6.31}$$

(We use the symbol $(d/d\zeta)\, \Phi(\zeta)$ to mean the matrix obtained by differentiating each element of $\Phi(\zeta)$.) Combining the above equation with (4.6.18), we obtain

$$\left[\frac{d}{d\zeta}\Phi(\zeta)\right]X(0) = AX(\zeta) \qquad (4.6.32)$$
$$= A\Phi(\zeta)X(0) \qquad \text{(by (4.6.30))}$$

and so, since $X(0)$ is arbitrary, we must have that

$$\frac{d}{d\zeta}\Phi(\zeta) = A\Phi(\zeta) \qquad (4.6.33)$$

Thus both the state-vector $X(t)$ *and the transition matrix* $\Phi(\zeta)$ *satisfy the same differential equation.* (See Ex. 4.13.)

Finally, consider the associated differential equation

$$\frac{d}{d\zeta}\Psi(\zeta) = -\Psi(\zeta)A \qquad (4.6.34)$$

Its solution is

$$\Psi(\zeta) = \Psi(0)\exp(-\zeta A) \qquad (4.6.35)$$

which can be verified as follows. By direct differentiation, (4.6.35) gives

$$\frac{d}{d\zeta}\Psi(\zeta) = -\Psi(0)\exp(-\zeta A)A \qquad (4.6.36)$$
$$= -\Psi(\zeta)A \qquad \text{(by (4.6.23))}$$

which is the same as (4.6.34), and so (4.6.35) is in fact the solution.

Suppose we now take, as initial conditions for (4.6.34),

$$\Psi(0) = I \qquad (4.6.37)$$

Then (4.6.35) shows that

$$\Psi(\zeta) = \exp(-\zeta A) = [\Phi(\zeta)]^{-1} \qquad (4.6.38)$$

i.e., (4.6.34) *is the differential equation for the inverse of* $\Phi(t)$. (See Ex. 4.13.) It is thus possible, if required, to obtain $\Phi(\zeta)$ and its inverse $\Psi(\zeta)$ from the matrix A, *by numerical integration of the system of differential equations*

$$\frac{d}{d\zeta}\Phi(\zeta) = A\Phi(\zeta) \qquad (4.6.39)$$

$$\frac{d}{d\zeta}\Psi(\zeta) = -\Psi(\zeta)A \tag{4.6.40}$$

with initial conditions

$$\Phi(0) = I = \Psi(0) \tag{4.6.41}$$

We shall return briefly to this question of numerical integration in the next section.

Consider the following simple example. Given the matrix

$$G(\zeta) = \begin{pmatrix} 1-\zeta & \zeta \\ -\zeta & 1+\zeta \end{pmatrix} \tag{4.6.42}$$

we wish to determine whether or not it is a transition matrix.

If it is, then by (4.6.27), there must be a matrix A so that

$$G(\zeta) = \exp(\zeta A) \tag{4.6.43}$$

i.e., we must be able to show that, for some A,

$$G(\zeta) = I + \zeta A + \frac{\zeta^2}{2!}A^2 + \frac{\zeta^3}{3!}A^3 + \cdots \tag{4.6.44}$$

We can determine the probable form of A by writing (4.6.42) as

$$G(\zeta) = \begin{pmatrix} 1 & 0 \\ 0 & 1 \end{pmatrix} + \zeta\begin{pmatrix} -1 & 1 \\ -1 & 1 \end{pmatrix} \tag{4.6.45}$$

which appears to be a truncated version of (4.6.44). From (4.6.45) the matrix A must be

$$A = \begin{pmatrix} -1 & 1 \\ -1 & 1 \end{pmatrix} \tag{4.6.46}$$

and since for this matrix, A^2 and all higher powers are null, we see that $G(\zeta)$ does indeed satisfy (4.6.44). Hence it is in fact a transition matrix.

We also verify that (4.6.39) and (4.6.41) hold true for $G(\zeta)$. Thus, from (4.6.42)

$$\frac{dG(\zeta)}{d\zeta} = \begin{pmatrix} -1 & 1 \\ -1 & 1 \end{pmatrix} \tag{4.6.47}$$

and from (4.6.46),

$$AG(\zeta) = \begin{pmatrix} -1 & 1 \\ -1 & 1 \end{pmatrix} \begin{pmatrix} 1-\zeta & \zeta \\ -\zeta & 1+\zeta \end{pmatrix} \tag{4.6.48}$$

$$= \begin{pmatrix} -1 & 1 \\ -1 & 1 \end{pmatrix}$$

Thus

$$\frac{dG(\zeta)}{d\zeta} = AG(\zeta) \tag{4.6.49}$$

and clearly also

$$G(0) = I \tag{4.6.50}$$

showing that $G(\zeta)$ does indeed satisfy (4.6.39) and (4.6.41). ◆◆

The reader can now verify, by the same argument, that

$$G(\zeta) \equiv \begin{pmatrix} \cos\zeta & \sin\zeta \\ -\sin\zeta & \cos\zeta \end{pmatrix} \tag{4.6.51}$$

is a transition matrix, whereas

$$G(\zeta) \equiv \begin{pmatrix} 1 & \zeta \\ \zeta & 1 \end{pmatrix} \tag{4.6.52}$$

is not.

4.7 TIME-VARYING LINEAR DIFFERENTIAL EQUATIONS

The next step in the generalization of the choice of models is to assume that the process satisfies a linear differential equation with *time-varying* coefficients, i.e.

$$\frac{d}{dt} X(t) = A(t)\, X(t) \tag{4.7.1}$$

(The preceding discussion in which $A(t)$ was a matrix of constants is, of course, subsumed by this more general model.)

We attempt to solve (4.7.1) using a vector Taylor's series as we did in (4.6.20). Thus we form the series

$$
\begin{aligned}
X(t) &= X(0) + t\left[DX(t)\Big|_{t=0}\right] + \frac{t^2}{2!}\left[D^2X(t)\Big|_{t=0}\right] + \cdots \\
&= \sum_{\nu=0}^{\infty} \frac{t^\nu}{\nu!}\left[D^\nu X(t)\Big|_{t=0}\right]
\end{aligned}
\tag{4.7.2}
$$

In this case, however, *the time-varying nature of the coefficient matrix* $A(t)$ *complicates matters considerably.* Thus, by (4.7.1)

$$
\begin{aligned}
DX &= AX \\
D^2X &= D(AX) = \dot{A}X + A\dot{X} = (\dot{A} + A^2)X \\
D^3X &= (\ddot{A} + A\dot{A} + \dot{A}A)X + (\dot{A} + A^2)AX \\
&= (\ddot{A} + A\dot{A} + 2\dot{A}A + A^3)X
\end{aligned}
\tag{4.7.3}
$$

and so the simple result obtained in (4.6.22) now fails to materialize.

However, under fairly relaxed conditions on $A(t)$ (see e.g. [4.6]), *by virtue of the linearity of* (4.7.1) *there also exists a linear relationship between* $X(t)$ *and its initial value* $X(0)$. Thus, it can be shown that there exists a matrix $P(t)$ so that $X(t)$ of (4.7.1) satisfies

$$
X(t) = P(t)X(0) \tag{4.7.4}
$$

and that the matrix $P(t)$ is nonsingular.

The above equation is suggestive of a transition relation, but on comparing it to (4.6.25), the essential difference between the time-varying and constant-coefficient models becomes very clear; in the present case the matrix $P(t)$ does not have the simple exponential form which it had previously. An immediate result is that, in general,

$$
[P(t)]^k \neq P(kt) \tag{4.7.5}
$$

equality holding *only* for the transition matrices which stem from constant-coefficient models (c/f (4.6.29)).

In order to examine further the properties of the transition matrices associated with time-varying linear systems, we first manipulate (4.7.4) into a more useful form. Thus (4.7.4) gives us the equations

$$
\begin{aligned}
X(t_n + \zeta) &= P(t_n + \zeta)X(0) \\
X(t_n) &= P(t_n)X(0)
\end{aligned}
\tag{4.7.6}
$$

and so, by combining them, we obtain

$$X(t_n + \zeta) = P(t_n + \zeta)[P(t_n)]^{-1} X(t_n) \tag{4.7.7}$$

in which we have used the fact that $P(t_n)$ is nonsingular.

Define

$$\Phi(t_n + \zeta, t_n) \equiv P(t_n + \zeta)[P(t_n)]^{-1} \tag{4.7.8}$$

(The double argument of Φ means that each of its elements is a function of *two* variables.) Then (4.7.7) can be written

$$X(t_n + \zeta) = \Phi(t_n + \zeta, t_n) X(t_n) \tag{4.7.9}$$

which is a transition relation in the more usual form.

This equation displays the time-varying nature of the process very clearly, since it shows that the transition matrix relating $X(t_n)$ and $X(t_n + \zeta)$ depends not only on the separation interval ζ, as it did formerly, *but also on* t_n, *the time at which the transition is occurring.* (See Ex. 4.14 and 4.15.) We note that (4.7.9) can be written

$$X(t_m) = \Phi(t_m, t_n) X(t_n) \tag{4.7.10}$$

and that the matrix Φ satisfies

$$\Phi(t_n, t_n) = I \tag{4.7.11}$$

The matrix $\Phi(t_n + \zeta, t_n)$ as defined by (4.7.8) is clearly not easily obtained from that equation, and we now consider how one does in fact generate $\Phi(t_n + \zeta, t_n)$ given the matrix $A(t)$ of (4.7.1). To do this, we examine the differential equation which Φ satisfies, and based on our experience with the constant-coefficient case, we suspect that it is of the same form as (4.7.1).

In forming (4.7.6), we set $t = t_n + \zeta$. Thus, differentiation with respect to t and ζ are equivalent. Differentiating (4.7.9) gives

$$\frac{\partial}{\partial \zeta} \Phi(t_n + \zeta, t_n) X(t_n) = \frac{\partial}{\partial \zeta} X(t_n + \zeta) \tag{4.7.12}$$

Then by (4.7.1) and (4.7.9)

$$\begin{aligned} \frac{\partial}{\partial \zeta} \Phi(t_n + \zeta, t_n) X(t_n) &= A(t_n + \zeta) X(t_n + \zeta) \\ &= A(t_n + \zeta) \Phi(t_n + \zeta, t_n) X(t_n) \end{aligned} \tag{4.7.13}$$

and since $X(t_n)$ can be made to assume an infinity of values depending on $X(0)$, we must have

$$\frac{\partial}{\partial \zeta} \Phi(t_n + \zeta, t_n) = A(t_n + \zeta)\Phi(t_n + \zeta, t_n) \tag{4.7.14}$$

Φ and X do in fact both satisfy the same differential equation.

We can obtain Φ by *numerical integration* of (4.7.14), using any chosen integration technique. One simple integration procedure can be obtained as follows. By Taylor's theorem

$$\Phi(t_n + h, t_n) = \Phi(t_n, t_n) + h \frac{\partial}{\partial \zeta} \Phi(t_n + \zeta, t_n)\Big|_{\zeta=0} + 0(h^2) \tag{4.7.15}$$

The term $0(h^2)$ is actually

$$0(h^2) \equiv \frac{h^2}{2!} \frac{\partial^2}{\partial \zeta^2} \Phi(t_n + \zeta, t_n)\Big|_{\zeta=\xi} \qquad 0 < \xi < h \tag{4.7.16}$$

and so goes to zero like h^2. Neglecting this error term and taking account of (4.7.11) and (4.7.14), we obtain

$$\Phi(t_n + h, t_n) = I + hA(t_n) \tag{4.7.17}$$

This represents the initial step of a numerical integration algorithm for $\Phi(t_n + \zeta, t_n)$.

First we note that if ζ is sufficiently small, we simply set $h = \zeta$ and obtain $\Phi(t_n + \zeta, t_n)$ by a single application of (4.7.17). The "smallness" of ζ is gauged by whether or not we are prepared to ignore the $0(h^2)$ term of (4.7.15) when $h = \zeta$. If this error is unacceptably large then we break ζ up into equal fractions, say $h = \zeta/m$ where m is an integer. We now use (4.7.17) followed by repeated applications of the same idea namely

$$\Phi(t_n + kh, t_n) = \Phi[t_n + (k-1)h, t_n] + h \frac{\partial}{\partial \zeta} \Phi(t_n + \zeta, t_n)\Big|_{\zeta=(k-1)h} \tag{4.7.18}$$

which, by the use of (4.7.14), reduces to

$$\Phi(t_n + kh, t_n) = \Big\{I + hA[t_n + (k-1)h]\Big\} \Phi[t_n + (k-1)h, t_n] \qquad k = 1, 2, 3, \ldots, m \tag{4.7.19}$$

The integer m is chosen so that the errors, which at each stage are $0(h^2)$, collectively give rise to an acceptably small error after the m cycles of the integration algorithm. Depending upon how much computation time we are prepared to use, this integration method can be made to yield $\Phi(t_n + \zeta, t_n)$ to any required degree of accuracy. Alternatively, we can use higher order numerical integration schemes (see e.g. [4.2] or [4.5]) to give $\Phi(t_n + \zeta, t_n)$ with improved accuracy but at the cost of more computation on each integration cycle.

In the case of the constant-coefficient model, the possibility of obtaining Φ by numerical integration (see p. 97) may have seemed somewhat gratuitous, since one expects that the direct evaluation of $\exp(tA)$ might be simpler. However, this is not in general true, since in order to derive $\exp(tA)$ the eigenvalues of A are required, and obtaining them poses a nontrivial problem. With certain exceptions numerical integration is, in fact, the most readily available way for actually obtaining the transition matrix of a model from its differential equation, whether constant-coefficient or time-varying.

Finally we consider the inverse of $\Phi(t_n + \zeta, t_n)$ a matrix which we shall make use of in a later development. By (4.7.6) and (4.7.8) we can write

$$\begin{aligned} X(t_n) &= P(t_n)[P(t_n + \zeta)]^{-1} X(t_n + \zeta) \\ &= \Phi(t_n, t_n + \zeta) X(t_n + \zeta) \end{aligned} \tag{4.7.20}$$

and so

$$[\Phi(t_n + \zeta, t_n)]^{-1} = \Phi(t_n, t_n + \zeta) \tag{4.7.21}$$

or

$$[\Phi(t_n, t_m)]^{-1} = \Phi(t_m, t_n) \tag{4.7.22}$$

We can thus obtain the inverse of Φ by simply interchanging its two functional arguments.

However, in practice we seldom expect to know the actual functional form of Φ, and so again we must rely on numerical integration. Thus, consider the differential equation

$$\frac{\partial}{\partial \zeta} \Psi(t_n + \zeta, t_n) = -\Psi(t_n + \zeta, t_n) A(t_n + \zeta) \tag{4.7.23}$$

and form the product $\Psi\Phi$. Then by differentiation and making use of (4.7.14) and (4.7.23) we see that

$$\frac{\partial}{\partial\zeta}[\Psi(t_n+\zeta,t_n)\Phi(t_n+\zeta,t_n)] = -\Psi(t_n+\zeta,t_n)A(t_n+\zeta)\Phi(t_n+\zeta,t_n)$$
$$+\Psi(t_n+\zeta,t_n)A(t_n+\zeta)\Phi(t_n+\zeta,t_n)$$
$$= \emptyset \qquad (4.7.24)$$

This means that $\Psi(t_n+\zeta,t_n)\Phi(t_n+\zeta,t_n)$ is independent of ζ and so we can write

$$\Psi(t_n+\zeta,t_n)\Phi(t_n+\zeta,t_n) = B(t_n) \qquad (4.7.25)$$

We now impose the initial conditions

$$\Psi(t_n,t_n) = I \qquad (4.7.26)$$

and setting $\zeta = 0$ in (4.7.25) shows that $B(t_n)$ is in fact the identity matrix, *and hence that* Ψ *is the inverse of* Φ.

This provides us with a way of obtaining Φ^{-1} by numerical integration. Thus by using (4.7.23) together with (4.7.26), we can follow the numerical integration scheme of (4.7.19) thereby obtaining the inverse of $\Phi(t_n+\zeta,t_n)$.

In conclusion, we observe that what is true for t_n is true for t generally, and so we summarize the theory developed in the present section by the following system of equations:

$$\frac{d}{dt}X(t) = A(t)X(t) \qquad (4.7.27)$$

$$X(t+\zeta) = \Phi(t+\zeta,t)X(t) \qquad (4.7.28)$$

$$\frac{\partial}{\partial\zeta}\Phi(t+\zeta,t) = A(t+\zeta)\Phi(t+\zeta,t) \qquad (4.7.29)$$

$$[\Phi(t+\zeta,t)]^{-1} = \Psi(t+\zeta,t) = \Phi(t,t+\zeta) \qquad (4.7.30)$$

$$\frac{\partial}{\partial\zeta}\Psi(t+\zeta,t) = -\Psi(t+\zeta,t)A(t+\zeta) \qquad (4.7.31)$$

$$\Phi(t,t) = I = \Psi(t,t) \qquad (4.7.32)$$

$$\Phi(t+\zeta,t+\mu)\Phi(t+\mu,t+\beta) = \Phi(t+\zeta,t+\beta) \qquad (4.7.33)$$

the last result following readily by a double application of (4.7.28).

4.8 NONLINEAR SYSTEMS

We come now to the final generalization in choosing the model, namely to the case of the *nonlinear* differential equation

$$\frac{d}{dt} X(t) = F[X(t), t] \qquad (4.8.1)$$

where X is the state-vector at time t and F is a vector of nonlinear functions of the variables of which X is comprised and possibly of t as well. To make our subsequent discussion more explicit, we shall actually follow an example through in this section, accompanying each of the theoretical statements with a corresponding further development of the example.

Example:

Consider a body moving under the action of gravity through the atmosphere. To a good approximation the magnitude of the drag force exerted by the atmosphere on the body is given by

$$f_d = \frac{1}{2} \rho v^2 \alpha \qquad (4.8.2)$$

where ρ is the atmospheric density, v is the speed of the body and α is assumed to be a *known constant.*

Define the x_0, x_1, x_2 coordinate system so that its origin is at some point on the Earth's surface, x_0 is *North,* x_1 is *West* and x_2 is *Up.* We assume the Earth's surface to be essentially flat. It is well known that to a good approximation, the atmospheric density obeys the exponential law

$$\rho = \rho_0 e^{-kx_2} \qquad (4.8.3)$$

which is consistent with the assumption of an isothermal atmosphere (k and ρ_0 are taken to be constants).

The total force acting on the body is the sum of drag and gravity forces, the former along the velocity vector, the latter along x_2. We resolve the drag-force into its components in the x_0, x_1, x_2 system by noting that the velocity vector is

$$V = v\hat{V} \qquad (4.8.4)$$

where v is its length and $\hat{V}$ is the unit vector along V. Thus (4.8.2) can be written in *vector* form as

$$F_d = -\frac{1}{2}\rho\alpha v^2 \hat{V} \tag{4.8.5}$$

Now, letting $\hat{i}$, $\hat{j}$, and $\hat{k}$ be unit vectors along the x_0, x_1, and x_2 axes respectively, we have

$$V = \dot{x}_0 \hat{i} + \dot{x}_1 \hat{j} + \dot{x}_2 \hat{k} \tag{4.8.6}$$

and so

$$\hat{V} = \frac{\dot{x}_0 \hat{i} + \dot{x}_1 \hat{j} + \dot{x}_2 \hat{k}}{v} \tag{4.8.7}$$

Hence

$$F_d = -\frac{1}{2}\rho\alpha v \left(\dot{x}_0 \hat{i} + \dot{x}_1 \hat{j} + \dot{x}_2 \hat{k}\right) \tag{4.8.8}$$

giving, as the differential equations of motion

$$\begin{aligned} m\ddot{x}_0 &= -\frac{1}{2}\rho\alpha v \dot{x}_0 \\ m\ddot{x}_1 &= -\frac{1}{2}\rho\alpha v \dot{x}_1 \\ m\ddot{x}_2 &= -\frac{1}{2}\rho\alpha v \dot{x}_2 - mg \end{aligned} \tag{4.8.9}$$

The term g is taken to be a constant although it is really a function of x_2, a refinement which is easily introduced.

We define the state-vector as

$$X \equiv \begin{pmatrix} x_0 \\ x_1 \\ x_2 \\ x_3 \\ x_4 \\ x_5 \end{pmatrix} \equiv \begin{pmatrix} x_0 \\ x_1 \\ x_2 \\ \dot{x}_0 \\ \dot{x}_1 \\ \dot{x}_2 \end{pmatrix} \tag{4.8.10}$$

Then forming $\dot{X}$ by the use of (4.8.9) and (4.8.3), we obtain

$$\begin{pmatrix} \dot{x}_0 \\ \dot{x}_1 \\ \dot{x}_2 \\ \dot{x}_3 \\ \dot{x}_4 \\ \dot{x}_5 \end{pmatrix} = \begin{pmatrix} x_3 \\ x_4 \\ x_5 \\ \dfrac{-\rho_0 \alpha}{2m} e^{-kx_2} \left(x_3^2 + x_4^2 + x_5^2\right)^{\frac{1}{2}} x_3 \\ \dfrac{-\rho_0 \alpha}{2m} e^{-kx_2} \left(x_3^2 + x_4^2 + x_5^2\right)^{\frac{1}{2}} x_4 \\ \dfrac{-\rho_0 \alpha}{2m} e^{-kx_2} \left(x_3^2 + x_4^2 + x_5^2\right)^{\frac{1}{2}} x_5 - g \end{pmatrix} \tag{4.8.11}$$

which is of the form (4.8.1). ◆◆

The treatment of such nonlinear systems will be essentially as follows. For a given set of initial conditions, $\overline{X}(0)$, we can numerically integrate (4.8.1), thereby obtaining the associated subsequent history of $\overline{X}(t)$ known as a *nominal trajectory*. We assume that we have such a trajectory,† based on some assumed initial conditions, but that the true initial conditions were in fact *slightly different*, by an unknown amount. Let the true trajectory, $X(t)$, differ from the nominal trajectory by an amount $\delta X(t)$, i.e.

$$X(t) = \overline{X}(t) + \delta X(t) \tag{4.8.12}$$

Then $\delta X(t)$ is a vector of time dependent functions *which are, by assumption, small in relation to the corresponding elements in* $\overline{X}(t)$. Differentiation of (4.8.12) gives us

$$\frac{d}{dt} X(t) = \frac{d}{dt} \overline{X}(t) + \frac{d}{dt} \delta X(t) \tag{4.8.13}$$

†At a later stage (in Chapter 8) we will describe explicitly how nominal trajectories are obtained.

and using (4.8.1) we obtain

$$\frac{d}{dt}\bar{X}(t) + \frac{d}{dt}\delta X(t) = F[\bar{X}(t) + \delta X(t)] \qquad \text{(See Note)} \tag{4.8.14}$$

Consider say the first equation of (4.8.14), namely

$$\frac{d}{dt}\bar{x}_0(t) + \frac{d}{dt}\delta x_0(t) = f_0\left[\bar{x}_0(t) + \delta x_0(t), \bar{x}_1(t) + \delta x_1(t)\right] \tag{4.8.15}$$

where we assume, for simplicity, that X is a 2-vector. *Let t be temporarily fixed.* Then the right-hand side of (4.8.15) is of the form

$$f_0(\bar{x}_0 + \zeta, \bar{x}_1 + \eta)$$

which, by a double application of Taylor's theorem, gives

$$f_0(\bar{x}_0 + \zeta, \bar{x}_1 + \eta) = f_0\Big|_{\substack{\bar{x}_0\\ \bar{x}_1}} + \zeta \frac{\partial f_0}{\partial x_0}\Big|_{\substack{\bar{x}_0\\ \bar{x}_1}} + \eta \frac{\partial f_0}{\partial x_1}\Big|_{\substack{\bar{x}_0\\ \bar{x}_1}}$$

$$+ \frac{1}{2!}\left(\zeta^2 \frac{\partial^2 f_0}{\partial x_0^{\,2}} + 2\zeta\eta \frac{\partial^2 f_0}{\partial x_0\,\partial x_1} + \eta^2 \frac{\partial^2 f_0}{\partial x_1^{\,2}}\right)\Bigg|_{\substack{\bar{x}_0 + \mu\\ \bar{x}_1 + \nu}} \tag{4.8.16}$$

where $0 < \mu < \zeta$, $0 < \nu < \eta$.

If we drop the final term, *as we shall henceforth do*, then we see that we are committing an error of the form

$$a\zeta^2 + b\zeta\eta + c\eta^2$$

and if ζ and η are themselves sufficiently small, this term is very small. *However, we should always bear this error term in mind, particularly if η and ζ are of questionably large size.* More will be said about this at a later stage.

Note: We have not shown that F may also depend explicitly on t, as we did in (4.8.1), but this is implied both here and henceforth. Omitting t as a possible functional argument has no effect on the discussion to be presented.

Returning to (4.8.15) we are now able to see that, *to first order*,

$$\frac{d}{dt}\bar{x}_0(t) + \frac{d}{dt}\delta x_0(t) = f_0(\bar{x}_0, \bar{x}_1) + \left.\frac{\partial f_0}{\partial x_0}\right|_{\substack{\bar{x}_0 \\ \bar{x}_1}} \cdot \delta x_0 + \left.\frac{\partial f_0}{\partial x_1}\right|_{\substack{\bar{x}_0 \\ \bar{x}_1}} \cdot \delta x_1 \tag{4.8.17}$$

and in a like manner

$$\frac{d}{dt}\bar{x}_1(t) + \frac{d}{dt}\delta x_1(t) = f_1(\bar{x}_0, \bar{x}_1) + \left.\frac{\partial f_1}{\partial x_0}\right|_{\substack{\bar{x}_0 \\ \bar{x}_1}} \cdot \delta x_0 + \left.\frac{\partial f_1}{\partial x_1}\right|_{\substack{\bar{x}_0 \\ \bar{x}_1}} \cdot \delta x_1 \tag{4.8.18}$$

But by assumption

$$\frac{d}{dt}\bar{X}(t) = F[\bar{X}(t)] \tag{4.8.19}$$

which means that (4.8.17) and (4.8.18) can be written

$$\begin{pmatrix} \dfrac{d}{dt}\delta x_0(t) \\ \\ \dfrac{d}{dt}\delta x_1(t) \end{pmatrix} = \left.\begin{pmatrix} \dfrac{\partial f_0}{\partial x_0} & \dfrac{\partial f_0}{\partial x_1} \\ \\ \dfrac{\partial f_1}{\partial x_0} & \dfrac{\partial f_1}{\partial x_1} \end{pmatrix}\right|_{\substack{\bar{x}_0(t) \\ \bar{x}_1(t)}} \begin{pmatrix} \delta x_0(t) \\ \\ \delta x_1(t) \end{pmatrix} \tag{4.8.20}$$

and in general we see that, to first order, $\delta X(t)$ satisfies the *time-varying linear* differential equation

$$\frac{d}{dt}\delta X(t) = A[\bar{X}(t)]\,\delta X(t) \tag{4.8.21}$$

where the matrix $A[\bar{X}(t)]$ is defined by

$$\left[A[\bar{X}(t)]\right]_{i,j} = \left.\frac{\partial f_i(X)}{\partial x_j}\right|_{X = \bar{X}(t)} \tag{4.8.22}$$

Example:

In order to make the above development more specific, we pause briefly to derive the A matrix for the system of equations (4.8.11). Thus, differentiating each term with respect to each variable gives the matrix $A[\bar{X}(t)]$, as follows:

$$A[\bar{X}(t)] = \left(\begin{array}{c|c|c|c|c|c} 0 & 0 & 0 & 1 & 0 & 0 \\ \hline 0 & 0 & 0 & 0 & 1 & 0 \\ \hline 0 & 0 & 0 & 0 & 0 & 1 \\ \hline 0 & 0 & cke^{-k\bar{x}_2}\bar{v}\bar{x}_3 & -ce^{-k\bar{x}_2}\left(\frac{\bar{v}^2 + \bar{x}_3^2}{\bar{v}}\right) & -ce^{-k\bar{x}_2}\left(\frac{\bar{x}_3\bar{x}_4}{\bar{v}}\right) & -ce^{-k\bar{x}_2}\left(\frac{\bar{x}_3\bar{x}_5}{\bar{v}}\right) \\ \hline 0 & 0 & cke^{-k\bar{x}_2}\bar{v}\bar{x}_4 & -ce^{-k\bar{x}_2}\left(\frac{\bar{x}_3\bar{x}_4}{\bar{v}}\right) & -ce^{-k\bar{x}_2}\left(\frac{\bar{v}^2 + \bar{x}_4^2}{\bar{v}}\right) & -ce^{-k\bar{x}_2}\left(\frac{\bar{x}_4\bar{x}_5}{\bar{v}}\right) \\ \hline 0 & 0 & cke^{-k\bar{x}_2}\bar{v}\bar{x}_5 & -ce^{-k\bar{x}_2}\left(\frac{\bar{x}_3\bar{x}_5}{\bar{v}}\right) & -ce^{-k\bar{x}_2}\left(\frac{\bar{x}_4\bar{x}_5}{\bar{v}}\right) & -ce^{-k\bar{x}_2}\left(\frac{\bar{v}^2 + \bar{x}_5^2}{\bar{v}}\right) \end{array}\right) \tag{4.8.23}$$

Note: $c \equiv \dfrac{\rho_0 \alpha}{2m}$ and $\bar{v} = \left(\bar{x}_3^2 + \bar{x}_4^2 + \bar{x}_5^2\right)^{1/2}$. By $\bar{x}_2$ we mean $\bar{x}_2(t)$, etc. ◆◆

Once the functions $\bar{x}_0(t), \bar{x}_1(t), \ldots, \bar{x}_5(t)$ are obtained by numerical integration of (4.8.11) starting from $\bar{X}(0)$, they can be inserted into $A[\bar{X}(t)]$ thereby providing a square matrix of numbers. Alternatively if $\bar{X}(t_n)$ has been obtained previously, we can integrate (4.8.11) forward starting from $\bar{X}(t_n)$ to obtain $\bar{X}(t_n + \zeta)$. These numbers, if inserted into A, then provide us with $A[\bar{X}(t_n + \zeta)]$ and we see that, in general, A is simply a *time-varying matrix* whose elements, instead of being given explicitly as functions of t, are given as functions of $\bar{x}_0(t), \ldots, \bar{x}_5(t)$ which are themselves defined by a differential equation rather than explicity.

Returning now to (4.8.21) we compare it to (4.7.1) and, as a result, *are able to see immediately that all of the theory of the preceding section on time-varying linear differential equations now applies,* where the state-vector $X(t)$ of (4.7.1) is now $\delta X(t)$ in (4.8.21), and the time-varying coefficient matrix is $A[\bar{X}(t)]$ rather than just $A(t)$.

Thus the transition equation for $\delta X(t)$ will be (c/f (4.7.9))

$$\delta X(t_n + \zeta) = \Phi(t_n + \zeta, t_n; \overline{X})\, \delta X(t_n) \tag{4.8.24}$$

where Φ is, as before, a matrix of functions of two variables as well as depending on the assumed reference trajectory $\overline{X}(t)$ about which $\delta X(t)$ is the perturbation.

The transition matrix Φ, and its inverse Ψ, satisfy (c/f (4.7.14) and (4.7.23))

$$\frac{\partial}{\partial \zeta} \Phi(t_n + \zeta, t_n; \overline{X}) = A[\overline{X}(t_n + \zeta)]\, \Phi(t_n + \zeta, t_n; \overline{X}) \tag{4.8.25}$$

$$\frac{\partial}{\partial \zeta} \Psi(t_n + \zeta, t_n; \overline{X}) = -\Psi(t_n + \zeta, t_n; \overline{X})\, A[\overline{X}(t_n + \zeta)] \tag{4.8.26}$$

with initial conditions

$$\Phi(t_n, t_n; \overline{X}) = I = \Psi(t_n, t_n; \overline{X}) \tag{4.8.27}$$

These can be integrated by the numerical schemes discussed in the previous section (or, of course, by any other scheme the user may wish to employ) to provide $\Phi(t_n + \zeta, t_n; \overline{X})$ and its inverse.

In this way we are able to reduce a model based on a *nonlinear* differential equation to a problem involving a *time-varying linear* differential equation. The reader is reminded that the linearization is based on the assumption that a nominal trajectory is available, and that the perturbation vector $\delta X(t)$ is sufficiently small, so that terms involving products and squares of the components of that vector can be safely ignored. How that nominal trajectory is actually obtained will be discussed later. We now consider a simple example in order to reinforce the above ideas.

Example:

Returning to (4.8.2), we assume one-dimensional motion along the x axis. Then (4.8.2) gives us the differential equation

$$\ddot{x}(t) = k[\dot{x}(t)]^2 \tag{4.8.28}$$

where we assume that k is a known constant. This is a nonlinear differential equation in the state-variables $x(t)$ and $\dot{x}(t)$.

First we reorganize (4.8.28) into the vector form of (4.8.1). Thus

$$\frac{d}{dt} x(t) = \dot{x}(t) \tag{4.8.29}$$

and by (4.8.28),

$$\frac{d}{dt} \dot{x}(t) = k[\dot{x}(t)]^2 \tag{4.8.30}$$

which means that we can write

$$\frac{d}{dt}\begin{pmatrix} x(t) \\ \dot{x}(t) \end{pmatrix} = \begin{pmatrix} f_0[x(t), \dot{x}(t)] \\ f_1[x(t), \dot{x}(t)] \end{pmatrix} \tag{4.8.31}$$

where

$$f_0[x(t), \dot{x}(t)] \equiv \dot{x}(t) \tag{4.8.32}$$

and

$$f_1[x(t), \dot{x}(t)] = k[\dot{x}(t)]^2 \tag{4.8.33}$$

The latter two equations define the nonlinear functions of the state-variables $x(t)$ and $\dot{x}(t)$, which make up the vector $F[X(t)]$ of (4.8.1).

We now linearize the above system about a nominal trajectory. Thus, letting

$$\overline{X}(t) \equiv \begin{pmatrix} \bar{x}(t) \\ \dot{\bar{x}}(t) \end{pmatrix} \tag{4.8.34}$$

be that trajectory, and assuming that

$$x(t) = \bar{x}(t) + \delta x(t) \tag{4.8.35}$$

$$\dot{x}(t) = \dot{\bar{x}}(t) + \delta \dot{x}(t) \tag{4.8.36}$$

we rewrite (4.8.31) as

$$\frac{d}{dt}\begin{pmatrix} \bar{x}(t) + \delta x(t) \\ \dot{\bar{x}}(t) + \delta \dot{x}(t) \end{pmatrix} = \begin{pmatrix} \dot{\bar{x}}(t) + \delta \dot{x}(t) \\ k[\dot{\bar{x}}(t) + \delta \dot{x}(t)]^2 \end{pmatrix} \tag{4.8.37}$$

Expanding this equation now gives us

$$\frac{d}{dt}\begin{pmatrix} \bar{x}(t) \\ \dot{\bar{x}}(t) \end{pmatrix} + \frac{d}{dt}\begin{pmatrix} \delta x(t) \\ \delta \dot{x}(t) \end{pmatrix} \tag{4.8.38}$$

$$= \begin{pmatrix} \dot{\bar{x}}(t) \\ k[\dot{\bar{x}}(t)]^2 \end{pmatrix} + \begin{pmatrix} \delta \dot{x}(t) \\ 2k\dot{\bar{x}}(t)\,\delta \dot{x}(t) \end{pmatrix} + \begin{pmatrix} 0 \\ k[\delta \dot{x}(t)]^2 \end{pmatrix}$$

and so we have, *precisely*, that

$$\frac{d}{dt}\begin{pmatrix} \delta x(t) \\ \delta \dot{x}(t) \end{pmatrix} = \begin{pmatrix} 0 & 1 \\ 0 & 2k\dot{\bar{x}}(t) \end{pmatrix}\begin{pmatrix} \delta x(t) \\ \delta \dot{x}(t) \end{pmatrix} + \begin{pmatrix} 0 \\ k[\delta \dot{x}(t)]^2 \end{pmatrix} \tag{4.8.39}$$

This is the nonlinear differential equation governing the evolution of the perturbation vector $\delta X(t) \equiv [\delta x(t), \delta \dot{x}(t)]^T$.

The *linearized* differential equation is obtained from (4.8.39) by dropping the higher order terms. Thus we obtain the *approximate* system

$$\frac{d}{dt}\begin{pmatrix} \delta x(t) \\ \delta \dot{x}(t) \end{pmatrix} = \begin{pmatrix} 0 & 1 \\ 0 & 2k\dot{\bar{x}}(t) \end{pmatrix}\begin{pmatrix} \delta x(t) \\ \delta \dot{x}(t) \end{pmatrix} \tag{4.8.40}$$

which can now be written in the form

$$\frac{d}{dt}\,\delta X(t) = A[\bar{X}(t)]\,\delta X(t) \tag{4.8.41}$$

where

$$A[\bar{X}(t)] \equiv \begin{pmatrix} 0 & 1 \\ 0 & 2k\dot{\bar{x}}(t) \end{pmatrix} \tag{4.8.42}$$

In this particular case A happens to depend on only *one* element of $\bar{X}$.

The matrix $A[\bar{X}(t)]$ above could also have been obtained by the use of (4.8.22). Thus, applying that equation to (4.8.32) and (4.8.33) gives us

$$a_{00}[\bar{X}(t)] \equiv \left.\frac{\partial f_0}{\partial x(t)}\right|_{X=\bar{X}} = 0 \tag{4.8.43}$$

$$a_{01}[\overline{X}(t)] \equiv \left.\frac{\partial f_0}{\partial \dot{x}(t)}\right|_{X=\overline{X}} = 1 \tag{4.8.44}$$

$$a_{10}[\overline{X}(t)] \equiv \left.\frac{\partial f_1}{\partial x(t)}\right|_{X=\overline{X}} = 0 \tag{4.8.45}$$

$$a_{11}[\overline{X}(t)] \equiv \left.\frac{\partial f_1}{\partial \dot{x}(t)}\right|_{X=\overline{X}} = 2k\dot{\bar{x}}(t) \tag{4.8.46}$$

which is in precise agreement with (4.8.42).

We conclude this example by integrating (4.8.25) to obtain the transition matrix. For our numerical integration algorithm we use the very simple rule (c/f (4.7.17))

$$\Phi(t_n + h, t_n; \overline{X}) = I + hA[\overline{X}(t_n)] \tag{4.8.47}$$

This gives us, by the use of (4.8.42) above,

$$\begin{aligned}\Phi(t_n + h, t_n; \overline{X}) &= \begin{pmatrix} 1 & 0 \\ 0 & 1 \end{pmatrix} + h\begin{pmatrix} 0 & 1 \\ 0 & 2k\dot{\bar{x}}(t) \end{pmatrix} \\ &= \begin{pmatrix} 1 & h \\ 0 & 1 + 2hk\dot{\bar{x}}(t) \end{pmatrix}\end{aligned} \tag{4.8.48}$$

which is a very elementary, but very often used method of approximating Φ. Given the numerical values of the state-variables $\bar{x}(t_n)$ and $\dot{\bar{x}}(t_n)$ we see that Φ above can be readily reduced to a 2 x 2 matrix of numbers. Of course the values of $\bar{x}(t_n)$ and $\dot{\bar{x}}(t_n)$ would have to be obtained by numerically integrating the differential equations given in (4.8.29) and (4.8.30), using some selected integration algorithm and a set of initial conditions. The latter defines the trajectory $\overline{X}(t)$. ◆◆

The reader is strongly urged to work Examples 4.20 through 4.22 where we consider further illustrations of the material discussed in this section.

We now close this chapter by considering the following very important interpretation of the transition matrix in the nonlinear case. Suppose that we define the matrix K whose i, j^{th} term is

$$\left[K(t_n + \zeta, t_n; \bar{X})\right]_{ij} \equiv \frac{\partial \bar{x}_i(t_n + \zeta)}{\partial \bar{x}_j(t_n)} \tag{4.8.49}$$

where $\bar{X}$ is governed by the differential equation

$$\frac{d}{dt} X(t) = F[X(t)] \tag{4.8.50}$$

Then it can easily be shown that K satisfies the differential equation

$$\frac{\partial}{\partial \zeta} K(t_n + \zeta, t_n; \bar{X}) = A[\bar{X}(t_n + \zeta)] K(t_n + \zeta, t_n; \bar{X}) \tag{4.8.51}$$

with initial conditions

$$K(t_n, t_n; \bar{X}) = I \tag{4.8.52}$$

and where the matrix A in (4.8.51) is obtained from (4.8.50) by

$$\left[A[\bar{X}(t)]\right]_{i,j} \equiv \left.\frac{\partial f_i(X)}{\partial x_j}\right|_{X = \bar{X}(t)} \tag{4.8.53}$$

But this means that K is equal to the matrix Φ, as defined by (4.8.25) and (4.8.27). We have thus arrived at an alternate expression for the transition matrix, namely

$$\left[\Phi(t_n + \zeta, t_n; \bar{X})\right]_{i,j} = \left.\frac{\partial x_i(t_n + \zeta)}{\partial x_j(t_n)}\right|_{X = \bar{X}} \tag{4.8.54}$$

The above result is an *exact* equation and there are no higher order terms being neglected. It states that the sensitivity matrix of $\bar{X}(t)$ at time $t_n + \zeta$, to its value at time t_n, is *precisely* equal to the transition matrix. Equation (4.8.54) has a very useful practical application in that it permits us to obtain closed-form approximations, of any desired degree of accuracy, for the transition matrix, without the need for numerical integration and without the need to obtain the matrix $A[\bar{X}(t_n)]$ of (4.8.22). This is a definite extension to the techniques that we have developed, and the reader is strongly urged to study Examples 4.16 through 4.19 where we prove (4.8.54) and then show how it can be applied. The results of Ex. 4.19 should be compared to (4.8.48) above.

NOTES

A recursion algorithm for the matrix S shown in Table 4.1 on p. 85 can be obtained from [4.5, p. 18] as

$$[S]_{i,j} = [S]_{i-1,j-1} + (j-1)[S]_{i,j-1}$$

with

$$[S]_{0,0} = 1 \quad \text{and} \quad [S]_{0,j} = 0 = [S]_{i,0} \quad \text{for} \quad i,j \geq 1$$

As an example, let $i = 5$, $j = 8$. Then the above recursion becomes

$$[S]_{5,8} = [S]_{4,7} + 7[S]_{5,7}$$

Table 4.1 shows that

$$[S]_{4,7} + 7[S]_{5,7} = 735 + 1225$$

giving

$$[S]_{5,8} = 1960$$

The corresponding recursion for S^{-1}(Table 4.2 on p. 86) is

$$[S^{-1}]_{i,j} = [S^{-1}]_{i-1,j-1} - i[S^{-1}]_{i,j-1}$$

$$[S^{-1}]_{0,0} = 1, \qquad [S^{-1}]_{0,i} = 0 = [S^{-1}]_{j,0}, \qquad \text{for} \quad i,j \geq 1$$

These two recursions should be compared to the recursion for S and S^{-1} given in (2.4.9) and (2.4.14) on pages 22 and 24 respectively.

EXERCISES

4.1 a) Given that $x(t) = t^2$, verify that for $\tau = 2$ seconds: $x_n = 4n^2$, $\dot{x}_n = 4n$, $\ddot{x}_n = 2$, $\nabla x_n = 8n - 4$, $\nabla^2 x_n = 8$. Also verify that $\tau\dot{x}_n = \nabla x_n +$

$(1/2!)\nabla^2 x_n$, and that $(\tau^2/2!)\ddot{x}_n = (1/2!)\nabla^2 x_n$. Are these consistent with (4.4.11)?

b) Given that $x(t) = \sin(\pi/2)t$, verify that for $\tau = 0.1$ seconds:

$$x_n = \sin\frac{n\pi}{20}, \quad \dot{x}_n = \frac{\pi}{2}\cos\frac{n\pi}{20}, \quad \ddot{x}_n = -\frac{\pi^2}{4}\sin\frac{n\pi}{20},$$

$$\nabla x_n = \left(1 - \cos\frac{\pi}{20}\right)\sin\frac{n\pi}{20} + \left(\sin\frac{\pi}{20}\right)\left(\cos\frac{n\pi}{20}\right).$$

Also verify that $\lim_{\tau \to 0} (1/\tau)\nabla x_n = (\pi/2)\cos(\pi/2)t = \dot{x}(t)$.

4.2 a) Given that $x_n = n^2$, what is $x(t)$ when $\tau = 1$, $\tau = 0.1$, $\tau = 2$? What is $\dot{x}_n$, $\ddot{x}_n$ for each of these cases?

b) Given that $x_n = \sin(n\pi/16)$, what is $\dot{x}_n$, $\ddot{x}_n$ when $\tau = 0.2, 2$?

4.3 For the matrices Φ of (4.2.5) and (4.2.15) verify (4.2.23) when the matrices are 4 x 4. Using $k = 2$, verify (4.2.24) for the same matrices.

4.4 Let $x(t) = t^3$. Form the state-vector $X(0) \equiv (x, \dot{x}, \ddot{x}, \dddot{x})^T_{t=0}$. Now using (4.2.6), verify that $X(\zeta) = \Phi(\zeta)X(0)$ where $\Phi(\zeta)$ is defined by (4.2.5). Verify that $X(2) = \Phi(1)X(1)$.

4.5 a) Let $x(t) = t^3$ and $\tau = 2$ seconds, and set up the vector X_n of (4.2.2).

b) Now form Z_n defined by (4.2.14), and verify that for Φ of (4.2.15), $\Phi(1)Z_1 = Z_2$ and $\Phi(h)Z_0 = Z_h$.

c) Next use (4.2.18) to obtain $X_2 \equiv D(\tau)Z_2$ and $X_h \equiv D(\tau)Z_h$ and compare these to X_2 and X_h obtained from part a) above.

4.6 Verify that (4.3.3) follows from (4.3.2). Obtain the alternate result

$$\nabla^i x_{n+h} = \sum_{j=i}^{m} (-1)^{j-i}\binom{-h}{j-i}\nabla^j x_n .$$

4.7 Using $x(t) = t^3$, set $t = n\tau$ with $\tau = 2$ and obtain the vector $U_n \equiv \left(x_n, \nabla x_n, \nabla^2 x_n, \nabla^3 x_n\right)^T$. Now verify that (4.3.6) gives

$$\Phi(h) = \begin{pmatrix} 1 & h & \dfrac{h^2 + h}{2} & \dfrac{h^3 + 3h^2 + 2h}{6} \\ & 1 & h & \dfrac{h^2 + h}{2} \\ & & 1 & h \\ & & & 1 \end{pmatrix}$$

Set $n = 0$ and verify (4.3.4). Repeat with $n = 1$.

4.8 Obtain (4.4.9) by direct computation on (4.4.5).

4.9 Using the vector U_n of Ex. 4.7, set up the vector V_n defined in (4.4.6). Now use (4.4.8) to find Z_n of (4.4.6), and using $\tau = 2$, obtain $\left(x_n, \dot{x}_n, \ddot{x}_n, \dddot{x}_n\right)^T$. Compare this to X_n of (4.2.2) for $x(t) = t^3$ with $\tau = 2$.

4.10 a) Obtain the Lagrange interpolator matrix W of (4.5.14) for a 1-step predictor, assuming a *cubic* polynomial and $\tau = 2$. Apply it to the function $x(t) = t^3$, and verify that it does in fact gives the 1-step predictions of x and its derivatives.

b) Using $x_n = \sin(n/8)(\pi/2)$, form the observation vector $Y_{(n)} \equiv \left(x_n, x_{n-1}, x_{n-2}, x_{n-3}\right)^T$ for $n = 3$, by the use of 5-figure tables. Now apply the first two rows of the Lagrange matrix from (a) above to $Y_{(3)}$ and obtain $x_4, \dot{x}_4$. Compare these to table values for $\sin(\pi/4)$ and $(\pi/32)\cos(\pi/4)$.

c) Repeat b) above using $n = 4$.

Note: $x_n = \sin[n\tau(1/8\tau)(\pi/2)]$ and so $x(t) = \sin[(\pi/16\tau)t]$. Thus $\dot{x}(t) = (\pi/16\tau)\cos[(\pi/16\tau)t]$ giving $\dot{x}_n = (\pi/16\tau)\cos(n/8)(\pi/2)$.

4.11 a) Write the differential equation for polynomials of degree 2 in the vector-matrix form of (4.6.7) using the state-vector of (4.6.5). Now form $\Phi(\zeta) = \exp(\zeta A)$ as defined by (4.6.28). Verify that $X(t+\zeta) = \Phi(\zeta)X(t)$. Compare $\Phi(\zeta)$ to (4.2.5).

b) Repeat, but using the state-vector Z_n of (4.2.14) instead. Obtain $\Phi(h)$ where $h = \zeta\tau$, and thus verify (4.2.15), (4.2.16) and (4.2.24).

4.12 Define the state-vector $V_n \equiv \left[x_n, \nabla x_n, (\nabla^2/2!)x_n\right]^T$ (c/f (4.4.6)). Now use (4.4.8) to write $Z(t) \equiv [x(t), \tau\dot{x}(t), (\tau^2/2)\ddot{x}(t)]^T$ in terms of $V(t)$.

Assuming that $x(t)$ is a quadratic, write the differential equation for $Z(t)$ in the form of (4.6.18), and hence (using Table 4.1 to obtain S), write the differential equation for $V(t)$ in the form $(d/dt)V(t) = AV(t)$. Now form $\exp(\zeta A)$ and reconcile it with $\Phi(h)$ of (4.3.6) where $\zeta = h\tau$.

4.13 a) For the matrix $\Phi(\zeta)$ of (4.2.5) form $(d/dt)\Phi(t)$. Now, using the associated A matrix for this Φ (see Ex. 4.11), verify that (4.6.33) holds.

b) Similarly verify (4.6.34) where $\Psi(t)$ is the inverse of $\Phi(t)$ above. (Use (4.2.23) to express $\Psi(t)$ in terms of $\Phi(t)$.)

4.14 Consider the *linear constant-coefficient* differential equation

$$\ddot{x}(t) = x(t)$$

a) Verify that its solutions are

$$\begin{pmatrix} x(t) \\ \dot{x}(t) \end{pmatrix} = \begin{pmatrix} e^{t} & e^{-t} \\ e^{t} & -e^{-t} \end{pmatrix} \begin{pmatrix} x(0) \\ \dot{x}(0) \end{pmatrix} \tag{II}$$

b) Verify that (I) can be written as $(d/dt)\,X(t) = AX(t)$ where $A \equiv \begin{pmatrix} 0 & 1 \\ 1 & 0 \end{pmatrix}$.

c) Write (II) in the form $X(t) = P(t)\,X(0)$ and find $[P(t)]^{-1}$. Now form $\Phi(t + \zeta, t) \equiv P(t + \zeta)[P(t)]^{-1}$. Verify that Φ is independent of t, and so we can write it as $\Phi(\zeta)$.

d) Find $\exp(\zeta A)$ and compare it to $\Phi(\zeta)$ of c) above.

4.15 Consider the following *linear time-varying differential* equation

$$(1 - 2t)\,\ddot{x}(t) + (1 + 4t^2)\,\dot{x}(t) - (2 - 2t + 4t^2)\,x(t) = 0 \tag{I}$$

a) Verify that its solutions are $x(t) = e^{t^2} x(0) + e^{t}\dot{x}(0)$.

b) Verify that (I) can be written in the form $(d/dt)\,X(t) = A(t)\,X(t)$, where $X(t) \equiv [x(t), \dot{x}(t)]^T$ and

$$A(t) \equiv \begin{pmatrix} 0 & 1 \\ \dfrac{2 - 2t + 4t^2}{1 - 2t} & \dfrac{-(1 + 4t^2)}{1 - 2t} \end{pmatrix}$$

c) Verify from a) above that we can write $X(t) = P(t)\,X(0)$ where

$$P(t) \equiv \begin{pmatrix} e^{t^2} & e^{t} \\ 2t\,e^{t^2} & e^{t} \end{pmatrix}$$

d) Find $[P(t)]^{-1}$ and hence form $\Phi(t + \zeta, t) \equiv P(t + \zeta)[P(t)]^{-1}$.

e) Finally verify that $X(t + \zeta) = \Phi(t + \zeta, t)\,X(t)$.

4.16 For the linear system $(d/dt)\,X(t) = A(t)\,X(t)$, show that if we define the matrix whose i, j^{th} terms is $\partial x_i(t_n + \zeta)/\partial x_j(t_n)$, then this matrix is precisely the associated transition matrix $\Phi(t_n + \zeta, t_n)$, i.e., $\left[\Phi(t_n + \zeta, t_n)\right]_{ij} = \partial x_i(t_n + \zeta)/\partial x_j(t_n)$.

4.17 a) Consider the nonlinear system

$$\dot{x}(t) = f[x(t)]$$

(where f and x are both scalars) and let the trajectory passing through $\bar{x}(t_n)$ be

$$\bar{x}(t_n + \zeta) = g\left[\bar{x}(t_n), \zeta\right] \tag{II}$$

where g depends on the form of f in (I) above. Define the functions

$$k(t_n + \zeta, t; \bar{x}) \equiv \frac{\partial \bar{x}(t_n + \zeta)}{\partial \bar{x}(t_n)} \tag{III}$$

and

$$a\left[\bar{x}(t_n)\right] \equiv \frac{\partial f\left[\bar{x}(t_n)\right]}{\partial \bar{x}(t_n)} \quad , \tag{IV}$$

Now show that k satisfies the differential equation

$$\frac{\partial}{\partial \zeta} k(t_n + \zeta, t_n; \bar{x}) = a\left[\bar{x}(t_n + \zeta)\right] k(t_n + \zeta, t_n; \bar{x})$$

with initial conditions $k(t_n, t_n; \bar{x}) = 1$. Hence infer that (III) is the (scalar) transition matrix for the linearized system using $\bar{x}(t_n)$ as the nominal trajectory.

b) Verify that the above is true in general for the *vector* system $(d/dt) X(t) = F[X(t)]$.

Comment: We see then that for any differential equation, whether linear or nonlinear we have precisely

$$\left[\Phi(t_n + \zeta, t_n; \bar{X})\right]_{ij} = \frac{\partial \bar{x}_i(t_n + \zeta)}{\partial \bar{x}_j(t_n)}$$

4.18 (This example follows Ex. 4.17 for a specific differential equation.)

a) For the nonlinear system $(d/dt) x(t) = f[x(t)]$, where $f[x(t)] \equiv [x(t)]^2$, show that the solutions are $x(t) = 1/(c - t)$, where c is an arbitrary constant. Hence prove that

$$\bar{x}(t_n + \zeta) = \frac{\bar{x}(t_n)}{1 - \zeta \bar{x}(t_n)} \tag{I}$$

b) Verify that for (I) above

$$\frac{\partial \bar{x}(t_n + \zeta)}{\partial \bar{x}(t_n)} = \frac{1}{\left[1 - \zeta \bar{x}(t_n)\right]^2} \tag{II}$$

c) Let the function (II) be called $k(t_n + \zeta, t_n; \bar{x})$, and define

$$a\left[\bar{x}(t_n)\right] \equiv \frac{\partial f\left[\bar{x}(t_n)\right]}{\partial \bar{x}(t_n)} \tag{III}$$

where $f\left[\bar{x}(t_n)\right]$ is given in a) above. Verify that $a\left[\bar{x}(t_n)\right] = 2\bar{x}(t_n)$, and that k satisfies the equation

$$\frac{\partial}{\partial \zeta} k(t_n + \zeta, t_n; \bar{x}) = a\left[\bar{x}(t_n + \zeta)\right] k(t_n + \zeta, t_n; \bar{x})$$

with initial conditions $k(t_n, t_n; \bar{x}) = 1$. Hence (II) is, in fact, the transition matrix, i.e.

$$\frac{\partial \bar{x}(t_n + \zeta)}{\partial \bar{x}(t_n)} = \varphi(t_n + \zeta, t_n; \bar{x})$$

4.19 Consider the differential equation

$$\ddot{x}(t) = k[\dot{x}(t)]^2 \tag{I}$$

(This could arise, for example, when a body is moving through a viscous medium whose drag is proportional to the speed squared.)

a) Using a two-term Taylor series, verify that

$$\begin{pmatrix} \bar{x}(t_n + \zeta) \\ \dot{\bar{x}}(t_n + \zeta) \end{pmatrix} = \begin{pmatrix} \bar{x}(t_n) + \zeta\dot{\bar{x}}(t_n) + \dfrac{\zeta^2}{2!} k\left[\dot{\bar{x}}(t_n)\right]^2 \\ \dot{\bar{x}}(t_n) + \zeta k\left[\dot{\bar{x}}(t_n)\right]^2 \end{pmatrix}$$

Now apply the results of Ex. (4.17b) to obtain an explicit approximation for the transition matrix by forming

$$\Phi(t_n + \zeta, t_n; \bar{X}) = \begin{pmatrix} \dfrac{\partial \bar{x}(t_n + \zeta)}{\partial \bar{x}(t_n)} & \dfrac{\partial \bar{x}(t_n + \zeta)}{\partial \dot{\bar{x}}(t_n)} \\ \dfrac{\partial \dot{\bar{x}}(t_n + \zeta)}{\partial \bar{x}(t_n)} & \dfrac{\partial \dot{\bar{x}}(t_n + \zeta)}{\partial \dot{\bar{x}}(t_n)} \end{pmatrix}$$

Compare the result to (4.8.48).

b) Repeat using a three-term Taylor series.

Comment: The above method avoids the need for numerical integration in obtaining the transition matrix.

4.20 Verify, for the differential equation of Ex. 4.18a), that the trajectory whose value is unity at $t = 1$ is given by $x(t) = 1/(2 - t)$.

a) Integrate $dx/dt = x^2$ numerically,† using a step-size $h = 0.1$, starting with $\bar{x}(1) = 1$ and continue to $\bar{x}(1.5)$. Use the integration algorithm $\bar{x}(t_n + h) = \bar{x}(t_n) + h\dot{\bar{x}}(t_n)$.

b) Repeat a) above, but use *Heun's Method* for the integration, viz.

$$\beta(t_n + h) = \bar{x}(t_n) + h\dot{\bar{x}}(t_n)$$

$$\gamma(t_n + h) = \bar{x}(t_n) + h\dot{\beta}(t_n + h)$$

$$\bar{x}(t_n + h) = \frac{1}{2}\left[\beta(t_n + h) + \gamma(t_n + h)\right]$$

Note: $\dot{\beta}(t_n + h) = \left[\beta(t_n + h)\right]^2$.

c) Compare the trajectories obtained from a) and b) above with the exact trajectory, $\bar{x}(t) = 1/(2 - t)$.

4.21 a) Assume that

$$\dot{x}(t) = [x(t)]^2 \tag{I}$$

and that $x(1) = 1.1$. Use the methods of Section 4.8 to linearize the above differential equation about the nominal trajectory $\bar{x}(t)$ where $\bar{x}(1) = 1$. Hence verify that

$$\frac{d}{dt}\delta x(t) = [2\bar{x}(t)]\,\delta x(t) \tag{II}$$

is the resultant linearized system.

b) The nominal trajectory $\bar{x}(t)$ has been obtained by numerical integration in Ex. 4.20b). Using those values for $\bar{x}(t)$, integrate (II) above using the algorithm $\delta x(t_n + h) = \delta x(t_n) + h\frac{d}{dt}\left[\delta x(t_n)\right]$. Start with $\delta x(1) = 0.1$ and continue to $t = 1.5$ with step-size $h = 0.1$.

c) Add $\delta x(t)$ obtained above to $\bar{x}(t)$, and so obtain the trajectory $x(t)$ passing through 1.1 at $t = 1$. Verify that the exact solution to $\dot{x}(t) = [x(t)]^2$ with initial condition $x(1) = 1.1$, is $x(t) = 1/(1.9091 - t)$, and compare this to $\delta x(t) + \bar{x}(t)$ obtained numerically above.

†i.e., with the aid of a desk-calculator or a computer.

4.22 a) For the linearized system (c/f Ex. 4.21)

$$\frac{d}{dt}\delta x(t) = [2\bar{x}(t)]\,\delta x(t) \tag{I}$$

verify that the transition relation is

$$\delta x(t_n + \zeta) = \varphi(t_n + \zeta, t_n; \bar{x})\,\delta x(t_n) \tag{II}$$

where φ is obtainable by integrating the differential equation

$$\frac{d}{d\zeta}\varphi(t_n + \zeta, t_n; \bar{x}) = \left[2\bar{x}(t_n + \zeta)\right]\varphi(t_n + \zeta, t_n; \bar{x}) \tag{III}$$

with initial conditions $\varphi(t_n, t_n; \bar{x}) = 1$.

b) Integrate (III) numerically, using the values for $\bar{x}(t)$ as obtained from Ex. (4.20b). Use step size $h = 0.1$ and start at $\varphi(1, 1; \bar{x}) = 1$, proceeding to $\varphi(1.5, 1; \bar{x})$, by the algorithm given in (4.7.19).

c) Apply the values of φ obtained in b) to (II) of a), to give $\delta x(1), \delta x(1.1), \ldots, \delta x(1.5)$. Compare the procedure and the results to Ex. 4.21.

REFERENCES

1. Rice, J. R., "The Approximation of Functions, Vol. 1," Addison-Wesley, 1964, p. 120 et seq.
2. Hildebrand, F. B., "Introduction to Numerical Analysis," McGraw-Hill Book Company, 1956.
3. Mirsky, L., "An Introduction to Linear Algebra," Oxford University Press, 1961.
4. Bellman, R., "Introduction to Matrix Analysis," McGraw-Hill Book Company, 1960.
5. Hamming, R. W., "Numerical Methods for Scientists and Engineers," McGraw-Hill Book Company, 1962.
6. Zadeh, L. A., and Desoer, C. A., "Linear System Theory," McGraw-Hill Book Company, 1963, p. 339.

5

STATISTICAL BACKGROUND

5.1 INTRODUCTION

The approach which we adopt in this book to the problem of smoothing and prediction is, briefly, as follows. A process, modelled by a given differential equation, is being observed in the presence of additive errors. The purpose of the algorithms which we propose to develop is to select a trajectory from the model equations which best fits the observations, in some specified sense.

The component of the observations which is truly related to the process will be referred to as the *signal,* and the additive error components will be called the *observation errors* or the *input errors.* The processes with which we shall concern ourselves are *deterministic* whereas the errors will be regarded as *random variables.*

In this chapter we review the statistical concepts needed to examine the errors both before and after smoothing. The background of the reader is assumed to be equivalent, at least, to the first four chapters of [5.1] or to the first four chapters of [5.2]. For a background in matrix theory, as required for this chapter, the reader is referred to [5.3].

5.2 CONCEPTS FROM PROBABILITY THEORY

Suppose at time $t = t_n$ we use a device, called a *transducer*, to make a single observation on a real-valued process. We obtain

$$y_n = x_n + \nu_n \tag{5.2.1}$$

where x_n is the value of the process at $t = t_n$, ν_n is the error introduced by the transducer, and y_n is the resultant observation in the presence of that error. In what follows, we use the terms *observation error* and *input error* interchangeably in reference to ν_n, and the component in (5.2.1) related to the process (in this case x_n) will be called the *signal.*

If we were to observe the process at *some fixed time* t_n *by simultaneously using a large ensemble of identical*† *transducers* (see Fig. 5.1), we would obtain a set of y_n's differing only in their respective ν_n's.

†The word "identical," as used here, implies only that the transducers are *macroscopically* identical. Thus, while each measures the process in precisely the same way, the individual corruptions added by the respective instruments are permitted to differ.

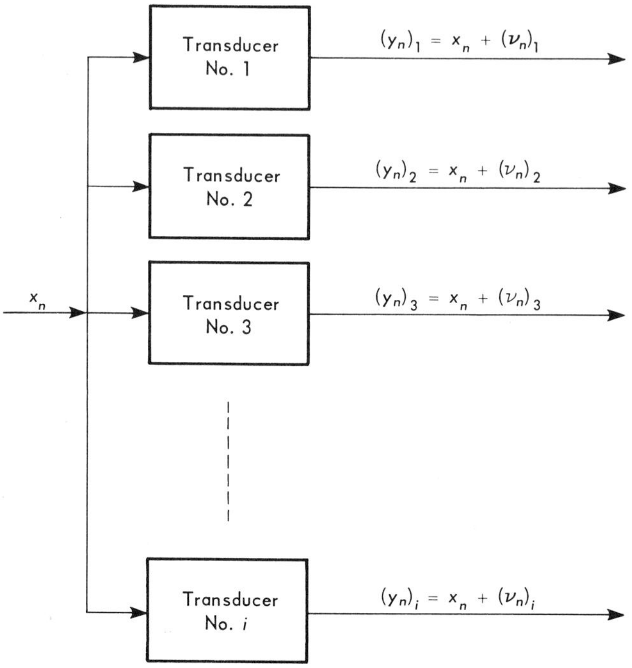

Fig. 5.1 *Simultaneous observations of a process by an ensemble of identical transducers.*

Since we are restricting ourselves to the domain of real numbers, ν_n is seen to be a real random variable which can range in value (from transducer to transducer at time t_n) over the continuous range $-\infty \leq \nu_n \leq +\infty$. Examination of this set of observations $\{y_n\}$, would show that, for a given real number λ, a certain fraction of the ν_n's satisfy

$$\nu_n \leq \lambda \tag{5.2.2}$$

That fraction is assumed to tend to a limit as the number of transducers used increases, and we obtain, as that limit, a well-defined function of λ, called the *probability distribution function* of the random variable ν_n. We symbolize it as $P_{\nu_n}(\lambda)$, by which we mean:

$$P_{\nu_n}(\lambda) = \left\{\text{The probability that } \nu_n \leq \lambda\right\} \tag{5.2.3}$$

We define the probability density function of ν_n as $p_{\nu_n}(\lambda)$ where

$$P_{\nu_n}(\lambda) = \int_{-\infty}^{\lambda} p_{\nu_n}(\xi)\, d\xi \tag{5.2.4}$$

It then follows immediately that $p_{\nu_n}(\lambda)$ is an integrable function of λ related to $P_{\nu_n}(\lambda)$ by

$$p_{\nu_n}(\lambda) = \frac{d}{d\lambda} P_{\nu_n}(\lambda) \tag{5.2.5}$$

i.e. the density function is simply the derivative of the distribution function.

The *ensemble expectation* of any function of the random variable ν_n is defined by

$$E\left\{f(\nu_n)\right\} \equiv \int_{-\infty}^{\infty} f(\lambda)\, p_{\nu_n}(\lambda)\, d\lambda \tag{5.2.6}$$

and in particular, when $f(\nu_n)$ is ν_n itself, then

$$E\left\{\nu_n\right\} \equiv \int_{-\infty}^{\infty} \lambda p_{\nu_n}(\lambda)\, d\lambda \tag{5.2.7}$$

The expectation of $\alpha\nu_n$, where α is a constant, is obtained from (5.2.7) as

$$E\{\alpha\nu_n\} = \alpha E\{\nu_n\} \tag{5.2.8}$$

In a like manner, by (5.2.6),

$$E\{\nu_n^2\} \equiv \int_{-\infty}^{\infty} \lambda^2 p_{\nu_n}(\lambda)\, d\lambda \tag{5.2.9}$$

and from this it follows immediately that

$$E\{(\alpha\nu_n)^2\} = \alpha^2 E\{\nu_n^2\} \tag{5.2.10}$$

The *variance* of the random variable ν_n is defined as

$$\text{Var}(\nu_n) \equiv \sigma^2(\nu_n) \equiv E\left\{\left(\nu_n - E\{\nu_n\}\right)^2\right\} \tag{5.2.11}$$

We say that the random variable ν_n is *zero-mean* if

$$E\{\nu_n\} = 0 \tag{5.2.12}$$

It thus follows that if ν_n is zero-mean, then

$$\sigma^2(\nu_n) = E\{\nu_n^2\} \tag{5.2.13}$$

We shall be dealing almost exclusively with zero-mean random variables and so $\sigma^2(\nu_n)$ will be used interchangeably with $E\{\nu_n^2\}$, the reader being asked to recall that $E\{\nu_n\}$ is zero by assumption.

Until now we have concerned ourselves with only a *single* random variable ν_n, which was introduced by examination of the errors in the ensemble of observations $\{y_n\}$ made at $t = t_n$. We now turn our attention to a *second* ensemble, $\{y_k\}$, made at $t = t_k$, where $k \neq n$.

The errors in y_k are symbolized as ν_k and as with ν_n, they can be seen to form a random variable in the range $-\infty \leq \nu_k \leq +\infty$. ν_k has a probability distribution function $P_{\nu_k}(\lambda)$ and a density function $p_{\nu_k}(\lambda)$. P and p of ν_k *may or may not* have the same functional form as they did for ν_n.

We now perform the following experiment. Assume that $t_n < t_k$. By using a *single transducer,* we obtain first a sample value of ν_n and then of ν_k, thereby providing the ordered pair of numbers (ν_n, ν_k). By the *simultaneous* use of a large number of such transducers, we can obtain many such ordered pairs

$$(\nu_n, \nu_k)_1, \quad (\nu_n, \nu_k)_2, \ldots, \quad (\nu_n, \nu_k)_i, \ldots$$

where the outer subscript, in each case, is identified with the transducer providing the pair. (See Figure 5.1.)

We now examine this ensemble of ordered pairs and determine the fraction of the overall set of pairs which satisfies, *simultaneously,*

$$\nu_n \leq \lambda, \quad \nu_k \leq \mu, \quad (\lambda, \mu \text{ real}) \tag{5.2.14}$$

As the number of transducers increases, this fraction is assumed to tend to a limit which depends on λ and μ. We call it the *bivariate* or *joint distribution function* of the random variables ν_n and ν_k, and symbolize it by

$$P_{\nu_n, \nu_k}(\lambda, \mu) \equiv \left\{ \begin{array}{l} \text{The fraction of the pairs } (\nu_n, \nu_k)_i \text{ which} \\ \text{satisfy } \nu_n \leq \lambda \text{ and } \nu_k \leq \mu \end{array} \right\} \tag{5.2.15}$$

We define the *joint probability density function* of ν_n and ν_k as $p_{\nu_n, \nu_k}(\lambda, \mu)$ where

$$P_{\nu_n, \nu_k}(\lambda, \mu) = \int_{-\infty}^{\lambda} \int_{-\infty}^{\mu} p_{\nu_n, \nu_k}(\xi, \eta)\, d\xi\, d\eta \tag{5.2.16}$$

and so it follows immediately that

$$p_{\nu_n, \nu_k}(\lambda, \mu) = \frac{\partial^2}{\partial\lambda\,\partial\mu} P_{\nu_n, \nu_k}(\lambda, \mu) \tag{5.2.17}$$

Define the *bivariate ensemble expectation* of the function $f(\nu_n, \nu_k)$ by

$$E\{f(\nu_n, \nu_k)\} \equiv \int_{-\infty}^{\infty}\int_{-\infty}^{\infty} f(\lambda, \mu)\, p_{\nu_n, \nu_k}(\lambda, \mu)\, d\lambda d\mu \tag{5.2.18}$$

One such $f(\nu_n, \nu_k)$ is the product $\nu_n \nu_k$. Thus

$$E\{\nu_n \nu_k\} \equiv \int_{-\infty}^{\infty}\int_{-\infty}^{\infty} \lambda\mu p_{\nu_n, \nu_k}(\lambda, \mu)\, d\lambda d\mu \tag{5.2.19}$$

The *covariance* of ν_n and ν_k is defined by

$$\mathrm{Cov}(\nu_n, \nu_k) \equiv E\left\{\left(\nu_n - E\{\nu_n\}\right)\left(\nu_k - E\{\nu_k\}\right)\right\} \tag{5.2.20}$$

For *zero-mean* random variables ν_n and ν_k, we see immediately that

$$\mathrm{Cov}(\nu_n, \nu_k) = E\{\nu_n \nu_k\} \tag{5.2.21}$$

The E operator has the very important *linearity* property. Thus, since by definition

$$E\{\alpha f(\nu_n) + \beta g(\nu_k)\} = \int_{-\infty}^{\infty}\int_{-\infty}^{\infty} [\alpha f(\lambda) + \beta g(\mu)]\, p_{\nu_n, \nu_k}(\lambda, \mu)\, d\lambda d\mu \tag{5.2.22}$$

It follows that (see Ex. 5.2)

$$E\{\alpha f(\nu_n) + \beta g(\nu_k)\} = \alpha E\{f(\nu_n)\} + \beta E\{g(\nu_k)\} \tag{5.2.23}$$

E is thus a linear operator. We now apply this linearity property of E to *vectors* and *matrices* of random variables.

Let N be given by

$$N \equiv \begin{pmatrix} \nu_0 \\ \nu_1 \\ \vdots \\ \nu_m \end{pmatrix} \tag{5.2.24}$$

i.e. N is a vector of $m + 1$ random variables. Then we define $E\{N\}$ as

$$E\{N\} \equiv \begin{pmatrix} E\{\nu_0\} \\ \vdots \\ E\{\nu_m\} \end{pmatrix} \tag{5.2.25}$$

Consider the matrix NN^T. By the formal rule for matrix multiplication

$$NN^T = \begin{pmatrix} \nu_0^2 & \nu_0\nu_1 & \cdots & \nu_0\nu_m \\ \nu_0\nu_1 & \nu_1^2 & \cdots & \nu_1\nu_m \\ \vdots & \vdots & & \vdots \\ \nu_0\nu_m & \nu_1\nu_m & \cdots & \nu_m^2 \end{pmatrix} \tag{5.2.26}$$

Then we define $E\{NN^T\}$ as

$$E\{NN^T\} = \begin{pmatrix} E\{\nu_0^2\} & E\{\nu_0\nu_1\} & \cdots & E\{\nu_0\nu_m\} \\ E\{\nu_0\nu_1\} & E\{\nu_1^2\} & \cdots & E\{\nu_1\nu_m\} \\ \vdots & \vdots & & \vdots \\ E\{\nu_0\nu_m\} & E\{\nu_1\nu_m\} & \cdots & E\{\nu_m^2\} \end{pmatrix} \tag{5.2.27}$$

Suppose now that U is a linear transformation on N, i.e., that

$$U = WN \tag{5.2.28}$$

where W is a matrix of constants, not necessarily square. We are interested in the first and second order statistics of U in terms of the statistics of N.

Consider first $E\{U\}$. By the linearity of the E operator, it follows immediately that

$$E\{U\} = WE\{N\} \tag{5.2.29}$$

Consider next the *pair* of linear transformations

$$\begin{aligned} U &= GN \\ V &= HN \end{aligned} \tag{5.2.30}$$

where U and V are not necessarily vectors of the same order. Then

$$\begin{aligned} UV^T &= GN(HN)^T \\ &= GNN^TH^T \\ &= G(NN^T)H^T \end{aligned} \tag{5.2.31}$$

Once again, the linearity of the E operator gives us

$$\begin{aligned} E\left\{UV^T\right\} &= E\left\{G(NN^T)H^T\right\} \\ &= GE\left\{NN^T\right\}H^T \end{aligned} \tag{5.2.32}$$

As an example, let G and H both be a matrix with a single row, say C^T. Then

$$U = V = C^TN \tag{5.2.33}$$

and so, by (5.2.32)

$$E\left\{C^TNN^TC\right\} = C^TE\left\{NN^T\right\}C \tag{5.2.34}$$

◆◆

More generally, letting $G = H = W$ say, the above discussion demonstrates that when

$$U = WN \tag{5.2.35}$$

then $E\{U\}$ and $E\left\{UU^T\right\}$ are given in terms of $E\{N\}$ and $E\left\{NN^T\right\}$ by

$$E\{U\} = E\{WN\} = WE\{N\} \tag{5.2.36}$$

and

$$E\{UU^T\} = E\{W(NN^T)W^T\} = WE\{NN^T\}W^T \tag{5.2.37}$$

The above two equations will be applied very frequently.

As an example, let

$$\begin{pmatrix} u_0 \\ u_1 \end{pmatrix} = \begin{pmatrix} 2 & -1 & 1 \\ 3 & -1 & -1 \end{pmatrix} \begin{pmatrix} \nu_0 \\ \nu_1 \\ \nu_2 \end{pmatrix} \tag{5.2.38}$$

Then

$$\begin{pmatrix} E\{u_0\} \\ E\{u_1\} \end{pmatrix} = \begin{pmatrix} 2E\{\nu_0\} - E\{\nu_1\} + E\{\nu_2\} \\ 3E\{\nu_0\} - E\{\nu_1\} - E\{\nu_2\} \end{pmatrix} \tag{5.2.39}$$

If ν_0, ν_1 and ν_2 are all zero-mean, then clearly so are both u_0 and u_1. Let this be the case, and suppose we are given that

$$E\{NN^T\} \equiv \begin{pmatrix} 1 & 1 & 1 \\ 1 & 2 & 1 \\ 1 & 1 & 3 \end{pmatrix} \tag{5.2.40}$$

Then by (5.2.38) and (5.2.37)

$$\begin{aligned} E\{UU^T\} &\equiv \begin{pmatrix} E\{u_0^2\} & E\{u_0 u_1\} \\ E\{u_0 u_1\} & E\{u_1^2\} \end{pmatrix} \\ &= \begin{pmatrix} 2 & -1 & 1 \\ 3 & -1 & -1 \end{pmatrix} \begin{pmatrix} 1 & 1 & 1 \\ 1 & 2 & 1 \\ 1 & 1 & 3 \end{pmatrix} \begin{pmatrix} 2 & 3 \\ -1 & -1 \\ 1 & -1 \end{pmatrix} \\ &= \begin{pmatrix} 7 & 1 \\ 1 & 4 \end{pmatrix} \end{aligned} \tag{5.2.41}$$

♦♦

Finally, we examine the notions of *correlation* and *statistical independence.*† By definition, two random variables ν_n and ν_k are said to be *statistically independent* if

$$p_{\nu_n,\nu_k}(\lambda,\mu) = p_{\nu_n}(\lambda)\,p_{\nu_k}(\mu) \tag{5.2.42}$$

i.e. if their joint density function equals the product of their univariate density functions. On the other hand ν_n and ν_k are said to be *uncorrelated* if

$$E\{\nu_n\nu_k\} = E\{\nu_n\}E\{\nu_k\} \tag{5.2.43}$$

Independence is a much stronger property and, as is well known, "independent" implies "uncorrelated," but the reverse is not in general true. (See Ex. 5.3 and 5.4.)

In practice, attempts to estimate the statistical properties of random variables from empirical data lead naturally to the estimation of expectations rather than of distribution (or density) functions. Expectations are, practically speaking, much easier to estimate than are distribution functions, this for the most part being because the expectation operator is a *functional* and so it leads in every case to a *single real number* as the sought for answer. On the other hand, distributions and densities are *functions,* and to estimate them from empirical data leads, at best, to a functional approximation process, and not to the true functions themselves. It thus turns out that in practice the notion of correlation is more frequently used. Two random variables ν_n and ν_k can be observed and empirical data can be gathered by drawing typical values. The expectations $E\{\nu_n\}$, $E\{\nu_k\}$, $E\{\nu_n^2\}$, $E\{\nu_k^2\}$ and $E\{\nu_n\nu_k\}$ can then be estimated, the accuracy of the estimates depending on the size of the samples.

Defining the vector N as in (5.2.24) we call the matrix

$$E\{(N - E\{N\})(N - E\{N\})^T\} \tag{5.2.44}$$

†The attention of the reader is directed to the fact that the term "independence" can be used in two distinct ways. First there is the *statistical* sense used above and defined by (5.2.42). Second there is the *linear* sense, i.e. the quantities f and g are said to be *linearly independent* if $\alpha f + \beta g = 0$ only when $\alpha = 0 = \beta$. Thus two random variables can be linearly independent and yet *not* statistically independent.

the *covariance matrix* of N. It will frequently be the case that the elements of N are zero-mean, i.e., that

$$E\{N\} = \emptyset \tag{5.2.45}$$

where $\emptyset$ symbolizes the null-vector of appropriate order. We see then that for two zero-mean random variables ν_n and ν_k, their covariance matrix is

$$E\{NN^T\} = \begin{pmatrix} E\{\nu_n^2\} & E\{\nu_n\nu_k\} \\ E\{\nu_n\nu_k\} & E\{\nu_k^2\} \end{pmatrix} \tag{5.2.46}$$

which by (5.2.13) means that for two zero-mean random variables we have, equivalently,

$$E\{NN^T\} = \begin{pmatrix} \sigma^2(\nu_n) & E\{\nu_n\nu_k\} \\ E\{\nu_n\nu_k\} & \sigma^2(\nu_k) \end{pmatrix} \tag{5.2.47}$$

5.3 PROPERTIES OF COVARIANCE MATRICES

For definiteness, we consider three random variables ν_0, ν_1 and ν_2, all of which are zero-mean and we define the vector

$$N \equiv \begin{pmatrix} \nu_0 \\ \nu_1 \\ \nu_2 \end{pmatrix} \tag{5.3.1}$$

Then the covariance matrix of N, namely $E\{NN^T\}$, is a real symmetric matrix. We investigate its properties.

Let C be any non-null vector of constants. Then by (5.2.34) we have that

$$\begin{aligned} C^T E\{NN^T\}C &= E\{C^T NN^T C\} \\ &= E\{(C^T N)(C^T N)\} \\ &= E\{(C^T N)^2\} \end{aligned} \tag{5.3.2}$$

But $C^T N$ is itself a scalar random variable, and clearly $E\{(C^T N)^2\}$ is *non-negative.* We have thus shown that for any non-null vector C,

$$C^T E\{NN^T\} C \geq 0 \tag{5.3.3}$$

which proves that the matrix $E\{NN^T\}$ is either positive definite or positive semidefinite.† We call a matrix which belongs to either of these two classes, a *nonnegative definite* matrix. Every covariance matrix is thus nonnegative definite.

In general, covariance matrices are positive definite. In examining the situation where a covariance matrix is positive semidefinite, we find that the following holds true.

Let N be a vector of m zero-mean random variables. Then the following two statements are equivalent.‡

A. $E\{NN^T\}$ *has a rank-defect of k.*§

B. *Precisely k of the random variables in N are linear combinations of the remaining $m - k$.*

The proof that B implies A is straightforward and is left as an exercise for the reader. As a demonstration of the fact that A implies B, we present the following numerical example. The reader can readily construct a general proof from our procedure.

Suppose that N has the covariance matrix¶

$$E\{NN^T\} = \begin{pmatrix} 1 & 1 & -1 & 2 \\ 1 & 2 & -2 & 3 \\ -1 & -2 & 2 & -3 \\ 2 & 3 & -3 & 5 \end{pmatrix} \tag{5.3.4}$$

We diagonalize this matrix by congruence, using the method of Gaussian elimination discussed in Ex. 5.11. Thus, let

$$B \equiv \begin{pmatrix} 1 & & & \\ -1 & 1 & & \\ 0 & 1 & 1 & \\ -1 & -2 & -1 & 1 \end{pmatrix} \tag{5.3.5}$$

†Whenever the term positive *semi*definite is used, we imply that the matrix in question is singular.

‡i.e., each implies the other.

§i.e., its rank is $m - k$.

¶We assume N to be zero-mean.

Then it is readily verified that

$$BE\{NN^T\}B^T = \begin{pmatrix} 1 & & & \\ & 1 & & \\ & & 0 & \\ & & & 0 \end{pmatrix} \tag{5.3.6}$$

which shows that $E\{NN^T\}$ is positive semidefinite with rank 2 and rank-defect 2. We thus suspect that two elements of N are linearly dependent on the other two which are independent of one another. This is now verified.

Define the vector U by the transformation

$$U = BN \tag{5.3.7}$$

where B is given in (5.3.5). Then

$$u_0 = \nu_0 \tag{5.3.8}$$

$$u_1 = -\nu_0 + \nu_1 \tag{5.3.9}$$

$$u_2 = \nu_1 + \nu_2 \tag{5.3.10}$$

$$u_3 = -\nu_0 - 2\nu_1 - \nu_2 + \nu_3 \tag{5.3.11}$$

Since N is zero-mean, so is U. Moreover, by (5.2.37)

$$E\{UU^T\} = BE\{NN^T\}B^T \tag{5.3.12}$$

which is displayed in (5.3.6). Thus u_2 and u_3 have zero variance, and being zero-mean, *they are themselves zero with probability one.* Hence from (5.3.10) and (5.3.11)

$$\begin{aligned} \nu_1 + \nu_2 &= 0 \\ -\nu_0 - 2\nu_1 - \nu_2 + \nu_3 &= 0 \end{aligned} \tag{5.3.13}$$

and so, solving for ν_2 and ν_3 gives us,

$$\begin{aligned} \nu_2 &= -\nu_1 \\ \nu_3 &= \nu_0 + \nu_1 \end{aligned} \tag{5.3.14}$$

This shows that ν_2 and ν_3 are indeed linearly dependent on ν_0 and ν_1. It remains only to show that the latter are linearly independent of each other.

From (5.3.12) and (5.3.6) we see that

$$E\left\{\begin{pmatrix}u_0\\u_1\end{pmatrix}(u_0, u_1)\right\} = \begin{pmatrix}1 & 0\\0 & 1\end{pmatrix} \tag{5.3.15}$$

which is a positive definite matrix. Thus for all nonzero constants c_0 and c_1,

$$(c_0, c_1)\, E\left\{\begin{pmatrix}u_0\\u_1\end{pmatrix}(u_0, u_1)\right\}\begin{pmatrix}c_0\\c_1\end{pmatrix} > 0$$

We now follow the reasoning employed in (5.3.2) and we thus obtain

$$E\left\{(c_0 u_0 + c_1 u_1)^2\right\} > 0 \tag{5.3.16}$$

and so it must also be true that the sum $c_0 u_0 + c_1 u_1$ cannot be identically zero, if c_0 and c_1 are nonzero. We have thus shown that u_0 and u_1 are linearly independent of each other. Finally then, by (5.3.8) and (5.3.9), ν_0 and ν_1 are also linearly independent of each other. ◆◆

In the above example, we saw that if $E\{NN^T\}$ has a rank-defect of 2 and a rank of 2, then two of the elements in N are linear combinations of the other two, and the latter are not linearly related to each other. We now summarize as follows: Let *$E\{NN^T\}$ be the $m \times m$ covariance matrix of N, a vector of m zero-mean random variables. Then*

a) *$E\{NN^T\}$ is symmetric nonnegative definite.*

b) *If k of the random variables in N are identically zero or are a linear combination of the remaining $m - k$, then $E\{NN^T\}$ is positive semidefinite with rank $m - k$ and vice versa.*

c) *If $k = 0$, i.e. if none of the random variables in N is linearly related to the others, then $E\{NN^T\}$ is positive definite and vice versa.*

In general, the latter will be the case in our subsequent discussion, and so, unless stated otherwise, a covariance matrix will always be positive definite. However, on specific occasions we will encounter situations where this is not the case, such as for example, when a perfect measurement is obtained, or when measurements with correlation coefficient† unity are introduced. However, these cases will be exceptional and will be clearly pointed out. The reader is referred to Example 5.6.

5.4 THE COVARIANCE MATRIX OF THE INPUT ERRORS

Suppose that at time t_n we make a vector of observations on a process. As an example we might choose to observe $x(t)$ and $\dot{x}(t)$ say, and in each case a zero-mean additive error is assumed to be present. Thus, letting

$$Y_n \equiv \begin{pmatrix} y_0 \\ y_1 \end{pmatrix}_n \quad \text{and} \quad N_n \equiv \begin{pmatrix} \nu_0 \\ \nu_1 \end{pmatrix}_n$$

be the vector of observations and errors respectively, we have

$$\begin{aligned} (y_0)_n &= x(t_n) + (\nu_0)_n \\ (y_1)_n &= \dot{x}(t_n) + (\nu_1)_n \end{aligned} \tag{5.4.1}$$

The covariance matrix of N_n will be written R_n, and since we are assuming zero-mean observation errors throughout,

$$R_n \equiv E\left\{N_n N_n^T\right\} \tag{5.4.2}$$

Thus,

$$R_n = \begin{pmatrix} E\left\{(\nu_0)_n^2\right\} & E\left\{(\nu_0 \nu_1)_n\right\} \\ E\left\{(\nu_1 \nu_0)_n\right\} & E\left\{(\nu_1)_n^2\right\} \end{pmatrix} \tag{5.4.3}$$

The numerical algorithms which we intend to develop will form estimates of the process based on sets of such vectors of observations, made

†The correlation coefficient of two zero-mean random variables ν_i and ν_j is defined by

$$\rho_{ij} \equiv \frac{E\{\nu_i \nu_j\}}{[\sigma^2(\nu_i)\,\sigma^2(\nu_j)]^{1/2}}$$

over the sequence of $L + 1$ *successive instants*

$$t_{n-L},\ t_{n-L+1},\ \ldots,\ t_{n-1},\ t_n$$

Accordingly we define the *total observation vector*

$$\mathsf{Y}_{(n)} \equiv \left(\begin{array}{c} Y_n \\ \hdashline Y_{n-1} \\ \hdashline \vdots \\ \hdashline Y_{n-L} \end{array}\right) \tag{5.4.4}$$

Y_{n-i} being the vector of observations made at time t_{n-i}. Associated with $\mathsf{Y}_{(n)}$ is the *total error vector*

$$\mathsf{N}_{(n)} \equiv \left(\begin{array}{c} N_n \\ \hdashline N_{n-1} \\ \hdashline \vdots \\ \hdashline N_{n-L} \end{array}\right) \tag{5.4.5}$$

N_{n-i} being the vector of errors in Y_{n-i}. We have assumed that the random variables forming $\mathsf{N}_{(n)}$ are all zero-mean, and we designate its covariance matrix as $\mathsf{R}_{(n)}$. Thus we define the *total covariance matrix*

$$\mathsf{R}_{(n)} \equiv \mathsf{E}\left\{\mathsf{N}_{(n)} \mathsf{N}_{(n)}{}^T\right\} \tag{5.4.6}$$

We pause briefly to comment on notation. The upper-case italic letters Y_n, N_n and R_n are used to signify observations, errors and their covariance matrix respectively, obtained at time t_n. The subscript shows the *single* instant at which they prevail, and is an essential piece of data along with the actual components of the vectors or matrix. On the other hand, the sans serif italic upper-case letters $\mathsf{Y}_{(n)}$, $\mathsf{N}_{(n)}$ and $\mathsf{R}_{(n)}$ will be used to represent what we call the *total* observation and error vectors, and covariance matrix, formed by a concatenation of the Y_n's and N_n's. The *parenthesized* subscript in $\mathsf{Y}_{(n)}$, $\mathsf{N}_{(n)}$ and $\mathsf{R}_{(n)}$ signifies that *the* n^{th} *is the most recent observation* of the sequence. The reader should be particularly aware of this distinction, and whenever Y_n, N_n or R_n are used, a *single* observation instant is implied, whereas $Y_{(n)}$, $N_{(n)}$, $R_{(n)}$ carry with them a *sequence* of observation instants.

We now examine the properties of $R_{(n)}$. For simplicity assume that we form $Y_{(n)}$ from observations at times t_n and t_{n-1}, and assume that we are observing two quantities in each case. (See e.g. (5.4.1).) Then

$$\begin{aligned} R_{(n)} &= E\{N_{(n)} N_{(n)}^T\} \\ &= E\left\{\begin{pmatrix} N_n \\ \hline N_{n-1} \end{pmatrix} \left(N_n^T \mid N_{n-1}^T\right)\right\} \\ &= \left(\begin{array}{c|c} E\{N_n N_n^T\} & E\{N_n N_{n-1}^T\} \\ \hline E\{N_{n-1} N_n^T\} & E\{N_{n-1} N_{n-1}^T\} \end{array}\right) \end{aligned} \tag{5.4.7}$$

and since, in this case,

$$N_n \equiv \begin{pmatrix} \nu_0 \\ \nu_1 \end{pmatrix}_n \qquad N_{n-1} \equiv \begin{pmatrix} \nu_0 \\ \nu_1 \end{pmatrix}_{n-1} \tag{5.4.8}$$

we see that

$$R_{(n)} = \left(\begin{array}{cc|cc} \sigma^2\left[(\nu_0)_n\right] & E\{(\nu_0 \nu_1)_n\} & E\{(\nu_0)_n(\nu_0)_{n-1}\} & E\{(\nu_0)_n(\nu_1)_{n-1}\} \\ E\{(\nu_1 \nu_0)_n\} & \sigma^2\left[(\nu_1)_n\right] & E\{(\nu_1)_n(\nu_0)_{n-1}\} & E\{(\nu_1)_n(\nu_1)_{n-1}\} \\ \hline E\{(\nu_0)_{n-1}(\nu_0)_n\} & E\{(\nu_0)_{n-1}(\nu_1)_n\} & \sigma^2\left[(\nu_0)_{n-1}\right] & E\{(\nu_0 \nu_1)_{n-1}\} \\ E\{(\nu_1)_{n-1}(\nu_0)_n\} & E\{(\nu_1)_{n-1}(\nu_1)_n\} & E\{(\nu_1 \nu_0)_{n-1}\} & \sigma^2\left[(\nu_1)_{n-1}\right] \end{array}\right) \tag{5.4.9}$$

Thus $R_{(n)}$ is seen to be a real symmetric matrix and by the arguments given in the previous section, it will be nonnegative definite. If none of the random variables $(\nu_0)_n$, $(\nu_1)_n$, $(\nu_0)_{n-1}$ or $(\nu_1)_{n-1}$ is a linear combination of the others, as will usually be the case in practice, then $R_{(n)}$ will be positive definite. As can be seen from (5.4.9), the *on-diagonal* blocks of $R_{(n)}$ contain only a *single* time-subscript whereas the *off-diagonal* ones contain *two*.

We address ourselves to three essential properties of $R_{(n)}$ in addition to its nonnegative definiteness:

1. Firstly, the random processes giving rise to the error vectors $\ldots, N_{n-L}, N_{n-L+1}, \ldots, N_{n-1}, N_n, \ldots$ may or may not be stationary, i.e., their statistics may or may not be changing with time. In the event that the second order statistics of these vectors are changing with time, then this information will be clearly conveyed by the successive total covariance matrices. Thus, we are not limited to stationary input-error statistics.

2. Secondly, the matrices $E\left\{N_{n-i}N_{n-j}^T\right\}$ for $i \neq j$ may or may not all be null. These matrices form the *off-diagonal blocks* in $R_{(n)}$. In the event that the observation errors made at *any one* time are uncorrelated with those made at *any other* time, then they *will* all be null. However, if some correlating action along the time-axis is present (e.g. electronic circuitry containing capacitors or inductors), then some or all of the off-diagonal blocks will be nonnull. If they are all null, we say that the errors are *stage-wise uncorrelated* and in such a case, all total covariance matrices obtained from that observation scheme will be *block-diagonal with a single time-subscript in each block.*

3. Finally, consider the covariance matrices $\ldots, R_{n-1}, R_n, \ldots$ which form the *on-diagonal blocks* of the total covariance matrices. The individual components of the error vector

$$N_n \equiv \begin{pmatrix} \nu_0 \\ \nu_1 \end{pmatrix}_{n-i} \tag{5.4.10}$$

may or may not be correlated with one another. (Note the single-time-subscript in (5.4.10).) If they are *not,* then its covariance matrix R_n will be a diagonal matrix. If this is true for all such vectors N making up $N_{(n)}$ of (5.4.5) then the errors are said to be *locally uncorrelated.*

Thus for example, if all of the total covariance matrices obtained from some observation scheme have the form

$$R_{(n)} = \alpha^2 I = \begin{pmatrix} \alpha^2 & & & \\ & \alpha^2 & & \\ & & \ddots & \\ & & & \alpha^2 \end{pmatrix} \tag{5.4.11}$$

where α is a constant, then clearly

a) the second order statistics are stationary,

b) the errors are stage-wise uncorrelated,

c) the errors are locally uncorrelated.

On the other hand, if $R_{(n)}$ has the form (assuming two observation instants in $Y_{(n)}$ and two measurements in each of Y_n and Y_{n-1})

$$R_{(n)} = \left(\begin{array}{cc|cc} \theta^n & \theta^n & & \\ \theta^n & 2\theta^n & & \\ \hline & & \theta^{n-1} & \theta^{n-1} \\ & & \theta^{n-1} & 2\theta^{n-1} \end{array}\right) \tag{5.4.12}$$

where θ is a real constant, then the errors are

a) nonstationary,

b) stage-wise uncorrelated,

c) locally correlated.

Since each of the three properties discussed above can be disjointly present or absent in any one case, there are clearly eight possibilities in all. While they can obviously all occur in practice, and our filtering algorithms will naturally be able to handle any of them, two cases will occur most frequently in our discussion. These are first, the case shown in (5.4.11) and second, the case of stage-wise uncorrelated errors of which (5.4.12) is a simple example.

5.5 THE COVARIANCE MATRIX OF THE OUTPUT ERRORS

The filters which we will develop will be of the form†

$$X^*_{n+h,n} = WY_{(n)} \tag{5.5.1}$$

where $X^*_{n+h,n}$ is a prediction of what the process state-vector will be at time t_{n+h} based on observations up to time t_n, W is a matrix, and $Y_{(n)}$ is the total observation vector on which the estimate is based.

For definiteness let $X^*_{n+h,n}$ be a 3-vector and let the row-vectors of the matrix W be called W_0, W_1, W_2. Then from (5.5.1)

$$\begin{aligned} \left(x^*_0\right)_{n+h,n} &= W_0 Y_{(n)} \\ \left(x^*_1\right)_{n+h,n} &= W_1 Y_{(n)} \\ \left(x^*_2\right)_{n+h,n} &= W_2 Y_{(n)} \end{aligned} \tag{5.5.2}$$

†See e.g. (4.5.14).

We now demonstrate that the rows of W will, in general, be linearly independent. This is so for the following reason. Suppose, to the contrary, that they were linearly related so that for example

$$W_2 = \alpha W_0 + \beta W_1 \tag{5.5.3}$$

where α and β are a pair of constants. Then it would be easier to compute $\left(x_2^*\right)_{n+h,n}$ from the relation

$$\left(x_2^*\right)_{n+h,n} = \alpha\left(x_0^*\right)_{n+h,n} + \beta\left(x_1^*\right)_{n+h,n} \tag{5.5.4}$$

rather than by the use of the final equation in (5.5.2). The matrix W would in fact have in it a *redundant* row, namely W_2, which we need never use. Thus, without loss of generality, the rows of W can be assumed to be *linearly independent* row-vectors. Hence, W will have rank equal to the number of its rows. We refer to this condition as *full row-rank*.

We now examine $Y_{(n)}$ of (5.5.1). We know that each of its components is a sum of a term due to the process, plus an additive observation error (see (5.4.1)). (We have referred to the former as the signal and to the latter as the input errors.) What is true, term by term for each component of $Y_{(n)}$, must be true for the entire vector. Suppose we now remove the signal. Then all that remains of $Y_{(n)}$ is the total error vector $N_{(n)}$ of (5.4.5), and by (5.5.1) we now see that the filter gives

$$X^*_{n+h,n} = WN_{(n)} \tag{5.5.5}$$

This shows how the observation errors are acted on by the algorithm. We call the above the *random output-error vector*, and to make matters completely explicit we designate it as

$$N^*_{n+h,n} \equiv WN_{(n)} \tag{5.5.6}$$

The symbol $N^*_{n+h,n}$ should be read as "the random errors in $X^*_{n+h,n}$", and the presence of the star and the subscripts are intended to link it very clearly to the vector $X^*_{n+h,n}$.

When the signal is not removed from $Y_{(n)}$ then by (5.5.1) we see that $X^*_{n+h,n}$ will have two components, one due to the signal in $Y_{(n)}$ and the other given by (5.5.6), which is due solely to the input errors in $Y_{(n)}$. At a later stage we will analyze the part of $X^*_{n+h,n}$ which is related to the signal and we will find that errors may be present in it too. Those errors will be called the *systematic* or *bias errors.*

We now examine the statistics of the random output-error vector $N^*_{n+h,n}$. By assumption, the input errors are zero-mean, i.e.,

$$E\{N_{(n)}\} = \emptyset \tag{5.5.7}$$

Hence, making use of (5.2.36) we see from (5.5.6) that $E\{N^*_{n+h,n}\} = \emptyset$ and so the random output errors will then also be zero-mean.

The covariance matrix of the random output-error vector $N^*_{n+h,n}$ is designated as $S^*_{n+h,n}$, i.e.

$$S^*_{n+h,n} \equiv E\{N^*_{n+h,n} N^{*T}_{n+h,n}\} \tag{5.5.8}$$

We can express $S^*_{n+h,n}$ in terms of the covariance matrix of the observation errors as follows. By (5.5.6)

$$\begin{aligned} S^*_{n+h,n} &= E\{WN_n (WN_n)^T\} \\ &= WE\{N_{(n)} N_{(n)}^T\} W^T \end{aligned} \tag{5.5.9}$$

and so by (5.4.6)

$$S^*_{n+h,n} = WR_{(n)} W^T \tag{5.5.10}$$

This equation forms a complementary pair with (5.5.1). In the earlier equation we saw how the filter, as exemplified by W, acts on the observation vector $Y_{(n)}$ to produce the estimate. In (5.5.10), on the other hand, we see how the filter acts on the covariance matrix of the observation errors to produce the covariance matrix of the errors in the estimate.

Assume now that we set up the filter which gives us updated estimates. We write (5.5.1) as

$$X^*_{n,n} = W(0)\, Y_{(n)} \tag{5.5.11}$$

and for an h-step prediction

$$X^*_{n+h,n} = W(h)\, Y_{(n)} \tag{5.5.12}$$

the functional argument of W being intended to show explicitly the amount of prediction being performed. Let Φ be the transition matrix for the process under consideration, assumed for simplicity to be a constant-coefficient linear differential equation. Then the assertion that $X^*_{n,n}$ and $X^*_{n+h,n}$ both define the same trajectory can be stated precisely by the equation

$$X^*_{n+h,n} = \Phi(h) X^*_{n,n} \tag{5.5.13}$$

Combining this with the previous two equations thus gives

$$W(h)\, \mathsf{Y}_{(n)} = \Phi(h)\, W(0)\, \mathsf{Y}_{(n)} \tag{5.5.14}$$

and clearly, since $\mathsf{Y}_{(n)}$ is arbitrary, we see then that $W(h)$ must satisfy

$$W(h) = \Phi(h)\, W(0) \tag{5.5.15}$$

This is in close analogy to (5.5.13) and is a direct consequence of the fact that the process is being modeled by a differential equation whose transition matrix is $\Phi(h)$.

Now the random output errors in $X^*_{n,n}$ and $X^*_{n+h,n}$ are, respectively,

$$N^*_{n,n} \equiv W(0)\, \mathsf{N}_{(n)} \tag{5.5.16}$$

and

$$N^*_{n+h,n} \equiv W(h)\, \mathsf{N}_{(n)} \tag{5.5.17}$$

Making use of (5.5.15) then shows that

$$N^*_{n+h,n} = \Phi(h) N^*_{n,n} \tag{5.5.18}$$

Thus, just as the prediction $X^*_{n+h,n}$ can be obtained from $X^*_{n,n}$ by the use of $\Phi(h)$ in (5.5.13), so the random errors in the prediction are related to those in the updated estimate by (5.5.18).

We examine the covariance matrices of $N^*_{n+h,n}$ and $N^*_{n,n}$. By (5.5.18),

$$\begin{aligned} S^*_{n+h,n} &\equiv \mathsf{E}\left\{N^*_{n+h,n} N^{*T}_{n+h,n}\right\} \\ &= \mathsf{E}\left\{\Phi(h) N^*_{n,n} \left[\Phi(h) N^*_{n,n}\right]^T\right\} \\ &= \Phi(h)\, \mathsf{E}\left\{N^*_{n,n} N^{*T}_{n,n}\right\} \Phi(h)^T \end{aligned} \tag{5.5.19}$$

i.e., $S^*_{n+h,n}$ and $S^*_{n,n}$ are related by the *congruence transformation*†

$$S^*_{n+h,n} = \Phi(h) S^*_{n,n} \Phi(h)^T \tag{5.5.20}$$

Thus the transition matrix $\Phi(h)$ is seen to have the ability to move the validity instant of the *estimate,* if used as in (5.5.13), or of the *covariance matrix,* if used as in (5.5.20). (Observe that in both cases only the *first* subscript of $X^*_{n+h,n}$ or $S^*_{n+h,n}$ is affected by Φ. The second subscript signifies that the most recent observations used in forming the estimate were obtained at t_n, and nothing can modify that short of changing the vector $Y_{(n)}$ in (5.5.1).

We pause briefly to examine (5.5.13) and (5.5.20) in somewhat further detail. This pair of equations will appear frequently in our subsequent discussion, and the reader will soon become accustomed to seeing filtering algorithms headed by the pair of equations (or their equivalent)

$$X^*_{n+h,n} = \Phi(h) X^*_{n,n} \tag{5.5.21}$$

$$S^*_{n+h,n} = \Phi(h) S^*_{n,n} \Phi(h)^T \tag{5.5.22}$$

Now, it is clear that given the model differential equations

$$\frac{d}{dt} X(t) = AX(t) \tag{5.5.23}$$

(assuming a linear constant-coefficient system) and given, as a set of initial conditions,

$$X_n \equiv X^*_{n,n} \tag{5.5.24}$$

a unique trajectory is implied. Moreover, it is also clear that we can move ourselves back and forth along that trajectory, by the use of backward or forward integration of (5.5.23) subject to (5.5.24). What equation (5.5.21) states is that, in addition to employing numerical integration, *we can also accomplish precisely the same result by an appropriate linear transformation.* Thus $X^*_{n,n}$ and $X^*_{n+h,n}$ define *one and the same* trajectory and are merely restatements of it. Given either, we can obtain the other by the use of (5.5.21) or by the integration of (5.5.23).

†In Ex. 5.9 we define the congruence transformation and examine some of its more important properties.

We must further bear in mind that the two vectors appearing in (5.5.21) are state-vectors of an estimated trajectory, and that estimation was based on a given set of data obtained up to t_n, as stated by the second of their dual subscripts. The symbol appearing in that position obviously cannot change as we move back and forth along the trajectory to which $X^*_{n,n}$ and $X^*_{n+h,n}$ belong, and this is in evidence in (5.5.21). In summary then, (5.5.21) connects two state-vectors of one and the same estimated trajectory based on a single given set of data.

In (5.5.22) we see how $S^*_{n,n}$ and $S^*_{n+h,n}$ are related. Thus, given that the errors in $X^*_{n,n}$ have as their covariance matrix $S^*_{n,n}$, a very natural question to ask is how those errors propagate if we integrate the differential equation using $X^*_{n,n}$ as initial conditions. The answer is given in (5.5.18), and (5.5.22) shows *how the second-order statistics propagate.* We accordingly think of (5.5.21) as the *state-prediction* equation, and of (5.5.22) as the *error-statistics propagation* equation. Of course, if we were only interested in moving the state-vector down the trajectory, it would be computationally easier to simply integrate the differential equations, and the transition matrix would not be needed. However, when we are interested in how the errors propagate, then the transition matrix is essential and must be obtained.† (Note that integrating the differential equations of the transition matrix is a much larger operation than integrating the differential equations of the state-vector.)

When the equations of motion are *nonlinear,* the problem is slightly more complicated. Given $X^*_{n,n}$ we can now no longer form $X^*_{n+h,n}$ by a relation of the form of (5.5.21), *since there is no transition equation for a nonlinear process.* Instead, we simply integrate the model equations

$$\frac{d}{dt} X(t) = F[X(t)] \tag{5.5.25}$$

using as initial conditions

$$X(t_n) = X^*_{n,n} \tag{5.5.26}$$

The truncation errors caused by the chosen numerical integration procedure can, in theory at any rate, be made as small as we please, and so $X^*_{n+h,n}$ can be obtained with any degree of precision. Movement along the unique trajectory specified by (5.5.25) and (5.5.26) can thus be implemented. *In the case of nonlinear systems, prediction or retrodiction is always carried out in this manner.*

†See Sections 4.7 and 4.8.

The error propagation problem is not solved quite so simply. Thus, assuming that $X^*_{n,n}$ were known to have as its covariance matrix $S^*_{n,n}$, the problem of determining *precisely* how those statistics propagate if we integrate (5.5.25) is, in general, very complex. In fact we usually require the entire density function of the errors, i.e. all of their statistics and not only their second-order ones. Even then, in all but the simplest cases, the situation is usually intractable.

However, if we are prepared to accept an *approximate solution,* we can obtain an answer quite readily. Thus, given that

$$\frac{d}{dt} X(t) = F[X(t)] \tag{5.5.27}$$

suppose that the true trajectory $X(t)$ were perturbed by a small amount to $X(t) + \epsilon[X(t)]$ where $\epsilon[X(t)]$ means "errors in $X(t)$." The vector $X(t) + \epsilon[X(t)]$ can be thought of as the estimate $X^*_{n,n}$, i.e.,

$$X^*_{n,n} = X(t_n) + \epsilon\left[X(t_n)\right] \tag{5.5.28}$$

Thus $\epsilon[X(t)]$ is actually the vector of errors in $X^*_{n,n}$, namely $N^*_{n,n}$. As a result of the perturbation, (5.5.27) gives us (c/f (4.8.14))

$$\frac{d}{dt} X(t) + \frac{d}{dt}\epsilon[X(t)] = F\left[X(t) + \epsilon[X(t)]\right] \tag{5.5.29}$$

and so (c/f (4.8.21)) we have, to *first order accuracy,*

$$\frac{d}{dt}\epsilon[X(t)] = A[X(t)]\,\epsilon[X(t)] \tag{5.5.30}$$

where

$$\left[A[X(t)]\right]_{ij} = \left.\frac{\partial f_i(X)}{\partial x_j}\right|_{X = X(t)} \tag{5.5.31}$$

Thus (5.5.30) gives us an *approximation* to the differential equation governing the evolution of the errors in $X(t)$, subject to (5.5.27).

We of course now recognize that the approximation we have obtained is a time-varying linear differential equation, and so we can write (c/f (4.8.24))

$$\epsilon\left[X(t_n + \zeta)\right] = \Phi(t_n + \zeta, t_n; X)\,\epsilon\left[X(t_n)\right] \tag{5.5.32}$$

This is exact relative to (5.5.30) but is of course only accurate to first order, relative to (5.5.27). Then the law by which the second-order statistics of $\epsilon[X(t)]$ will propagate is given from (5.5.32) as (c/f (5.5.10))

$$S^*_{t_n+\zeta, t_n} = \Phi(t_n + \zeta, t_n; X)\, S^*_{t_n, t_n}\, \Phi(t_n + \zeta, t_n; X)^T \tag{5.5.33}$$

where, of course, we define $S^*_{t_n, t_n}$ by

$$\begin{aligned} S^*_{t_n, t_n} &\equiv E\left\{\epsilon\left[X(t_n)\right]\epsilon\left[X(t_n)\right]^T\right\} \\ &\equiv E\left\{N^*_{n,n} N^{*T}_{n,n}\right\} \end{aligned} \tag{5.5.34}$$

The transition matrix is thus required, and it is obtainable either by the method of integrating (4.8.25), i.e.

$$\frac{\partial}{\partial\zeta}\Phi(t_n + \zeta, t_n; X) = A\left[X(t_n + \zeta)\right]\Phi(t_n + \zeta, t_n; X) \tag{5.5.35}$$

subject to

$$\Phi(t_n, t_n; X) = I \tag{5.5.36}$$

or else directly from (4.8.54), namely

$$\left[\Phi(t_n + \zeta, t_n; X\right]_{ij} = \frac{\partial x_i(t_n + \zeta)}{\partial x_j(t_n)} \tag{5.5.37}$$

Equation (5.5.33) is accurate *only to second order terms* involving components of $\epsilon\left[X(t_n)\right]$. If $\epsilon\left[X(t_n)\right]$ involves only small numbers, i.e., if the errors in $X^*_{n,n}$ are small relative to $X(t_n)$, then (5.5.33) shows, to good precision, how the second-order statistics of the errors in $X^*_{n,n}$ will propagate. However, if those errors are not small relative to $X(t_n)$ then (5.5.33) may in fact be a very poor approximation to the true value of $S^*_{t_n+\zeta, t_n}$ given $S^*_{t_n, t_n}$.

As the nominal trajectory in the above procedure, we can now use the state-vector $X^*_{n,n}$, since, after all, (5.5.28) could also be written as

$$X_n = X^*_{n,n} - \epsilon(X_n) \tag{5.5.38}$$

i.e. we can also think of the true trajectory X_n as being a perturbed version of the estimated trajectory $X^*_{n,n}$. In this case we would write (5.5.35) as

$$\frac{\partial}{\partial \zeta}\Phi(t_n + \zeta, t_n; X^*) = A\left(X^*_{t_n+\zeta, t_n}\right)\Phi(t_n + \zeta, t_n; X^*) \tag{5.5.39}$$

where, by (5.5.31)

$$\left[A\left(X^*_{t_n+\zeta, t_n}\right)\right]_{ij} \equiv \left.\frac{\partial f_i(X)}{\partial x_j}\right|_{X = X^*_{t_n+\zeta, t_n}} \tag{5.5.40}$$

The vector $X^*_{t_n+\zeta, t_n}$ is obtained by integrating (5.5.27) from t_n to $t_n + \zeta$, using $X^*_{n,n}$ as initial conditions. Thus in the nonlinear case, to integrate the differential equations of the transition matrix, we must *also* integrate the state-equations. We note that if we are only moving the state-vector along the trajectory then we accomplish this without the transition matrix in both the linear and nonlinear case, by simply integrating the state-equations. However, if we also wish to study the error propagation, then the transition matrix is required.

Finally we examine the rank of the covariance matrices of the output errors. By (5.5.10)

$$\begin{aligned} S^*_{n,n} &\equiv E\left(N^*_{n,n} N^{*T}_{n,n}\right) \\ &= W(0)\, R_{(n)}\, W(0)^T \end{aligned} \tag{5.5.41}$$

and since $R_{(n)}$ is symmetric, so then is $S^*_{n,n}$. Moreover, we have already demonstrated that, without loss of generality, $W(0)$ can be assumed always to have full row-rank. Thus from (5.5.41) it follows that *if $R_{(n)}$ is positive definite, then so is $S^*_{n,n}$* (see Ex. 5.7).

From (5.5.20) we recall that $S^*_{n+h,n}$ is a congruence transformation on $S^*_{n,n}$. Hence, if the latter is positive definite, then so is the former.† Thus for any amount of prediction or retrodiction, the covariance matrix of the output errors will be positive definite if the covariance matrix of the input errors is positive definite.

†Positive definiteness is preserved under congruence. (See Ex. 5.9.)

The reader is also referred to Examples 5.8 through 5.12 which contain much useful information related to covariance matrices and congruence transformations.

We consider a simple example. Let $X^*_{n+h,n}$ be a 2-vector with the transition matrix

$$\Phi(\tau) = \begin{pmatrix} 1 & \tau \\ 0 & 1 \end{pmatrix} \tag{5.5.42}$$

and let the filtering algorithm be

$$X^*_{n,n} = W(0)\,Y_{(n)} \tag{5.5.43}$$

where

$$W(0) = \frac{1}{10}\begin{pmatrix} 7 & 4 & 1 & -2 \\ 3 & 1 & -1 & -3 \end{pmatrix} \tag{5.5.44}$$

we wish to find $W(1)$.

Clearly,

$$\begin{aligned} X^*_{n+1,n} &= \Phi(1)\,X^*_{n,n} \\ &= \Phi(1)\,W(0)\,Y_{(n)} \\ &= W(1)\,Y_{(n)} \end{aligned} \tag{5.5.45}$$

Hence [c/f (5.5.15)]

$$\begin{aligned} W(1) &= \Phi(1)\,W(0) \\ &= \frac{1}{10}\begin{pmatrix} 10 & 5 & 0 & -5 \\ 3 & 1 & -1 & -3 \end{pmatrix} \end{aligned} \tag{5.5.46}$$

Thus if $W(0)$ of (5.5.44) gives an *updated* estimate, then $W(1)$ of (5.5.46) will give the 1-*step prediction* down the same trajectory.

Assume next that $R_{(n)} = I$ where I is 4×4. Then by (5.5.44) and (5.5.10)

$$\begin{aligned} S^*_{n,n} &= W(0)\,R_{(n)}\,W(0)^T \\ &= \frac{1}{100}\begin{pmatrix} 70 & 30 \\ 30 & 20 \end{pmatrix} \end{aligned} \tag{5.5.47}$$

Likewise, by (5.5.46) and (5.5.10),

$$\begin{aligned} S^*_{n+1,n} &= W(1)\,R_{(n)}\,W(1)^T \\ &= \frac{1}{100}\begin{pmatrix} 150 & 50 \\ 50 & 20 \end{pmatrix} \end{aligned} \tag{5.5.48}$$

Moreover, by (5.5.20)

$$\begin{aligned} S^*_{n+1,n} &= \Phi(1)\,S^*_{n,n}\,\Phi(1)^T \\ &= \frac{1}{100}\begin{pmatrix} 150 & 50 \\ 50 & 20 \end{pmatrix} \end{aligned} \tag{5.5.49}$$

which obviously agrees with (5.5.48).

Note that $W(0)$ of (5.5.44) has two independent rows and hence it has rank 2, i.e. full row-rank. Thus $S^*_{n,n}$ and $S^*_{n+1,n}$ should be positive definite and from (5.5.47) and (5.5.48) this is in fact the case. In general,

$$\begin{aligned} S^*_{n+h,n} &= \Phi(h)\,S^*_{n,n}\,\Phi(h)^T \\ &= \frac{1}{100}\begin{pmatrix} 70 + 60h + 20h^2 & 30 + 20h \\ 30 + 20h & 20 \end{pmatrix} \end{aligned} \tag{5.5.50}$$

and, as is easily verified, this is positive definite for all h.

The smallest diagonal elements in $S^*_{n+h,n}$ are obtained when $h = -3/2$, as we see by setting

$$\frac{d}{dh}(70 + 60h + 20h^2) = 0 \tag{5.5.51}$$

Thus, at $h = -3/2$,

$$S^*_{n-3/2,n} = \frac{1}{100}\begin{pmatrix} 25 & 0 \\ 0 & 20 \end{pmatrix} \tag{5.5.52}$$

and so for the case of 3/2-second retrodiction, the errors in the estimate have the smallest variances and are uncorrelated. ◆◆

5.6 THE GAUSSIAN AND CHI-SQUARED DENSITY FUNCTIONS

In the discussion to come on various smoothing algorithms, we will constantly be interested in the statistics of the random output errors, given the statistics of the observation errors. It will be assumed that the latter have zero mean and that they have covariance matrices with bounded elements, all of which are known. One of the primary tasks in the analysis of the filtering schemes will then be to develop expressions which give the covariance matrix of the random output errors as a function of the covariance matrix of the input errors.

Concerning the form of the probability density functions for the observation errors, however, *we will make no restrictions whatever.* All we will require is that they be zero-mean and have finite second-order statistics, which are known. Within these limits, any density function will be acceptable, and the derivation of the algorithms or the analysis of their properties will be completely unrelated to the particular forms which those density functions might assume.

Now, it is a known fact that one particular density function occurs very frequently in practice, and this is the *normal* or *Gaussian* probability density function. It is the purpose of this section to consider that function briefly, and to itemize some of its salient properties. We also examine briefly the *Chi-squared* density function and consider some extremely valuable applications of the latter to our present discussion. Our approach will be very elementary, and for further detail the reader is referred to [5.2], [5.4] or [5.5].

A random variable ν with expectation b and variance r is said to be *normal* or *Gaussian* if its density function has the form

$$p_\nu(\zeta) = \frac{1}{(2\pi r)^{1/2}} \exp\left\{-\frac{1}{2}\left[\frac{(\zeta - b)^2}{r}\right]\right\} \tag{5.6.1}$$

This is the very familiar bell-shaped curve that occurs so frequently in probability and statistics. If ν has zero mean and unity variance, then we obtain what is called the *standard* normal distribution

$$p_\nu(\zeta) = \frac{1}{(2\pi)^{1/2}} \exp\left\{-\frac{1}{2}\zeta^2\right\} \tag{5.6.2}$$

Assume now that N is a vector of m Gaussian random variables with mean B and positive-definite covariance matrix R. Then its multivariate density function is given by

$$p_N(Z) = \frac{1}{(2\pi)^{m/2}(\det R)^{1/2}} \exp\left[-\frac{1}{2}(Z - B)^T R^{-1}(Z - B)\right] \tag{5.6.3}$$

We see that if N is multivariate normal, then its density function, and hence all of its statistics, are completely determined by its expectation B and covariance matrix R.† (See Ex. 5.14.)

One important property of Gaussian random variables is that *any linear combination of them is again Gaussian.* This fact is of direct interest to us for the following reason. As pointed in the preceding section, all of our smoothing algorithms will essentially be in the form of a linear transformation on the observations, and from (5.5.6) we thus see that each element of the output-error vector $N^*_{n+h,n}$ is a linear combination of the input errors in $N_{(n)}$. It thus follows that if the input errors are Gaussian with mean zero and covariance matrix $R_{(n)}$, then the output errors are also Gaussian with mean zero. From (5.5.10) we obtain the covariance matrix of the output errors as $WR_{(n)}W^T$, and so the statistics of the output errors are, in this case, completely known.

We now examine a further very important aspect of the Gaussian density function. Thus, suppose that we equate (5.6.3) to a constant. Then clearly we are considering *contours* or *surfaces* in the Z-space, made up of all the points which give rise to a certain value of $p_N(Z)$. These contours are of great interest in performing system analyses, and we now examine their shapes.

To within a constant (i.e. -1/2), the exponent of the multivariate Gaussian density function of (5.6.3) is the quadratic form

$$f(Z) \equiv (Z - B)^T R^{-1}(Z - B) \tag{5.6.4}$$

Thus, if we equate $p_N(Z)$ of that equation to a constant, then we are also equating the above scalar $f(Z)$ to a constant. Consider first the bivariate case. Setting $f(Z)$ to a constant, say k^2, gives us the equation

$$(z_0 - b_0,\ z_1 - b_1)\begin{pmatrix} r_{00} & r_{01} \\ r_{10} & r_{11} \end{pmatrix}^{-1}\begin{pmatrix} z_0 - b_0 \\ z_1 - b_1 \end{pmatrix} = k^2 \tag{5.6.5}$$

†This is also true even if R is singular. In this case the density function for N can be given in terms of its characteristic function. (See e.g. [5.2, p. 130].)

and as is well known, if the matrix R is positive definite, *then this defines a family of ellipses* with respect to z_0 and z_1, which are possibly rotated, depending on R, and whose centers are displaced by the vector B from the system origin. In this way we see that if the bivariate normal density function is set equal to a constant, then the sample values of the random variables ν_0 and ν_1 will be constrained to lie on ellipses whose properties are given completely by R, B and k^2. In higher dimensions, the analogy goes over to families of hyper-ellipsoids in the hyper-spaces of four and five dimensions, etc.

An examination of the covariance matrix of a Gaussian density function can thus be an extremely fruitful undertaking, if viewed in this light. It gives us the general shape of the constant probability contours for the errors, telling us for example whether they are flat and pancake-like, cigar-shaped or spherical. To perform this study, the Z-space is best transformed to the coordinate system in which the ellipsoids are symmetrically oriented with respect to the axes, and centered at the origin.

It can be shown quite simply (see e.g. [5.3]) that if R is a positive-definite matrix and if Q is the orthogonal matrix formed from its orthonormalized eigenvectors, then

$$Q^T R^{-1} Q = \Lambda^{-1} \tag{5.6.6}$$

where Λ^{-1} is a diagonal matrix, made up of the eigenvalues of R^{-1}. By inverting both sides of the above equation, we note that these are the reciprocals of the eigenvalues of R.

Suppose then that we have the ellipsoid

$$(Z - B)^T R^{-1} (Z - B) = k^2 \tag{5.6.7}$$

and that we wish to study its shape. Let Q be defined as above, and define the transformation

$$Z - B = QJ \tag{5.6.8}$$

Then the vector J is an orthogonal transformation on the vector Z-B, and so *shapes have been preserved* (since orthogonal transformations only rotate and do not stretch). Using (5.6.8) in (5.6.7) gives

$$J^T Q^T R^{-1} Q J = k^2 \tag{5.6.9}$$

which, by (5.5.6) becomes

$$J^T \Lambda^{-1} J = k^2 \tag{5.6.10}$$

But Λ^{-1} is diagonal and so (5.6.10) reduces to

$$\frac{j_0^2}{\lambda_0} + \frac{j_1^2}{\lambda_1} + \cdots + \frac{j_m^2}{\lambda_m} = k^2 \tag{5.6.11}$$

showing that the ellipsoids are now symmetrically oriented in the J-space and centered on the J-origin. Moreover for $k = 1$, *the numbers* $\lambda_0^{1/2}$, $\lambda_1^{1/2}, \ldots, \lambda_m^{1/2}$ *are the intercepts which the ellipsoid makes along the axes of the J-system.*

Thus in this new coordinate system, the eigenvalues of the matrix R give us all the information we require to form a picture of the shape of the ellipsoids, and so to study the shapes of the error surfaces. All that we in fact require are those eigenvalues. While we do not propose to continue this discussion further, it is readily seen that the shape of the error ellipsoids is a most useful concept in performing a system analysis, and in determining major sources of error. (See Ex. (5.15).)

A density function which is very closely related to the Gaussian one is the *Chi-squared* function. Assume that N is an m-vector of *independent standard normal variables* (i.e. zero-mean and covariance matrix given by an identity matrix). Then the scalar $N^T N$, i.e. the sum of the squares of those random variables, has a density function known as the *Chi-squared function with m degrees of freedom.* The details of its form need not concern us here (see e.g. [5.5, p. 98]), and all that we require are tables of its cumulative distribution function (see e.g. [5.6]).

In the event that N has a covariance matrix R (not necessarily an identity matrix), then it is readily shown (see Ex. 5.16) that the scalar

$$\chi^2 \equiv N^T R^{-1} N \tag{5.6.12}$$

is also Chi-squared.†

The Chi-squared function provides us with an extremely sensitive test for verifying, to within confidence limits, the hypothesis that a given vector is a likely member of an ensemble whose covariance matrix is known. Thus, assume that we were told that a Gaussian ensemble has as its covariance

†The vector N is still assumed to be Gaussian with zero-mean.

matrix

$$R \equiv \begin{pmatrix} 1 & -1 \\ -1 & 1.01 \end{pmatrix} \tag{5.6.13}$$

and suppose that the vector $N \equiv \left(\frac{1}{2}, \frac{1}{2}\right)^T$ were presented. We wish to test the hypothesis that *this N is a likely member from that ensemble.* To do this, we compute the value of χ^2 for N relative to R, thereby obtaining:

$$\chi^2 \equiv N^T R^{-1} N = 100.25 \tag{5.6.14}$$

Then reference to tables (Chi-squared with two degrees of freedom) shows that

$$P(\chi^2 \leq 100.25) = 0.99999 \ldots \tag{5.6.15}$$

which means that of all possible points in the 2-space in which N lies, 99.999% of them have smaller values of χ^2 than the N given above. If our confidence limit is established at, say 95%, *then we infer that our hypothesis is false,* i.e. that N is not a likely member of our ensemble. Alternatively we can infer that R is not its covariance matrix.

On the other hand, as is readily verified, the vector $N \equiv \left(\frac{1}{2}, -\frac{1}{2}\right)^T$ has a χ^2, relative to the above R of 0.5, and from tables

$$P(\chi^2 \leq 0.5) = 0.22120 \tag{5.6.16}$$

This, by our chosen confidence limit, does nothing to disprove our hypothesis, and so we can assume that this N *is* a likely member.

While the reader may well feel that in the above case the likelihood of the two N's could have been judged by inspection from the sign of the off-diagonal element in R, *in higher dimensions this is not possible.* But in all cases, regardless of the dimension of the space, the Chi-squared test proves to be an extremely powerful and very easily implemented criterion.

In much of the work to come, our algorithms will produce both an estimate $X^*_{n,n}$ and a covariance matrix $S^*_{n,n}$. Needless to say, the algorithm is not of much use if the errors in the former are inconsistent with the latter.

During the testing and de-bugging phase, when the inputs to the filter are being generated by the addition of synthetic errors to a known trajectory, it should be possible to compute the actual errors in $X^*_{n,n}$, namely $N^*_{n,n}$.

Then, by computing the χ^2 of the latter, relative to the matrix $S^*_{n,n}$, it is possible to verify whether or not inconsistencies are present. If they are, the matter should be further investigated *until the reason for those inconsistencies is found.* This could lie either in mathematical errors in setting up the algorithm, or in programming errors. But regardless of the cause, the problem should always be rectified before proceeding further.

In many cases the algorithms also contain a prediction phase, in which $X^*_{n+1,n}$ is computed from $X^*_{n,n}$ by forward-integration of the model, and $S^*_{n+1,n}$ is computed from $S^*_{n,n}$ by the use of (5.5.33). Again possible inconsistencies between the errors in $X^*_{n+1,n}$ and the matrix $S^*_{n+1,n}$ should be checked for using the Chi-squared criterion with a specified confidence limit. For if inconsistencies do exist, the algorithm will be of questionable value.

In Ex. 5.15 it is shown that the Chi-squared value of the errors in $X^*_{n+1,n}$ relative to $S^*_{n+1,n}$ should be equal to the Chi-squared for the errors in $X^*_{n,n}$ relative to $S^*_{n,n}$. This equality should also be checked for, and in this way the consistency of the algorithm which computes $X^*_{n+1,n}$ from $X^*_{n,n}$ with that of the algorithm for computing $S^*_{n+1,n}$ from $S^*_{n,n}$ can be established.

EXERCISES

5.1 Assume that a random variable x has mean μ and variance σ^2. We sample x by making observations on it, thereby obtaining the values $(x)_1, (x)_2, \ldots, (x)_i$ from which we form

$$m \equiv \frac{1}{k} \sum_{i=1}^{k} (x)_i \tag{I}$$

$$s^2 \equiv \frac{1}{k-1} \sum_{i=1}^{k} \left[(x)_i - m\right]^2 \tag{II}$$

Assuming that the observations are statistically independent, verify that $E\{m\} = \mu$, $E\{s^2\} = \sigma^2$ thereby showing that μ and σ^2 can be estimated by the above averaging schemes. Show also the $E\left\{(m - \mu)^2\right\} = \sigma^2/k$ and so the variance of the estimate of μ goes to zero as $k \to \infty$.

Hint: Set $(x)_i = \mu + \nu_i$ where $E\left\{\nu_i\right\} = 0$, and $E\left\{\nu_i \nu_j\right\} = \sigma^2$ if $i = j$ and zero if $i \neq j$.

5.2 Show formally, using (5.2.18), that

a) $$E\{\alpha\nu_n + \beta\nu_k\} = \alpha E\{\nu_n\} + \beta E\{\nu_k\}$$

b) $$E\{(\alpha\nu_n + \beta\nu_k)^2\} = \alpha^2 E\{\nu_n^2\} + 2\alpha\beta E\{\nu_n\nu_k\} + \beta^2 E\{\nu_k^2\}$$

c) Hence verify that (c/f (5.2.34))

$$E\{(\alpha\nu_n + \beta\nu_k)^2\} = (\alpha, \beta) E\left\{\begin{pmatrix}\nu_n\\ \nu_k\end{pmatrix}(\nu_n, \nu_k)\right\}\begin{pmatrix}\alpha\\ \beta\end{pmatrix}$$

5.3 Verify that if ν_n and ν_k are statistically independent then

$$E\{\nu_n^{\alpha}\nu_k^{\beta}\} = E\{\nu_n^{\alpha}\}E\{\nu_n^{\beta}\} \tag{I}$$

for any exponents α and β. For uncorrelated random variables (I) holds in general only for $\alpha, \beta = 1$.

5.4 Let ν_n, ν_k be two *discrete* random variables which assume values $\{-1, 0, 2\}$ and $\{0, 1, 2\}$ respectively. Their joint density function is given by

ν_k \ ν_n	−1	0	2
0	$\frac{1}{24}$	$\frac{1}{6}$	$\frac{1}{24}$
1	$\frac{1}{6}$	$\frac{1}{6}$	$\frac{1}{6}$
2	$\frac{1}{24}$	$\frac{1}{6}$	$\frac{1}{24}$

Compute

a) the probability that $\nu_n = j$, for $j = -1, 0, 2$,

b) the probability that $\nu_k = i$, for $i = 0, 1, 2$,

c) $E\{\nu_n\}$, $E\{\nu_k\}$, $E\{\nu_n\nu_k\}$

d) Hence verify that ν_n and ν_k are uncorrelated but are not indepenent.

5.5 Prove that if all of the random variables in

$$N \equiv (\nu_0, \nu_1, \ldots, \nu_m)^T$$

are uncorrelated, then, whether they are zero-mean or not, the covariance matrix of N is diagonal.

5.6 Let r_{ij} be the i, j^{th} element of R, the covariance matrix of m random variables. Show that

$$r_{ii} r_{jj} \geq r_{ij}^2 \qquad 1 \leq i, j \leq m \tag{I}$$

What happens to the above inequality in the event that every one of the m random variables is linearly independent of the others? If (I) has a strict inequality, can we infer that all m random variables are linearly independent?

5.7 Let A be an $m \times k$ matrix and R a $k \times k$ matrix. Show that if R is positive definite then ARA^T is positive definite if and only if A has full row-rank.

5.8 Prove that if R is nonnegative definite then its diagonal elements are all nonnegative.

5.9 Let B and M be real square matrices with M symmetric and B nonsingular. Then BMB^T is said to be a *congruence transformation* on M. Verify

a) that since M is symmetric then so is any congruence transformation on M and

b) that each of the properties
i) positive definite, ii) positive semidefinite, and iii) nonnegative definite, is preserved under congruence. (In fact the *signature* of a matrix is also preserved under congruence. The signature is the triplet of numbers α, β, γ where α is the number of *positive* eigenvalues, β the multiplicity with which *zero* is an eigenvalue and γ the number of *negative* eigenvalues. This accounts for b) above but is considerably harder to prove.)

5.10 Let D be a congruence transformation on M, i.e. $D = BMB^T$, all matrices being $m \times m$. Verify

a) that if M is positive definite then all of D's diagonal elements are positive.

b) If D is diagonal with positive diagonal elements then M is positive definite.

c) Hence infer that if D is diagonal then its diagonal elements are all positive *if and only if* M is positive definite.

d) If D is diagonal with $m - k$ positive elements and k zeros on the diagonal, verify that M is positive semidefinite with rank $m - k$.

5.11 Let

$$M \equiv \begin{pmatrix} 1 & 2 & 2 \\ 2 & 5 & 5 \\ 2 & 5 & 9 \end{pmatrix}$$

a) Verify that if

$$B_1 \equiv \begin{pmatrix} 1 & & \\ -2 & 1 & \\ 0 & 0 & 1 \end{pmatrix}$$

then $B_1 M B_1^T$ has zeros in the 1,2 and 2,1 positions. We have, in fact, performed one step of Gaussian elimination. We call a matrix such as B_1 an *elementary* matrix, which by definition is an identiy matrix with a single nonzero off-diagonal element.

b) Construct the elementary matrix B_2 such that $B_2 \left(B_1 M B_1^T\right) B_2^T$ has zeros in 1,3 and 3,1 (in addition to 1,2 and 2,1). Finally construct the elementary matrix B_3 which then also puts zeros in 2,3 and 3,2, thus making $B_3 \left[B_2 \left(B_1 M B_1^T\right) B_2^T\right] B_3^T$ diagonal.

c) Using part c) of Ex. 5.10, infer that M above is positive definite.

d) Verify that $B \equiv B_3 B_2 B_1$ is lower triangular.

5.12 Diagonalize

$$M_1 \equiv \begin{pmatrix} 2 & 1 & 3 \\ 1 & 5 & 0 \\ 3 & 0 & 5 \end{pmatrix}$$

by congruence using the method of Ex. 5.11. From that diagonal matrix, what is the rank of M_1? Is M_1 nonnegative definite? Repeat using

$$M_2 = \begin{pmatrix} 1 & 2 & 3 \\ 2 & 3 & 4 \\ 3 & 4 & 5 \end{pmatrix}$$

Is M_2 nonnegative definite?
Hint: See Ex. 5.10.

5.13 Given that the covariance matrix of the zero-mean random variables ν_0, ν_1, ν_2 and ν_3 is

$$E\{NN^T\} = \begin{pmatrix} 1 & -2 & 2 & 3 \\ -2 & 5 & -2 & -10 \\ 2 & -2 & 6 & 2 \\ 3 & -10 & 7 & 17 \end{pmatrix}$$

show that two of the random variables in N are linearly dependent on the other two which are independent of one another.
Hint: Diagonalize the above matrix by congruence.

5.14 Verify that normally distributed random variables are independent *if and only if* they are uncorrelated.
Hint: What form does the covariance matrix for uncorrelated random variables assume?

5.15 Assume that ν_0 and ν_1 are two normally distributed zero-mean random variables which have as their covariance matrix

$$R_n = \begin{pmatrix} 3 & 1 \\ 1 & 3 \end{pmatrix}$$

a) Verify that the eigenvalues of R_n are 2 and 4 and that the associated normalized eigenvectors are, respectively

$$V_1 \equiv \frac{1}{\sqrt{2}} \begin{pmatrix} 1 \\ -1 \end{pmatrix} \quad \text{and} \quad V_2 \equiv \frac{1}{\sqrt{2}} \begin{pmatrix} 1 \\ 1 \end{pmatrix}$$

b) Verify that the matrix

$$Q \equiv (V_1, V_2)$$

is orthogonal, i.e. $Q^T = Q^{-1}$ and that

$$Q^T R_n Q = \begin{pmatrix} 2 & 0 \\ 0 & 4 \end{pmatrix}$$

i.e. Q is the orthogonal transformation which diagonalizes R_n to its eigenvalues.

c) Infer that

$$Q^T R_n^{-1} Q = \begin{pmatrix} \frac{1}{2} & 0 \\ 0 & \frac{1}{4} \end{pmatrix}$$

d) The density function of ν_0 and ν_1 is

$$p_{\nu_0, \nu_1}(z_0, z_1) \equiv \frac{1}{2\pi\sqrt{8}} \exp\left\{ -\frac{1}{2}(z_0, z_1) R_n^{-1} \begin{pmatrix} z_0 \\ z_1 \end{pmatrix} \right\}$$

Infer that the contours of equal probability are ellipses centered on the z_0, z_1 origin, their semimajor axes rotated by 45° with respect to the z_0 axis, and with the lengths of the semimajor and semiminor axes in the ratio $\sqrt{2} : 1$.

5.16 a) Prove that if N is an m-vector of Gaussian random variables† with $E\{N\} \equiv B$ and $E\left\{(N - B)(N - B)^T\right\} = R$, then $R^{-1/2}(N - B)$ is a vector of independent standard Gaussian variables.

b) Infer from a) above that the scalar χ^2 defined by

$$\chi^2 \equiv (N - B)^T R^{-1} (N - B)$$

is a random variable which has a Chi-squared distribution function.

5.17 a) Define

$$\chi^2_{n,n} \equiv N^{*T}_{n,n} S^{*-1}_{n,n} N^*_{n,n}$$

$$\chi^2_{n+1,n} \equiv N^{*T}_{n+1,n} S^{*-1}_{n+1,n} N^*_{n+1,n}$$

Making use of (5.5.18) and (5.5.20) prove that

$$\chi^2_{n+1,n} = \chi^2_{n,n}$$

b) Assume that $\Phi = \begin{pmatrix} 1 & 1 \\ 0 & 1 \end{pmatrix}$, $N^*_{n,n} = (1, 1)^T$ and $S^*_{n,n} = \begin{pmatrix} 2 & 1 \\ 1 & 1 \end{pmatrix}$. Compute the χ^2 for $N^*_{n,n}$. Now derive $N^*_{n+1,n}$ and $S^*_{n+1,n}$. Compute the χ^2 of $N^*_{n+1,n}$ and compare this to the χ^2 for $N^*_{n,n}$.

† See p. 154 for a definition.

REFERENCES

1. Davenport, W. B., and Root, W. L., "An Introduction to the Theory of Random Signals and Noise," McGraw-Hill Book Company, 1958.
2. Lindgren, B. W., "Statistical Theory," MacMillan, 1962.
3. Hildebrand, F. B., "Methods of Applied Mathematics," Prentice-Hall, 1961, Chapter 1.
4. Miller, K. S., "Multidimensional Gaussian Processes," John Wiley & Sons, New York, 1964.
5. Kenney, J. F., and Keeping, E. S., "Mathematics of Statistics" (Part Two), Van Nostrand, 1956, Chapter 5.
6. Abramowitz, M., and Stegun, I. A., (Eds.) "Handbook of Mathematical Functions," Nat. Bureau of Stds., App. Math. Ser. 55, U.S. Govt. Printer, 1965.

6

EXACTNESS, LEAST-SQUARES AND MINIMUM-VARIANCE

6.1 INTRODUCTION

This chapter is concerned with the general form of linear estimators, i.e.,

$$X^*_{n,n} = W_n Y_{(n)} \tag{6.1.1}$$

in which the estimate is obtained as a linear transformation on the observations.

We show that the matrix W_n must satisfy a fundamental condition, called the *exactness constraint.* This is followed by a statement and proof of two theorems by which W_n is chosen, the first being called the *least-squares* method and the second the *minimum-variance* method. All of our subsequent work will be merely implementations of W chosen either by least-squares or by minimum-variance.

The chapter then proceeds with an analysis of the properties of these two concepts. The minimum-variance procedure is shown to be very far-reaching, and will be seen to be the cornerstone on which our most powerful smoothing algorithms are based.

6.2 THE OBSERVATION SCHEME

Assume that we are observing a process and that we wish to smooth our data or make predictions based on them. As a first step, we will have to make an *a priori* assumption concerning the mathematical form which we believe will adequately describe that process. As was shown in Chapter 4, this means the assumption of a state-vector and a differential equation, and the consequent derivation of a transition matrix.

Suppose then that a model has been selected, based on the state-vector $X(t)$. In Chapter 4 we showed how, depending on the type of differential equation on which that model is based, three successively more general situations arise. These can be summarized by stating the form of the transition equation for each case. Thus

Constant-Coefficient Linear Model

$$X(t_n + \zeta) = \Phi(\zeta)X(t_n) \tag{6.2.1}$$

Time-Varying Linear Model

$$X(t_n + \zeta) = \Phi(t_n + \zeta, t_n)X(t_n) \tag{6.2.2}$$

Nonlinear Model

$$\begin{aligned} X(t_n) &= \overline{X}(t_n) + \delta X(t_n) \\ \delta X(t_n + \zeta) &= \Phi\left(t_n + \zeta, t_n; \overline{X}\right)\delta X(t_n) \end{aligned} \tag{6.2.3}$$

In the first case, the transition matrix depends on only *one* parameter, namely ζ, the amount by which time is shifted. In the second case, *two* parameters enter into Φ, the amount of time-shift, ζ, and the time at which the shift takes place, t_n. Finally in the nonlinear case, *three* facts are required for the formation of Φ. In addition to the amount of time-shift and the time at which it takes place, a specific trajectory is involved, as specified by $\overline{X}$ in (6.2.3). In all three cases, however, a linear transition relation can always be written.

We now examine the possible ways in which the *observations* can be related to the state-vector, and show that a very close analogy to the above three cases also exists here.

Consider first the very simple situation where we have, say, a quadratic model, and where we are observing $x(t)$ at each sampling instant. Then the observation y_n is related to x_n by

$$y_n = x_n + \nu_n \tag{6.2.4}$$

where ν_n is the nth observation error. This can also be written

$$y_n = (1, 0, 0) \begin{pmatrix} x \\ \dot{x} \\ \ddot{x} \end{pmatrix}_n + \nu_n \tag{6.2.5}$$

which is of the form

$$Y_n = MX_n + N_n \tag{6.2.6}$$

where Y_n is the vector of observations taken at time t_n, and N_n is the associated error vector. (In this case, $Y_n = y_n$, $N_n = \nu_n$.)

In the above example, the matrix M assumes a particularly simple form, but it is easy to generalize slightly and to assume that M is *any constant matrix* of appropriate dimensions. (See Ex. 6.1.)

Suppose next that the matrix M changes with n. Then we would write the observation relation as

$$Y_n = M_n X_n + N_n \tag{6.2.7}$$

Consider, finally, the most general observation model which could (and frequently does) arise, namely,

$$Y_n = G(X_n) + N_n \tag{6.2.8}$$

where G is a vector of *nonlinear* functions of the state-variables.

A common example of such a situation is the following. Let $X(t)$ be the state-vector consisting of the Cartesian coordinates of the position of an object and the first derivatives of position, i.e.

$$X(t) = \left(x_0(t), x_1(t), x_2(t), \dot{x}_0(t), \dot{x}_1(t), \dot{x}_2(t)\right)^T \tag{6.2.9}$$

The reader is referred to Figure 6.1.

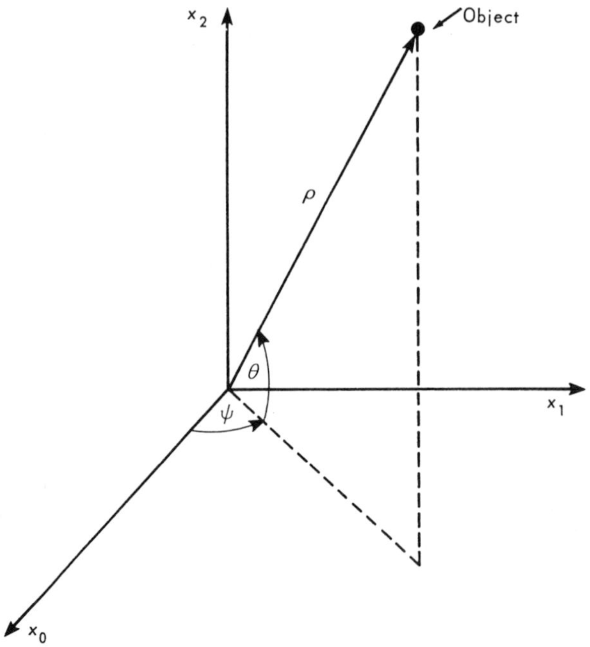

Fig. 6.1 *Cartesian and polar coordinates.*

Suppose now, that a radar is being used to observe the object and that the radar puts out the triplet of numbers

$$Y_n \equiv \begin{pmatrix} \rho \\ \psi \\ \theta \end{pmatrix}_n + \begin{pmatrix} \nu_\rho \\ \nu_\psi \\ \nu_\theta \end{pmatrix}_n \tag{6.2.10}$$

where ρ, ψ and θ are the *polar coordinates* of position

$$\begin{aligned} \rho &\equiv \left(x_0^2 + x_1^2 + x_2^2\right)^{1/2} && \text{(range)} \\ \psi &\equiv \tan^{-1}\left(\frac{x_1}{x_0}\right) && \text{(azimuth)} \\ \theta &\equiv \tan^{-1}\frac{x_2}{\left(x_0^2 + x_1^2\right)^{1/2}} && \text{(elevation)} \end{aligned} \tag{6.2.11}$$

and $\nu_\rho, \nu_\psi, \nu_\theta$ are the respective observation errors. This is a nonlinear observation system, and if (6.2.11) is used in (6.2.10) we obtain an equation of the form of (6.2.8). ◆◆

While we shall return to the nonlinear case in a later chapter, and discuss it in further detail, we outline here briefly how we will handle such situations.

Assume that we have a *known* nominal trajectory $\overline{X}(t)$, close to the true trajectory, i.e.

$$X(t) = \overline{X}(t) + \delta X(t) \tag{6.2.12}$$

where the components of $\delta X(t)$ are small compared to those of $X(t)$. We compute

$$\overline{Y}_n \equiv G\left(\overline{X}_n\right) \tag{6.2.13}$$

Now subtract (6.2.13) from (6.2.8), and letting

$$\delta Y_n \equiv Y_n - \overline{Y}_n \tag{6.2.14}$$

this gives us

$$\delta Y_n = G\left(\overline{X}_n + \delta X_n\right) - G\left(\overline{X}_n\right) + N_n \tag{6.2.15}$$

We now expand the first two terms on the right, using the Taylor series method discussed on p. 108, thereby obtaining *to first order,*

$$\delta Y_n = M\left(\overline{X}_n\right)\delta X_n + N_n \tag{6.2.16}$$

where

$$\left[M\left(\overline{X}_n\right)\right]_{ij} = \left.\frac{\partial g_i(X)}{\partial x_j}\right|_{X = \overline{X}_n} \tag{6.2.17}$$

For the example considered in (6.2.11) $M\left(\overline{X}_n\right)$ is given on the following page.

$$M\left(\bar{X}_n\right) = \begin{pmatrix} \dfrac{\bar{x}_0}{\bar{\rho}} & \dfrac{\bar{x}_1}{\bar{\rho}} & \dfrac{\bar{x}_2}{\bar{\rho}} & 0 & 0 & 0 \\ -\dfrac{\bar{x}_1}{\bar{s}^2} & \dfrac{\bar{x}_0}{\bar{s}^2} & 0 & 0 & 0 & 0 \\ -\dfrac{\bar{x}_0\bar{x}_2}{\bar{\rho}^2\bar{s}} & -\dfrac{\bar{x}_1\bar{x}_2}{\bar{\rho}^2\bar{s}} & \dfrac{\bar{s}}{\bar{\rho}^2} & 0 & 0 & 0 \end{pmatrix}_n$$

where

$$\bar{\rho} \equiv \left(\bar{x}_0^{\,2} + \bar{x}_1^{\,2} + \bar{x}_2^{\,2}\right)^{1/2}$$

and

$$\bar{s} \equiv \left(\bar{x}_0^{\,2} + \bar{x}_1^{\,2}\right)^{1/2}$$

Equation (6.2.16) is seen to provide an *approximate but linear relation,* with which we can replace the nonlinear system of (6.2.8). The linearized replacement is, of course, subject to errors due to the truncation of the Taylor series. More will be said about this later.

We see then, that in close analogy to (6.2.1), (6.2.2) and (6.2.3), we can summarize the above analysis on the possible observation methods by the following sets of equations.

Constant-Coefficient Linear Observation System

$$Y_n = MX_n + N_n \tag{6.2.18}$$

Time-Varying Linear Observation System

$$Y_n = M_n X_n + N_n \tag{6.2.19}$$

Nonlinear Observation System

$$\begin{aligned} X_n &= \bar{X}_n + \delta X_n \\ \bar{Y}_n &\equiv G\left(\bar{X}_n\right) \\ \delta Y_n &\equiv Y_n - \bar{Y}_n \\ \delta Y_n &= M\left(\bar{X}_n\right)\delta X_n + N_n \end{aligned} \tag{6.2.20}$$

In practice, systems may arise which have any of the three model choices mated with any of the three types of observation systems. Thus, for example, we could have a nonlinear set of observation equations used in conjunction with a linear model. This will pose no problems. However, for the remainder of this chapter we shall assume, essentially, that both the model and the observation equations are time-varying linear, and this naturally covers the constant-coefficient linear cases as well. The nonlinear cases will be discussed, *per se*, in later chapters. However, as we shall then see, very little changes, and the results of this chapter continue to apply with only very minor modifications.

6.3 THE EXACTNESS CONSTRAINT

Suppose that at time t_n we make a vector of observations on a process. Let the chosen state-vector be the $(m+1)$-vector

$$X(t) \equiv \begin{pmatrix} x_0(t) \\ x_1(t) \\ \vdots \\ x_m(t) \end{pmatrix} \tag{6.3.1}$$

and let the observation and error vectors be, respectively, the $(r+1)$-vectors

$$Y_n \equiv \begin{pmatrix} y_0 \\ y_1 \\ \vdots \\ y_r \end{pmatrix}_n, \qquad N_n \equiv \begin{pmatrix} \nu_0 \\ \nu_1 \\ \vdots \\ \nu_r \end{pmatrix}_n \tag{6.3.2}$$

where r and m are not necessarily equal. Then, assuming a time-varying linear observation scheme of the form shown in (6.2.19), we have that

$$Y_n = M_n X_n + N_n \tag{6.3.3}$$

On page 139 in Chapter 5, we define the total observation and error vectors as

$$Y_{(n)} \equiv \begin{pmatrix} Y_n \\ \hline Y_{n-1} \\ \hline \vdots \\ \hline Y_{n-L} \end{pmatrix}, \qquad N_{(n)} \equiv \begin{pmatrix} N_n \\ \hline N_{n-1} \\ \hline \vdots \\ \hline N_{n-L} \end{pmatrix} \tag{6.3.4}$$

formed from the concatenation of $L + 1$ successive vectors of the form of (6.3.2). The total observation vector $Y_{(n)}$ will form the input to our filtering algorithms, and we investigate how it is related to the state-vector X_n. Accordingly we use (6.3.3) for each of the instants t_{n-L} to t_n and obtain

$$\begin{pmatrix} Y_n \\ \hline Y_{n-1} \\ \hline \vdots \\ \hline Y_{n-L} \end{pmatrix} = \begin{pmatrix} M_n X_n \\ \hline M_{n-1} X_{n-1} \\ \hline \vdots \\ \hline M_{n-L} X_{n-L} \end{pmatrix} + \begin{pmatrix} N_n \\ \hline N_{n-1} \\ \hline \vdots \\ \hline N_{n-L} \end{pmatrix} \tag{6.3.5}$$

Assume now, that the transition relation for the chosen model has the form

$$X(t_{n-k}) = \Phi(t_{n-k}, t_n) X(t_n) \tag{6.3.6}$$

(which means that the model is a time-varying linear differential equation). Then (6.3.5) can be written

$$\begin{pmatrix} Y_n \\ \hline Y_{n-1} \\ \hline \vdots \\ \hline Y_{n-L} \end{pmatrix} = \begin{pmatrix} M_n X_n \\ \hline M_{n-1} \Phi(t_{n-1}, t_n) X_n \\ \hline \vdots \\ \hline M_{n-L} \Phi(t_{n-L}, t_n) X_n \end{pmatrix} + \begin{pmatrix} N_n \\ \hline N_{n-1} \\ \hline \vdots \\ \hline N_{n-L} \end{pmatrix} \tag{6.3.7}$$

showing that the entire set of observations, *made over the extended time span* $t_{n-L} \leq t \leq t_n$ can be related to the model state-vector *evaluated at the single instant* t_n.

We now factor X_n out of the right-hand side of (6.3.7) and obtain

$$\begin{pmatrix} Y_n \\ \hline Y_{n-1} \\ \hline \vdots \\ \hline Y_{n-L} \end{pmatrix} = \begin{pmatrix} M_n \\ \hline M_{n-1}\Phi(t_{n-1}, t_n) \\ \hline \vdots \\ \hline M_{n-L}\Phi(t_{n-L}, t_n) \end{pmatrix} X_n + \begin{pmatrix} N_n \\ \hline N_{n-1} \\ \hline \vdots \\ \hline N_{n-L} \end{pmatrix} \tag{6.3.8}$$

Define the matrix

$$T_n \equiv \begin{pmatrix} M_n \\ \hline M_{n-1}\Phi(t_{n-1}, t_n) \\ \hline \vdots \\ \hline M_{n-L}\Phi(t_{n-L}, t_n) \end{pmatrix} \tag{6.3.9}$$

Then (6.3.8) can be written as

$$Y_{(n)} = T_n X_n + N_{(n)} \tag{6.3.10}$$

This compact equation is of crucial importance in what follows and will recur over and over again. It is the most general form of a linear observation scheme, combined with a linear model,† and shows that the total observation vector can be related, by the use of the matrix T_n, to a single value of the state-vector. (See Ex. 6.2.) There are essentially three types of information which are incorporated into T_n.

1. As is evident from (6.3.9), *the instants at which the observations were taken* is recorded in T_n, both indirectly by the subscripts of the M's as well as explicitly, in the dual arguments of the transition matrices. Note that at no stage have we restricted the observation times $t_{n-L}, \ldots, t_n$ in any way whatever. They can be concurrent, some or all distinct, equally or unequally spaced, etc., and in fact t_n need not even be the most recent, although for convenience we take it to be so. They also need not follow

†Discussion of the nonlinear cases is deferred to Chapter 8.

one another according to their subscripts, although this is again taken to be the case for convenience.

2. *The choice of observation schemes* is recorded clearly in T_n by the elements of the matrices $M_{n-L}, \ldots, M_n$. Complete freedom is possible, and the dimensionality of the vectors $Y_{n-L}, \ldots, Y_n$ can differ from one another. Moreover the relationships of those vectors to the state-vectors $X_{n-L}, \ldots, X_n$ need not all be the same, provided only that they are all linear. The M matrices will convey that information in an appropriate way.

3. Finally, *the specific choice of model* is firmly contained in T_n by the presence of the transition matrices. The only restriction, at this stage, is that the model be a linear one, although this too will be relaxed later on.

Thus all of the *a priori* decisions which we make for our smoothing scheme – *when* we observe, *how* we observe and *what* we believe we are observing – enter into the formation of T_n. It is not surprising that this matrix occurs so frequently in what lies ahead.

The decision is now made *that we shall limit ourselves exclusively to linear estimation schemes.* Thus all of our algorithms will either be of the form, or reducible to the form,

$$X^*_{n,n} = W_n Y_{(n)} \tag{6.3.11}$$

As yet the matrix W_n is completely arbitrary in all respects, other than its array size, which is, of course, fixed by the orders of the vectors $Y_{(n)}$ and X_n. It will be our aim, in what follows, to develop various schemes for deriving W so that certain objectives are met.

We now combine (6.3.10) and (6.3.11), thereby obtaining

$$X^*_{n,n} = W_n T_n X_n + W_n N_{(n)} \tag{6.3.12}$$

We also make the further restriction *that the only matrices W which are of interest to us, are those which guarantee that their associated $X^*_{n,n}$ is an unbiased estimate of X_n*, i.e. that

$$E\left\{X^*_{n,n}\right\} = X_n \tag{6.3.13}$$

Taking the expectation of both sides of (6.2.13) gives us†

$$E\left\{X^*_{n,n}\right\} = W_n T_n X_n \tag{6.3.14}$$

† $N_{(n)}$ is zero-mean, by assumption.

and so if (6.3.13) is to hold, then we must have

$$X_n = W_n T_n X_n \tag{6.3.15}$$

However, since X_n is arbitrary, this can only mean that W_n must satisfy

$$W_n T_n = I \tag{6.3.16}$$

where T_n is given. *Only those W's which do satisfy* (6.3.16) *will be acceptable, since only for those cases will $X^*_{n,n}$ be an unbiased estimate of X_n.*

We refer to (6.3.16) as the *exactness constraint* and as we shall see, it forms the basis of all of our further developments.† We examine it in further detail.

Suppose, first, that the order of the vector $Y_{(n)}$ is *less than* the order of X_n. For definiteness let $Y_{(n)}$ be a 2-vector and X_n a 3-vector. Then (6.3.10) has the form

$$\begin{pmatrix} y_0 \\ y_1 \end{pmatrix}_{(n)} = \begin{pmatrix} t_{00} & t_{01} & t_{02} \\ t_{10} & t_{11} & t_{12} \end{pmatrix} \begin{pmatrix} x_0 \\ x_1 \\ x_2 \end{pmatrix}_n + \begin{pmatrix} \nu_0 \\ \nu_1 \end{pmatrix}_n \tag{6.3.17}$$

and (6.3.11) appears as

$$\begin{pmatrix} x_0^* \\ x_1^* \\ x_2^* \end{pmatrix}_{n,n} = \begin{pmatrix} w_{00} & w_{01} \\ w_{10} & w_{11} \\ w_{20} & w_{21} \end{pmatrix}_n \begin{pmatrix} y_0 \\ y_1 \end{pmatrix}_{(n)} \tag{6.3.18}$$

Hence the exactness constraint (6.3.16) gives

$$\begin{pmatrix} w_{00} & w_{01} \\ w_{10} & w_{11} \\ w_{20} & w_{21} \end{pmatrix}_n \begin{pmatrix} t_{00} & t_{01} & t_{02} \\ t_{10} & t_{11} & t_{12} \end{pmatrix}_n = \begin{pmatrix} 1 & 0 & 0 \\ 0 & 1 & 0 \\ 0 & 0 & 1 \end{pmatrix} \tag{6.3.19}$$

†See chart at the end of Chapter 1.

which is seen to be 9 equations for the 6 unknown w_{ij}'s. A solution for such an overdetermined system does not in general exist *and so we shall exclude the case where the order of* $Y_{(n)}$ *is less than the order of* X_n.

Suppose next that $Y_{(n)}$ and X_n are vectors of the *same* order. Then both W_n and T_n are square and since (6.3.16) implies that T_n is nonsingular, we have the solution

$$W_n = T_n^{-1} \tag{6.3.20}$$

In this case W_n is uniquely determined *and the resulting filter turns out to be simply an interpolation on the data,* in which the trajectory is forced to pass through every observation precisely. This case will only be of marginal interest to us, since it leaves us no freedom to attempt to do something about the observation errors.

Finally then, we consider the case where the order of $Y_{(n)}$ *exceeds* the order of X_n, e.g. $Y_{(n)}$ is a 4-vector and X_n a 2-vector. The exactness constraint now gives us

$$\begin{pmatrix} w_{00} & w_{01} & w_{02} & w_{03} \\ w_{10} & w_{11} & w_{12} & w_{13} \end{pmatrix}_n \begin{pmatrix} t_{00} & t_{01} \\ t_{10} & t_{11} \\ t_{20} & t_{21} \\ t_{30} & t_{31} \end{pmatrix}_n = \begin{pmatrix} 1 & 0 \\ 0 & 1 \end{pmatrix} \tag{6.3.21}$$

which is an underdetermined system. There exists *an infinity* of W's which will satisfy it for a given T, and yet all of them will lead to unbiased estimates. Within that underdetermined structure we will be free to maneuver, in an attempt to offset the observational errors, *and so this is the case of real interest to us.*

As we shall soon show, by setting up further constraints, we will arrive at two choices for W, the first being called the *least-squares* filter matrix, symbolized $\hat{W}_n$, and the second the *minimum-variance* filter matrix, symbolized $\mathring{W}_n$.

Example:† We take as our model a first-degree polynomial, and as the state-vector we choose (c/f (4.2.14))

$$Z(t) \equiv \begin{pmatrix} x(t) \\ \tau\dot{x}(t) \end{pmatrix} \tag{6.3.22}$$

†This example is continued in each of the sections of this chapter to illustrate the successive developments.

where τ is the constant inter-sample time. The transition matrix is (c/f (4.2.15))

$$\Phi(h) = \begin{pmatrix} 1 & h \\ 0 & 1 \end{pmatrix} \tag{6.3.23}$$

and, assuming that we are observing only the position variable, (c/f (6.2.5)) the observation matrix is

$$M = (1, 0) \tag{6.3.24}$$

For the case, where three observations go into each estimate, we have

$$Y_{(n)} = \begin{pmatrix} y_n \\ y_{n-1} \\ y_{n-2} \end{pmatrix} \tag{6.3.25}$$

and so by (6.3.9)

$$T_n = \begin{pmatrix} (1, \ 0) \\ \hline (1, \ 0)\begin{pmatrix} 1 & -1 \\ 0 & 1 \end{pmatrix} \\ \hline (1, \ 0)\begin{pmatrix} 1 & -2 \\ 0 & 1 \end{pmatrix} \end{pmatrix} \tag{6.3.26}$$

which reduces to

$$T_n = \begin{pmatrix} 1 & 0 \\ 1 & -1 \\ 1 & -2 \end{pmatrix} \tag{6.3.27}$$

Thus (6.3.10) becomes

$$\begin{pmatrix} y_n \\ y_{n-1} \\ y_{n-2} \end{pmatrix} = \begin{pmatrix} 1 & 0 \\ 1 & -1 \\ 1 & -2 \end{pmatrix} \begin{pmatrix} x \\ \tau\dot{x} \end{pmatrix}_n + \begin{pmatrix} \nu_n \\ \nu_{n-1} \\ \nu_{n-2} \end{pmatrix} \tag{6.3.28}$$

The exactness constraint now gives us

$$\begin{pmatrix} w_{00} & w_{01} & w_{02} \\ w_{10} & w_{11} & w_{12} \end{pmatrix} \begin{pmatrix} 1 & 0 \\ 1 & -1 \\ 1 & -2 \end{pmatrix} = \begin{pmatrix} 1 & 0 \\ 0 & 1 \end{pmatrix} \tag{6.3.29}$$

which is four equations in six unknowns, namely

$$\begin{aligned} w_{00} + w_{01} + w_{02} &= 1 \\ w_{01} + 2w_{02} &= 0 \\ w_{10} + w_{11} + w_{12} &= 0 \\ w_{11} + 2w_{12} &= -1 \end{aligned} \tag{6.3.30}$$

♦♦

6.4 FURTHER RELATIONSHIPS

Prior to embarking on a discussion of the least-squares and minimum-variance techniques for selecting W, a few additional results are required.

Returning to (6.3.12) we see that $X^*_{n,n}$ is composed of the sum of two parts, a deterministic vector $W_n T_n X_n$ plus a random vector $W_n N_{(n)}$. Since only W's satisfying the exactness constraint will be considered, the former reduces simply to X_n, and the latter we recognize, of course, as $N^*_{n,n}$ defined on p. 143. Thus (6.3.12) can actually be written

$$X^*_{n,n} = X_n + N^*_{n,n} \tag{6.4.1}$$

This very simple form shows that $X^*_{n,n}$ will equal X_n to within a vector of zero-mean, random errors, $N^*_{n,n}$.

Unfortunately *another error component enters* for the following reason. We have assumed, somewhat naively perhaps, that the chosen model *exactly* describes the true process. What if it does not? As we shall see later, a further vector of *deterministic* errors will then be added to (6.4.1). However, for the present we ignore that possibility, and we continue to assume a perfect match between the chosen model and the true process.

From (6.4.1) we solve for X_n, and inserting the result so obtained in (6.3.10) gives us

$$Y_{(n)} - T_n X^*_{n,n} = N_{(n)} - T_n N^*_{n,n} \tag{6.4.2}$$

Then, by taking the expectation of both sides, we see that

$$E\{Y_{(n)} - T_n X^*_{n,n}\} = \emptyset \tag{6.4.3}$$

The vector $Y_{(n)} - T_n X^*_{n,n}$ is thus zero-mean if the exactness constraint holds. We now examine the *rank* of the matrix T_n, a quantity of considerable importance.

The matrix T_n first arose in (6.3.10) which we repeat here, i.e.

$$Y_{(n)} = T_n X_n + N_{(n)} \tag{6.4.4}$$

In general, the order of $Y_{(n)}$ will exceed that of X_n, and so T_n will be a rectangular matrix with more rows than columns. The maximum rank which T_n could possibly have is thus equal to the number of its columns. *We show that, without loss of generality, only* T's *with maximum rank, i.e. full column-rank, need ever be considered.*

Suppose, for definiteness, that $Y_{(n)}$ is a 4-vector and X_n is a 3-vector. Then (6.4.4) would become

$$\begin{pmatrix} y_0 \\ y_1 \\ y_2 \\ y_3 \end{pmatrix}_{(n)} = \begin{pmatrix} t_{00} & t_{01} & t_{02} \\ t_{10} & t_{11} & t_{12} \\ t_{20} & t_{21} & t_{22} \\ t_{30} & t_{31} & t_{32} \end{pmatrix}_n \begin{pmatrix} x_0 \\ x_1 \\ x_2 \end{pmatrix}_n + \begin{pmatrix} \nu_0 \\ \nu_1 \\ \nu_2 \\ \nu_3 \end{pmatrix}_{(n)} \tag{6.4.5}$$

Assume now that the three column vectors of T_n *are* linearly related. For example, let the first column be α-times the second plus β-times the third, where α and β are some scalars. Then (6.4.5) can be written as

$$\begin{pmatrix} y_0 \\ y_1 \\ y_2 \\ y_3 \end{pmatrix}_n = \begin{pmatrix} t_{01} & t_{02} \\ t_{11} & t_{12} \\ t_{21} & t_{22} \\ t_{31} & t_{32} \end{pmatrix} \begin{pmatrix} \alpha & 1 & 0 \\ \beta & 0 & 1 \end{pmatrix} \begin{pmatrix} x_0 \\ x_1 \\ x_2 \end{pmatrix}_n + \begin{pmatrix} \nu_0 \\ \nu_1 \\ \nu_2 \\ \nu_3 \end{pmatrix}_n \tag{6.4.6}$$

But this is now of the form

$$Y_{(n)} = T'_n X'_n + N_n \tag{6.4.7}$$

where T'_n has two columns and where

$$X'_n \equiv \begin{pmatrix} \alpha x_0 + x_1 \\ \beta x_0 + x_2 \end{pmatrix}_n \tag{6.4.8}$$

is a 2-vector. We started out with the assumption that the state-vector should be a 3-vector and we have arrived at a situation where we are observing a 2-vector. It will *not be possible* to estimate the required *three* quantities $(x_0, x_1, x_2)_n$ from an estimate of (6.4.8), and so we are forced to conclude that we cannot let the columns of T_n be linearly related. *Thus only T's with full column-rank will be considered.*†

As we show in the next section, it is precisely that assumption that T has full column-rank which enables us, from a matrix formalism standpoint, to obtain estimates at all. Without it we encounter operations which call for the inverses of singular matrices.

Example: Returning to T_n of (6.3.27) we see clearly that it has rank 2, i.e. full column-rank. ◆◆

6.5 LEAST-SQUARES ESTIMATION

In the preceding section we showed how the total observation vector could be related to the true state-vector by

$$Y_{(n)} = T_n X_n + N_{(n)} \tag{6.5.1}$$

We now introduce the idea of a *simulated observation vector.*

Thus, let the state-vector X_n in (6.5.1) have some *given numerical value,* say V. Then the simulated observation vector is that vector of numbers *computed* from (6.5.1) as though an *error-free* total observation vector were being made on V. Thus the simulated observation vector on V, based on the observation relation (6.5.1), is

$$Y \equiv T_n V \tag{6.5.2}$$

†This question of the rank of T is of course governed by a relationship between M and Φ (see (6.3.9)). In Section 8.9 we pursue this matter further.

As another example of a simulated observation vector the reader is referred to (6.2.13).

The next concept which we introduce is the idea of the *residual vector*. Thus, suppose $Y_{(n)}$ is a total observation vector related to the true state-vector by (6.5.1), and let $X^*_{n,n}$ be some estimate of X_n, extracted from $Y_{(n)}$ by the estimation scheme

$$X^*_{n,n} = W_n Y_{(n)} \tag{6.5.3}$$

We now form the simulated observation vector based on $X^*_{n,n}$, namely $T_n X^*_{n,n}$. Then the *total residual vector* is defined as the true observation vector minus the simulated observation vector based on that estimate. We designate the residual vector as E, and so we have

$$E\left(X^*_{n,n}\right) \equiv Y_{(n)} - T_n X^*_{n,n} \tag{6.5.4}$$

where we display clearly the functional dependence of E on the estimate $X^*_{n,n}$.

At this stage we are in a position to embark on a general discussion of the *least-squares principle*. Of the infinite number of ways in which $X^*_{n,n}$ can be chosen, the principle of least-squares states, simply, *that it should be chosen so that the sum of the squared components of its residual vector is least.* Thus, least-squares calls for the vector $X^*_{n,n}$ to be chosen so that the scalar inner-product,

$$e\left(X^*_{n,n}\right) \equiv \left[E\left(X^*_{n,n}\right)\right]^T E\left(X^*_{n,n}\right) \tag{6.5.5}$$

shall be smallest. We now show how $e\left(X^*_{n,n}\right)$ is minimized over $X^*_{n,n}$.

First, we recall from the previous section that, without loss of generality, we can always assume the matrix T_n to have full column rank. That being the case, the matrix $T_n^T T_n$ will be symmetric and positive definite (see Ex. 6.4), and so it is quite permissible to talk about its inverse $\left(T_n^T T_n\right)^{-1}$, as well as about the matrices $\left(T_n^T T_n\right)^{1/2}$ and $\left(T_n^T T_n\right)^{-1/2}$. (See Ex. 6.5.)

We now insert (6.5.4) into (6.5.5), obtaining

$$e\left(X^*_{n,n}\right) = \left(T_n X^*_{n,n} - Y_{(n)}\right)^T \left(T_n X^*_{n,n} - Y_{(n)}\right) \tag{6.5.6}$$

which, by a rearrangement of terms analogous to "completing squares," becomes

$$e\left(X^*_{n,n}\right) = \left[(T^TT)^{1/2}X^* - (T^TT)^{-1/2}T^TY\right]^T\left[(T^TT)^{1/2}X^* - (T^TT)^{-1/2}T^TY\right]$$
$$- Y^TT(T^TT)^{-1}T^TY \tag{6.5.7}$$
$$+ Y^TY$$

Proof follows quite readily by direct expansion. (See Ex. 6.9.)

Examination of (6.5.7) shows that, of the three terms on the right, *only one of them contains* $X^*_{n,n}$. We have no control over the other two, and so if e is to be least *then* $X^*_{n,n}$ *must be chosen so as to minimize the first of those terms.* This is readily accomplished if we set

$$\left(T_n^T T_n\right)^{1/2} X^*_{n,n} = \left(T_n^T T_n\right)^{-1/2} T_n^T Y_{(n)} \tag{6.5.8}$$

for then, that first term becomes zero. Since $e\left(X^*_{n,n}\right)$ is a nonnegative scalar, this is clearly the correct way to minimize it.

From (6.5.8) *we thus get the least-squares estimate*

$$\hat{X}^*_{n,n} = \left(T_n^T T_n\right)^{-1} T_n^T Y_{(n)} \tag{6.5.9}$$

and under these circumstances, by (6.5.7),

$$e\left(\hat{X}^*_{n,n}\right) = Y^T_{(n)}\left[I - T_n\left(T_n^T T_n\right)^{-1} T_n^T\right] Y_{(n)} \tag{6.5.10}$$

(Henceforth when we place a hat over $X^*_{n,n}$, i.e. $\hat{X}^*_{n,n}$, then we are referring *specifically* to the unique estimate shown in (6.5.9), obtained by the least-squares principle.)

Equation (6.5.9) is seen to be of the same form as (6.5.3), the weight matrix being

$$\hat{W}_n \equiv \left(T_n^T T_n\right)^{-1} T_n^T \tag{6.5.11}$$

The reader can verify immediately that $\hat{W}_n$, so defined, satisfies the exactness constraint (6.3.16) and so $\hat{W}_n$ is an unbiased linear estimate of X_n. This is a direct consequence of choice by least-squares.

In Section 5.5 we pointed out that W_n can be taken, without loss of generality, to always have full row-rank. Likewise in Section 6.4 we showed that T_n can be assumed to always have full column-rank. The reader can easily verify, from (6.5.11), that $\hat{W}_n$ will have the same rank as T_n^T, and so the full column-rank assumption on T_n *implies* full row-rank for W_n.

Consider now the covariance matrix of the random errors in $\hat{X}^*_{n,n}$, namely $\hat{S}^*_{n,n}$. By (5.5.10) and (6.5.11) this will be

$$\begin{aligned} \hat{S}^*_{n,n} &\equiv \hat{W}_n R_{(n)} \hat{W}_n^T \\ &= \left(T_n^T T_n\right)^{-1} T_n^T R_{(n)} T_n \left(T_n^T T_n\right)^{-1} \end{aligned} \tag{6.5.12}$$

and so, if the covariance matrix of the random input errors in $Y_{(n)}$ is known, then the covariance matrix of the random errors in $\hat{X}^*_{n,n}$ is obtainable from this expression.

One case of particular interest is the following. Suppose that a single instrument is being used to observe the process and that the observation errors have constant variance and are uncorrelated with each other. Then $R_{(n)}$ would have the form

$$R_{(n)} = \sigma_\nu^2 I \tag{6.5.13}$$

where I is the identity matrix and σ_ν^2 the variance of the errors. For this case, (6.5.12) reduces quite readily to

$$\hat{S}^*_{n,n} = \sigma_\nu^2 \left(T_n^T T_n\right)^{-1} \tag{6.5.14}$$

The matrix $\left(T_n^T T_n\right)^{-1}$ plays a very significant role in least-squares estimation, as can be seen from (6.5.11), (6.5.12) or (6.5.14). Obtaining that inverse is, in fact, the major problem of least-squares estimation from a computational standpoint. We refer to (6.5.9) as the *classical formula.*

In Chapters 7, 9 and 13 we will derive least-squares algorithms which avoid the need for a matrix inversion. This is accomplished by the use of the orthogonal polynomials developed in Chapter 3, and it will be seen that, as a result of the orthogonality property, the matrix $T_n^T T_n$ will, in effect, be replaced by a *diagonal matrix* whose inversion is trivial. However, there are many practical situations where the classical formula must be used and we cannot stress its value sufficiently.

As long as both the model and the observation scheme are linear, the foregoing discussion shows how the principle of least-squares can be used to provide an estimate. If either of them is nonlinear, however, then a further analysis will have to be undertaken before an estimation scheme is obtained. In Chapter 8 we treat this question and arrive at what is known as *the method of iterative differential-correction.*

We note in conclusion that $\hat{W}$ can also be obtained by using the differential calculus to minimize $e\left(X^*_{n,n}\right)$ of (6.5.6). The reader is referred to Ex. 6.6 and 6.7 and to the important comment following Ex. 6.7.

Example: Using the observation equation (6.3.28) we obtain the least-squares weight matrix. Thus, by (6.3.27)

$$T_n^T T_n = \begin{pmatrix} 3 & -3 \\ -3 & 5 \end{pmatrix} \tag{6.5.15}$$

and so by (6.5.11)

$$\begin{aligned} \hat{W} &= \left(T_n^T T_n\right)^{-1} T_n^T \\ &= \frac{1}{6}\begin{pmatrix} 5 & 2 & -1 \\ 3 & 0 & -3 \end{pmatrix} \end{aligned} \tag{6.5.16}$$

Note that $\hat{W}T_n = I$, showing that the exactness constraint is satisfied.

We recall, from (6.3.23), that the model which gave rise to T_n used above, was a first-degree polynomial.† Assume now that the observations on that polynomial are

$$Y_{(n)} = \begin{pmatrix} 3.1 \\ 2.2 \\ 0.9 \end{pmatrix} \tag{6.5.17}$$

Then (6.5.9) gives us

$$\begin{aligned} \hat{X}^*_{n,n} &= \frac{1}{6}\begin{pmatrix} 5 & 2 & -1 \\ 3 & 0 & -3 \end{pmatrix}\begin{pmatrix} 3.1 \\ 2.2 \\ 0.9 \end{pmatrix} \\ &= \begin{pmatrix} 3.167 \\ 1.1 \end{pmatrix} \end{aligned} \tag{6.5.18}$$

†See also (6.3.22) where its state-vector was defined.

Thus the estimate of $x(t)$, at the time that the observation $y_n = 3.1$ was obtained, is

$$x^*_{n,n} = 3.167 \tag{6.5.19}$$

and the estimate of $\tau\dot{x}(t)$ at that time is

$$\tau\dot{x}^*_{n,n} = 1{\cdot}1 \tag{6.5.20}$$

Assuming that $\tau = 1/10$ second, we see that

$$\dot{x}^*_{n,n} = \frac{1{\cdot}1}{\tau} = 11 \tag{6.5.21}$$

Let the covariance matrix of the errors in $Y_{(n)}$ of (6.5.17) be

$$R_{(n)} = \begin{pmatrix} 1 & & \\ & 2 & \\ & & 2 \end{pmatrix} \tag{6.5.22}$$

Then by (6.5.12)

$$\begin{aligned} \hat{S}^*_{n,n} &= \hat{W}_n R_{(n)} \hat{W}_n^T \\ &= \begin{pmatrix} \frac{35}{36} & \frac{21}{36} \\ \frac{21}{36} & \frac{27}{36} \end{pmatrix} \end{aligned} \tag{6.5.23}$$

◆◆

6.6 MINIMUM-VARIANCE ESTIMATION

In the least-squares estimation approach discussed above, we directed our attention towards the summed squared residuals. Reference to (6.4.3) shows that the total residual vector is a vector of zero-mean random variables, and what we did, by the least-squares principle, was to set up the algorithm which minimizes the sum of the squares of those random variables obtained on each successive sample draw from the observation process.

We now direct our attention, instead, to the covariance matrix of the output errors. Assume that the matrix T_n is given, and that the covariance matrix of the input errors, $R_{(n)}$, is known. This last assumption is basic to the minimum-variance estimation procedure.† *Restricting ourselves to those matrices W which satisfy the exactness constraint* (6.3.16), we conceptually form all possible estimates

$$X^*_{n,n} = W_n Y_{(n)} \tag{6.6.1}$$

Their respective covariance matrices will be

$$S^*_{n,n} = W_n R_{(n)} W_n^T \tag{6.6.2}$$

We now set about sorting through this infinity of covariance matrices, generated by the infinity of W's which satisfy the exactness constraint. Among those covariance matrices, one in particular is of interest to us. We call it the *minimum-variance covariance matrix,* designated $\mathring{S}^*_{n,n}$. Likewise the W which produced it, in (6.6.2), is called the *minimum-variance weight matrix,* designated $\mathring{W}_n$, and the resultant estimate is called the *minimum-variance estimate* and designated $\mathring{X}^*_{n,n}$.

The property of this matrix $\mathring{S}^*_{n,n}$ which interests us is that *each of its diagonal elements is individually the smallest* if compared to the corresponding diagonal elements of all of the other matrices $S^*_{n,n}$ generated by (6.6.2). Specifically, assuming 3 × 3 matrices say, then the diagonal elements of $\mathring{S}^*_{n,n}$ satisfy

$$\begin{aligned} \left(\mathring{s}^*_{0,0}\right)_{n,n} &\le \left(s^*_{0,0}\right)_{n,n} \\ \left(\mathring{s}^*_{1,1}\right)_{n,n} &\le \left(s^*_{1,1}\right)_{n,n} \\ \left(\mathring{s}^*_{2,2}\right)_{n,n} &\le \left(s^*_{2,2}\right)_{n,n} \end{aligned} \tag{6.6.3}$$

for *every* matrix $S^*_{n,n}$ generated by (6.6.2).‡

The matrix $\mathring{S}^*_{n,n}$ so selected exists, it is unique, and it constitutes the crux of the minimum-variance estimation procedure. While we have singled it out by examination and comparison of diagonal elements only, we shall see, in the succeeding sections, that the minimum-variance property actually reaches further than just the diagonal terms. It in fact covers the entire matrix and has a number of very far-reaching implications.

†By contrast, we did not require $R_{(n)}$ in order to perform least-squares estimation. See (6.5.9).
‡Always assuming that W_n satisfies the exactness constraint (6.3.16).

We now embark on solving the minimum-variance estimation problem, namely: Given $Y_{(n)}$, $R_{(n)}$ and T_n, find $\mathring{W}_n$, $\mathring{X}^*_{n,n}$ and $\mathring{S}^*_{n,n}$. The problem can be restated more formally as follows:

Given the observation relation

$$Y_{(n)} = T_n X_n + N_{(n)} \tag{6.6.4}$$

and given that $R_{(n)}$ is the positive definite covariance matrix of the zero-mean error vector $N_{(n)}$, find the weight matrix $\mathring{W}_n$ so that

a) *If we estimate X_n by*

$$\mathring{X}^*_{n,n} = \mathring{W}_n Y_{(n)} \tag{6.6.5}$$

then the errors in $\mathring{X}^*_{n,n}$ have the smallest variances, and

b) *The matrix $\mathring{W}_n$ satisfies*

$$\mathring{W}_n T_n = I \tag{6.6.6}$$

The problem is well suited to an application of Lagrange's *method of undetermined multipliers* (see e.g. [6.1]). Thus, given the positive definite matrix $R_{(n)}$ and the full-column-rank rectangular matrix T_n, find the rectangular matrix W_n so that each of the diagonal elements of

$$S^*_{n,n} \equiv W_n R_n W_n^T \tag{6.6.7}$$

is minimized, subject to the constraint equations

$$W_n T_n = I \tag{6.6.8}$$

For definiteness let $R_{(n)}$ be 3×3 and T_n be 3×2. Then W_n will have to be 2×3. Define its *rows* by the row-vectors

$$\begin{aligned} W_0 &\equiv (w_{00},\ w_{01},\ w_{02}) \\ W_1 &\equiv (w_{10},\ w_{11},\ w_{12}) \end{aligned} \tag{6.6.9}$$

and likewise, define the *column*-vectors of T_n by

$$T_0 \equiv \begin{pmatrix} t_{00} \\ t_{10} \\ t_{20} \end{pmatrix}, \qquad T_1 \equiv \begin{pmatrix} t_{01} \\ t_{11} \\ t_{21} \end{pmatrix} \tag{6.6.10}$$

Also define a 2×2 matrix of (as yet undetermined) multipliers by

$$\Lambda \equiv \begin{pmatrix} \lambda_{00} & \lambda_{01} \\ \lambda_{10} & \lambda_{11} \end{pmatrix} \equiv (\Lambda_0 \mid \Lambda_1) \tag{6.6.11}$$

Thus Λ_0 is the first column of Λ and Λ_1 the second.

The covariance matrix $S^*_{n,n}$ is now 2×2 and becomes, by (6.6.7),

$$S^*_{n,n} = \begin{pmatrix} W_0 R W_0^T & W_0 R W_1^T \\ W_1 R W_0^T & W_1 R W_1^T \end{pmatrix} \tag{6.6.12}$$

The constraint equation (6.6.8), on the other hand, gives us the four scalar equations

$$\begin{aligned} W_0 T_0 &= 1, \qquad & W_0 T_1 &= 0, \\ W_1 T_0 &= 0, \qquad & W_1 T_1 &= 1, \end{aligned} \tag{6.6.13}$$

and it is now clear that the Lagrangian minimization problem can be separated into two completely disconnected problems. Thus:

Problem I

Minimize the scalar $W_0 R W_0^T$ *over the row-vector of numbers* W_0, *subject to the constraints*

$$W_0 T_0 = 1 \qquad W_0 T_1 = 0 \tag{6.6.14}$$

Problem II

Minimize the scalar $W_1 R W_1^T$ *over the row-vector of numbers* W_1, *subject to the constraints*

$$W_1 T_0 = 0 \qquad W_1 T_1 = 1 \tag{6.6.15}$$

Lagrange's method requires that we form the *Lagrangian function* L, which is the sum of the function to be minimized and each of the constraint equations multiplied by an "undetermined multiplier." Thus for Problem I above we form the scalar L_0, defined by

$$L_0 \equiv W_0 R W_0^T - 2(W_0 T_0 - 1)\lambda_{00} - 2(W_0 T_1 - 0)\lambda_{10} \tag{6.6.16}$$

where, for convenience we use $-2\lambda_{00}$ and $-2\lambda_{10}$ as the multipliers. L_0 is a function of the vector $(W_0 \mid \lambda_{00}, \lambda_{10})$.

Following Lagrange's method we now differentiate L_0 with respect to each of the components of W_0, setting the results to zero. This gives the *vector* of equations (see Ex. 6.10)

$$2RW_0^T - 2T_0\lambda_{00} - 2T_1\lambda_{10} = \emptyset \qquad \text{(see Note)} \tag{6.6.17}$$

which we reorganize as

$$RW_0^T = (T_0 \mid T_1)\begin{pmatrix}\lambda_{00}\\ \lambda_{10}\end{pmatrix} \tag{6.6.18}$$

Lagrange's method also requires that we differentiate L_0 with respect to each of the multipliers λ_{00} and λ_{10}. This gives

$$W_0T_0 = 1 \quad \text{and} \quad W_0T_1 = 0 \tag{6.6.19}$$

We now have exactly the right number of equations to solve for λ_{00}, λ_{10} and the numbers in W_0, and we could, if we wish, solve for W_0 from the above system of equations.

The situation is simplified, however, if we solve Problems I and II above *simultaneously*. Applying the same operations to Problem II as we did to Problem I, leads to the set of equations

$$RW_1^T = (T_0 \mid T_1)\begin{pmatrix}\lambda_{01}\\ \lambda_{11}\end{pmatrix} \tag{6.6.20}$$

$$W_1T_0 = 0 \quad \text{and} \quad W_1T_1 = 1 \tag{6.6.21}$$

which can be combined with (6.6.18) and (6.6.19) to give

$$RW^T = T\Lambda \tag{6.6.22}$$

$$WT = I \tag{6.6.23}$$

It is easily inferred that this pair would also result if we were solving the *general case* rather than the 2×2 case considered above.

Note: By the symbol $\emptyset$, both here and henceforth, we mean a null-vector or a null-matrix, whichever is appropriate.

The matrix W is now obtained as follows. By transposing (6.6.22) we get

$$WR^T = \Lambda^T T^T \tag{6.6.24}$$

but since R is a (symmetric) positive definite matrix (6.6.24) can be written as

$$W = \Lambda^T T^T R^{-1} \tag{6.6.25}$$

Post-multiplying by T then gives

$$\begin{aligned} WT &= \Lambda^T T^T R^{-1} T \\ &= I \qquad \text{by (6.6.23).} \end{aligned} \tag{6.6.26}$$

It is shown in Ex. 6.4 that $T^T R^{-1} T$ is positive definite if R is positive definite and T has full column-rank,† and since by assumption, this is the case, $T^T R^{-1} T$ can then be inverted. Thus by (6.6.26)

$$\Lambda^T = (T^T R^{-1} T)^{-1} \tag{6.6.27}$$

This is now used in (6.6.25) to give, finally,

$$W_n = \left(T_n^T R_{(n)}^{-1} T_n\right)^{-1} T_n^T R_{(n)}^{-1} \tag{6.6.28}$$

Strictly speaking, Lagrange's method leads to a *stationary* point, i.e. the above W may have made the diagonal elements of $S^*_{n,n}$ to be maximum, minimum, or at a saddle point. However, by (6.6.12) we see that each of those elements is a quadratic form on a positive definite matrix which rules out the possibility of a maximum or a saddle point. Thus we have arrived at the matrix W which *minimizes,* individually, each of the diagonal elements of $S^*_{n,n}$ and we write, as the resultant filter

$$\overset{\circ}{X}{}^*_{n,n} = \overset{\circ}{W}_n Y_{(n)} \tag{6.6.29}$$

where

$$\overset{\circ}{W}_n \equiv \left(T_n^T R_{(n)}^{-1} T_n\right)^{-1} T_n^T R_{(n)}^{-1} \tag{6.6.30}$$

This is called the *minimum-variance unbiased linear estimator.*

†This fact also follows directly from (6.6.26), i.e., $\Lambda^T T^T R^{-1} T = I$ implies Λ^T and $T^T R^{-1} T$ have nonzero determinants.

The resultant covariance matrix $\overset{\circ}{S}{}^{*}_{n,n}$ is, by (6.6.7)

$$\overset{\circ}{S}{}^{*}_{n,n} = \overset{\circ}{W}_n R_{(n)} \overset{\circ}{W}{}_n^T \tag{6.6.31}$$

and it is easily verified (see Ex. 6.11) that this reduces to

$$\overset{\circ}{S}{}^{*}_{n,n} = \left(T_n^T R_{(n)}^{-1} T_n\right)^{-1} \tag{6.6.32}$$

The reader is also referred to Examples 6.12 through 6.14.

It is particularly interesting to examine the case where the input errors are uncorrelated, stationary and have covariance matrix (c/f (6.5.13))

$$R_{(n)} = \sigma_\nu^2 I \tag{6.6.33}$$

The minimum-variance filter matrix and the output-error covariance matrix then become

$$\overset{\circ}{W}_n = \left(T_n^T T_n\right)^{-1} T_n^T \tag{6.6.34}$$

$$\overset{\circ}{S}{}^{*}_{n,n} = \sigma_\nu^2 \left(T_n^T T_n\right)^{-1} \tag{6.6.35}$$

which follows readily by using (6.6.33) in (6.6.30) and (6.6.32). Comparison of the above two equations with (6.5.11) and (6.5.14) thus shows that if the input errors have a covariance matrix given by (6.6.33) *then the least-squares filter is also the minimum-variance filter.* For any other case of input errors, however, the diagonal elements of $\overset{\circ}{S}{}^{*}_{n,n}$ will be less than or equal, term by term, to corresponding elements in all covariance matrices of unbiased linear estimates on the same data, including the least-squares estimate.

In this way, then, we have solved the minimum-variance estimation problem. Again we see from (6.6.30) that matrix inversions are called for – not one, as in the least-squares case, but two. We have also seen that, under certain circumstances, the minimum-variance and least-squares estimators are one and the same.

The minimum-variance estimate has been derived on the basis of complete linearity, i.e. a linear model and linear observation relations have been assumed throughout. In many cases of practical interest however, linearity does not prevail, and in Chapter 8 we will show how minimum-variance estimation can be applied to nonlinear situations by the use of the method of iterative differential-correction.

The remainder of this chapter is devoted to a study of the minimum-variance property in greater depth. Many relationships will be revealed which will play important roles in our future discussion.

Example: Using the observation equation (6.3.28) and the observation-error covariance matrix (6.5.22), we now derive the minimum-variance matrices $\mathring{W}_n$ and $\mathring{S}^*_{n,n}$. Thus

$$\left(T_n^T R_{(n)}^{-1} T_n\right)^{-1} = \begin{pmatrix} \frac{10}{11} & \frac{6}{11} \\ \frac{6}{11} & \frac{8}{11} \end{pmatrix} \tag{6.6.36}$$

which, by (6.6.32), is the matrix $\mathring{S}^*_{n,n}$. By (6.6.30) the minimum-variance weight matrix $\mathring{W}$ becomes

$$\mathring{W} = \frac{1}{11}\begin{pmatrix} 10 & 2 & -1 \\ 6 & -1 & -5 \end{pmatrix} \tag{6.6.37}$$

As a check, we note that $\mathring{W}T_n = I$, and so $\mathring{W}$ of (6.6.37) does satisfy the exactness constraint.

We compare the covariance matrix $\mathring{S}^*_{n,n}$ in (6.6.36) to the matrix $\hat{S}^*_{n,n}$ of (6.5.23). Reducing them to their common denominator, namely 396, we obtain from (6.6.36),

$$\mathring{S}^*_{n,n} = \frac{1}{396}\begin{pmatrix} 360 & 216 \\ 216 & 288 \end{pmatrix} \tag{6.6.38}$$

and from (6.5.23)

$$\hat{S}^*_{n,n} = \frac{1}{396}\begin{pmatrix} 385 & 231 \\ 231 & 297 \end{pmatrix} \tag{6.6.39}$$

showing clearly that the diagonal elements of $\mathring{S}^*_{n,n}$ are smaller than those of $\hat{S}^*_{n,n}$. ◆◆

6.7 GENERAL ASPECTS OF MINIMUM-VARIANCE

The approach we adopted in the preceding section to the minimum-variance estimator was through the diagonal elements of the output-error covariance matrix. We now show that the minimum-variance property actually goes far deeper, only one aspect being the minimal nature of those diagonal elements.

Suppose, as we did in the previous section, that we consider all covariance matrices

$$S^*_{n,n} \equiv W_n R_{(n)} W_n^T \tag{6.7.1}$$

which arise from weight matrices W_n which satisfy the exactness constraint

$$W_n T_n = I \tag{6.7.2}$$

where $R_{(n)}$ and T_n are given.† Each of those matrices $S^*_{n,n}$ is positive definite if $R_{(n)}$ is positive definite and T_n has full column rank.

Suppose that we now subtract from each matrix $S^*_{n,n}$ the minimum-variance matrix $\overset{\circ}{S}{}^*_{n,n}$, as obtained in Section 6.6. In general, the difference between two positive definite matrices may or may not be a definite matrix. However, as we now prove, *the matrix* $\overset{\circ}{S}{}^*_{n,n}$ *is such that every one of the matrices C defined by*

$$C \equiv S^*_{n,n} - \overset{\circ}{S}{}^*_{n,n} \tag{6.7.3}$$

is a nonnegative definite matrix.

To prove this we require the following.

Lemma

Let

$$C \equiv BB^T - BG(G^T G)^{-1} G^T B^T \tag{6.7.4}$$

where B and G^T *are any* $m \times k$ *matrices with* $k \geq m$, *and* $G^T G$ *is nonsingular. Then C is nonnegative definite.*

Proof

Returning to (6.5.10) we recall that $e\left(\hat{X}^*_{n,n}\right)$ is nonnegative, and so since $Y_{(n)}$ of that equation is arbitrary, it must be true that the matrix $I - T(T^T T)^{-1} T^T$ is nonnegative definite. Hence so is $I - G(G^T G)^{-1} G^T$ for *any*

†Throughout this section, when we write $S^*_{n,n}$, we mean a covariance matrix of the form of (6.7.1) where W_n satisfies (6.7.2).

matrix G for which $(G^T G)^{-1}$ exists. Finally then for any matrix B, the matrix

$$C \equiv B\left[I - G(G^T G)^{-1} G^T\right] B^T \tag{6.7.5}$$

is nonnegative definite and so the lemma is proved. ◆◆

We now have the following:

Theorem 6.1

Let R be a positive definite $k \times k$ matrix and T a $k \times m$ matrix $(k \geq m)$ with full column rank. Let W be any $m \times k$ matrix satisfying

$$WT = I \tag{6.7.6}$$

Then the matrix

$$C \equiv WRW^T - (T^T R^{-1} T)^{-1} \tag{6.7.7}$$

is nonnegative definite.

Proof

Using the previous lemma we set

$$B = WR^{1/2} \qquad G = R^{-1/2} T \tag{6.7.8}$$

Then (6.7.4) shows that the matrix

$$WRW^T - WT(T^T R^{-1} T)^{-1} T^T W^T$$

is nonnegative definite. But by (6.7.6) this now reduces to

$$WRW^T - (T^T R^{-1} T)^{-1}$$

and so the theorem is proved. ◆◆

Of course, it is now evident that the following is also true.

Corollary 6.1.1

For $S^*_{n,n}$ and $\mathring{S}^*_{n,n}$ defined by (6.6.2) *and* (6.6.31) *respectively, the matrix*

$$C \equiv S^*_{n,n} - \mathring{S}^*_{n,n} \tag{6.7.9}$$

is nonnegative definite.

Proof follows directly from Theorem 6.1 above. ◆◆

This corollary is the very fundamental aspect of the minimum-variance concept to which we have been alluding, and all of the properties of the minimum-variance covariance matrix can be obtained from it.

As a start, if $S^*_{n,n} - \mathring{S}^*_{n,n}$ is nonnegative definite, then its diagonal elements are all nonnegative,† i.e.

$$[C]_{i,i} \equiv \left[S^*_{n,n}\right]_{i,i} - \left[\mathring{S}^*_{n,n}\right]_{i,i} \geq 0 \qquad (0 \leq i \leq m) \tag{6.7.10}$$

We thus have

Corollary 6.1.2

$$\left[S^*_{n,n}\right]_{i,i} \geq \left[\mathring{S}^*_{n,n}\right]_{i,i} \qquad (0 \leq i \leq m) \tag{6.7.11}$$

◆◆

The above proves that each of the diagonal elements of $\mathring{S}^*_{n,n}$ are individually less than, or equal to, corresponding elements in any of the other matrices $S^*_{n,n}$. This formed the basis of the preceding section and now emerges simply as a corollary of the minimum-variance property as stated in Theorem 6.1. Henceforth, when we talk of the *minimum-variance property* of the matrix $\mathring{S}^*_{n,n}$, then we are referring specifically to the fact that the matrix

$$C \equiv S^*_{n,n} - \mathring{S}^*_{n,n}$$

is nonnegative definite.

As a further important corollary we have the following. Suppose that $\mathring{X}^*_{n,n}$ is the unbiased minimum-variance updated estimate of X_n, based on $Y_{(n)}$, $R_{(n)}$ and T_n, and assume that we form the *prediction*

$$U \equiv \Phi(h)\,\mathring{X}^*_{n,n} \tag{6.7.12}$$

Clearly U is an unbiased estimate of X_{n+h}, since

$$\begin{aligned} E\left\{U - X_{n+h}\right\} &= \Phi(h)\, E\left\{\mathring{X}^*_{n,n} - X_n\right\} \\ &= \emptyset \end{aligned} \tag{6.7.13}$$

†See Ex. 5.8.

The question we now ask is, if U is compared with any other unbiased estimate of X_{n+h} based on $Y_{(n)}$ and T_n, *will the covariance matrix of U have the smallest diagonal elements?* The answer is yes, as we now propose to show.

Thus, let $X^*_{n,n}$ be *any* unbiased estimate of X_n based on $Y_{(n)}$ and T_n, i.e. let

$$X^*_{n,n} = W_n Y_{(n)} \tag{6.7.14}$$

where $W_n T_n = I$. Form the prediction

$$V \equiv \Phi(h) X^*_{n,n} \tag{6.7.15}$$

Then by the reasoning of (6.7.13), V is an unbiased estimate of X_{n+h} based on $Y_{(n)}$ and T_n.

The covariance matrix of V is, by (6.7.15) and (5.5.20),

$$S^*_V \equiv \Phi(h) S^*_{n,n} \Phi(h)^T \tag{6.7.16}$$

Likewise the covariance matrix of U is seen from (6.7.12) to be

$$S^*_U \equiv \Phi(h) \overset{\circ}{S}{}^*_{n,n} \Phi(h)^T \tag{6.7.17}$$

Hence

$$S^*_V - S^*_U = \Phi(h) \left(S^*_{n,n} - \overset{\circ}{S}{}^*_{n,n}\right) \Phi(h)^T \tag{6.7.18}$$

But by Corollary 6.1.1, the right-hand side is now seen to be a congruence transformation on a nonnegative definite matrix. This property is preserved under congruence (see Ex. 5.9) and so the left-hand side of (6.7.18) is then also nonnegative definite. Thus the diagonal elements of $S^*_V - S^*_U$ are nonnegative and so we have proved the following:

Corollary 6.1.3

Each of the diagonal elements of S^*_U and S^*_V satisfy

$$\left[S^*_V\right]_{i,i} \geq \left[S^*_U\right]_{i,i} \tag{6.7.19}$$

◆◆

By equation (6.7.18) we now have the final corollary,

Corollary 6.1.4

The minimum-variance property is preserved under prediction or retrodiction, i.e. for any h, the matrix

$$C \equiv S^*_{n+h,n} - \mathring{S}^*_{n+h,n} \tag{6.7.20}$$

is nonnegative definite. ◆◆

We shall accordingly write (6.7.12) and (6.7.17) as

$$\mathring{X}^*_{n+h,n} = \Phi(h)\,\mathring{X}^*_{n,n} \tag{6.7.21}$$

$$\mathring{S}^*_{n+h,n} = \Phi(h)\,\mathring{S}^*_{n,n}\,\Phi(h)^T \tag{6.7.22}$$

showing that if a prediction is made on the basis of a minimum-variance estimate, then that prediction is also minimum-variance by comparison to all other unbiased predictions to the same time-instant, based on the same data.

The minimum-variance property does indeed contain some important ramifications. They will feature more and more strongly as our discussion proceeds.

Example: In (6.6.38) we obtained the minimum-variance covariance matrix $\mathring{S}^*_{n,n}$ whereas in (6.6.39) we have the least-squares matrix $\hat{S}^*_{n,n}$ for the same model and the same data. Their difference is

$$\hat{S}^*_{n,n} - \mathring{S}^*_{n,n} = \frac{1}{396}\begin{pmatrix} 25 & 15 \\ 15 & 9 \end{pmatrix} \tag{6.7.23}$$

which is quite clearly nonnegative definite, as Theorem 6.1 predicts. Moreover, if we now form, say, a 1-step prediction using $\Phi(1)$ from (6.3.23), then

$$\begin{aligned} \mathring{S}^*_{n+1,n} &= \Phi(1)\,\mathring{S}^*_{n,n}\,\Phi(1)^T \\ &= \begin{pmatrix} 1 & 1 \\ 0 & 1 \end{pmatrix} \frac{1}{396}\begin{pmatrix} 360 & 216 \\ 216 & 288 \end{pmatrix}\begin{pmatrix} 1 & 0 \\ 1 & 1 \end{pmatrix} \\ &= \frac{1}{396}\begin{pmatrix} 1080 & 504 \\ 504 & 288 \end{pmatrix} \end{aligned} \tag{6.7.24}$$

whereas

$$\hat{S}^*_{n+1,n} = \Phi(1)\hat{S}^*_{n,n}\Phi(1)^T$$
$$= \frac{1}{396}\begin{pmatrix} 1144 & 528 \\ 528 & 297 \end{pmatrix} \qquad (6.7.25)$$

Then,

$$\hat{S}^*_{n+1,n} - \mathring{S}^*_{n+1,n} = \frac{1}{396}\begin{pmatrix} 64 & 24 \\ 24 & 9 \end{pmatrix} \qquad (6.7.26)$$

which is also nonnegative definite as Corollary 6.1.4 predicts. ◆◆

6.8 UNIQUENESS OF LEAST-SQUARES AND MINIMUM-VARIANCE

We now raise the question of *uniqueness* of the minimum-variance and least-squares estimates. It is clear that given $Y_{(n)}$, T_n and $R_{(n)}$, there can be only *one* minimum-variance and *one* least-squares estimate, and these are clearly spelled out by (6.6.29) and (6.5.9). However, the following question arises.

Suppose that the *units,* in which the components of $Y_{(n)}$ are expressed, were changed – feet to inches, etc. Would this have any effect on the resultant estimate? The problem goes somewhat further yet, since we must also recognize that the elements of $Y_{(n)}$ may be of mixed *dimensions* – length, velocity, mass, etc., and we must also examine whether or not it is legitimate to mix such measurements.

It is now shown that alteration of units and the mixing of dimensions *constitutes no problem whatever* for the case of minimum-variance estimation. However, the same is *not* true for the least-squares case.

To see this, consider the following hypothetical situation. Two observers, A and B, are assumed to be observing a process, simultaneously using the *same* set of observation instruments. *A reads the measurements in one set of units and B in another.* Letting Q be the diagonal matrix of scale factors relating A to B (e.g. 12 inches per foot), then A's and B's total observation vectors Y_A and Y_B will be numerically related by

$$Y_A = QY_B \qquad (6.8.1)$$

Note that Q is diagonal with positive entries.

The errors in Y_A and Y_B are N_A and N_B, related by

$$N_A = QN_B \tag{6.8.2}$$

and so clearly the respective covariance matrices are connected by

$$R_A = QR_B Q^T \tag{6.8.3}$$

For simplicity, we suppose that A and B both use the same state-vector X_n (although we could also assume a difference in units, with an equation similar to (6.8.1) connecting X_A and X_B). Thus, assume that the equations

$$Y_A = T_A X_n + N_A \tag{6.8.4}$$

and

$$Y_B = T_B X_n + N_B \tag{6.8.5}$$

constitute the observation relations which A and B are using. Then by virtue of (6.8.1) and (6.8.2) we must also have

$$T_A = QT_B \tag{6.8.6}$$

Observers A and B now form their respective minimum-variance estimates obtaining

$$\mathring{X}_A^* = \left(T_A^T R_A^{-1} T_A\right)^{-1} T_A^T R_A^{-1} Y_A \tag{6.8.7}$$

and

$$\mathring{X}_B^* = \left(T_B^T R_B^{-1} T_B\right)^{-1} T_B^T R_B^{-1} Y_B \tag{6.8.8}$$

But if we now apply (6.8.1), (6.8.3) and (6.8.6) to (6.8.7), we obtain

$$\begin{aligned}\mathring{X}_A^* &= \left[(QT_B)^T\left(QR_B Q^T\right)^{-1}(QT_B)\right]^{-1}(QT_B)^T\left(QR_B Q^T\right)^{-1} QY_B \\ &= \left(T_B^T R_B^{-1} T_B\right)^{-1} T_B^T R_B^{-1} Y_B \\ &= \mathring{X}_B^* \qquad \text{(by (6.8.8))}\end{aligned} \tag{6.8.9}$$

and so, *regardless of the choice of units, A and B obtain precisely the same estimate vector.*

The case of least-squares is, however, somewhat different. Based on Y_A and Y_B our two observers obtain their respective least-squares estimates as

$$\hat{X}_A^* = \left(T_A^T T_A\right)^{-1} T_A^T Y_A \tag{6.8.10}$$

and

$$\hat{X}_B^* = \left(T_B^T T_B\right)^{-1} T_B^T Y_B \tag{6.8.11}$$

Then by (6.8.1) and (6.8.6)

$$\begin{aligned} \hat{X}_A^* &= \left[(QT_B)^T QT_B\right]^{-1} (QT_B)^T QY_B \\ &= \left(T_B^T Q^2 T_B\right)^{-1} T_B^T Q^2 Y_B \end{aligned} \tag{6.8.12}$$

and now in general, only if

$$Q^2 = k^2 I \tag{6.8.13}$$

where k^2 is some scalar, is $\hat{X}_A^*$ equal to $\hat{X}_B^*$. Hence (6.8.13) means that in general only when $Q = kI$ is $\hat{X}_A^* = \hat{X}_B^*$. Since a change of units does not generally amount to such a Q when the dimensions in the observation vector are mixed, we thus see that a least-squares estimate *can definitely be influenced, and in fact modified, by choice of units* (see Ex. 6.15).

The above must then serve to cast some very serious doubts on the value of least-squares estimation *when the data have mixed dimensions.* Thus, suppose that distances and speeds are observed for the purpose of estimating the state of a process. What units shall be chosen? Recasting the distances from microns to miles or the speeds from light years per microsecond to feet per century would have a definite effect on the estimate vector, and certainly units such as these are every bit as valid as any others that we might choose.

It is for this reason that least-squares estimation, based on mixed dimension observations is *seldom, if ever, performed.* In our case, least-squares will *always* be restricted to situations where a *single* aspect of a process (i.e. a single element of the state-vector X_n) is being observed. We will also assume that all observations are scaled in the same way. Thus $Y_{(n)}$ will be a vector of measurements, all on a single quantity in X_n, all of the same dimensions and all scaled in the same units.

If mixed dimensional information is to be combined, as is very often the case, then minimum-variance estimation *must* be used. The presence of the covariance matrix $R_{(n)}$ serves as a normalizer, permitting measurements of differing dimensions to be combined. Moreover, if units are changed in $Y_{(n)}$, then as we saw in (6.8.9) they are also *changed in* $R_{(n)}$ *in precisely the required way to nullify the effects of that change.* Minimum-variance estimation is again seen to be a very profound concept. (See Ex. 6.16.)

6.9 WEIGHTED LEAST-SQUARES

In the preceding section we showed that, in the case of minimum-variance estimation, the covariance matrix of the input errors serves as a normalizer which compensates precisely for any changes in the units of the observations, and permits us to combine observations with mixed dimensions. We now show that it plays another exceedingly important role, and in so doing we shall be able to demonstrate yet another aspect of the power of the minimum-variance approach.

Thus, suppose that we have a total observation vector $Y_{(n)}$ related to the true state-vector X_n by

$$Y_{(n)} = T_n X_n + N_{(n)} \tag{6.9.1}$$

and that we wish to form a *least-squares* estimate. According to the least-squares criterion, the summed squared residuals must be minimized over the estimate vector $X^*_{n,n}$. Thus $X^*_{n,n}$ must be chosen so that (c/f (6.5.6)) the scalar

$$e\left(X^*_{n,n}\right) \equiv \left(Y_{(n)} - T_n X^*_{n,n}\right)^T \left(Y_{(n)} - T_n X^*_{n,n}\right) \tag{6.9.2}$$

is least. Note the dependence of $e\left(X^*_{n,n}\right)$ on $X^*_{n,n}$.

For definiteness, let $Y_{(n)} \equiv \left(y_0, y_1, y_2, y_3\right)^T_{(n)}$ and call the rows of T_n the row-vectors T_0, T_1, T_2 and T_3. Then (6.9.2) can be written in the form:

$$e\left(X^*_{n,n}\right) = \sum_{k=0}^{3} \left[\left(y_k\right)_{(n)} - T_k X^*_{n,n}\right]^2 \tag{6.9.3}$$

This shows clearly that $e\left(X^*_{n,n}\right)$ is a sum of squares, each comprised of the difference between an actual observation and its simulated counterpart based on the trajectory $X^*_{n,n}$.

Now, the closeness with which a simulated observation approaches the actual one *is a measure of the credence which the selected trajectory is giving to that observation.* In the case of interpolation the trajectory passes through all of the observations precisely, and so complete credence is given to every element of $Y_{(n)}$. However in the least-squares process, the chosen trajectory does not necessarily pass through all or even any of the observations.

Suppose that we are able, somehow, to determine that one particular observation is more precise than the rest, and decide that a higher than average amount of credence should be given to it. This means that we would somehow want to force the selected trajectory to pass somewhat closer to that observation.

To accomplish this we examine the possibility of *applying weights to the squared residuals* in (6.9.3). One form of obtaining the weights, which presents itself very naturally, is to use the *inverse of the variance* of the observation on which the residual is based. In this way if the variance is large, the weight is small and that particular residual will be played down. If the variance is small, then the residual is stressed more heavily, *as it should be.* Equation (6.9.3) is thus made to read as follows:

$$e\left(X^*_{n,n}\right) = \sum_{k=0}^{3}\left[\left(y_k\right)_{(n)} - T_k X^*_{n,n}\right]^2 \frac{1}{\left(\sigma_k^2\right)_{(n)}} \tag{6.9.4}$$

where $\left(\sigma_k^2\right)_{(n)}$ is the variance of $\left(y_k\right)_{(n)}$ and is obtained by taking the appropriate diagonal element out of $R_{(n)}$, the covariance matrix of the errors in $Y_{(n)}$.

The criterion (6.9.4) is essentially what Gauss chose to minimize in his estimation procedures, discussed in [6.2]. It remained until 1934 when Aitken (see [6.3]) proposed using the *entire* covariance matrix (and not just its diagonal elements) in the criterion. Thus, Aitken proposed replacing (6.9.2) by the *weighted* criterion

$$e\left(X^*_{n,n}\right) = \left(Y_{(n)} - T_n X^*_{n,n}\right)^T R_{(n)}^{-1} \left(Y_{(n)} - T_n X^*_{n,n}\right) \tag{6.9.5}$$

and, as is readily verified, when $R_{(n)}$ is diagonal, this immediately reduces to Gauss' criterion (6.9.4). The criterion (6.9.5) is known as *weighted least-squares.*

If we now minimize (6.9.5) over $X^*_{n,n}$, we obtain (see Ex. 6.17)

$$X^*_{n,n} = \left(T_n^T R_{(n)}^{-1} T_n\right)^{-1} T_n^T R_{(n)}^{-1} Y_{(n)} \tag{6.9.6}$$

which we immediately recognize as the minimum-variance estimate.

We have thus shown that the minimum-variance estimate can be arrived at through the concept of weighted least-squares, and in so doing we have thrown further light on the importance of the covariance matrix of the input errors. Hence, in addition to performing the normalizing role pointed out in the previous section, *its inverse also serves as a weight matrix* for selection of the trajectory, stressing the contributions which the more precise observations make and playing down the contributions of the less precise ones.†

The minimum-variance estimate has thus been shown to have some very wide implications. Our approach was initially to choose W in order to minimize the diagonal elements of the covariance matrix, but we soon showed that the minimum-variance property in reality extends deep into the entire matrix through the fact that $C \equiv S^*_{n,n} - \mathring{S}^*_{n,n}$, is nonnegative definite. This then showed us that the minimum-variance property also extends to the entire trajectory, i.e. if $\mathring{X}^*_{n,n}$ is minimum-variance then so is $\Phi(h)\mathring{X}^*_{n,n}$ for all h. In the preceding section we showed that mixed dimensions and arbitrary units can be used, and in this section we have shown, further, how the minimum-variance estimate stresses the observations in accordance with their quality as exemplified by their second order statistics.

In passing, we point out that if all squared residuals are weighted equally, i.e. if the error covariance matrix has the form $R_{(n)} \equiv \alpha^2 I$, then minimum-variance and least-squares, as seen from (6.9.6), would coincide.‡ The weighted least-squares criterion is thus an important unifying concept between the two techniques.

Finally, we mention that if the errors in $Y_{(n)}$ are multivariate Gaussian, then the minimum-variance estimate is also the *maximum-likelihood estimate* (see e.g. [6.4]). This is discussed in Ex. 6.19.

6.10 THE RESIDUALS

In this, the final section of the present chapter, we show that the residuals can be utilized to very great advantage in a method for calibrating or monitoring the accuracy of the observation instruments.

†The reader is referred to the historical note at the end of this chapter.

‡See also p. 191 where we also discussed the relationship between least-squares and minimum-variance.

We suppose that $R_{(n)}$ is known *to within a scale-factor,* i.e. we assume that

$$R_{(n)} = \lambda^2 P_{(n)} \tag{6.10.1}$$

where $P_{(n)}$ *is known but* λ^2 *is unknown.* We pause briefly to demonstrate that such a situation could easily arise in practice.

As a first example, suppose that a single instrument is observing a process and that the observation errors are completely uncorrelated and stationary. (This is certainly a reasonable situation.) Then a total covariance matrix, formed from those errors, would be of the form

$$R_{(n)} = \sigma_\nu^2 I \tag{6.10.2}$$

where σ_ν^2 is the variance of the errors on any one measurement. If the value of σ_ν^2 is *unknown,* then this corresponds precisely to the situation conjectured by (6.10.1). ◆◆

As a second example, suppose that the entries in $Y_{(n)}$ are obtained *by the consolidation of other measurements.* As a simple demonstration, let

$$\begin{aligned} y_n &\equiv \frac{1}{5}\sum_{i=1}^{5} g_i \\ y_{n-1} &\equiv \frac{1}{3}\sum_{i=1}^{3} h_i \\ y_{n-2} &\equiv \frac{1}{7}\sum_{i=1}^{7} k_i \end{aligned} \tag{6.10.3}$$

where the g's, h's and k's are drawn from parent bodies of data.† If the situation is such that the g, h and k measurements are all of equal variance σ_ν^2, then, assuming an absence of correlation, the covariance matrix of the errors in

$$Y_{(n)} \equiv \begin{pmatrix} y_n \\ y_{n-1} \\ y_{n-2} \end{pmatrix} \tag{6.10.4}$$

†Such situations occur frequently in astronomical work in the formation of what are called "normal places" from raw data.

would be of the form

$$R_{(n)} = \sigma_\nu^2 \begin{pmatrix} \frac{1}{5} & & \\ & \frac{1}{3} & \\ & & \frac{1}{7} \end{pmatrix} \tag{6.10.5}$$

If σ_ν^2 is unknown, then again this is of the form of (6.10.1).

We now show that λ^2 in (6.10.1) can be estimated from the residuals.

The minimum-variance estimate is obtained from $Y_{(n)}$, $R_{(n)}$ and T_n by

$$\mathring{X}^*_{n,n} = \left(T_n^T R_{(n)}^{-1} T_n\right)^{-1} T_n^T R_{(n)}^{-1} Y_{(n)} \tag{6.10.6}$$

and making use of (6.10.1), this becomes

$$\mathring{X}^*_{n,n} = \left(T_n^T P_{(n)}^{-1} T_n\right)^{-1} T_n^T P_{(n)}^{-1} Y_{(n)} \tag{6.10.7}$$

which means that the minimum-variance estimate can be obtained *even though* λ^2 *is unknown.* We are thus also able to compute the residual vector using $Y_{(n)}$ and (6.10.7), i.e.

$$E\left(\mathring{X}^*_{n,n}\right) \equiv Y_{(n)} - T\mathring{X}^*_{n,n} \tag{6.10.8}$$

from which we form the *weighted* summed squared residuals

$$\mathring{e}_n \equiv \left[E\left(\mathring{X}^*_{n,n}\right)\right]^T P_{(n)}^{-1} E\left(\mathring{X}^*_{n,n}\right) \tag{6.10.9}$$

This quantity can be computed in any given situation.

Returning now to (6.4.2), we see that (6.10.9) can be written as

$$\mathring{e}_n = \left(N_{(n)} - T_n\mathring{N}^*_{n,n}\right)^T P_{(n)}^{-1} \left(N_{(n)} - T_n\mathring{N}^*_{n,n}\right) \tag{6.10.10}$$

However we recall that

$$\mathring{N}^*_{n,n} \equiv \mathring{W}_n N_{(n)} \tag{6.10.11}$$

and so (6.10.10) reduces to

$$\overset{\circ}{e}_n = \left[\left(I - T_n \overset{\circ}{W}_n\right) N_{(n)}\right]^T P_{(n)}^{-1} \left(I - T_n \overset{\circ}{W}_n\right) N_{(n)} \tag{6.10.12}$$
$$= N_{(n)}^T G N_{(n)}$$

where

$$G \equiv \left(I - T_n \overset{\circ}{W}_n\right)^T P_{(n)}^{-1} \left(I - T_n \overset{\circ}{W}_n\right) \tag{6.10.13}$$

If we now make the substitution (c/f (6.10.7))

$$\overset{\circ}{W}_n = \left(T_n^T P_{(n)}^{-1} T_n\right)^{-1} T_n^T P_n^{-1} \tag{6.10.14}$$

then it is readily verified (see Ex. 6.20) that (6.10.13) reduces to

$$G = P_{(n)}^{-1} \left(I - T_n \overset{\circ}{W}_n\right) \tag{6.10.15}$$

Moreover, by Ex. 6.23

$$N_{(n)}^T G N_{(n)} = \mathrm{Tr}\left(G^T N_{(n)} N_{(n)}^T\right) \tag{6.10.16}$$

(where Tr means *trace*). Combining the above equations then reduces (6.10.12) to

$$\overset{\circ}{e}_n = \mathrm{Tr}\left[\left(I - T_n \overset{\circ}{W}_n\right)^T P_{(n)}^{-1} N_{(n)} N_{(n)}^T\right] \tag{6.10.17}$$

We now take the expectation of both sides. Then since the expectation of a trace equals the trace of the expectation, (6.10.17) gives us

$$E\left\{\overset{\circ}{e}_n\right\} = E\left\{\mathrm{Tr}\left[\left(I - T_n \overset{\circ}{W}_n\right)^T P_{(n)}^{-1} N_{(n)} N_{(n)}^T\right]\right\}$$
$$= \mathrm{Tr}\left[\left(I - T_n \overset{\circ}{W}_n\right)^T P_{(n)}^{-1} R_{(n)}\right] \tag{6.10.18}$$
$$= \lambda^2 \,\mathrm{Tr}\left(I - T_n \overset{\circ}{W}_n\right)^T \qquad \text{by (6.10.1).}$$

The identity matrix in the above equation is of the same order as $Y_{(n)}$. Call that number N_O. Then (6.10.18) becomes

$$E\left\{\overset{\circ}{e}_n\right\} = \lambda^2\left[N_O - \operatorname{Tr}\left(\overset{\circ}{W}_n{}^T T_n{}^T\right)\right] \tag{6.10.19}$$

Moreover, as is shown in Ex. 6.22

$$\operatorname{Tr}\left(\overset{\circ}{W}_n{}^T T_n{}^T\right) = \operatorname{Tr}\left(\overset{\circ}{W}_n T_n\right) \tag{6.10.20}$$

and since $\overset{\circ}{W}_n$ satisfies the exactness constraint, $\overset{\circ}{W}_n T_n$ equals an identity matrix of the same order as the state-vector. Call that number N_X. We thus write (6.10.19) as

$$E\left\{\overset{\circ}{e}_n\right\} = \lambda^2 (N_O - N_X) \tag{6.10.21}$$

This gives us, finally,

$$\lambda^2 = \frac{E\left\{\overset{\circ}{e}_n\right\}}{N_O - N_X} \tag{6.10.22}$$

In practice we do not know $E\left\{\overset{\circ}{e}_n\right\}$. All we have is a *single sample value* of the random variable $\overset{\circ}{e}_n$. However, in lieu of $E\left\{\overset{\circ}{e}_n\right\}$ we decide simply to use that sample value, i.e. we assert that

$$E\left\{\overset{\circ}{e}_n\right\} \approx e_n \qquad \text{(See Note)} \tag{6.10.23}$$

and so by (6.10.22)

$$\lambda^2 \approx \frac{\overset{\circ}{e}_n}{N_O - N_X} \tag{6.10.24}$$

from which λ^2 can be estimated. The number $\overset{\circ}{e}_n$ is computed from (6.10.9).

Note: When $N_O \gg N_X$, then we can expect $\overset{\circ}{e}_n$ to have a small variance, making (6.10.23) a more valid approximation.

By the use of this method, $R_{(n)}$ is now completely known to an accuracy depending upon how well (6.10.23) holds true. Thus the residuals can be used, very effectively, *either as a means of calibrating the observation instruments or else as a method of keeping a check on their precision.*

One caveat of importance should be pointed out. All of the above arguments originated with (6.10.8), which by (6.4.2) is equivalent to

$$E\left(\overset{\circ}{X}{}^*_{n,n}\right) = N_{(n)} - T_n \overset{\circ}{N}{}^*_{n,n} \tag{6.10.25}$$

and shows that the residual vector is founded on the input and output random error vectors. However, if the assumed model does *not* match the true process precisely, *then there exists another set of errors,* called the bias errors.† These will show up in $\overset{\circ}{N}{}^*_{n,n}$ and cause an inflated reading.

In many situations it is possible to match the model to the true process very accurately. For example in celestial or orbital mechanics, differential equations of motion are usually known very precisely. In other situations however, the match is often not so good, as for example in process control systems. The user must accordingly exercise judgment in applying the results of this section.

NOTES

For a historical note on the method of least-squares, weighted least-squares and minimum-variance estimation (sometimes also known as Markoff estimation), the reader is referred to Plackett [6.5], who concludes that it was Gauss who first justified the use of weighted least-squares, and showed that it led to the minimum-variance estimate. Gauss restricted himself to *diagonal* covariance matrices and it was Aitken [6.3] who developed the generalization given on p. 202.

In [6.2], Gauss states the following. "Our principle,‡ which we have made use of since the year 1795, has lately been published by Legendre in the work *Nouvelles methodes pour la determination des orbites des cometes,'* Paris, 1806 . . .".

The fact that [6.2] was published in 1809, i.e. three years after Legendre's work was published, yet claiming to have ante-dated him by eleven years, led to a one-sided feud between them (Gauss simply ignored Legendre). An account of this battle is given by Bell in [6.6], in which Gauss is cited as having replied, simply: "I communicated the whole matter to Olbers in 1802." Apparently Legendre was able to contact Olbers who still had the manuscript at the time the dispute erupted.

†To be discussed later.
‡i.e. weighted least-squares.

EXERCISES

6.1 a) Suppose that the observation vector is $Y_n \equiv (y_0, y_1)_n^T$, where $(y_0)_n$ and $(y_1)_n$ are both observations on $x(t)$ made simultaneously by two separate transducers. Assume that $x(t)$ is a cubic, characterized by $X_n \equiv (x, \dot{x}, \ddot{x}, \dddot{x})_n^T$. Verify that if we wish to write $Y_n = MX_n + N_n$, then we must use

$$M = \begin{pmatrix} 1 & 0 & 0 & 0 \\ 1 & 0 & 0 & 0 \end{pmatrix}$$

b) Let the observation vector be $Y_n \equiv (y_0, y_1)_n^T$ where y_0 and y_1 are observations on the *quadratic* process $x(t)$ and its derivative. We choose as the state-vector $U_n \equiv (x, \nabla x, \nabla^2 x)_n^T$.

Show that if we wish to write $Y_n = MU_n + N_n$, then we must use

$$M = \begin{pmatrix} 1 & 0 & 0 \\ 0 & \frac{1}{\tau} & \frac{1}{2\tau} \end{pmatrix}$$

Hint: See (4.4.8).

6.2 Assume that $Y_n = y_n$, where y_n is an observation on $x(t)$, assumed to be a quadratic, with state-vector $Z_n \equiv \left(x, \tau\dot{x}, \frac{\tau^2}{2}\ddot{x}\right)_n^T$ (τ is the period between samples).

a) Write out the matrix T of (6.3.9) for $L = 10$ and verify that it has rank 3, i.e. full column rank.

b) Show that

$$T^T T = \begin{pmatrix} \sum_{i=0}^{10} i^0 & -\sum_{i=0}^{10} i^1 & \sum_{i=0}^{10} i^2 \\ -\sum_{i=0}^{10} i^1 & \sum_{i=0}^{10} i^2 & -\sum_{i=0}^{10} i^3 \\ \sum_{i=0}^{10} i^2 & -\sum_{i=0}^{10} i^3 & \sum_{i=0}^{10} i^4 \end{pmatrix}$$

c) Using the summation formulae on page 40 and the method of Ex. 5.12, verify that $T^T T$ is positive definite.

6.3 Repeat Ex. 6.2 above for $L = 2$ to obtain the 3×3 matrix T.

a) Solve (by inverting T or otherwise) for the matrix W which satisfies (6.3.16) i.e., $WT = I$. Is W unique in this case and if so why?

b) Verify that the algorithm

$$\begin{pmatrix} z_0^* \\ z_1^* \\ z_2^* \end{pmatrix}_{n,n} = W \begin{pmatrix} y_n \\ y_{n-1} \\ y_{n-2} \end{pmatrix} \qquad \text{(I)}$$

makes $Z_{n,n}^*$ the derivative-vector of (4.5.9) and that (I) is in fact a Lagrange estimator. (Compare (I) to the filter obtained from (4.5.17) setting $h = 0$ and removing $D(\tau)$.)

6.4 Let T be the $k \times (m+1)$ matrix

$$T = \left(T_0 \,\vdots\, T_1 \,\vdots\, \cdots \,\vdots\, T_m\right)$$

where the T_i are k-vectors and $k \geq m + 1$.

a) Show that $T^T T$ is positive definite if and only if the vectors T_i are all linearly independent.

b) Verify that if R is positive definite then $T^T R^{-1} T$ is positive definite, if and only if T has full column rank.

6.5 Let A be a real symmetric matrix. Then, as is well known,† there exists an orthogonal matrix Q, made up of the normalized eigenvectors of A such that

$$Q^T AQ = \Lambda$$

where Λ is a diagonal matrix made up of A's eigenvalues. If A is positive definite then all of its eigenvalues are positive

a) Verify that

$$Q\Lambda^k Q^T = A^k$$

†See e.g. [6.7] or [6.8].

where k is a rational number and hence write the expression for $A^{1/2}$ and $A^{-1/2}$ in terms of Q and Λ.

b) Letting

$$A = \begin{pmatrix} 3 & 1 \\ 1 & 3 \end{pmatrix}$$

find $A^{1/2}$ and $A^{-1/2}$.

6.6 a) Let

$$e(x_0, x_1) \equiv (x_0, x_1) \begin{pmatrix} a_{00} & a_{01} \\ a_{01} & a_{11} \end{pmatrix} \begin{pmatrix} x_0 \\ x_1 \end{pmatrix}$$

Verify that

$$\frac{\partial e}{\partial x_0} = 2(1, 0) \begin{pmatrix} a_{00} & a_{01} \\ a_{01} & a_{11} \end{pmatrix} \begin{pmatrix} x_0 \\ x_1 \end{pmatrix}$$

and that

$$\frac{\partial e}{\partial x_1} = 2(0, 1) \begin{pmatrix} a_{00} & a_{01} \\ a_{01} & a_{11} \end{pmatrix} \begin{pmatrix} x_0 \\ x_1 \end{pmatrix}$$

and hence that

$$\begin{pmatrix} \dfrac{\partial e}{\partial x_0} \\[2ex] \dfrac{\partial e}{\partial x_1} \end{pmatrix} = 2 \begin{pmatrix} a_{00} & a_{01} \\ a_{01} & a_{11} \end{pmatrix} \begin{pmatrix} x_0 \\ x_1 \end{pmatrix}$$

The vector on the left is usually abbreviated as $\partial e/\partial X$. Infer that if $e(x) \equiv X^T AX$, where A is symmetric and X is an $(m + 1)$-vector, then

$$\frac{\partial e}{\partial X} \equiv \begin{pmatrix} \frac{\partial e}{\partial x_0} \\ \frac{\partial e}{\partial x_1} \\ \vdots \\ \frac{\partial e}{\partial x_m} \end{pmatrix} = 2AX$$

b) Let $e(X) \equiv X^T Y$ where X and Y are both $(m+1)$-vectors. Verify that

$$\frac{\partial e}{\partial X} = Y \tag{I}$$

c) Verify that if $e(X) \equiv Y^T X$, then (I) above still holds.

6.7 Using the results of Ex. 6.6, verify that if

$$e\left(X^*_{n,n}\right) \equiv \left(Y_{(n)} - T_n X^*_{n,n}\right)^T \left(Y_{(n)} - T_n X^*_{n,n}\right) \tag{I}$$

then deriving $\partial e/\partial\left(x^*_j\right)_{n,n}$ for each element in $X^*_{n,n}$ gives the vector of equations

$$\frac{\partial e}{\partial X^*_{n,n}} = -2T_n^T\left(Y_{(n)} - T_n X^*_{n,n}\right) \tag{II}$$

Now infer that if T_n has full column rank, then setting $\partial e/\partial X^*_{n,n}$ equal to a null-vector gives

$$\hat{X}^*_{n,n} = \left(T_n^T T_n\right)^{-1} T_n^T Y_{(n)}$$

which is the same as the least-squares estimate. Comment: As yet we have only shown that this $X^*_{n,n}$ is a *stationary point* of the surface $e\left(X^*_{n,n}\right)$ of (I). To show that it is a *minimum* (and not a maximum or a saddle point) we must also show that in its neighborhood

$$e\left(\hat{X}^*_{n,n} + \Delta\right) > e_n\left(\hat{X}^*_{n,n}\right)$$

where Δ is any small vector added to $\hat{X}^*_{n,n}$. This inequality will hold if and only if the matrix of second partials of $e\left(\hat{X}^*_{n,n}\right)$ (i.e. the matrix whose i, j^{th} element is $\partial^2 e/\partial x^*_i \partial x^*_j$), is positive definite in the neighborhood of $\hat{X}^*_{n,n}$. (See e.g. [6.8].) But from (II) above, we see that this matrix is precisely $2T_n^{\,T} T_n$ which, if T_n has full column rank, is positive definite. Thus $\hat{X}^*_{n,n}$ is in fact a *minimum* of $e\left(X^*_{n,n}\right)$ of (I), and so the least-squares estimate can be derived using the differential calculus.

6.8 a) Using the observation scheme implied in Ex. 6.2, show that

$$e_n \equiv \left(Y_{(n)} - TX^*_{n,n}\right)^T \left(Y_{(n)} - TX^*_{n,n}\right)$$

can be written

$$e_n = \sum_{k=0}^{10} \left[y_{n-k} - \left(x^*_0 - kx^*_1 + k^2 x^*_2\right)_{n,n}\right]^2$$

b) Set $\partial e_n / \partial \left(x^*_i\right)_{n,n} = 0$ for $0 \le i \le 2$, to obtain a set of linear equations of the form

$$A\hat{X}^*_{n,n} = B \tag{I}$$

where A is a 3×3 matrix and B is a 3-vector.

c) Verify that

$$A = T^T T$$

$$B = T^T Y_{(n)}$$

By Ex. 6.2, $T^T T$ is positive definite. Infer that (I) above gives

$$\hat{X}^*_{n,n} = (T^T T)^{-1} T^T Y_{(n)}$$

6.9 By direct expansion on (6.5.7), verify that it follows from (6.5.6).

6.10 For the scalar (c/f (6.6.16))

$$L_0 \equiv W_0 R W_0^{\,T} - 2(W_0 T_0 - 1)\lambda_{00} - 2W_0 T_1 \lambda_{10}$$

use the results of Ex. 6.6 to verify that

$$\frac{\partial L_0}{\partial W_0} = 2RW_0{}^T - 2T_0\lambda_{00} - 2T_1\lambda_{10}$$

thereby proving (6.6.17).

6.11 Verify that (6.6.30) and (6.6.31) give (6.6.32), i.e.

$$\mathring{S}^*_{n,n} = \left(T_n{}^T R^{-1}_{(n)} T_n\right)^{-1}$$

6.12 Assume that the state-vector

$$Z_n = \begin{pmatrix} x \\ \tau\dot{x} \end{pmatrix}_n$$

defines a polynomial process sampled at equally spaced instants. Let the observation vector be Y_n consisting of the scalar y_n which is an observation on x_n. Let the covariance matrix of the errors in three successive observations be

$$R_{(n)} = \begin{pmatrix} 2 & -1 & 0 \\ -1 & 2 & -1 \\ 0 & -1 & 2 \end{pmatrix}$$

a) Write the transition matrix for Z_n, the observation matrix M of (6.2.18), and hence the matrix T of (6.3.9) using $L = 2$.

b) By the use of (6.6.30), (6.6.32) and (6.7.22) find $\mathring{W}$, $\mathring{S}^*_{n,n}$ and $\mathring{S}^*_{n+h,n}$.

c) Verify that as we vary the prediction instant h, the diagonal elements of $\mathring{S}^*_{n+h,n}$ are least when $h = -1$, i.e. for retrodiction to the center of the observation interval, $t = n\tau$, $(n-1)\tau$, $(n-2)\tau$.

6.13 a) For the above example, find $\hat{W}$, $\hat{S}^*_{n,n}$ and $\hat{S}^*_{n+h,n}$ by the use of (6.5.11), (6.5.12) and (5.5.22).

b) Verify that the diagonal elements of $\hat{S}^*_{n+h,n}$ are also least when $h = -1$.

c) Compare the covariance matrices $\overset{\circ}{S}{}^*$ of Ex. 6.12 and $\hat{S}^*$ of this example, and verify that *for any value* of the prediction instant h,

$$\left[\overset{\circ}{S}{}^*_{n+h,n}\right]_{i,i} \le \left[\hat{S}^*_{n+h,n}\right]_{i,i} \qquad i = 0, 1$$

Thus the diagonal elements of $\overset{\circ}{S}{}^*_{n+h,n}$ are less than or equal to those of $\hat{S}^*_{n+h,n}$. (In fact the diagonal elements of $\overset{\circ}{S}{}^*_{n+h,n}$ are less than or equal to those of any other covariance matrix $S^*_{n+h,n}$ resulting from an unbiased linear estimation when the errors have the covariance matrix $R_{(n)}$ of Ex. 6.12.)

6.14 Repeat Ex. 6.12 and Ex. 6.13 using $R_{(n)} = \sigma_\nu^2 I$.

6.15 The following example demonstrates how the choice of units affects the least-squares estimate.

A body is in one-dimensional, unaccelerated motion. It is observed on two occasions, one second apart, the measurements being

$$\begin{pmatrix} \text{position (ft)} \\ \text{velocity (ft/sec)} \end{pmatrix}_n = \begin{pmatrix} 2 \\ 2 \end{pmatrix}$$

$$\begin{pmatrix} \text{position (ft)} \\ \text{velocity (ft/sec)} \end{pmatrix}_{n-1} = \begin{pmatrix} 1 \\ 2 \end{pmatrix}$$

(Neglecting observation errors, the position measurements suggest a speed of 1 ft/sec but the velocity measurements suggest 2 ft/sec.)

a) Obtain the least-squares estimate for X based on the above four observations *but with the position measurements rescaled in inches,* leaving the velocity measurements scaled as given in feet per second. Verify that

$$\hat{X}^*_{n,n} = \begin{pmatrix} 1.986 \text{ ft} \\ 0.972 \text{ ft/sec} \end{pmatrix}$$

b) Now repeat *but rescale the velocity measurements in inches/sec* and leave the position measurements scaled in feet. Verify that the least-squares estimate becomes

$$\hat{X}^*_{n,n} = \begin{pmatrix} 2.499 \text{ ft} \\ 1.998 \text{ ft/sec} \end{pmatrix}$$

Thus, rescaling the observations into smaller units has emphasized them much more strongly, and shows that a least-squares estimate is not unique.

6.16 Using the data of the previous example *plus* the following covariance matrix of the observations:

$$R = \begin{pmatrix} 1\ \text{ft}^2 & & & \\ & 1\ \text{ft}^2/\text{sec}^2 & & \\ & & 1\ \text{ft}^2 & \\ & & & 1\ \text{ft}^2/\text{sec}^2 \end{pmatrix}$$

verify that rescaling of the observations has no effect on the minimum-variance estimate.

6.17 By setting $\partial e/\partial X^*_{n,n} = \emptyset$ for $e\left(X^*_{n,n}\right)$ of (6.9.5), show that (6.9.6) results.

6.18 Assume that the covariance matrix $R_{(n)}$ of (6.9.5) is block-diagonal. (The errors are stage-wise uncorrelated.)

a) Using (6.3.4) for $Y_{(n)}$ and (6.3.9) for T_n, verify that for such an $R_{(n)}$, e_n of (6.9.5) can be written

$$e_n = \sum_{k=0}^{L} \left[Y_{n-k} - M\Phi(-k)\,X^*_{n,n}\right]^T R^{-1}_{n-k} \left[Y_{n-k} - M\Phi(-k)\,X^*_{n,n}\right]$$

where M is the observation matrix, $X^*_{n,n}$ is an estimate vector and $\Phi(h)$ is the transition matrix of the process.

b) By setting $\partial e_n/\partial x^*_{n,n} = 0$ for each component of $X^*_{n,n}$, (see Ex. 6.6), verify that this leads to the estimator

$$\mathring{X}^*_{n,n} = \mathring{S}^*_{n,n} \sum_{k=0}^{L} [M\Phi(-k)]^T R^{-1}_{n-k} Y_{n-k}$$

where

$$\mathring{S}^*_{n,n} = \left(\sum_{k=0}^{L} [M\Phi(-k)]^T R^{-1}_{n-k} M\Phi(-k)\right)^{-1}$$

c) Reconcile the above derivation with the minimum-variance filter given in (6.6.29), (6.6.30) and (6.6.32).

6.19 Referring to (6.3.10), i.e.

$$Y_{(n)} = T_n X_n + N_{(n)} \tag{I}$$

we see that if $N_{(n)}$ is zero-mean, then $T_n X_n$ is the mean of $Y_{(n)}$. Suppose that the variables in $N_{(n)}$ are all *normally distributed.* Then their multivariate density function is (see (5.6.3))

$$p(N_{(n)}) = \frac{1}{(2\pi)^{m/2} (\det R)^{1/2}} \exp\left[-\frac{1}{2}\left(N_{(n)}^T R_{(n)}^{-1} N_{(n)}\right)\right]$$

where

$$R_{(n)} \equiv E\left\{N_{(n)} N_{(n)}^T\right\}$$

and m is the order of N. Verify that the random variables $Y_{(n)}$ of (I) have distribution

$$p(Y_{(n)}) = \frac{1}{(2\pi)^{m/2} (\det R)^{1/2}} \exp\left[-\frac{1}{2}(Y_{(n)} - T_n X_n)^T R_{(n)}^{-1} (Y_{(n)} - T_n X_n)\right]$$

Since X_n is in general unknown, we estimate it by a vector $X^*_{n,n}$. The *Principle of Maximum-Likelihood* [6.4] states that the unknown vector $X^*_{n,n}$ shall be chosen so that the value of the density function $p(Y_{(n)})$ is maximized for the particular draw of $Y_{(n)}$. Verify that when $p(Y_{(n)})$ is Gaussian, then this leads to (6.9.5) and hence, via Ex. 6.17 to the minimum-variance filter.

6.20 Verify that (6.10.13) reduces to (6.10.15) when we insert

$$\mathring{W}_n = \left(T_n^T P_{(n)}^{-1} T_n\right)^{-1} T_n^T P_{(n)}^{-1}$$

6.21 Let N be a vector, G a matrix. Prove that

$$N^T G N = \operatorname{Tr}(G N N^T)$$

where Tr ≡ Trace means the sum of the diagonal elements.

6.22 Assuming that AB, BA, and A^TB^T are all defined (A and B are matrices), prove that

$$\mathrm{Tr}(AB) = \mathrm{Tr}(BA) = \mathrm{Tr}(A^TB^T)$$

6.23 Using Ex. 6.21 and 6.22 above, infer that

$$N^TGN = \mathrm{Tr}(G^TNN^T)$$

and hence verify that

$$E\{N^TGN\} = \mathrm{Tr}\left(G^T E\{NN^T\}\right)$$

6.24 Suppose we wish to calibrate an instrument by making a series of uncorrelated measurements on a *constant* quantity. The statistics of the instrument's errors are stationary (i.e. the variances of all observations are equal).

a) Using a least-squares approach set up the observation relation as

$$Y_{(n)} = Tx_n + N_{(n)}$$

where x_n is a scalar and $Y_{(n)}$ and $N_{(n)}$ are k-vectors. Verify that

$$T = (1, 1, \ldots, 1)^T$$

b) Verify that least-squares estimation of x_n gives

$$\hat{x}^*_{n,n} = \frac{1}{k}\sum_{i=1}^{k} y_{n-i+1} \tag{I}$$

c) Using the results of Section 6.10 (see (6.10.24)) to estimate the variance of the instrument errors, verify that

$$\sigma_\nu^2 \approx \frac{1}{k-1}\sum_{i=1}^{k}\left(y_{n-i+1} - \hat{x}^*_{n,n}\right)^2 \tag{II}$$

d) Are (I) and)II) above consistent with (I) and (II) of Ex. 5.1?

REFERENCES

1. Fulks, W., "Advanced Calculus," John Wiley & Sons, 1962, p. 266 et. seq.
2. Gauss, K. F., "Theory of the Motion of the Heavenly Bodies Moving About the Sun in Conic Sections," 1809. English translation, Dover Publications, Inc., N.Y., 1963.
3. Aitken, A. C., "On Least-Squares and Linear Combinations of Observations," Proc. Roy. Soc. Edinb. A, 55, 42-7, 1934.
4. Lindgren, B. W., "Statistical Theory," MacMillan, N.Y., 1962, p. 222 et. seq.
5. Plackett, R. L., "A Historical Note on the Method of Least-Squares," Biometrika, 36, 458-460, 1949.
6. Bell, E. T., "Men of Mathematics," Simon and Schuster, 1937, pp. 259-60 (paperback edition).
7. Hildebrand, F. B., "Methods of Applied Mathematics," Prentice-Hall, 1961.
8. Bellman, R., "Introduction to Matrix Analysis," McGraw-Hill Book Company, 1960, pp. 2 and 3.
9. Mirsky, L., "Introduction to Linear Algebra," Oxford University Press, 1961, p. 159.

See Also

10. Hamilton, W. C., "Statistics in Physical Science," Ronald Press, New York, 1964.

PART 2

FIXED–MEMORY FILTERING

Two essential concepts have been reviewed, namely, the idea of a *model* based on a differential equation, and the idea of *trajectory selection,* based on the application of least-squares or minimum-variance to a sequence of observations. We are now in a position to commence our discussion of the main topic.

This, the second part of the book,† is devoted to one aspect of that problem and covers what we term the Fixed-Memory Filters. Their mode of operation is, briefly, as follows.

Observations are being repeatedly taken on a process and are streaming into a computer where they are received and stored upon arrival. They are placed into, what is called, a *push-down table.* This is a memory storage area in which the most recent observations are entered at the top, while all of their predecessors are moved down to make room for them. Each time new observations arrive, the procedure is repeated, all previous observations being moved down the table and the most recent ones again being entered at the top. In this way each observation occupies a place in the table in the same *spatial* sequence as the *time* sequence of arrival.

The push-down table is of *fixed length,* and so each observation eventually reaches the bottom of the table. Upon the next receipt of new data, the bottom-most entry is then simply discarded or forgotten.

†Chapters 7 and 8.

The algorithms to be developed in the next two chapters base their estimates on the current entries in the push-down table discussed above. This means that a record of *fixed* length is being operated on to produce an estimate, and for this reason we refer to these algorithms as Fixed-Memory smoothing and prediction schemes. They will be contrasted, in the remaining two parts of the book, with the Expanding-Memory and the Fading-Memory approaches.

As in the preceding chapters we continue to provide examples at the end of each chapter, but by necessity these now become progressively less theoretical and more applied. For the serious reader, however, there is no substitute for actually running the algorithms on a computer, and every effort should be made to program as many as possible of the filters and to feed them with artificially generated data. Only then will the true subtleties of the various schemes become apparent.

7

THE FIXED-MEMORY POLYNOMIAL FILTER

7.1 INTRODUCTION

The first of the filtering schemes to be considered is chosen for its conceptual simplicity and is entitled the *Fixed-Memory Polynomial Filter,* the reason for this name soon to become apparent.

We assume that a process is under observation and that, *at equally spaced instants of time,* a single aspect of that process is being observed. This gives rise to a sequence of *scalar* observations and we retain the most recent $L + 1$ of them in a push-down table, calling them

$$y_{n-L},\ y_{n-L+1},\ \ldots,\ y_{n-1},\ y_n$$

The purpose of the Fixed-Memory Polynomial Filter is to fit a polynomial to those $L + 1$ numbers, *in the sense of least-squares.* Thereafter the resulting polynomial is taken to be an estimate of the process which is giving rise to this sequence of observations. If derivatives of the process are needed, then the derivatives of the estimating polynomial are used, and if prediction is called for then it is based entirely on the selected polynomial.

As an example, suppose that an object in flight is being observed by the use of a radar, pulses being transmitted at equally spaced instants. Three observations are obtained on the basis of each pulse, these being the object's range (ρ), azimuth (ψ) and elevation (θ) (see p. 168). This gives rise to three sequences of numbers

$$\ldots \rho_{n-2}, \rho_{n-1}, \rho_n, \ldots$$

$$\ldots \psi_{n-2}, \psi_{n-1}, \psi_n, \ldots$$

$$\ldots \theta_{n-2}, \theta_{n-1}, \theta_n, \ldots$$

Three Fixed-Memory Polynomial Filters can now be used to fit polynomials to *each* of these sequences. On the basis of those polynomials we can, if we wish, estimate $(d/dt)\rho(t)$, $(d/dt)\psi(t)$, $(d/dt)\theta(t)$ etc., or we can estimate what future values of ρ, ψ, and θ will be. *The three filters are completely unconnected, and operate entirely without any reference to each other.*

There are two essential ideas underlying the Fixed-Memory Polynomial Filter. *First,* by using a polynomial, we are not required to know much about the true process, since if the polynomial is of adequate degree, it will automatically seek out the signal in the presence of the observational errors and give us a reasonable estimate of it.

Second, by using least-squares, we are not forcing the polynomial to equal any of the observations precisely (as was the case with the Lagrange interpolation scheme of Section 4.5). As a result, the polynomial positions itself in relation to those observations without necessarily passing through any of them, and as we shall see this results in a certain amount of smoothing.

A considerable amount of attention will be given to analyzing the statistical properties of the random errors in the estimate, caused by the random observation errors. We will also consider a second type of error in the estimate, known as the *bias errors* or *systematic errors.* These arise when the true process is not adequately approximated by the polynomial model we have chosen.

The case where the observations are not equally spaced, or where the model is other than a polynomial, will be discussed in Chapter 8. For the present we restrict ourselves to *equally spaced, scalar observations and to polynomial models.*

7.2 CLASSICAL LEAST-SQUARES

Suppose that we are making scalar observations on a process every τ seconds. Let the most recent observation be y_n and its predecessors $y_{n-1}, y_{n-2} \ldots$ etc. In Figure 7.1 we depict the situation, showing the time-axis, quantized by the index n every τ seconds. The r-axis is now drawn, r increasing positively with time, with its origin located at $t = (n - L)\tau$, i.e. $L\tau$ seconds before the most recent observation, (where L is a chosen number whose significance will soon become apparent).

We choose, as our model, a polynomial in r, designated $\left[p^*(r)\right]_n$, where the star signifies *estimation,* and the subscript n shows that this is the polynomial based on the total observation vector

$$Y_{(n)} \equiv (y_n, y_{n-1}, \ldots, y_{n-L})^T \tag{7.2.1}$$

in which y_n is the most recent observation. When a further τ seconds elapse and y_{n+1} is observed, we shall recompute p^* using the vector

$$Y_{(n+1)} \equiv (y_{n+1}, y_n, \ldots, y_{n-L+1})^T \tag{7.2.2}$$

The new polynomial will be designated $\left[p^*(r)\right]_{n+1}$.†

Assume for the moment that $\left[p^*(r)\right]_n$ has been obtained. Then by assigning various values to r, we are able to use it to estimate the process at various points. As an example, setting $r = L$ (see Figure 7.1) gives us the *updated estimate* $\left[p^*(L)\right]_n$. Setting $r = L + 1$ gives the 1-step prediction, i.e., $\left[p^*(L+1)\right]_n$.

We now address ourselves to the selection of p^*. For definiteness, let it be of first degree, and write it as

$$\left[p^*(r)\right]_n = \left(\beta_0\right)_n + \left(\beta_1\right)_n r \tag{7.2.3}$$

where the β's are as yet arbitrary. (Note that they are subscripted with n for the same reason that p^* is subscripted with n.)‡

At this stage we make the decision that the β's shall be chosen by the *least-squares* criterion. Thus the vector of residuals is (c/f (6.5.4)):

†For $\left[p^*(r)\right]_n$ we read "polynomial in r at time n."
‡For $\left(\beta_j\right)_n$ we read "β_j at time n."

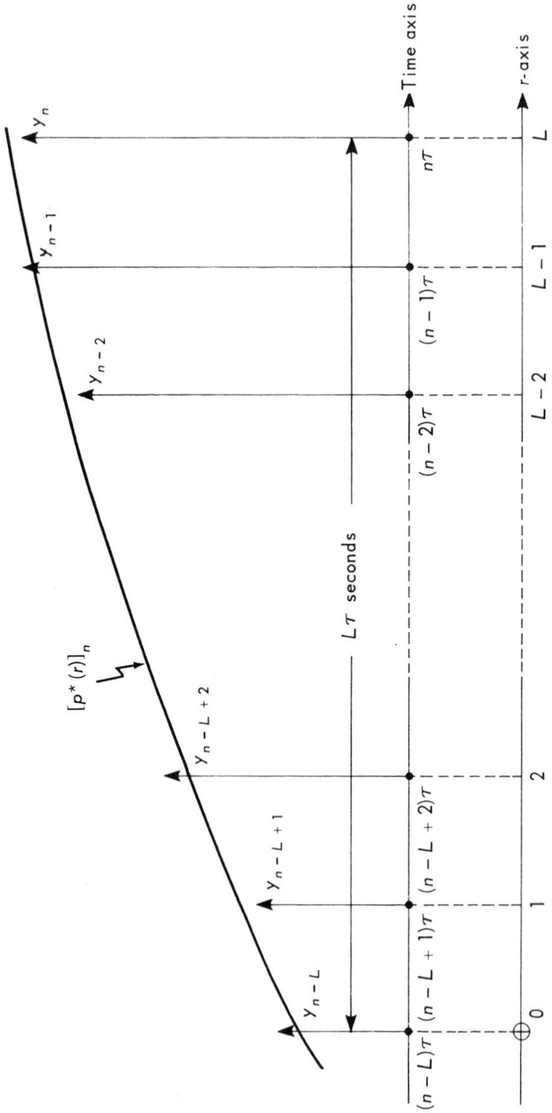

Fig. 7.1 *Definition of the r-axis.*

$$E_{(n)} \equiv \begin{pmatrix} y_n - \left[p^*(L)\right]_n \\ y_{n-1} - \left[p^*(L-1)\right]_n \\ \vdots \\ y_{n-L} - \left[p^*(0)\right]_n \end{pmatrix} \tag{7.2.4}$$

and so the sum of the squared residuals becomes

$$e_n = \sum_{r=0}^{L} \left\{ y_{n-L+r} - \left[p^*(r)\right]_n \right\}^2 \tag{7.2.5}$$

Combining this with (7.2.3) then gives us

$$e_n = \sum_{r=0}^{L} \left[y_{n-L+r} - \sum_{j=0}^{1} \left(\beta_j\right)_n r^j \right]^2 \tag{7.2.6}$$

which is seen to be a scalar function of the β's.

Following the least-squares approach, we now minimize e_n by setting $\partial e_n / \partial \left(\beta_i\right)_n = 0$ for $i = 0, 1$. This results in

$$\sum_{j=0}^{1} \left(\beta_j\right)_n \left(\sum_{r=0}^{L} r^{i+j} \right) = \sum_{r=0}^{L} r^i y_{n-L+r} \qquad i = 0, 1 \tag{7.2.7}$$

which is equivalent to the system of linear algebraic equations

$$\begin{pmatrix} \sum_{r=0}^{L} 1 & \sum_{r=0}^{L} r \\ \sum_{r=0}^{L} r & \sum_{r=0}^{L} r^2 \end{pmatrix} \begin{pmatrix} \beta_0 \\ \beta_1 \end{pmatrix}_n = \begin{pmatrix} \sum_{r=0}^{L} y_{n-L+r} \\ \sum_{r=0}^{L} r y_{n-L+r} \end{pmatrix} \tag{7.2.8}$$

By the use of the formulae on p. 40 the above now reduces to

$$\begin{pmatrix} L+1 & \frac{L}{2}(L+1) \\ \\ \frac{L}{2}(L+1) & \frac{L}{2}(L+1)(2L+1) \end{pmatrix} \begin{pmatrix} \beta_0 \\ \\ \beta_1 \end{pmatrix}_n = \begin{pmatrix} \sum_{r=0}^{L} y_{n-L+r} \\ \\ \sum_{r=0}^{L} r y_{n-L+r} \end{pmatrix} \tag{7.2.9}$$

and since we wish to know the values of the β's in terms of the observations, it is necessary that the matrix on the left be inverted.

While this is a trivial task, *if L is assigned a definite numerical value,* it is by no means so easy *if left in functional form.* For the 2 x 2 case the inverse in *functional form* would be

$$\frac{\begin{pmatrix} 2L(2L+1) & -6L \\ -6L & 12 \end{pmatrix}}{(L+2)^{(3)}}$$

but obtaining the *functional* form of the inverse of anything much beyond a 2 x 2 is obviously out of the question. Although *numerical* inverses would give correct answers, resorting to numbers would, at this stage, force the analysis to terminate, and the analysis has as yet hardly begun.

It is this impasse which makes us abandon the present approach. In the next section we show that *if* $\left[p^*(r)\right]_n$ *is written as a linear combination of the discrete Legendre polynomials,* then the matrix which comes about is the identity matrix and so no inversion problem arises.

7.3 THE ORTHOGONAL POLYNOMIAL APPROACH

In Section 3.2 we derived the polynomials which satisfy the orthogonality condition

$$\sum_{r=0}^{L} p(r;i,L)\,p(r;j,L) = 0 \qquad i \neq j \tag{7.3.1}$$

The polynomials $p(r;j,L)$ are given in (3.2.20) as

$$p(r;j,L) = \sum_{\nu=0}^{j} (-1)^{\nu} \binom{j}{\nu} \binom{j+\nu}{\nu} \frac{r^{(\nu)}}{L^{(\nu)}} \tag{7.3.2}$$

Also, the quantity $c(j,L)$, defined by

$$[c(j,L)]^2 = \sum_{r=0}^{L} [p(r;j,L)]^2 \tag{7.3.3}$$

is given in (3.2.30) as

$$[c(j,L)]^2 = \frac{(L+j+1)^{(j+1)}}{(2j+1)L^{(j)}} \tag{7.3.4}$$

For the remainder of this chapter we abbreviate as follows:

$$\begin{aligned} p_j(r) &\equiv p(r;j,L) \\ c_j &\equiv c(j,L) \end{aligned} \tag{7.3.5}$$

the L being implicit throughout.

We now define the *normalized* discrete Legendre polynomial of degree j by

$$\varphi_j(r) \equiv \frac{1}{c_j} p_j(r) \tag{7.3.6}$$

Then it is easily verified that

$$\sum_{r=0}^{L} \varphi_i(r)\varphi_j(r) = \delta_{ij} \tag{7.3.7}$$

where δ_{ij} is the Kronecker delta. At this stage we return to the derivation of the estimating polynomial $\left[p^*(r)\right]_n$.

In the preceding section we started with $\left[p^*(r)\right]_n$ as a power series in r (see (7.2.3)). This led directly to the impasse arising out of the difficulty of obtaining functional matrix inverses. Instead, suppose we write p^* as a *linear combination of the discrete Legendre polynomials.* Thus, for a polynomial of degree m,† we write

†The letter m will be reserved exclusively for the *degree* of the estimating polynomial.

$$\left[p^*(r)\right]_n = \sum_{j=0}^{m} \left(\beta_j\right)_n \varphi_j(r) \tag{7.3.8}$$

where the β's are as yet arbitrary.

Once again we set up the least-squares error functional [c/f (7.2.5)],

$$e_n = \sum_{r=0}^{L} \left\{ y_{n-L+r} - \left[p^*(r)\right]_n \right\}^2 \tag{7.3.9}$$

and then, using (7.3.8), this becomes

$$e_n = \sum_{r=0}^{L} \left[y_{n-L+r} - \sum_{j=0}^{m} \left(\beta_j\right)_n \varphi_j(r) \right]^2 \tag{7.3.10}$$

which is a scalar function of the β's.

We minimize e_n by setting $\partial e_n/\partial\beta_i = 0$, for $i = 0, 1, \ldots, m$, which now results in the $m + 1$ equations,

$$\sum_{k=0}^{L} \sum_{j=0}^{m} \left(\beta_j\right)_n \varphi_i(k) \varphi_j(k) = \sum_{k=0}^{L} y_{n-L+k} \varphi_i(k) \qquad i = 0, 1, \ldots, m \tag{7.3.11}$$

where r has been replaced by k for convenience. Interchanging the order of summation on the left, we obtain

$$\sum_{j=0}^{m} \left(\beta_j\right)_n \sum_{k=0}^{L} \varphi_i(k) \varphi_j(k) = \sum_{k=0}^{L} y_{n-L+k} \varphi_i(k) \tag{7.3.12}$$

to which we apply (7.3.7), giving

$$\left(\beta_j\right)_n = \sum_{k=0}^{L} y_{n-L+k} \varphi_j(k) \qquad j = 0, 1, \ldots, m \tag{7.3.13}$$

Then by (7.3.8)

$$\left[p^*(r)\right]_n = \sum_{j=0}^{m} \left[\sum_{k=0}^{L} y_{n-L+k}\varphi_j(k) \right] \varphi_j(r) \tag{7.3.14}$$

This is the required expression for the polynomial $\left[p^*(r)\right]_n$, which best fits the data vector $Y_{(n)}$ of (7.2.1) in the sense of least-squares. The orthogonality property of the polynomials $\varphi_j(r)$ has thus enabled us to overcome completely the matrix inversion impasse of the previous section. No matrix inversion problems arose this time, because, as we see by comparing (7.3.12) with (7.2.7), the matrix is now the *identity matrix* rather than the one shown in (7.2.9).

We obtain the *time-derivatives* of the estimating polynomial form (7.3.14) as follows. From Figure 7.1, we see that, *to within a constant,* $t = r\tau$, and so $dt = \tau dr$, i.e.

$$\frac{d}{dt} = \frac{1}{\tau}\frac{d}{dr} \tag{7.3.15}$$

Applying this to (7.3.14) gives us

$$\left[\frac{d}{dt}p^*(r)\right]_n = \frac{1}{\tau} \sum_{j=0}^{m} \left[\sum_{k=0}^{L} y_{n-L+k}\varphi_j(k) \right] \frac{d}{dr}\varphi_j(r) \tag{7.3.16}$$

and in general (letting $D \equiv d/dt$) we have, for the ith time-derivative of $\left[p^*(r)\right]_n$:

$$\left[D^i p^*(r)\right]_n = \frac{1}{\tau^i} \sum_{j=0}^{m} \left[\sum_{k=0}^{L} y_{n-L+k}\varphi_j(k) \right] \frac{d^i}{dr^i}\varphi_j(r) \tag{7.3.17}$$

Just as we did in the case of the Lagrange interpolator, *we have decided that* $\left[p^*(r)\right]_n$ *shall be used as an estimate of the true process.* This estimation is good or bad, depending upon how well the process can be approximated by a polynomial of degree m, and depending on how badly corrupted were the observations on which $\left[p^*(r)\right]_n$ is based. However, for better or for worse, henceforth $\left[p^*(r)\right]_n$ will be used as the estimate. Moreover, just as the process is estimated by $\left[p^*(r)\right]_n$, so its time-derivatives will be estimated by the use of the time-derivatives of $\left[p^*(r)\right]_n$.

By varying r, we can estimate the process at various instants in the future or the past, based on observations up to the present. Thus $\left[p^*(r)\right]_n$ is the estimate of what the process is at time $t = (n - L + r)\tau$, based on observations up to $t = n\tau$. Setting $r = L$, say, gives the updated estimate, and the 1-step prediction of the ith derivative of the process is obtained from (7.3.17) by setting $r = L + 1$.

At this stage we pause for a small example. Let $m = 1$ and $L = 4$. Then by (7.3.6), (3.2.21) and (3.2.31),

$$\varphi_0(r) = \frac{1}{\sqrt{L+1}} = \frac{1}{\sqrt{5}} \tag{7.3.18}$$

and

$$\begin{aligned} \varphi_1(r) &= \sqrt{\frac{3L}{(L+1)(L+2)}}\left(1 - 2\frac{r}{L}\right) \\ &= \sqrt{\frac{2}{5}}\left(1 - \frac{r}{2}\right) \end{aligned} \tag{7.3.19}$$

We rearrange (7.3.14) as

$$\left[p^*(r)\right]_n = \sum_{k=0}^{L}\left[\sum_{j=0}^{m} \varphi_j(k)\,\varphi_j(r)\right] y_{n-L+k} \tag{7.3.20}$$

and likewise (7.3.16) is written

$$\tau\left[\frac{d}{dt}\,p^*(r)\right]_n = \sum_{k=0}^{L}\left[\sum_{j=0}^{m} \varphi_j(k)\,\frac{d}{dr}\,\varphi_j(r)\right] y_{n-L+k} \tag{7.3.21}$$

in which the factor τ has been moved to the left-hand side. Then using (7.3.18) and (7.3.19)

$$\begin{pmatrix} p^*(r) \\ \\ \tau D p^*(r) \end{pmatrix}_n = \begin{pmatrix} \sum_{k=0}^{4} \left[\frac{1}{5} + \frac{2}{5}\left(1 - \frac{k}{2}\right)\left(1 - \frac{r}{2}\right) \right] y_{n-L+k} \\ \\ \sum_{k=0}^{4} \left[-\frac{1}{5}\left(1 - \frac{k}{2}\right) \right] y_{n-L+k} \end{pmatrix} \tag{7.3.22}$$

If, as an example, we wish to obtain the 1-*step prediction* of the process, then we must set $r = L + 1$. For the above case, where $L = 4$, this results in

$$\begin{pmatrix} p^*(5) \\ \tau D p^*(5) \end{pmatrix}_n = \frac{1}{10} \begin{pmatrix} 8 & 5 & 2 & -1 & -4 \\ 2 & 1 & 0 & -1 & -2 \end{pmatrix} \begin{pmatrix} y_n \\ y_{n-1} \\ y_{n-2} \\ y_{n-3} \\ y_{n-4} \end{pmatrix} \tag{7.3.23}$$

We now define the vector $Z^*_{n+1,n}$ by

$$\begin{pmatrix} z_0^* \\ z_1^* \end{pmatrix}_{n+1,n} \equiv \begin{pmatrix} p^*(5) \\ \tau D p^*(5) \end{pmatrix}_n \tag{7.3.24}$$

(Note that z_1^* is τ times the derivative of p^* and so it is a *scaled* version of the first derivative of the latter.) Also, the total observation vector is designated as

$$Y_{(n)} \equiv \begin{pmatrix} y_n \\ y_{n-1} \\ y_{n-2} \\ y_{n-3} \\ y_{n-4} \end{pmatrix} \tag{7.3.25}$$

Then (7.3.23) can be written in matrix form as

$$Z^*_{n+1,n} = W\,Y_{(n)} \tag{7.3.26}$$

The 0-step prediction is obtained from (7.3.22), by setting $r = 4$. This gives

$$Z^*_{n,n} = \frac{1}{10}\begin{pmatrix} 6 & 4 & 2 & 0 & -2 \\ 2 & 1 & 0 & -1 & -2 \end{pmatrix} Y_{(n)} \tag{7.3.27}$$

Alternatively, it can be derived from (7.3.23), using (4.2.16), i.e.,

$$Z^*_{n,n} = \Phi(-1)\,W Y_{(n)} \tag{7.3.28}$$

where in this case, by (4.2.15)

$$\Phi(-1) = \begin{pmatrix} 1 & -1 \\ 0 & 1 \end{pmatrix} \tag{7.3.29}$$

The actual algorithms for the 1-step position and scaled-velocity predictors, based on $L = 4$, would be respectively, [c/f (7.3.23)]

$$\left(z^*_0\right)_{n+1,n} = \frac{1}{10}\left(8y_n + 5y_{n-1} + 2y_{n-2} - y_{n-3} - 4y_{n-4}\right) \tag{7.3.30}$$

$$\left(z^*_1\right)_{n+1,n} = \frac{1}{10}\left(2y_n + y_{n-1} - y_{n-3} - 2y_{n-4}\right) \tag{7.3.31}$$

As each new observation y_n arrives, we simply "push down" the previous ones, discarding y_{n-5} and computing the latest 1-step predictions using the above. This illustrates the fixed-length nature of the memory. The filter stores in memory, and uses only the most recent five observations, and any which are older or staler than 5τ seconds are completely eradicated or "forgotten." The "memory" of the Fixed-Memory Polynomial Filter has the graphic form depicted in Figure 7.2 where all observations from the most recent, y_n, to its Lth predecessor, y_{n-L}, are weighted by unity and the remainder by zero.

We recall that $\left(z^*_1\right)_{n+1,n}$ of (7.3.24) is $\left[\tau(d/dt)\,p^*(L+1)\right]_n$ and so it is an estimate of τ-times the first derivative of the process. To obtain an estimate

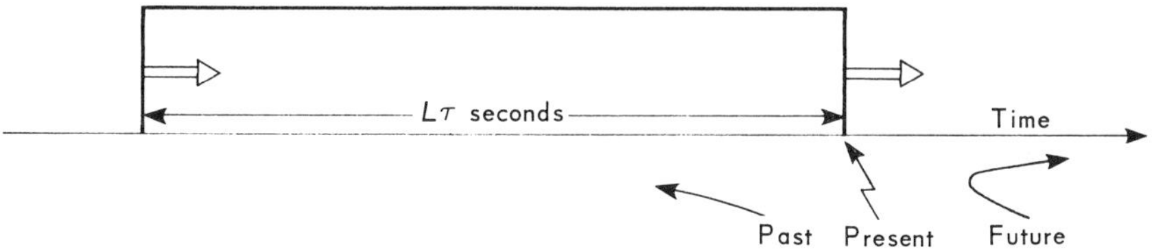

Fig. 7.2 *Memory window of fixed-memory filter.*

of the derivative itself, (7.3.31) must be divided by τ. If y_n is in feet, then the units of (7.3.31) are feet. If however we divide it by τ, and τ is in seconds, the result will then be velocity in ft/sec. Thus, from (7.3.30) and (7.3.31) we obtain the algorithms

$$(x^*)_{n+1,n} = \frac{1}{10}(8y_n + 5y_{n-1} + 2y_{n-2} - y_{n-3} - 4y_{n-4}) \tag{7.3.32}$$

$$(\dot{x}^*)_{n+1,n} = \frac{1}{10\tau}(2y_n + y_{n-1} - y_{n-3} - 2y_{n-4}) \tag{7.3.33}$$

We can think of the above algorithms in the form shown in Figure 7.3 where the observations are streaming in, and where two outputs are available, namely x^* and $\dot{x}^*$. We call these the *position* and *velocity* outputs respectively, and of course, if only one is required, we simply operate only that channel. ◆◆

7.4 THE GENERAL FORM

In the preceding section we obtained the filter matrices for the 1-step and 0-step, first-degree estimators, based on fixed-memory least-squares. We

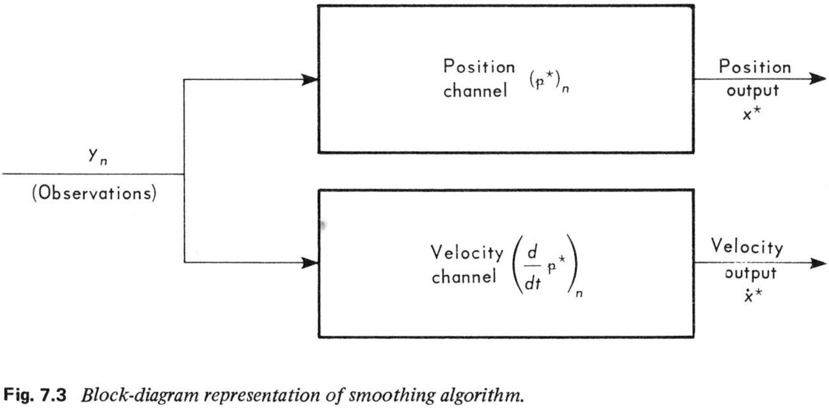

Fig. 7.3 *Block-diagram representation of smoothing algorithm.*

now develop a *general expression* which gives the weight matrix W for any values of L, m and prediction interval h. For ease of notation, we continue to use the abbreviations given in (7.3.5).

Returning to (7.3.17), we see that it can be written

$$\left[\frac{\tau^i}{i!} D^i p^*(r)\right]_n = \sum_{j=0}^{m} \frac{1}{i!} \frac{d^i}{dr^i} p_j(r) \sum_{k=0}^{L} \frac{p_j(k)}{c_j^2} y_{n-L+k} \tag{7.4.1}$$

Assume that $m = 1$ (i.e. p^* is first degree) and set $r = L$ (i.e. we are considering an updated estimation). Then (7.4.1) is equivalent to the matrix equation

$$\begin{pmatrix} p^*(L) \\ \\ \tau \dfrac{d}{dt} p^*(L) \end{pmatrix}_n = \begin{pmatrix} p_0(L) & p_1(L) \\ \\ 0 & \dfrac{d}{dr} p_1(L) \end{pmatrix} \begin{pmatrix} \displaystyle\sum_{k=0}^{L} \frac{p_0(k)}{c_0^2} y_{n-L+k} \\ \\ \displaystyle\sum_{k=0}^{L} \frac{p_1(k)}{c_1^2} y_{n-L+k} \end{pmatrix} \tag{7.4.2}$$

where $(d/dr)\, p_1(L)$ means $(d/dr)\, p_1(r)\big|_{r=L}$.

We now observe that the vector on the extreme right of (7.4.2) can be factored as follows:

$$\begin{pmatrix} \displaystyle\sum_{k=0}^{L} \frac{p_0(k)}{c_0^2} y_{n-L+k} \\ \\ \displaystyle\sum_{k=0}^{L} \frac{p_1(k)}{c_1^2} y_{n-L+k} \end{pmatrix} = \begin{pmatrix} \dfrac{1}{c_0^2} & 0 \\ \\ 0 & \dfrac{1}{c_1^2} \end{pmatrix} \begin{pmatrix} p_0(L),\ p_0(L-1),\ \ldots,\ p_0(0) \\ \\ p_1(L),\ p_1(L-1),\ \ldots,\ p_1(0) \end{pmatrix} \begin{pmatrix} y_n \\ y_{n-1} \\ \vdots \\ y_{n-L} \end{pmatrix} \tag{7.4.3}$$

We designate the first matrix on the right of (7.4.2) as P, which, for an estimator of degree m, has as its i, j^{th} element

$$[P]_{ij} \equiv \frac{1}{i!} \frac{d^i}{dr^i} p_j(r) \Bigg|_{r=L} \qquad 0 \le i, j \le m \tag{7.4.4}$$

Also, the first matrix on the right of (7.4.3), will be called C. It is diagonal with i, j^{th} element

$$[C]_{ij} \equiv \frac{1}{c_j^2}\delta_{ij} \qquad 0 \le i, j \le m \tag{7.4.5}$$

The second matrix on the right of (7.4.3) will be called B. The i, j^{th} element of B is

$$[B]_{ij} \equiv p_i(r)\Big|_{r = L - j} \qquad \left\{\begin{matrix} 0 \le i \le m \\ 0 \le j \le L \end{matrix}\right\} \tag{7.4.6}$$

Note that P and C above are square $(m + 1) \times (m + 1)$ matrices, whereas B is rectangular, having $m + 1$ rows and $L + 1$ columns. Finally we define the state-vectors $Z^*_{n+h,n}$ and $X^*_{n+h,n}$ by

$$Z^*_{n+h,n} \equiv \begin{pmatrix} p^*(L + h) \\ \tau Dp^*(L + h) \\ \dfrac{\tau^2}{2!} D^2 p^*(L + h) \\ \vdots \\ \dfrac{\tau^m}{m!} D^m p^*(L + h) \end{pmatrix}_n \qquad X^*_{n+h,n} \equiv \begin{pmatrix} p^*(L + h) \\ Dp^*(L + h) \\ \cdot \\ \cdot \\ \cdot \\ D^m p^*(L + h) \end{pmatrix}_n \tag{7.4.7}$$

We shall refer to $X^*_{n+h,n}$ as the *unscaled* estimate vector and to $Z^*_{n+h,n}$ as the *scaled* estimate vector.

Combining all of the above, we see that for $r = L$, (7.4.1) is equivalent to the matrix equation

$$Z^*_{n,n} = PCB\,Y_{(n)} \tag{7.4.8}$$

It thus follows immediately that we can write

$$Z^*_{n+h,n} = W(h)\,Y_{(n)} \tag{7.4.9}$$

where

$$W(h) \equiv \Phi(h)PCB \tag{7.4.10}$$

and where $\Phi(h)$ is defined on p. 76 by (4.2.15). Equation (7.4.10) is the general form of the weight matrix for obtaining the scaled estimate vector $Z^*_{n+h,n}$, based on a total observation vector $Y_{(n)}$.

Finally we obtain $X^*_{n+h,n}$, the vector of unscaled derivatives, by the use of the matrix $D(\tau)$ also defined on p. 76. Thus we have

$$X^*_{n+h,n} = W(h,\tau)\,Y_{(n)} \tag{7.4.11}$$

where

$$W(h,\tau) \equiv D(\tau)\Phi(h)PCB \tag{7.4.12}$$

This is the general form of the weight-matrix for estimating the unscaled derivatives of the process for any values of τ, h, m and L.

There remains one small problem, namely the form of the i, j^{th} term of the matrix P of (7.4.4). By its definition this is

$$\frac{1}{i!}\frac{d^i}{dr^i}p_j(r)\Bigg|_{r=L}$$

However, a small amount of exploration on (3.2.21), shows that the discrete Legendre polynomials do *not* easily permit of differentiation. Any attempt to differentiate them completely fractures their structure, and all order is thereafter irretrievably lost. There are two solutions to this problem.

In Appendix I we present a recursion formula for the derivatives of $p_j(r)$, and that can be used, in conjunction with a computer, to fill in the numbers of the matrix P, once L is chosen.

An alternate, and as we shall see in subsequent developments, very useful approach, is as follows. It is also shown in Appendix I that

$$\frac{1}{i!}\nabla^i p_j(r)\Bigg|_{r=L} = (-1)^j\binom{j}{i}\binom{j+i}{i}\frac{1}{L^{(i)}} \tag{7.4.13}$$

This is indeed a compact and easily computed form. We now define the matrix G, whose i, j^{th} term is precisely (7.4.13), i.e.

$$[G]_{ij} \equiv (-1)^i \binom{j}{i} \binom{j+i}{i} \frac{1}{L^{(i)}} \tag{7.4.14}$$

The jth column of G is seen (from the left-hand side of (7.4.13)) to be the vector

$$\begin{pmatrix} p_j(L) \\ \nabla p_j(L) \\ \frac{1}{2!} \nabla^2 p_j(L) \\ \vdots \\ \frac{1}{j!} \nabla^j p_j(L) \end{pmatrix}$$

i.e. a polynomial of degree j and its scaled backward differences of successively higher order, all evaluated at $r = L$. Comparison with (4.4.6) shows that this is of the same form as V_n of that equation.

On the other hand, the jth column of the matrix P [see (7.4.4)], is of the form of Z_n in (4.4.6) with τ of that equation set equal to unity. We can thus obtain P from G by the use of (4.4.8) i.e.,

$$P = SG \tag{7.4.15}$$

S being the associate Stirling matrix of the first kind, given on p. 85. We have thus shown that the matrix $W(h)$ of (7.4.10) can be written in the alternate form

$$W(h) = \Phi(h)\, SGCB \tag{7.4.16}$$

As an example, we check (7.3.23) by the use of (7.4.16). Since (7.3.23) was a 1-step predictor, we must use $h = 1$. Also (7.3.23) was based on $L = 4$ and $m = 1$. Thus (7.4.16) gives

$$W(1) = \Phi(1)\,SGCB$$

$$= \begin{pmatrix} 1 & 1 \\ 0 & 1 \end{pmatrix}\begin{pmatrix} 1 & 0 \\ 0 & 1 \end{pmatrix}\begin{pmatrix} 1 & -1 \\ 0 & -\frac{1}{2} \end{pmatrix}\begin{pmatrix} \frac{1}{5} & 0 \\ 0 & \frac{2}{5} \end{pmatrix}\begin{pmatrix} 1 & 1 & 1 & 1 & 1 \\ -1 & -\frac{1}{2} & 0 & \frac{1}{2} & 1 \end{pmatrix}$$

$$= \frac{1}{10}\begin{pmatrix} 8 & 5 & 2 & -1 & -4 \\ 2 & 1 & 0 & -1 & -2 \end{pmatrix} \tag{7.4.17}$$

which of course agrees precisely with (7.3.23). ◆◆

Equation (7.4.16) can be used in conjunction with a computer to generate the weight-matrices for the Fixed-Memory Polynomial Filter, for any degree and any value of h and L. If unscaled derivatives are required, we pre-multiply $W(h)$ by $D(\tau)$ as shown in (7.4.12).

For convenience, the matrix G is given in Table 7.1 up to $i, j = 10$, although in any computer program which computes W from (7.4.16) the general term of each of the matrices should be programmed. The general form for S is given as a recursion in the notes following Chapter 4. (See p. 116.)

7.5 VARIANCE REDUCTION

In the previous section we derived the general form of the weight-matrices for obtaining the scaled and unscaled estimate-vectors, $Z^*_{n+h,n}$ and $X^*_{n+h,n}$, from the vector of observations $\mathsf{Y}_{(n)}$. We now examine the statistical properties of the random errors in those estimates.

The observational errors in $\mathsf{Y}_{(n)}$ are denoted by the vector $\mathsf{N}_{(n)}$, and so $\mathsf{Y}_{(n)}$ is related to the model state-vector, X_n, by, an equation of the form

$$\mathsf{Y}_{(n)} = \mathsf{T}X_n + \mathsf{N}_{(n)} \tag{7.5.1}$$

This was discussed in Section 6.3 where T was first defined. We now use (7.5.1) in the smoothing and prediction algorithm (7.4.9), thereby obtaining

$$Z^*_{n+h,n} = W(h)\,\mathsf{T}X_n + W(h)\,\mathsf{N}_{(n)} \tag{7.5.2}$$

Thus the vector $\mathsf{N}_{(n)}$ gives rise to a vector of random errors in the estimate defined by

$$\mathsf{N}^*_{n+h,n} \equiv W(h)\,\mathsf{N}_{(n)} \tag{7.5.3}$$

Table 7.1 Numerical Coefficients of G Matrix

$i \backslash j$	0	1	2	3	4	5	6	7	8	9	10
0	1	−1	1	−1	1	−1	1	−1	1	−1	1
1		−2	6	−12	20	−30	42	−56	72	−90	110
2			6	−30	90	−210	420	−756	1260	−1980	2970
3				−20	140	−560	1680	−4200	9240	−18480	34320
4					70	−630	3150	−11550	34650	−90090	210210
5						−252	2772	−16632	72072	−252252	756756
6							924	−12012	84084	−420420	1681680
7								−3432	51480	−411840	2333760
8									12870	−218790	1969110
9										−48620	923780
10											184756

Note 1. The G matrix is defined on p. 239 by $[G]_{ij} = (-1)^j \binom{j}{i}\binom{j+i}{i}\frac{1}{L^{(i)}}$. Displayed above is $(-1)^j \binom{j}{i}\binom{j+i}{i}$. Ex: $[G]_{3,5} = \frac{-560}{L^{(3)}}$, $[G]_{7,8} = \frac{51480}{L^{(7)}}$.

Note 2. With appropriate adjustment of signs, the above matrix also gives the coefficients of the discrete Legendre polynomials (c/f p. 61).

If $N_{(n)}$ is a vector of zero-mean random variables then (7.5.3) shows that $N^*_{n+h,n}$ is also zero-mean. Assuming this to be the case, the covariance matrix of the latter is given by

$$S^*_{n+h,n} = W(h)\,R_{(n)}\,W(h)^T \tag{7.5.4}$$

where, of course, $R_{(n)}$ is the covariance matrix of $N_{(n)}$. Given $R_{(n)}$, we can use the above relationship to compute the covariance matrix of the random errors in the scaled-derivative estimate vector, $Z^*_{n+h,n}$.

An assumption which is very frequently made is that all of the observation errors are zero-mean and uncorrelated, with equal variance σ_ν^2. This means that $R_{(n)}$ will have the form

$$R_{(n)} = \sigma_\nu^2 I \tag{7.5.5}$$

For the remainder of this chapter, we take this to be the case. Applying (7.5.5) to (7.5.4), we accordingly obtain

$$S^*_{n,n} = \sigma_\nu^2\, W(0)\, W(0)^T \tag{7.5.6}$$

and then, by (7.4.10),

$$S^*_{n,n} = \sigma_\nu^2\, PCBB^TC^TP^T \tag{7.5.7}$$

This expression simplifies considerably.

First, it is easily verified (see Ex. 7.12) that

$$BB^T = C^{-1} \tag{7.5.8}$$

and so (since $C = C^T$), we see that (7.5.7) is equivalent to

$$S^*_{n,n} = \sigma_\nu^2\, PCP^T \tag{7.5.9}$$

Next, we factor C into $C^{1/2}C^{1/2}$. (Since C is diagonal, so is $C^{1/2}$, and the elements of the latter are simply the square-roots of those of the former.) It is then easily verified that $PC^{1/2}$ has as its i, j^{th} element

$$\left[PC^{1/2}\right]_{ij} = \frac{1}{i!}\,\frac{d^i}{dr^i}\,\varphi_j(r)\bigg|_{r=L} \qquad 0 \le i, j \le m \tag{7.5.10}$$

where $\varphi_j(r)$ is the normalized discrete Legendre polynomial, defined in (7.3.6). (See Ex. 7.13.)

We now define the matrix $Q(x)$, whose i, j^{th} element is

$$\left[Q(x)\right]_{ij} = \frac{1}{i!}\frac{d^i}{dr^i}\varphi_j(r)\bigg|_{r=x} \qquad 0 \le i, j \le m \tag{7.5.11}$$

Then (7.5.10) can be written

$$PC^{1/2} = Q(L) \tag{7.5.12}$$

and so (7.5.9) becomes, finally,

$$S^*_{n,n} = \sigma_\nu^2\, Q(L)\, Q(L)^T \tag{7.5.13}$$

The covariance matrix $S^*_{n,n}$ thus has, as its i, j^{th} element,

$$\begin{aligned}\left[S^*_{n,n}\right]_{ij} &= \sigma_\nu^2 \sum_{k=0}^{m} \left[Q(L)\right]_{ik}\left[Q(L)^T\right]_{kj} \\ &= \sigma_\nu^2 \sum_{k=0}^{m}\left[\frac{1}{i!}\frac{d^i}{dr^i}\varphi_k(r)\,\frac{1}{j!}\frac{d^j}{dr^j}\varphi_k(r)\right]\bigg|_{r=L}\end{aligned} \tag{7.5.14}$$

As an example, for $m = 1$, the covariance matrix of the random output errors in $Z^*_{n,n}$ of (7.4.8) becomes, using (7.5.14),

$$S^*_{n,n} = \sigma_\nu^2 \begin{pmatrix} \varphi_0^2(r) + \varphi_1^2(r) & \varphi_1(r)\dfrac{d}{dr}\varphi_1(r) \\ \varphi_1(r)\dfrac{d}{dr}\varphi_1(r) & \left[\dfrac{d}{dr}\varphi_1(r)\right]^2 \end{pmatrix}_{r=L} \tag{7.5.15}$$

Using (7.3.18) and (7.3.19), this gives (see Ex. 7.14),

$$S^*_{n,n} = \sigma_\nu^2 \begin{pmatrix} \dfrac{2(2L+1)}{(L+2)(L+1)} & \dfrac{6}{(L+2)(L+1)} \\ \dfrac{6}{(L+2)(L+1)} & \dfrac{12}{(L+2)(L+1)L} \end{pmatrix} \tag{7.5.16}$$

and so *for very large L*

$$S^*_{n,n} \approx \sigma_\nu^2 \begin{pmatrix} \frac{4}{L} & \frac{6}{L^2} \\ \frac{6}{L^2} & \frac{12}{L^3} \end{pmatrix} \tag{7.5.17}$$

showing that $S^*_{n,n}$ goes to a null matrix as $L \to \infty$. ◆◆

Consider next the covariance matrix of $N^*_{n+h,n}$, i.e. the random output-error vector of an h-step prediction. By (7.4.9)

$$Z^*_{n+h,n} = \Phi(h)\,PCB\,Y_{(n)} \tag{7.5.18}$$

and so,

$$S^*_{n+h,n} = \Phi(h)\,PCB\,R_{(n)}\,B^T C^T P^T \Phi(h)^T \tag{7.5.19}$$

When $R_{(n)} = \sigma_\nu^2 I$, this reduces, by the arguments given above, to

$$S^*_{n+h,n} = \sigma_\nu^2 \left[\Phi(h)\,Q(L)\right]\left[\Phi(h)\,Q(L)\right]^T \tag{7.5.20}$$

But (7.5.11) shows that

$$\Phi(h)\,Q(L) = Q(L+h) \tag{7.5.21}$$

and so (7.5.20) gives us

$$S^*_{n+h,n} = \sigma_\nu^2 Q(L+h)\,Q(L+h)^T \tag{7.5.22}$$

Comparison with (7.5.14) then shows that the i,j^{th} term of $S^*_{n+h,n}$ will be

$$\left[S^*_{n+h,n}\right]_{ij} = \sigma_\nu^2 \sum_{k=0}^{m} \frac{1}{i!}\frac{d^i}{dr^i}\varphi_k(r)\,\frac{1}{j!}\frac{d^j}{dr^j}\varphi_k(r)\Bigg|_{r=L+h} \tag{7.5.23}$$

As with the example of (7.5.15), when $m = 1$ this is just

$$S^*_{n+h,n} = \sigma_\nu^2 \begin{pmatrix} \varphi_0^2(r) + \varphi_1^2(r) & \varphi_1(r)\dfrac{d}{dr}\varphi_1(r) \\ \varphi_1(r)\dfrac{d}{dr}\varphi_1(r) & \left[\dfrac{d}{dr}\varphi_1(r)\right]^2 \end{pmatrix}_{r=L+h} \tag{7.5.24}$$

and so, for the 1-step predictor say, we set $r = L + 1$ in (7.3.18) and (7.3.19), and obtain

$$S^*_{n+1,n} = \sigma_\nu^2 \begin{pmatrix} \dfrac{2(2L+3)}{(L+1)L} & \dfrac{6}{(L+1)L} \\ \dfrac{6}{(L+1)L} & \dfrac{12}{(L+2)(L+1)L} \end{pmatrix} \tag{7.5.25}$$

♦♦

Equation (7.5.23) is the general analytic expression for $S^*_{n+h,n}$, assuming that (7.5.5) holds true. In the next section we shall discuss its properties.

It was pointed out earlier that a computer program can readily be written which generates the matrix $W(h)$ of (7.4.10) for any h, L and m. It is then a trivial matter to add the additional program to generate $S^*_{n+h,n}$. By (7.5.4) we obtain

$$S^*_{n+h,n} = \sigma_\nu^2 W(h) W(h)^T \tag{7.5.26}$$

and in practice, the best way of obtaining the actual numerical values of $S^*_{n+h,n}$ is directly from this expression.

Until now we have assumed that the output of the filter is $Z^*_{n+h,n}$, the vector of *scaled* estimates. If the filter output is to be the *unscaled* vector $X^*_{n+h,n}$ of (7.4.7) then the covariance matrix of the random output errors is $\mathsf{S}^*_{n+h,n}$, where (see (7.4.12))

$$\mathsf{S}^*_{n+h,n} \equiv D(\tau) S^*_{n+h,n} D(\tau) \tag{7.5.27}$$

It is easily shown (see Ex. 7.17) that this together with (7.5.23) gives

$$\left[\mathsf{S}^*_{n+h,n}\right]_{ij} = \frac{i!\,j!}{\tau^{i+j}}\left[S^*_{n+h,n}\right]_{ij}$$
$$= \frac{\sigma_\nu^2}{\tau^{i+j}} \sum_{k=0}^{m} \frac{d^i}{dr^i}\varphi_k(r)\,\frac{d^j}{dr^j}\varphi_k(r)\Bigg|_{r=L+h} \tag{7.5.28}$$

The attention of the reader is directed to the use of an upper-case italic S for the covariance matrix of the *scaled* estimate vector Z^*, and a sans serif italic S for the covariance matrix of the *unscaled* vector X^*. The matrix $S^*_{n+h,n}$ is independent of τ, as seen from (7.5.23), whereas $\mathsf{S}^*_{n+h,n}$ contains τ *and is really the matrix of interest.*

7.6 DEPENDENCE OF THE COVARIANCE MATRIX ON L AND τ

In any practical situation, the user of a filtering algorithm faces the following problem: Given the statistical properties of the observation errors, how should one select the parameters of the filter so that the outputs satisfy a given set of requirements? As an example, suppose that the observation errors are zero-mean, uncorrelated and of standard-deviation $\sigma_\nu = 10$ feet. What first-degree fixed-memory smoothing algorithm should be selected so that errors in the position output have a standard deviation of 3 feet, assuming 1-step prediction?

From (7.5.25) we see that

$$\left[S^*_{n+1,n}\right]_{0,0} = \sigma_\nu^2 \frac{2(2L+3)}{(L+1)L} \tag{7.6.1}$$

and so we require that

$$9 = 100 \frac{2(2L+3)}{(L+1)L} \tag{7.6.2}$$

giving $L = 45$. The vector $Y_{(n)}$ must thus contain 46 entries.

What will be the standard deviation of the random errors in the velocity estimate?

Again by (7.5.25) we have

$$\left[S^*_{n+1,n}\right]_{1,1} = \sigma_\nu^2 \frac{12}{(L+2)(L+1)L} \tag{7.6.3}$$

To obtain the variance of the errors in the unscaled velocity estimate, we must take into account the scale-factor τ. Thus by (7.5.28), using $L = 45$,

$$\begin{aligned}\left[\mathsf{S}^*_{n+1,n}\right]_{1,1} &= \frac{1}{\tau^2}\left[S^*_{n+1,n}\right]_{1,1} \\ &= \frac{1}{\tau^2}\sigma_\nu^2 \frac{12}{(L+2)(L+1)L} \\ &= \frac{0.0123}{\tau^2}\end{aligned} \tag{7.6.4}$$

This is the variance of the random output errors in the velocity estimate.

It now remains to fix $\left[S^*_{n+1,n}\right]_{1,1}$ in order to fix τ. Thus, if we require the random velocity errors to have a σ of say 4 ft/sec, then

$$16 = \frac{1}{\tau^2}(0.0123)$$

giving $\tau = 0.029$ seconds, i.e. about 36 observations per second.

In summary, assuming a first-degree, 1-step predictor, if the standard-deviation of the observation errors is 10 feet, $\tau = 0.029$ secs and $L = 45$, then the standard-deviations of the errors in the position and velocity estimates will be, respectively:

$$\begin{aligned} \sigma(x^*) &= 3 \text{ feet} \\ \sigma(\dot{x}^*) &= 4 \text{ ft/sec} \end{aligned} \tag{7.6.5}$$

◆◆

The above problem had a simple solution because it was deliberately constructed that way. Note also that it was based on the assumption that (7.5.5) is true. In practice things are seldom so easy, and what we are usually required to do is to fix parameters by trade-off, i.e. we are given a set of conditions which cannot all be met and the only thing we can hope to do is to find the best compromise.

For this reason, it is essential that we gain some insight into the behavior of the terms of the covariance matrices as we vary the parameters L, m, h and τ. In this section we consider their behavior as L and τ are varied. The effects of varying m and h are examined in the next section. *The entire analysis assumes that* (7.5.5) *holds true.*

Consider first the matrices of (7.5.16) and (7.5.25). In both cases we see that *all of their elements go to zero as L increases.* This is true, as we now show, for *any degree* (m), and *any prediction interval* (h), and means that we can make the covariance matrices arbitrarily close to null matrices by taking L sufficiently large.

The proof is as follows. By (7.5.9) (letting $\sigma_\nu^2 = 1$),

$$\begin{aligned} S^*_{n+h,n} &= \Phi(h)\, PCP^T\, \Phi(h)^T \\ &= \Phi(h)\, SGCG^T S^T\, \Phi(h)^T \qquad \text{[by (7.4.15)]} \end{aligned} \tag{7.6.6}$$

When L is large, we see from (7.4.5) and (3.2.31), that

$$C \approx \frac{1}{L}\begin{pmatrix} 1 & & & \\ & 3 & & \\ & & 5 & \\ & & & \ddots \end{pmatrix} \tag{7.6.7}$$

Moreover, from (7.4.14) and Table 7.1 on p. 241, when L is large:

$$G \approx \begin{pmatrix} 1 & -1 & 1 & \cdots \\ & -\frac{2}{L} & \frac{6}{L} & \cdots \\ & & \frac{6}{L^2} & \cdots \\ & & & \ddots \end{pmatrix}$$

$$= \begin{pmatrix} 1 & & & \\ & \frac{1}{L} & & \\ & & \frac{1}{L^2} & \\ & & & \ddots \end{pmatrix}\begin{pmatrix} 1 & -1 & 1 & \cdots \\ & -2 & 6 & \cdots \\ & & 6 & \cdots \\ & & & \ddots \end{pmatrix} \tag{7.6.8}$$

Thus (for a 3×3),

$$GCG^T =$$

$$= \frac{1}{L}\begin{pmatrix} 1 & & \\ & \frac{1}{L} & \\ & & \frac{1}{L^2} \end{pmatrix}\begin{pmatrix} 1 & -1 & 1 \\ & -2 & 6 \\ & & 6 \end{pmatrix}\begin{pmatrix} 1 & & \\ & 3 & \\ & & 5 \end{pmatrix}\begin{pmatrix} 1 & & \\ -1 & -2 & \\ 1 & 6 & 6 \end{pmatrix}\begin{pmatrix} 1 & & \\ & \frac{1}{L} & \\ & & \frac{1}{L^2} \end{pmatrix}$$

$$= \frac{1}{L}\begin{pmatrix} 1 & & \\ & \frac{1}{L} & \\ & & \frac{1}{L^2} \end{pmatrix}\begin{pmatrix} 9 & 36 & 30 \\ 36 & 192 & 180 \\ 30 & 180 & 180 \end{pmatrix}\begin{pmatrix} 1 & & \\ & \frac{1}{L} & \\ & & \frac{1}{L^2} \end{pmatrix}$$

$$= \begin{pmatrix} \frac{9}{L} & \frac{36}{L^2} & \frac{30}{L^3} \\ \frac{36}{L^2} & \frac{192}{L^3} & \frac{180}{L^4} \\ \frac{30}{L^3} & \frac{180}{L^4} & \frac{180}{L^5} \end{pmatrix} \tag{7.6.9}$$

which goes to a null matrix as $L \to \infty$. *Thus every element of* $S^*_{n+h,n}$ *goes to zero as L goes to infinity.* (Although we proved this for the 3×3 case, it is easily seen that the method of proof is quite general, and can be applied to matrices of any size.) ◆◆

Since $\Phi(h)S$ [on the right of (7.6.6)] is a matrix which does not depend on L, we see that (for large L),

$$S^*_{n+h,n} \equiv [\Phi(h)S](GCG^T)\left[S^T\Phi(h)^T\right]$$

$$\approx \begin{pmatrix} \frac{\alpha_{00}}{L} & \frac{\alpha_{01}}{L^2} & \frac{\alpha_{02}}{L^3} \\ \frac{\alpha_{10}}{L^2} & \frac{\alpha_{11}}{L^3} & \frac{\alpha_{12}}{L^4} \\ \frac{\alpha_{20}}{L^3} & \frac{\alpha_{21}}{L^4} & \frac{\alpha_{22}}{L^5} \end{pmatrix} \tag{7.6.10}$$

We are thus able to see from (7.6.10), just *how* the elements of S^* go to zero as L increases. Thus the i, j^{th} element of $S^*_{n+h,n}$ will be approximately

$$\left[S^*_{n+h,n}\right]_{ij} \approx \frac{\alpha_{ij}}{L^{i+j+1}}\sigma_\nu^2 \tag{7.6.11}$$

where α_{ij} is a constant, depending on h and m, but not on L. From (7.5.28) we see further, that for large L, the i, j^{th} element of the matrix $S^*_{n+h,n}$ will be

$$\left[S^*_{n+h,n}\right]_{ij} \approx \frac{\lambda_{ij}}{\tau^{i+j}L^{i+j+1}}\sigma_\nu^2 \tag{7.6.12}$$

where λ_{ij} is a constant depending on h and m. (In fact, $\lambda_{ij} = i!\,j!\,\alpha_{ij}$.)

It thus follows that any element of $S^*_{n+h,n}$ can be made as small as we wish, by simply taking the number of observations $(L + 1)$ sufficiently large. This is, of course, consistent with the heuristic idea that *by averaging over a sufficiently large number of unbiased errors, the net error can be made arbitrarily small.*

Consider next, the behavior of $S^*_{n+h,n}$ as τ is varied. Define the *smoothing time* T by

$$T \equiv L\tau \tag{7.6.13}$$

Thus T is the total time between the most recent and the oldest observations used in the filter for any given estimate.† Then by (7.6.12),

$$\left[S^*_{n+h,n}\right]_{ij} \approx \frac{\lambda_{ij}\tau}{T^{i+j+1}}\sigma_\nu^2 \tag{7.6.14}$$

We can vary τ in two ways.

CASE A

Reduce τ keeping L fixed. Then (7.6.12) shows that (for $i, j > 0$), $\left[S^*_{n+h,n}\right]_{ij}$ will increase. *This means that if we make a fixed number of observations at a greater rate, the variances of the random output errors increase.*

For $i = j = 0$ we see from (7.6.12) that the variance of the random errors in the position estimate is

$$\left[S^*_{n+h,n}\right]_{0,0} \approx \frac{\lambda_{00}}{L}\sigma_\nu^2 \tag{7.6.15}$$

which does not depend on τ and so remains constant as we reduce τ while keeping L fixed.

†Equivalently T is the length of the memory (see Figure 7.2 on p. 235).

CASE B

Reduce τ keeping T fixed. This means that the observations are being made over an interval of fixed duration but that their number is increasing. By (7.6.14) we see that the elements of $S^*_{n+h,n}$ diminish as we reduce τ.

It would appear that in the limit, as $\tau \to 0$, then $S^*_{n+h,n}$ tends to a null matrix if T is kept fixed, but this is true only in theory. We recall that all of the above arguments were based on the assumption [see (7.5.5)] that the observation errors were uncorrelated and of constant variance. As $\tau \to 0$ it is difficult to envisage this situation persisting, and sooner or later, in a physical case, the errors would become correlated and then (7.6.14) would cease to hold. The reader may feel that, as the sampling interval diminishes, the requisite absence of correlation could prevail if, in the limit, the random errors assumed the form of *white noise.* However, ideal white noise (which is uncorrelated regardless of how frequently sampled) has *infinite* variance (autocorrelation function equal to a Dirac delta), and so this violates the assumption that σ_ν^2 remains constant.

In summary then, as $\tau \to 0$ with T fixed, $S^*_{n+h,n}$ diminishes to a null matrix according to (7.6.14), provided that (7.5.5) continues to be true and σ_ν^2 remains constant.

We have discussed the behavior of S^* as we vary L and τ. In the next section we examine how that matrix depends on m and h.

7.7 DEPENDENCE OF THE COVARIANCE MATRIX ON m AND h

Some of the properties of the covariance matrix of the Fixed-Memory Polynomial Filter were analyzed in the preceding section. We showed there that for any degree and prediction interval, *when L is large,*

$$\left[S^*_{n+h,n}\right]_{ij} \approx \frac{\lambda_{ij}}{\tau^{i+j}L^{i+j+1}}\sigma_\nu^2 \tag{7.7.1}$$

We now examine the behavior of S^* as we increase the degree, m, while keeping L fixed.

When m is 1 say, then S^* is 2×2. Let $h = 0$. The diagonal elements of S^* are

$$\begin{aligned} \left[S^*_{n,n}\right]_{0,0} &\equiv \text{Var (random errors in position)}^\dagger \\ \left[S^*_{n,n}\right]_{1,1} &\equiv \text{Var (random errors in velocity)} \end{aligned} \tag{7.7.2}$$

†Where Var means variance.

For $m = 2$, then S^* is 3×3 with diagonal elements

$$\begin{aligned} \left[S^*_{n,n}\right]_{0,0} &= \text{Var (random errors in position)} \\ \left[S^*_{n,n}\right]_{1,1} &= \text{Var (random errors in velocity)} \\ \left[S^*_{n,n}\right]_{2,2} &= \text{Var (random errors in acceleration)} \end{aligned} \tag{7.7.3}$$

We can array these diagonal members of $S^*_{n,n}$ as follows,

Degree

	$i \backslash m$	0	1	2	3
position	0	X	X	X	X
velocity	1		X	X	X
acceleration	2			X	X
jerk	3				X

and for $i = 0$, say, we see that the first row of the table gives us the variances in position $\left[S^*_{n,n}\right]_{0,0}$, *as m increases through successive values.* (The above table applies *only* for $h = 0$, say, and for some other values of h we can derive the corresponding tables.)

Suppose $h = 0$, By (7.5.28) the ith *diagonal* term of S^*, for an mth-degree estimator, is

$$\left[S^*_{n,n}\right]_{ii} = \frac{\sigma_\nu^2}{\tau^{2i}} \sum_{k=0}^{m} \left[\frac{d^i}{dr^i}\varphi_k(r)\right]^2 \Bigg|_{r=L} \tag{7.7.4}$$

and so this is the i, m^{th} term of the above table.

Now, it is a known fact that all of the polynomials $p_j(r)$ and their derivatives have all of their zeros *strictly inside* the interval $0 < r < L$.† Thus it follows that at $r = L$ and outside of the interval, the polynomials and their derivatives are nonzero, and so, for any i and k:

$$\left[\frac{d^i}{dr^i}\varphi_k(r)\right]^2 \Bigg|_{r=L} > 0 \tag{7.7.5}$$

†This applies more generally to *any* set of orthogonal polynomials. Thus (see e.g. [7.1]) it can be shown that any set of such polynomials has all of its zeros *strictly inside* the orthogonality interval and that all of those zeros are *distinct.* Moreover this is also true for the zeros of all of the derivatives of the entire set.

This means that for each increase in m, a positive number is added to the right hand side of (7.7.4), and so moving across any row of the table on p. 252, the entries will be *strictly increasing functions of m.*

We know from (7.7.1), that for large L,

$$\left[S^*_{n,n}\right]_{i,i} \approx \frac{\lambda_{ii}(m)}{\tau^{2i}L^{2i+1}}\sigma_\nu^2 \tag{7.7.6}$$

where we now show explicitly that λ depends on m. In Table 7.2 (p. 258), we show the values of $\lambda_{ii}(m)$ for $0 \leq i, m \leq 10$. Moving along the zeroth row of that table, we see that as the degree increases, the position-output variance increases through the values

m	0	1	2	...
$\left[S^*_{n,n}\right]_{0,0}$	$\frac{1}{L}\sigma_\nu^2$	$\frac{4}{L}\sigma_\nu^2$	$\frac{9}{L}\sigma_\nu^2$	...

The velocity-output variance likewise increases through the values

m	0	1	2	...
$\left[S^*_{n,n}\right]_{1,1}$	-	$\frac{12}{\tau^2L^3}\sigma_\nu^2$	$\frac{192}{\tau^2L^3}\sigma_\nu^2$	...

For other values of h, the corresponding tables show a similar increase with m. Table 7.3 on p. 259 gives the table for $h = -L/2$ (center of the observation interval), when L is large. Note how the values increase *step-wise* along any row. This is a direct consequence of the fact that

$$\left.\frac{d^i}{dr^i}\varphi_j(r)\right|_{r=\frac{L}{2}} = 0 \qquad (i + j \text{ odd}) \tag{7.7.7}$$

For large L, the table for $h = 0$ can be used equally well for $h = 1$, i.e. Table 7.2 *applies to both 0-step and 1-step prediction.*

It is this increase of $\left[S^*_{n+h,n}\right]_{ii}$ with m *that makes it imperative that we keep the degree of the estimator as low as possible.* (This problem will be discussed again in Section 7.13.)

We now consider what happens to $\left[S^*_{n+h,n}\right]_{ii}$ *when we vary* h and keep m (and of course, L) fixed. Returning to (7.5.28), we see that, since $(d^i/dr^i)\varphi_k(r)$ is a polynomial in r of degree $k-i$, the term

$$\left[S^*_{n+h,n}\right]_{ii} \equiv \frac{\sigma_\nu^2}{\tau^{2i}} \sum_{k=0}^{m} \left[\frac{d^i}{dr^i}\varphi_k(r)\right]^2 \Bigg|_{r=L+h} \tag{7.7.8}$$

will be a polynomial in h of degree $2(m-i)$. Moreover, since $(d^i/dr^i)\varphi_k(r)$ has all of its zeros *inside* the observation interval, $\left[S^*_{n+h,n}\right]_{ii}$ will be *strictly increasing* with h when we are predicting or retrodicting to any points *outside* the observation interval. Thus, as an example, for $m=3$, $i=1$ say, $\left[S^*_{n+h,n}\right]_{ii}$ is a polynomial in h of degree 4 which increases strictly when $h>0$.

Figure 7.4 depicts the behavior of $\left[S^*_{n+h,n}\right]_{ii}$ outside the smoothing interval, and shows how rapidly the variances of the random output errors increase when $h>0$ or $h<-L$. *It is clear that long predictions outside of the smoothing interval are always to be avoided,* particularly when m is larger than 1 or 2.

Inside the smoothing interval, the zeros of $(d^i/dr^i)\varphi_k(r)$ come into play and cause the diagonal elements of $S^*_{n+h,n}$ to remain relatively small. In Figures 7.5 and 7.6 we show typical plots of such elements, and the reader will note that they do in fact remain relatively small in the interval. He will also note how rapidly they are increasing at the end points. Outside the interval they continue to increase as depicted in Figure 7.4.

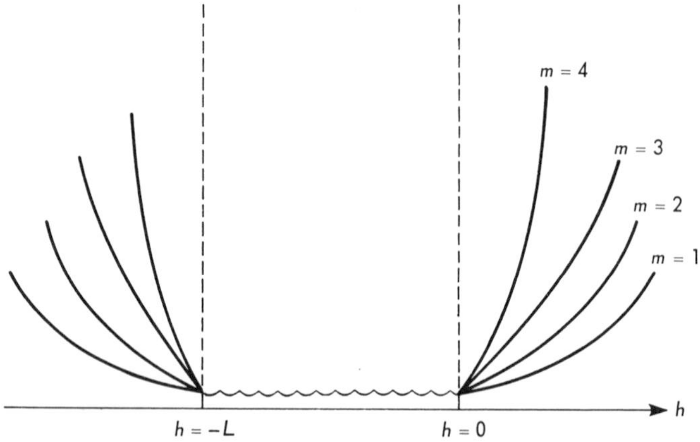

Fig. 7.4 *Variance of the random output errors as a function of h.*

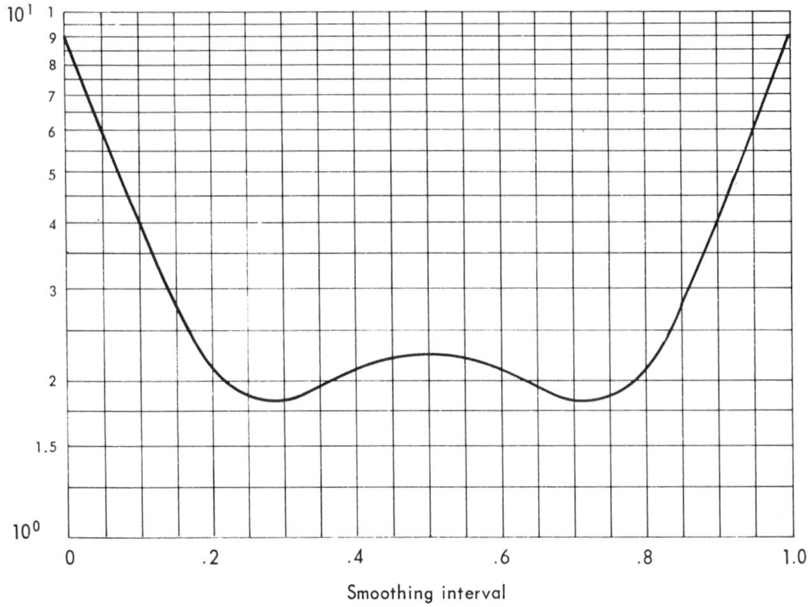

Fig. 7.5 *Asymptotic curve of* $L \times \left[S^*_{n+h,n}\right]_{0,0}$ *for* $m = 2$ *as* $L \to \infty$.

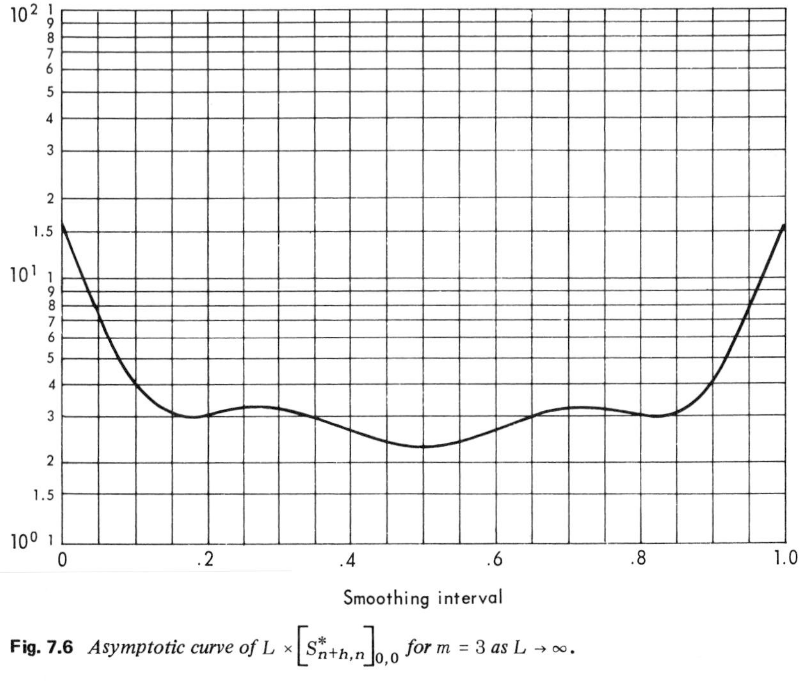

Fig. 7.6 *Asymptotic curve of* $L \times \left[S^*_{n+h,n}\right]_{0,0}$ *for* $m = 3$ *as* $L \to \infty$.

In summary then, the zeros of the polynomials and their derivatives, being all located inside the smoothing interval, will cause the curve of $\left[S^*_{n+h,n}\right]_{ii}$ to remain *relatively small in the smoothing interval,* whereas the complete absence of zeros, *outside* the observation interval, will result in a *rapid increase* in that region.

By virtue of their symmetry over the orthogonality interval, one half of all of the polynomials and their derivatives have a zero at the center of the interval. This very high preponderance of zeros at that point causes the curves such as Figures 7.5 and 7.6 to be minimal in one half of all cases at the center and close to minimal in the remainder. For this reason (and others to follow) *the center of the observation interval should always be chosen as the estimation point* unless the latter is fixed by prior constraints to lie elsewhere.

7.8 THE VARIANCE REDUCTION FACTORS

In the preceding section, reference was made to Tables 7.2 and 7.3. These tables give us what we call the Variance Reduction Factors (VRF's), for any degree, and any derivative up to 10, assuming that L is large.†

We demonstrate their use by an example. Given that we are observing a process at intervals of τ seconds, and assuming that the observation errors are zero-mean, uncorrelated and stationary with variance σ_ν^2, find the variance of the random errors in $x^*_{n+h,n}$, $\dot{x}^*_{n+h,n}$ and $\ddot{x}^*_{n+h,n}$ for $h = 0$, and $h = -L/2$. The degree of the estimator is 4 and L is very large.

Table 7.2 applies for h = 1 or 0. For $m = 4$, we obtain the $\lambda_{ii}(m)$ of (7.7.6). Thus (see the fourth column on p. 258)

$$\begin{aligned} \lambda_{00} &= 25 \\ \lambda_{11} &= 4800 \\ \lambda_{22} &= 317500 \end{aligned} \qquad (7.8.1)$$

Then by (7.7.6), for $\sigma_\nu^2 = 1$, $\left[S^*\right]_{ii} = \lambda_{ii}(m)/\tau^{2i}L^{2i+1}$, and so

$$\begin{aligned} \left[S^*\right]_{0,0} &= \frac{25}{L} \\ \left[S^*\right]_{1,1} &= \frac{4800}{\tau^2 L^3} \\ \left[S^*\right]_{2,2} &= \frac{317500}{\tau^4 L^5} \end{aligned} \qquad (7.8.2)$$

†We continue to assume, throughout, that (7.5.5) is true.

We call these the *Variance Reduction Factors.* They are the ratio of the output-error variance on each channel to the variance of the input errors.

For an input variance of σ_ν^2 other than unity, the output variance equals the input variance times the VRF. Thus,

$$\begin{aligned} \operatorname{Var}(x^*_{n,n}) &= \frac{25}{L}\sigma_\nu^2 \\ \operatorname{Var}(\dot{x}^*_{n,n}) &= \frac{4800}{\tau^2 L^3}\sigma_\nu^2 \\ \operatorname{Var}(\ddot{x}^*_{n,n}) &= \frac{317500}{\tau^4 L^5}\sigma_\nu^2 \end{aligned} \tag{7.8.3}$$

For smoothing to the center, $(h = -L/2)$ the VRF's are obtained from Table 7.3. Thus, from the fourth column of that table,

$$\begin{aligned} \operatorname{VRF}\left(x^*_{n-\frac{L}{2},n}\right) &= \frac{3.516}{L} \\ \operatorname{VRF}\left(\dot{x}^*_{n-\frac{L}{2},n}\right) &= \frac{75}{\tau^2 L^3} \\ \operatorname{VRF}\left(\ddot{x}^*_{n-\frac{L}{2},n}\right) &= \frac{720}{\tau^4 L^5} \end{aligned} \tag{7.8.4}$$

Thus, given that the observation errors have variance σ_ν^2, the random output errors in the estimate will have variances

$$\begin{aligned} \operatorname{Var}\left(x^*_{n-\frac{L}{2},n}\right) &= \frac{3.516}{L}\sigma_\nu^2 \\ \operatorname{Var}\left(\dot{x}^*_{n-\frac{L}{2},n}\right) &= \frac{75}{\tau^2 L^3}\sigma_\nu^2 \\ \operatorname{Var}\left(\ddot{x}^*_{n-\frac{L}{2},n}\right) &= \frac{720}{\tau^4 L^5}\sigma_\nu^2 \end{aligned} \tag{7.8.5}$$

♦♦

The reader should compare corresponding entries in Tables 7.2 and 7.3 and observe

Table 7.2 Numerical Constants in the VRF Formula (λ_{ii}) (L Large)

Smoothing to the End Point (h = 0) or 1-Step Prediction (h = 1)

i \ m	0	1	2	3	4	5	6	7	8	9	10
0	1.0(0)	4.0(0)	9.00(0)	1.600(1)	2.500(1)	3.600(1)	4.900(1)	6.400(1)	8.100(1)	1.000(2)	1.210(2)
1		1.2(1)	1.92(2)	1.200(3)	4.800(3)	1.470(4)	3.763(4)	8.467(4)	1.728(5)	3.267(5)	5.808(5)
2			7.2 (2)	2.592(4)	3.175(5)	2.258(6)	1.143(7)	4.572(7)	1.537(8)	4.516(8)	1.193(9)
3				1.008(5)	6.451(6)	1.306(8)	1.452(9)	1.098(10)	6.323(10)	2.968(11)	1.187(12)
4					2.540(7)	2.540(9)	7.684(10)	1.229(12)	1.299(13)	1.018(14)	6.363(14)
5						1.006(10)	1.449(12)	6.120(13)	1.333(15)	1.874(16)	1.919(17)
6							5.754(12)	1.128(15)	6.344(16)	1.804(18)	3.259(19)
7								4.488(15)	1.149(18)	8.301(19)	2.988(21)
8									4.578(18)	1.483(21)	1.339(23)
9										5.914(21)	2.366(24)
10											9.439(24)

Note: 1.449(12) means 1.449×10^{12} Ex: $i = 2$, $m = 3$, $\text{VRF} = \dfrac{2.592 \times 10^4}{\tau^4 L^5}$ (see (7.7.6)).

Table 7.3 Numerical Constants in the VRF Formula (λ_{ii}) (L Large) Smoothing to the Center $\left(h = -\frac{L}{2}\right)$

i \ m	0	1	2	3	4	5	6	7	8	9	10
0	1.0(0)	1.0(0)	2.25(0)	2.250(0)	3.516(0)	3.516(0)	4.785(0)	4.785(0)	6.056(0)	6.056(0)	7.328(0)
1		1.2(1)	1.20(1)	7.500(1)	7.500(1)	2.297(2)	2.297(2)	5.168(2)	5.168(2)	9.771(2)	9.771(2)
2			7.20(2)	7.200(2)	(8.820(3)	8.820(3)	4.465(4)	4.465(4)	1.501(5)	1.501(5)	3.963(5)
3				1.008(5)	1.008(5)	2.041(6)	2.041(6)	1.544(7)	1.544(7)	7.247(7)	7.247(7)
4					2.540(7)	2.540(7)	7.684(8)	7.684(8)	8.116(9)	8.116(9)	5.073(10)
5						1.006(10)	1.006(10)	4.250(11)	4.250(11)	5.977(12)	5.977(12)
6							5.754(12)	5.754(12)	3.237(14)	3.237(14)	5.846(15)
7								4.488(15)	4.488(15)	3.243(17)	3.243(17)
8									4.578(18)	4.578(18)	4.131(20)
9										5.914(21)	5.914(21)
10											9.439(24)

Note: 2.540(7) means 2.540 x 10^7 Ex: $i = 3$, $m = 5$, $\text{VRF} = \dfrac{2.041 \times 10^6}{\tau^6 L^7}$ (see (7.7.6)).

a. How much more slowly the VRF's increase with degree when one is smoothing to the center vs. the end point.

b. How much smaller (sometimes by three orders of magnitude) the VRF's are at the center, vs. the end point.

It is for both of these reasons and others to follow *that we always prefer to smooth to the center,* if we have the choice.

When L is not large (in relation to the degree) then we must use the exact expressions for the VRF's. These are tabulated on p. 371 where the symbol n is used in place of L, for the 1-step predictors up to degree 3. For other cases, use must be made of (7.5.28).

7.9 A SIMPLE EXAMPLE

In order to reinforce some of the ideas developed above, we consider the following simple example.

A radar is making equally spaced measurements, ρ_n (range), ψ_n (azimuth) and θ_n (elevation),† on an object in orbit. The observation errors are zero-mean, uncorrelated, and have variances σ_ρ^2, σ_ψ^2 and σ_θ^2.

Assuming a Fixed-Memory Polynomial Filter of degree $m = 2$, and observation length $L\tau$ seconds, how shall the filter be organized so as to

(I) Make 1-step predictions of ρ, ψ and θ to assist the radar in making further measurements.

(II) Make extended predictions into the future.

Part (I) requires that we compute, repeatedly, the 1-step predictions of ρ, ψ and θ. We obtain these from three identical Fixed-Memory Polynomial Filters, as follows:

$$
\begin{aligned}
\rho^*_{n+1,n} &= \sum_{j=0}^{L} w_{0j}\rho_{n-j} \\
\psi^*_{n+1,n} &= \sum_{j=0}^{L} w_{0j}\psi_{n-j} \\
\theta^*_{n+1,n} &= \sum_{j=0}^{L} w_{0j}\theta_{n-j}
\end{aligned}
\qquad (7.9.1)
$$

†For a definition of these quantities, see p. 168.

where the weights

$$(w_{00}, w_{01}, \ldots, w_{0L})$$

are the zeroth row of the $W(h)$ matrix of (7.4.10), using $h = 1$, $m = 2$ and L as selected. These predictions can be used to help point the radar at the object in order to make further observations.

We also usually require the variance of the random errors in these predictions. From Table 7.2, we see that the VRF in each case is (assuming L is large),

$$\text{VRF} = \frac{9}{L} \tag{7.9.2}$$

Hence

$$\begin{aligned} \sigma^2(\rho^*_{n+1,n}) &= \frac{9}{L}\sigma_\rho{}^2 \\ \sigma^2(\psi^*_{n+1,n}) &= \frac{9}{L}\sigma_\psi{}^2 \\ \sigma^2(\theta^*_{n+1,n}) &= \frac{9}{L}\sigma_\theta{}^2 \end{aligned} \tag{7.9.3}$$

For Part (II), we know, from the theory of Section 7.7, that it is undesirable to use the estimating polynomial to make long-term predictions because of the rapid growth in the VRF's with h (see Figure 7.4). (A second important reason will be discussed under *Systematic Errors* in the next section.) Accordingly, we decide instead to determine the *orbital parameters,* and then to make our predictions from them. Moreover, to get the best possible estimates of those parameters, we decide to estimate at the *center* of the observation interval. (This is desirable, *both* from a variance reduction standpoint as we have already shown, as well as for keeping the systematic errors down, as we shall show in the next section.)

To get the *center-smoothed values* we use the filter equations:

$$\begin{aligned} \rho^*_{n-\frac{L}{2},n} &= \sum_{j=0}^{L} w_{0j}\rho_{n-j} \\ \dot{\rho}^*_{n-\frac{L}{2},n} &= \tau\sum_{j=0}^{L} w_{1j}\rho_{n-j} \end{aligned} \tag{7.9.4}$$

where in this case the w's are respectively the zeroth and first rows of (7.4.10) with $h = -L/2$, $m = 2$ and L as selected. In the same way, using the same weights, we obtain $\left(\psi^*, \dot{\psi}^*, \theta^*, \dot{\theta}^*\right)_{n-\frac{L}{2},n}$, *and these six numbers can then be used to compute estimates of the six orbital parameters.* Long-term predictions can now be made with considerably more confidence by integrating the true equations of motion.

The covariance matrix of the random errors in the estimated orbital parameters can be estimated as follows. Define the vectors

$$X^*_{n-\frac{L}{2},n} \equiv \left(\rho^*, \dot{\rho}^*, \psi^*, \dot{\psi}^*, \theta^*, \dot{\theta}^*\right)^T_{n-\frac{L}{2},n}$$

$$K^* \equiv \left(k_0^*, k_1^*, k_2^*, k_3^*, k_4^*, k_5^*\right)^T \tag{7.9.5}$$

where the k's are the six orbital parameters. Let G be the vector of non-linear functions by which K^* is computed from X^*, i.e.,

$$K^* = G(X^*) \tag{7.9.6}$$

Then if δX^* is a small perturbation on X^*, and δK^* is the resultant change in K^*, we have [see (6.2.20) and its derivation], to first order,

$$\delta K^* \approx \mathrm{M}\left(X^*_{n-\frac{L}{2},n}\right)\delta X^* \tag{7.9.7}$$

where the matrix $\mathrm{M}\left(X^*_{n-\frac{L}{2},n}\right)$ is defined by

$$\left[\mathrm{M}\left(X^*_{n-\frac{L}{2},n}\right)\right]_{ij} = \left.\frac{\partial g_i}{\partial x_j}\right|_{X = X^*_{n-\frac{L}{2},n}} \tag{7.9.8}$$

g_i being the ith element of G and x_j the jth independent variable in the right-hand side of (7.9.6).

Now assume that the perturbations in δX^* are the random output errors, i.e., $N^*_{n-\frac{L}{2},n}$. Then δK^* is the corresponding vector of random errors in K^*. From (7.9.7) we have, as the covariance matrix of the errors in K^*,

$$
\begin{aligned}
Z^* &\equiv E\left\{\delta K^*(\delta K^*)^T\right\} \\
&\approx M E\left\{N^*_{n-\frac{L}{2},n} N^{*T}_{n-\frac{L}{2},n}\right\} M^T \\
&= M S^*_{n-\frac{L}{2},n} M^T
\end{aligned} \tag{7.9.9}
$$

where $S^*_{n-\frac{L}{2},n}$ is the covariance matrix of the random output errors in $X^*_{n-\frac{L}{2},n}$. S^* can be obtained as follows.

Define the observation vectors

$$
\mathrm{P}_{(n)} \equiv \begin{pmatrix} \rho_n \\ \vdots \\ \rho_{n-L} \end{pmatrix}, \qquad \Psi_{(n)} \equiv \begin{pmatrix} \psi_n \\ \vdots \\ \psi_{n-L} \end{pmatrix}, \qquad \Theta_{(n)} \equiv \begin{pmatrix} \theta_n \\ \vdots \\ \theta_{n-L} \end{pmatrix} \tag{7.9.10}
$$

Then from (7.9.4) we see that the filter which gives $X^*_{n-\frac{L}{2},n}$ from the observations can be written as

$$
\begin{pmatrix} \rho^* \\ \dot{\rho}^* \\ \psi^* \\ \dot{\psi}^* \\ \theta^* \\ \dot{\theta}^* \end{pmatrix}_{n-\frac{L}{2},n} = \left(\begin{array}{c|c|c} W & \emptyset & \emptyset \\ \hline \emptyset & W & \emptyset \\ \hline \emptyset & \emptyset & W \end{array}\right) \left(\begin{array}{c} \mathrm{P}_{(n)} \\ \hline \Psi_{(n)} \\ \hline \Theta_{(n)} \end{array}\right) \tag{7.9.11}
$$

where W is the $2 \times (L+1)$ matrix of weights used in (7.9.4).

The covariance matrix of the vector on the extreme right of (7.9.11) is

$$
R_{(n)} = \left(\begin{array}{c|c|c} \sigma_\rho^2 I & \emptyset & \emptyset \\ \hline \emptyset & \sigma_\psi^2 I & \emptyset \\ \hline \emptyset & \emptyset & \sigma_\theta^2 I \end{array}\right) \tag{7.9.12}
$$

where I is the identity matrix of order $L + 1$. Thus from (7.9.11) the covariance matrix of $X^*_{n-\frac{L}{2},n}$ is

$$S^*_{n-\frac{L}{2},n} = \left(\begin{array}{c|c|c} \sigma_\rho^2 WW^T & \emptyset & \emptyset \\ \hline \emptyset & \sigma_\psi^2 WW^T & \emptyset \\ \hline \emptyset & \emptyset & \sigma_\theta^2 WW^T \end{array}\right) \tag{7.9.13}$$

and so by (7.9.9) the covariance matrix of the orbital parameters can now be computed. Note that this is only an approximation because of the higher-order terms neglected in going from (7.9.6) to (7.9.7). Note also that it depends both on the variance of the original observations $\left(\sigma_\rho^2, \sigma_\psi^2, \sigma_\theta^2\right)$, as well as on $X^*_{n-\frac{L}{2},n}$, since M of (7.9.7) is a matrix function of the latter. ♦♦

7.10 THE SYSTEMATIC ERRORS

Until now we have concerned ourselves only with the random errors in the output. A second very important type of error, known as the *systematic errors*, arises as follows.

Suppose that we were able to observe the true process without errors, thereby forming the observation vector $\tilde{Y}_{(n)} \equiv (\tilde{y}_n, \ldots, \tilde{y}_{n-L})^T$, which is then presented to a Fixed-Memory Polynomial Filter. If the process were a polynomial of degree m or less (m is the degree of the filter), then the estimating polynomial $\left[p^*(r)\right]_n$ would agree precisely with this error-free observation vector at every point. However, when it is a polynomial of degree greater than m, or is not a polynomial, then the polynomial p^* differs from these observations in a precisely predictable way. This difference is called the *systematic error in position*. In a like manner, the derivatives of p^* will not exactly equal the derivatives of the process, and the differences are termed the *systematic errors in the derivatives*. Assembling all of these errors into vector form gives us the *vector of systematic errors*.

Let $B^*_{n,n}$ symbolize the vector of systematic errors in the unscaled output vector, $X^*_{n,n}$ [see (7.4.7)], and assume that the input to the polynomial filter is the error-free vector $\tilde{Y}_{(n)}$ above. Then we define B^* by the equation

$$X^*_{n,n} = \Pi_n - B^*_{n,n} \tag{7.10.1}$$

where we have introduced the state-vector

$$\Pi_n \equiv \begin{pmatrix} \pi(t) \\ D\pi(t) \\ \vdots \\ D^m\pi(t) \end{pmatrix}_{t=nT} \qquad (7.10.2)$$

to represent the *true* process.

Suppose, now, that we add a vector of zero-mean observation errors to $\tilde{Y}_{(n)}$. The filter output will be

$$X^*_{n,n} = W\left(Y_{(n)} + N_{(n)}\right) \qquad (7.10.3)$$

$$\left(\Pi_n - B^*_{n,n}\right) + N^*_{n,n} \qquad \text{(by (7.10.1))}$$

Define the *total error* in the estimate as

$$H^*_{n,n} \equiv \Pi_n - X^*_{n,n} \qquad (7.10.4)$$

Then we see that

$$H^*_{n,n} = B^*_{n,n} - N^*_{n,n} \qquad (7.10.5)$$

and so H^* is a sum of the random output errors and the systematic errors.†

H^* is a *biased* random variable, because B^* is a deterministic function of $\pi(t)$, the true process. Thus, since $N^*_{n,n}$ is zero-mean, we see from the above that

$$E\left\{H^*_{n,n}\right\} = B^*_{n,n} \qquad (7.10.6)$$

The systematic error vector is thus the mean of the total output errors, and for this reason the systematic errors are also frequently called the *bias errors.* We can obtain the precise form of $B^*_{n,n}$ if $\pi(t)$ is known. This is done as follows.

†The attention of the reader is drawn to the fact that the term "random output errors" refers to $N^*_{n,n}$, which results from the vector of random observation errors in the input, namely $N_{(n)}$. The "systematic errors" on the other hand are generated by the mismatch between the model and the true process.

First suppose that $\pi(t)$ is a polynomial of degree d. If $\pi(t)$ is not a polynomial, then we assume that it is defined by an infinite power-series. That being the case, there is an integer d, large enough, so that we can write

$$\pi(t) = \sum_{k=0}^{d} a_k t^k + \epsilon(t) \tag{7.10.7}$$

where the error $\epsilon(t)$ is *completely negligible.* $\pi(t)$ is thus always assumed to be a polynomial of degree d, and since we can take d as large as we please, no generality is lost.

As before, assume that we can observe $\pi(t)$ without errors, letting the observations be

$$\tilde{y}_{n-L+r} \equiv \pi_{n-L+r} \qquad r = 0, 1, 2, \ldots, L \tag{7.10.8}$$

Thus $\tilde{y}_{n-L+r}$ is a polynomial of degree d in r, and so we can express it as a sum of the discrete Legendre polynomials, i.e.

$$\tilde{y}_{n-L+r} = \sum_{j=0}^{d} \left(\tilde{\gamma}_j\right)_n p_j(r) \tag{7.10.9}$$

(See Note)

where the r-origin is chosen to be at $t = (n - L)\tau$. (See Figure 7.1.) The $\tilde{\gamma}_j$'s can be obtained by forming

$$\begin{aligned} \sum_{k=0}^{L} \tilde{y}_{n-L+k} p_i(k) &= \sum_{k=0}^{L} \left[\sum_{j=0}^{d} \left(\tilde{\gamma}_j\right)_n p_j(k) \right] p_i(k) \\ &= \sum_{j=0}^{d} \left(\tilde{\gamma}_j\right)_n c_j^2 \delta_{ij} \\ &= \left(\tilde{\gamma}_i\right)_n c_i^2 \end{aligned} \tag{7.10.10}$$

Hence

$$\left(\tilde{\gamma}_j\right)_n = \frac{1}{c_j^2} \sum_{k=0}^{L} \tilde{y}_{n-L+k} p_j(k) \qquad (0 \le j \le d) \tag{7.10.11}$$

Note: $p_j(r)$ and c_j below were defined in (7.3.5).

Suppose, next, that the error-free total observation vector, $\tilde{Y}_{(n)}$, is used as the observation vector for a Fixed-Memory Polynomial Filter of degree m. As in (7.3.8), we write the estimating polynomial as a combination of the discrete Legendre polynomials, i.e.

$$\left[p^*(r)\right]_n = \sum_{j=0}^{m} \left(\gamma_j\right)_n p_j(r) \tag{7.10.12}$$

Then, either by the reasoning of (7.10.10), or else by that used in Section 7.3,

$$\left(\gamma_j\right)_n = \frac{1}{c_j^2} \sum_{k=0}^{L} \tilde{y}_{n-L+k} p_j(k) \qquad (0 \le j \le m) \tag{7.10.13}$$

We now form the systematic error defined in (7.10.1). The zeroth element of the vector B^* will be

$$\begin{aligned} \left[b^*(r)\right]_n &\equiv \left(\pi_0\right)_{n-L+r} - \left(x_0^*\right)_{n-L+r,n} \\ &= \tilde{y}_{n-L+r} - \left[p^*(r)\right]_n \end{aligned} \tag{7.10.14}$$

and by (7.10.9) and (7.10.12) this can now be seen to be

$$\begin{aligned} \left[b^*(r)\right]_n &= \sum_{j=0}^{d} \left(\tilde{\gamma}_j\right)_n p_j(r) - \sum_{j=0}^{m} \left(\gamma_j\right)_n p_j(r) \\ &= \sum_{j=0}^{m} \left[\left(\tilde{\gamma}_j\right)_n - \left(\gamma_j\right)_n\right] p_j(r) + \sum_{j=m+1}^{d} \left(\tilde{\gamma}_j\right)_n p_j(r) \end{aligned} \tag{7.10.15}$$

But comparison of (7.10.11) and (7.10.13) shows that

$$\left(\tilde{\gamma}_j\right)_n = \left(\gamma_j\right)_n \qquad (0 \le j \le m) \tag{7.10.16}$$

Hence for $d > m$, we see from (7.10.15) that the systematic error is

$$\left[b^*(r)\right]_n = \sum_{j=m+1}^{d} \left(\tilde{\gamma}_j\right)_n p_j(r) \tag{7.10.17}$$

Note that $\left[b^*(r)\right]_n$ *is a function of* r, *if* n *is fixed, and a function of* n, *if* r *is fixed.* It is thus important, when discussing systematic errors, to decide whether r or n is to be regarded as the independent variable.

In exactly the same way, the systematic error in the estimate of the ith derivative is easily shown to be

$$\begin{aligned}\left(b_i^*\right)_{n+h,n} &\equiv \left[\frac{d^i}{dt^i} b^*(r)\right]_n \Bigg|_{r=L+h} \\ &= \frac{1}{\tau^i} \sum_{j=m+1}^{d} \left(\tilde{\gamma}_j\right)_n \frac{d^i}{dr^i} p_j(r) \Bigg|_{r=L+h}\end{aligned} \tag{7.10.18}$$

This is then the general form of the ith element of the systematic error vector $B^*_{n+h,n}$.

7.11 BEHAVIOR OF THE SYSTEMATIC ERRORS WITH h AND m

From (7.10.18), we see that the systematic error in the estimate of the ith derivative of $\pi(t)$ is given by

$$\left[\frac{d^i}{dt^i} b^*(r)\right]_n = \frac{1}{\tau^i} \sum_{j=m+1}^{d} \left(\tilde{\gamma}_j\right)_n \frac{d^i}{dr^i} p_j(r) \tag{7.11.1}$$

This is simply a sum of the orthogonal polynomials and their derivatives, and so any behavior exhibited by them is also exhibited in (7.11.1).

Inside the observation interval the polynomials and their derivatives fluctuate about zero. All of their zeros are inside this interval, and from plots of these functions (see Figures 7.7 and 7.8), we see that the fluctuations away from zero are surprisingly small. The polynomials are characterized as being very *quiet* inside the smoothing interval. Taking sums of these quiet functions gives a relatively quiet function for (7.11.1), and as a result, *the systematic errors show relatively small oscillations about zero, inside the smoothing interval.*

Outside the observation interval, however, the orthogonal polynomials and their derivatives are devoid of zeros, and they are strictly increasing or decreasing with r. Thus in general as we leave the observation interval, the expression in (7.11.1) shows a tendency to increase rapidly. The general

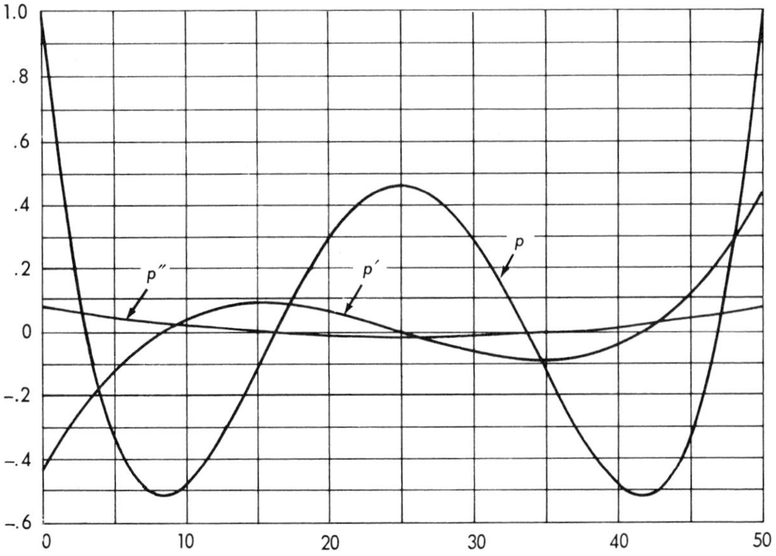

Fig. 7.7 $p(x; 4{,}50)$ *and its first two derivatives.*

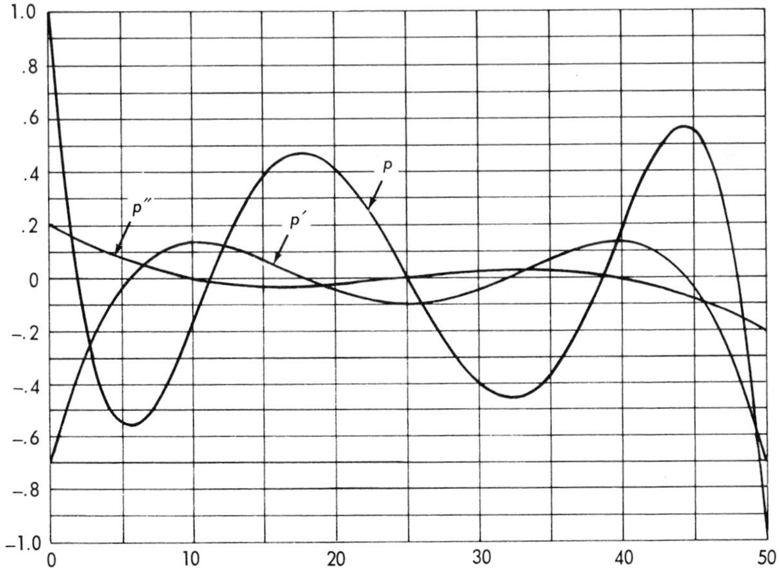

Fig. 7.8 $p(x; 5{,}50)$ *and its first two derivatives.*

picture is then one of small systematic errors inside the interval and large ones outside. This is, after all, what we would expect, *since* $\left[p^*(r)\right]_n$ *was chosen so as to approximate* $\pi(t)$ *well, inside the interval, with no conditions on how they were to agree outside of it.*

The zeros of the systematic errors inside the interval (if they exist), are related to the zeros of the orthogonal polynomials as shown by (7.11.1). Since half of the polynomials or their derivatives have a zero at the center, it follows then that the systematic error vector $\left[B^*(r)\right]_n$ should reflect this very high preponderance of zeros at that point. This is in fact the case, and so *smoothing to the center of the interval is highly desirable* from a systematic error standpoint.

In Section 7.7 we pointed out that the VRF's are close to minimum, if not actually minimum, at $r = L/2$, also because of the high preponderance of zeros at that point. *Thus for both reasons, smoothing to the center is always to be preferred,* if any freedom of choice exists.

Consider next the behavior of $b^*(r)$ *as we increase the degree of the estimating polynomial.* By assumption, the true signal was [see (7.10.7)],

$$\pi(t) = \sum_{k=0}^{d} a_k t^k + \epsilon(t) \tag{7.11.2}$$

where $|\epsilon(t)|$ can be made as small as we please by choosing d sufficiently large. From (7.11.1), we see that if we increase m, then $\left[b^*(r)\right]_n$ is reduced. Up to a point we can make the bias as small as we please by increasing the degree of the filter.

However, on p. 61, we pointed out that the set of discrete Legendre polynomials with parameter L form a finite set. Thus our attempts to increase the degree of the filter may, in fact, be foreshortened if L is not sufficiently large, since m cannot exceed L. Thus $m = L$ will be the highest degree filter we can implement, and if this is not adequate, we will be unable to approximate $\pi(t)$ to better than a certain precision unless we increase L.

In practice we seldom approach this limit, and by *increasing the degree of the filter the systematic errors are reduced.* But as we increase the degree, the variances of the random errors *increase* as we have shown in Sections 7.7 and 7.8. These conflicting facts mean that a balance must be struck in the choice of m between the systematic errors and the random errors.

7.12 BEHAVIOR OF THE SYSTEMATIC ERRORS WITH L AND τ

Suppose that the power series expansion for the true process $\pi(t)$ is

$$\pi(t) = \sum_{k=0}^{d} \left(a_k\right)_n t^k \tag{7.12.1}$$

where for each succeeding value of n the t-origin is chosen to lie at the beginning of an observation interval of our Fixed-Memory Polynomial Filter.† (See Figure 7.9.) Let $\tilde{y}_n$ be an error-free observation made on $\pi(t)$ at time $t = n\tau$. Then

$$\tilde{y}_{n-L+r} = \sum_{k=0}^{d} \left(a_k\right)_n \tau^k r^k \tag{7.12.2}$$

where the r-axis is also shown in Figure 7.9.

Using the Stirling numbers of the second kind (see (2.4.13) on p. 24), we know that

$$r^k = \sum_{i=0}^{k} \left[S^{-1}\right]_{ki} r^{(i)} \tag{7.12.3}$$

Thus (7.12.2) can be written as

$$\tilde{y}_{n-L+r} = \sum_{k=0}^{d} \left(a_k\right)_n \tau^k \sum_{i=0}^{k} \left[S^{-1}\right]_{ki} r^{(i)} \tag{7.12.4}$$

†It is because the t-origin is redefined for each value of n that the constants a_k in (7.12.1) are subscripted with n.

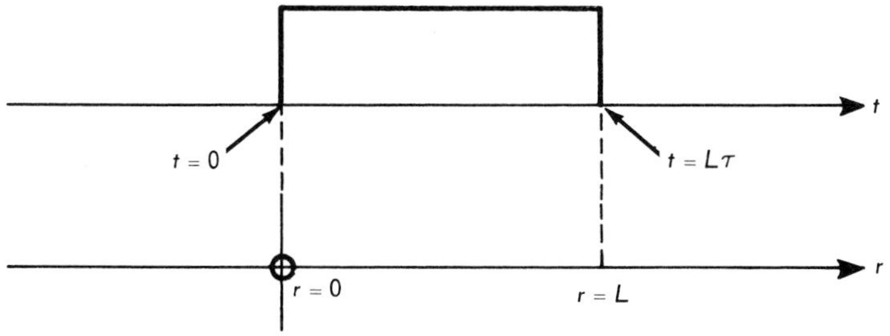

Fig. 7.9 *The t and r axes.*

We interchange the order of summation. Then

$$\tilde{y}_{n-L+r} = \sum_{i=0}^{d} r^{(i)} \left[\sum_{k=i}^{d} \left(a_k\right)_n \tau^k \left[S^{-1}\right]_{ki} \right] \tag{7.12.5}$$

$$= \sum_{i=0}^{d} \lambda_i(\tau, n)\, r^{(i)}$$

where we define

$$\lambda_i(\tau, n) = \sum_{k=i}^{d} \left(a_k\right)_n \tau^k \left[S^{-1}\right]_{ki} \tag{7.12.6}$$

Observe that if $i > 0$, then

$$\lim_{\tau \to 0} \lambda_i(\tau, n) = 0 \tag{7.12.7}$$

In Appendix I we prove that we can express $r^{(i)}$ as a linear combination of the discrete Legendre polynomials as follows:

$$r^{(i)} = \frac{L^{(i)}}{\binom{2i}{i}} \sum_{j=0}^{i} (-1)^j \left(\frac{2j+1}{2i+1}\right) \binom{2i+1}{i-j} p_j(r) \tag{7.12.8}$$

Define

$$g_{ij} \equiv \frac{(-1)^j \left(\dfrac{2j+1}{2i+1}\right) \dbinom{2i+1}{i-j}}{\dbinom{2i}{i}} \tag{7.12.9}$$

Then (7.12.8) can be written

$$r^{(i)} = L^{(i)} \sum_{j=0}^{i} g_{ij}\, p_j(r) \tag{7.12.10}$$

and so (7.12.5) becomes

$$\tilde{y}_{n-L+r} = \sum_{i=0}^{d} \lambda_i(\tau, n) L^{(i)} \sum_{j=0}^{i} g_{ij} p_j(r) \tag{7.12.11}$$

Interchanging the order of summation once more,

$$\begin{aligned} \tilde{y}_{n-L+r} &= \sum_{j=0}^{d} p_j(r) \left[\sum_{i=j}^{d} \lambda_i(\tau, n) L^{(i)} g_{ij} \right] \\ &= \sum_{j=0}^{d} \left(\tilde{\gamma}_j\right)_n p_j(r) \end{aligned} \tag{7.12.12}$$

where

$$\left(\tilde{\gamma}_j\right)_n \equiv \sum_{i=j}^{d} \lambda_i(\tau, n) L^{(i)} g_{ij} \tag{7.12.13}$$

We have thus expressed $\tilde{y}_{n-L+r}$ as a sum of the discrete Legendre polynomials in the form of (7.10.9), *and so* (7.12.13) *is the explicit form for* $\left(\tilde{\gamma}_j\right)_n$, *given the power series expansion for* $\pi(t)$ *about the beginning of the observation interval.* We are now in a position to see how $\left[b^*(r)\right]_n$ behaves as we vary L.

From (7.10.17) and (7.12.13), we have

$$\left[b^*(r)\right]_n = \sum_{j=m+1}^{d} \left[\sum_{i=j}^{d} \lambda_i(\tau, n) L^{(i)} g_{ij} \right] p_j(r) \tag{7.12.14}$$

and so when L is large the coefficients multiplying the $p_j(r)$'s increase rapidly due to the $L^{(i)}$ term. Outside the observation interval, the $p_j(r)$'s are all nonzero. Thus in that region $\left[b^*(r)\right]_n$ consists of a sum of nonzero numbers, *all of which are increasing in magnitude with* L. Although there may be cancellations at discrete points, for various L and r, in general it is easy to see that *as* L *increases the magnitude of the systematic errors will increase.*

Consider next how $\left[b^*(r)\right]_n$ behaves as we vary τ. As with variance reduction (see p. 250), there are two cases.

CASE A

τ is decreased while L is held constant.

Then from (7.12.7), all of the $\lambda_i(\tau, n)$ appearing in (7.12.14) will decrease, and so, in the limit, for fixed r, $\left[b^*(r)\right]_n$ is seen to go to zero as $\tau \to 0$. This means that any power series can be approximated arbitrarily well by a polynomial of given degree if the interval over which the approximation is made is taken to be sufficiently small.

CASE B

τ is decreased while L is increased so that $T = L\tau$ is constant.

From (7.12.6), as $\tau \to 0$,

$$\lambda_i(\tau, n) \to \left(a_i\right)_n \tau^i \tag{7.12.15}$$

and so as $\tau \to 0$ and $L \to \infty$, (7.12.13) gives

$$\begin{aligned}\left(\tilde{\gamma}_j\right)_n &\to \sum_{i=j}^{d} \left(a_i\right)_n \tau^i L^i g_{ij} \\ &= \sum_{i=j}^{d} \left(a_i\right)_n T g_{ij}\end{aligned} \tag{7.12.16}$$

which is a constant. The systematic errors will then tend to an asymptotic value which would be the same as the errors of an approximation of $\pi(t)$, over the *continuous* range $0 \leq t \leq T$, using the *continuous* Legendre polynomials up to degree m, defined in (3.1.2).

From the above arguments it is clearly desirable, from a systematic error standpoint, that we keep T as small as possible. For τ fixed, *this means that L should be kept as small as possible.* This is in direct contradiction to the choice we make to minimize the random output errors which is to make L as large as possible (see p. 250).

Clearly then, as with the choice of m, a compromise must be struck in the choice of L so that the systematic errors are balanced, in some sense, against the random output errors. We now present an analytical way of achieving this compromise.

7.13 BALANCING THE SYSTEMATIC AND RANDOM ERRORS

In practice we seldom know the true form of $\pi(t)$. (If we do, so much the better!) What we usually know are the *maximum* values, or *bounds* on the coefficients of the power series, in the general time range over which the filter is to operate. We now make the assumption that, at all times, $\pi(t)$ has a power series which converges rapidly enough so that if truncated, *then the truncation error is essentially dominated by the first neglected term.*

Thus (restricting ourselves to, say, the position output of the filter) by (7.10.17), if $\left[p^*(r)\right]_n$ is of degree m, then

$$\left[b^*(r)\right]_n \approx \left(\tilde{\gamma}_{m+1}\right)_n p_{m+1}(r) \tag{7.13.1}$$

and by (7.12.13), setting $d = m + 1$, we have

$$\left(\tilde{\gamma}_{m+1}\right)_n = \lambda_{m+1}(\tau, n)\, L^{(m+1)} g_{m+1,\, m+1} \tag{7.13.2}$$

From (7.12.6), setting $d = m + 1$,

$$\lambda_{m+1}(\tau, n) = \left(a_{m+1}\right)_n \tau^{m+1} \tag{7.13.3}$$

since $\left[S^{-1}\right]_{m+1,\, m+1} = 1$. Moreover, by (7.12.9), to within sign,

$$g_{m+1,\, m+1} = \frac{1}{\binom{2m+2}{m+1}} \tag{7.13.4}$$

Hence, we obtain (to within sign)

$$\left[b^*(r)\right]_n \approx \frac{\left(a_{m+1}\right)_n \tau^{m+1} L^{(m+1)}}{\binom{2m+2}{m+1}} p_{m+1}(r) \tag{7.13.5}$$

and in general (to within sign),

$$\left[D^i b^*(r)\right]_n \approx \frac{1}{\tau^i} \frac{\left(a_{m+1}\right)_n \tau^{m+1} L^{(m+1)}}{\binom{2m+2}{m+1}} \frac{d^i}{dr^i} p_{m+1}(r) \tag{7.13.6}$$

Equation (7.13.5) [and its generalization in (7.13.6)] is an extremely useful result. *It enables us to compute the systematic error anywhere in the observation interval for any* L *and* n, *on the assumption that* $\pi(t)$ *is a polynomial of degree one more than the model.*

As an example, assume that $m = 1$. (i.e. we are fitting a first-degree polynomial to a quadratic.) Then (7.13.5) gives

$$\left[b^*(r)\right]_n = \frac{a_2 \tau^2 L(L-1)}{6} p_2(r) \tag{7.13.7}$$

where we have used the fact that a_2 is a constant in this case. This then is the error in the approximation which results when we perform such a fit.

Perhaps the most useful application of (7.13.5) *is that it enables us to fix* L *and* m *so that the systematic errors are balanced relative to the random ones.* This is done as follows.

From (7.7.6) we recall that the variance of the random errors in the position estimate is

$$\left[S^*_{n,n}\right]_{0,0} = \frac{\lambda_{0,0}(m)}{L} \sigma_\nu^2 \tag{7.13.8}$$

where $\lambda_{0,0}(m)$ is tabulated in either Table 7.2 or Table 7.3, depending on whether we are smoothing to the end point or the center. Assume that the former is the case. Then we obtain the *bias error* at the end point by setting $r = L$ in (7.13.5). As is readily verified from (3.2.20), when $r = L$, then $|p_{m+1}(r)| = 1$ for *any* m, and so (7.13.5) becomes

$$\left[b(r)\right]_n \approx \frac{\left(a_{m+1}\right)_n \tau^{m+1} L^{(m+1)}}{\binom{2m+2}{m+1}} \tag{7.13.9}$$

We now make the assumption that we know enough about $\pi(t)$ so that a bound can be placed on $\left(a_{m+1}\right)_n$. This being the case we replace $\left(a_{m+1}\right)_n$ by that bound, obtaining

$$\left|\left[b^*(r)\right]_n\right| \leq \frac{|a_{m+1}| \tau^{m+1} L^{(m+1)}}{\binom{2m+2}{m+1}} \tag{7.13.10}$$

From this expression *we see how the bias increases with* L, whereas in (7.13.8) *we see how the random-error variance decreases with* L.

In practice we can tolerate a certain amount of bias in our answers. Just how much depends on each specific case, but suppose that we decide that we can accept a bias error equal to the one-σ value of the random errors. We thus equate (7.13.10) to the square-root of (7.13.8), and for *fixed* m, we obtain a value for L. *This then is the memory length which achieves our balance* for that value of m. Repeating this procedure for $m = 0, 1, 2, \ldots$ we then select the best overall situation *and in this way the balance between the random and systematic errors can be optimized.*

The above method works very well in practice and can generally be used very effectively to select specific values for L and m in each given situation.

7.14 TREND REMOVAL

It is sometimes the case that a nominal trajectory exists which is close to the data being filtered by a polynomial smoothing algorithm. This can be used to great advantage.

Suppose, specifically, that our model is given by the nonlinear differential equation

$$\frac{d}{dt} X(t) = F[X(t)] \tag{7.14.1}$$

where $X(t)$ is, for example, the state-vector of a body in orbit. Thus,

$$X(t) = \begin{pmatrix} x(t) \\ y(t) \\ z(t) \\ \dot{x}(t) \\ \dot{y}(t) \\ \dot{z}(t) \end{pmatrix} \tag{7.14.2}$$

where the x, y, z axes form some Cartesian reference frame.

We observe say range, azimuth and elevation (see p. 168) and obtain the sequences of observation

$$\begin{aligned} &\rho_{n-L}, \ldots, \rho_n \\ &\psi_{n-L}, \ldots, \psi_n \\ &\theta_{n-L}, \ldots, \theta_n \end{aligned} \tag{7.14.3}$$

Based on the nominal trajectory $\overline{X}(t)$ that we assumed to be available, we now also compute the *simulated observations*†

$$\begin{aligned} &\bar{\rho}_{n-L}, \ldots, \bar{\rho}_n \\ &\bar{\psi}_{n-L}, \ldots, \bar{\psi}_n \\ &\bar{\theta}_{n-L}, \ldots, \bar{\theta}_n \end{aligned} \tag{7.14.4}$$

and we then form the differences

$$\delta\rho_{n-k} \equiv \rho_{n-k} - \bar{\rho}_{n-k} \qquad 0 \le k \le L \tag{7.14.5}$$

$$\delta\psi_{n-k} \equiv \psi_{n-k} - \bar{\psi}_{n-k} \qquad 0 \le k \le L \tag{7.14.6}$$

$$\delta\theta_{n-k} \equiv \theta_{n-k} - \bar{\theta}_{n-k} \qquad 0 \le k \le L \tag{7.14.7}$$

We then send these differences to three Fixed-Memory Polynomial Filters, and obtain the center-smoothed values

$$\delta\rho^*_{c,n},\ \delta\dot{\rho}^*_{c,n},\ \delta\psi^*_{c,n},\ \delta\dot{\psi}^*_{c,n},\ \delta\theta^*_{c,n},\ \delta\dot{\theta}^*_{c,n}$$

to which we add back the simulated observations

$$\bar{\rho}_c,\ \bar{\dot{\rho}}_c,\ \bar{\psi}_c,\ \bar{\dot{\psi}}_c,\ \bar{\theta}_c,\ \bar{\dot{\theta}}_c$$

computed at the center of the observation interval. This gives us the outputs

$$\begin{aligned} \rho^*_{c,n} &\equiv \delta\rho^*_{c,n} + \bar{\rho}_c \\ \dot{\rho}^*_{c,n} &\equiv \delta\dot{\rho}^*_{c,n} + \bar{\dot{\rho}}_c \\ \psi^*_{c,n} &\equiv \delta\psi^*_{c,n} + \bar{\psi}_c \\ \dot{\psi}^*_{c,n} &\equiv \delta\dot{\psi}^*_{c,n} + \bar{\dot{\psi}}_c \\ \theta^*_{c,n} &\equiv \delta\theta^*_{c,n} + \bar{\theta}_c \\ \dot{\theta}^*_{c,n} &\equiv \delta\dot{\theta}^*_{c,n} + \bar{\dot{\theta}}_c \end{aligned} \tag{7.14.8}$$

The scheme is depicted in Figure 7.10 in which we show the portion associated with the ρ-data, assuming a quadratic filter. A similar system would be used to smooth the ψ and θ observations.

†See (6.2.13).

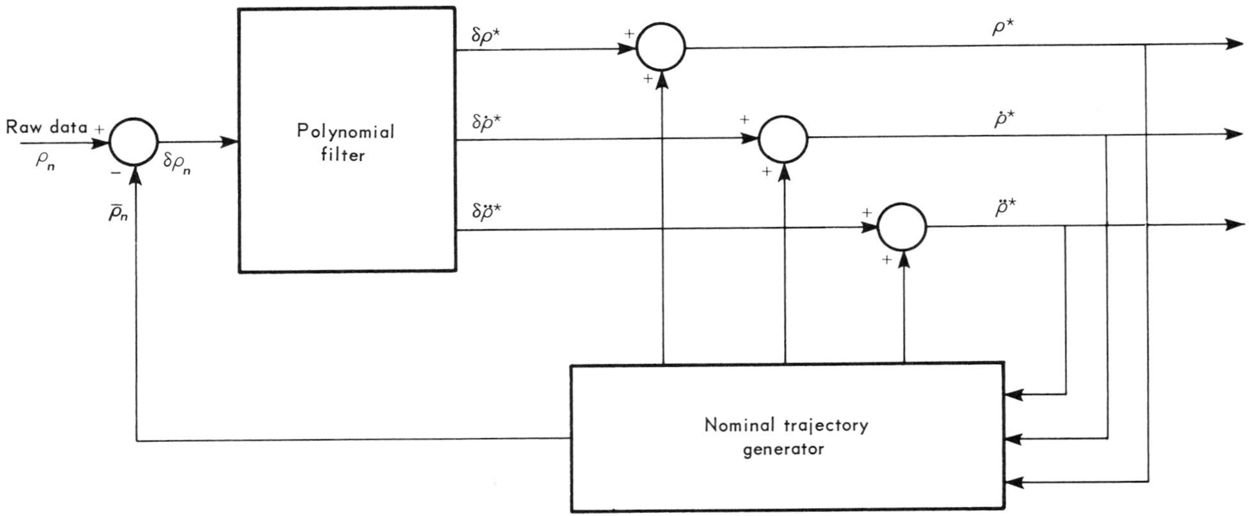

Fig. 7.10 *Polynomial filter with trend removal.*

From the 6-vector in (7.14.8) we can, if we so choose, compute† a state-vector $X^*_{c,n}$ which we now use in place of our nominal trajectory, thereby setting up an iteration procedure. We can also use this $X^*_{c,n}$ as the nominal trajectory *on the next cycle,* when further observations become available.

The rationale underlying the trend-removal procedure described above is that the differences in (7.14.5), (7.14.6) and (7.14.7), being formed by subtracting two close trajectories from one another will have power series with *reduced coefficients.* Thus let the true range-function be

$$\rho(t) = \alpha_0 + \alpha_1 t + \alpha_2 t^2 + \cdots \tag{7.14.9}$$

and let the nominal range-function be

$$\bar{\rho}(t) = \bar{\alpha}_0 + \bar{\alpha}_1 t + \bar{\alpha}_2 t^2 + \cdots \tag{7.14.10}$$

Then the $\delta\rho$ function of (7.14.5) is (neglecting observation errors)

$$\delta\rho(t) = (\alpha_0 - \bar{\alpha}_0) + (\alpha_1 - \bar{\alpha}_1)t + (\alpha_2 - \bar{\alpha}_2)t^2 + \cdots \tag{7.14.11}$$

which is a power series whose coefficients are small, relative to those of (7.14.9).

The systematic errors, we have seen, *depend linearly on the coefficients of the power series* [see e.g. (7.13.6)]. Assuming a first-degree polynomial filter, the systematic-errors, *if we filter the total ρ observations,* would be proportional‡ to α_2, whereas if we filter the $\delta\rho$'s then they are proportional to $\alpha_2 - \bar{\alpha}_2$. Thus, in the former case the bias error is, say

$$|b(\rho^*)| = k|\alpha_2| \tag{7.14.12}$$

whereas in the latter case it is

$$|b(\delta\rho^*)| = k|\alpha_2 - \bar{\alpha}_2| \tag{7.14.13}$$

However, we now add the nominal trajectory back into $\delta\rho^*$ [see (7.14.8)] *and this does not lead to any further systematic errors, since we are merely adding back in what we subtracted out earlier.* Hence the systematic errors in the trend removal case (7.14.13) become

$$|b(\delta\rho^* + \bar{\rho})| = k|\alpha_2 - \bar{\alpha}_2|$$

which is clearly smaller than (7.14.12) if α_2 and $\bar{\alpha}_2$ are close.

†This is discussed in further detail in Section 8.5.

‡We are assuming, as we did on p. 275 that the first neglected term in the truncated power series dominates the entire truncation error.

Thus by trend removal we are able to reduce (significantly in many cases) the systematic errors, thereby enabling us to 1) reduce the degree of the filter and 2) lengthen the observation interval. Both of the latter lead to reductions in the random-error variances.

Trend removal is thus strongly recommended whenever a polynomial filter is being applied. When a nominal trajectory is not available initially, we can derive one based on the first set of outputs of the filter. This is discussed in Section 8.5 in further detail.

Examples on the technique of trend removal are included in the exercises for Chapter 8.

7.15 CASCADED SIMPLE AVERAGING

A very effective method of reducing the amount of computation for the Fixed-Memory Polynomial Filters was devised by R. B. Blackman, and is entitled *Cascaded Simple Averaging.* It is described in [7.2] in a somewhat condensed way, but the reader who is seriously interested in these types of filters is strongly advised to make the investment needed to understand this technique.

Basically, what Blackman shows is that, to a very close approximation, the operations of the Fixed-Memory Polynomial Filters can be reduced to sequences comprised *only* of push-down, addition and subtraction, together with an extremely small number of multiplies. The weights of the filter matrix disappear, and are, in effect, almost entirely replaced by +1's, −1's and zeros. This significantly reduces the computational load and also decreases the memory-space requirements.

Blackman's scheme is very well suited to implementation in special purpose digital computers involving delay-line elements. It would be relatively easy to implement fairly large numbers of these filters in a single such machine, and the overall amount of equipment required would be quite small, this by comparison to a machine in which an equivalent number of Fixed-Memory Polynomial Filters, as described earlier in this chapter, were being operated.

EXERCISES

7.1 a) Starting from

$$\left[p^*(r)\right]_n = \left(\beta_0\right)_n + \left(\beta_1\right)_n r$$

obtain the expressions for β_0 and β_1 which best fit $p^*(r)$ to the observation sequence $y_{n-L}, \ldots, y_n$ in the sense of least-squares.

b) Hence set up $p^*(r)$ in the form

$$\left[p^*(r)\right]_n = \sum_{k=0}^{L} w(k, L, r) y_{n-L+k}$$

and verify that the multiplier $w(k, L, r)$ is a first-degree polynomial function of r.

c) Using b) above, show that

$$x^*_{n,n} = \sum_{k=0}^{L} \frac{2 - 2L + 6k}{(L+2)(L+1)} y_{n-L+k}$$

Verify that for $L = 4$,

$$x^*_{n,n} = \frac{1}{10}(6y_n + 4y_{n-1} + 2y_{n-2} - 2y_{n-4})$$

d) Attempt to repeat the above for $p^*(r)$ a quadratic in r.

7.2 a) Starting from

$$\left[p^*(r)\right]_n = \left(\beta_0\right)_n \varphi_0(r) + \left(\beta_1\right)_n \varphi_1(r) + \left(\beta_2\right)_n \varphi_2(r)$$

where the φ's are the normalized discrete Legendre polynomials, obtain the expressions for the β's which best fit $p^*(r)$ to the sequence $y_{n-L}, \ldots, y_n$ in the sense of least-squares.

b) Hence set up $p^*(r)$ in the form

$$\left[p^*(r)\right]_n = \sum_{k=0}^{L} w(k, L, r) y_{n-L+k}$$

and note that $w(k, L, r)$ is a quadratic in r.

c) Compare the above with the difficulties encountered in part d) of Ex. 7.1.

7.3 a) Verify that for a degree-1 estimator

$$\dot{x}^*_{n,n} = \frac{1}{\tau} \sum_{k=0}^{L} \frac{12k - 6L}{(L+2)(L+1)L} y_{n-L+k}$$

and reconcile this with (7.3.30) when $L = 4$.

b) Verify that for $m = 1$, $L = 4$,

$$Z^*_{n-2,n} = \frac{1}{10}\begin{pmatrix} 2 & 2 & 2 & 2 & 2 \\ 2 & 1 & 0 & -1 & -2 \end{pmatrix} Y_{(n)}$$

is the Fixed-Memory Polynomial Filter which estimates at the center of the observation interval.

c) Prove that for any L, when $m = 1$,

$$\left(z^*_0\right)_{c,n} = \frac{1}{L+1} \sum_{k=0}^{L} y_{n-k}$$

where c means the center of the interval.

d) Compare c) above to the zeroth-degree polynomial estimator.

7.4 a) Set up the Fixed-Memory Polynomial Filter

$$Z^*_{n,n} = W(0)\, Y_{(n)}$$

for $m = 2$, $L = 4$, making use of (7.4.10) using P as defined in (7.4.4).

b) Repeat a) above, but this time use P as defined in (7.4.15).

c) From either a) or b) obtain the 1-step predictor matrix $W(1)$ and the center-smoothing matrix $W(-L/2)$.

7.5 a) On p. 46 we defined the *impulse response* of a system as its output when the input is a Kronecker delta $\delta_{n,0}$. Verify that for the Fixed-Memory Polynomial Filter, $Z^*_{n,n} = WY_{(n)}$, the impulse response of the $\left(z^*_i\right)_{n,n}$ channel is simply the sequence of numbers given in the ith row of W. Hence infer that *any* Fixed-Memory Filter has an impulse response which becomes *precisely zero* after a finite time equal to the memory length.

b) Plot the impulse responses for each of the elements of $Z^*_{n,n}$ for the cases $m = 1$, $m = 2$ with $L = 10$ [make use of (7.4.10)].

7.6 For the case where $R = I$, (7.5.4) gives us

$$\left[S^*_{n,n}\right]_{ij} = \sum_{k=0}^{L} \left[W(0)\right]_{ik} \left[W(0)\right]_{jk}$$

Let the input to the filter

$$Z^*_{n,n} = W(0)\, Y_{(n)}$$

be a Kronecker delta $\delta_{n,0}$.

a) Prove that if we square the sequence of outputs from the z_i^* channel and add them together, then we obtain $\left[S^*_{n,n}\right]_{i,i}$ above.

b) Prove that if we multiply each output from the z_i^* channel by the concurrent one from the z_j^* channel and then add these products, we obtain $\left[S^*_{n,n}\right]_{i,j}$ above.

In this way we can, with the aid of a computer, obtain numerical values for $S^*_{n,n}$ for the case $R = I$.

c) Using the matrix $W(0)$ of (7.3.27) carry out the above operations either by hand or preferably with the aid of a computer, and reconcile the result with (7.5.16).

7.7 Verify that the product PC in (7.4.10) is nonsingular and that the rows of B are linearly independent. Hence infer that the matrix $W(h)$ of (7.4.10) has full row-rank and that by Ex. 6.4, WW^T is positive definite.

7.8 a) Obtain the quadratic least-squares estimating polynomial $p^*(r)$ for the observation vector

$$Y = \begin{pmatrix} \sin 90^\circ \\ \sin 67\tfrac{1}{2}^\circ \\ \sin 45^\circ \\ \sin 22\tfrac{1}{2}^\circ \\ \sin 0^\circ \end{pmatrix}$$

by the use of (7.3.20).

b) Now tabulate $p^*(r)$ for 1° increments (using a computer) and compare the results to the true values of $\sin\theta$ $(0 \le \theta \le 90^\circ)$.

c) Note that the estimate $p^*(r)$ above is a linear combination of $\varphi_0(r)$, $\varphi_1(r)$ and $\varphi_2(r)$ [c/f (7.3.14)] and so it can be rearranged into a power series

$$p^*(r) = \alpha_0 + \alpha_1 r + \alpha_2 r^2$$

in which form it could be programmed to serve as the basis for a sine/cosine subroutine.

Note: A slightly better quadratic approximation to $\sin\theta$ can be obtained by *minimizing the maximum error* rather than minimizing the *squared* residuals. (See e.g. [7.3].)

7.9 a) Using a computer, generate the sequence of observations

$$y_n = \frac{n}{100} + \nu_n$$

where ν_n is obtained from a random-noise subroutine within the computer. (Set the subroutine so that $\sigma_\nu = 10^{-3}$.)

b) Now apply this sequence to a first-degree Fixed-Memory Polynomial Filter using first $L = 4$, then $L = 10$, then $L = 100$. Compute the errors in the outputs.

c) Observe that those errors diminish as L is increased.

d) In each case also compute the elements of the covariance matrix by the procedure outlined in Ex. 7.6.

7.10 Given that

$$R_{(n)} = \begin{pmatrix} 2 & 1 & 0 & 0 & 0 \\ 1 & 2 & 1 & 0 & 0 \\ 0 & 1 & 2 & 1 & 0 \\ 0 & 0 & 1 & 2 & 1 \\ 0 & 0 & 0 & 1 & 2 \end{pmatrix}$$

use the weight-matrix W from (7.3.27) in (7.5.4) to obtain $S^*_{n,n}$. Verify that the diagonal terms of $S^*_{n,n}$ are 1.76 and 0.28.

7.11 Using (7.5.22), obtain the 2×2 (first-degree) matrix $S^*_{c,n}$ where c means "at the center of the observation interval." Verify that $S^*_{c,n}$ is diagonal, i.e. the position and velocity estimates are uncorrelated. Note: In general for higher degree filters, $S^*_{c,n}$ is not diagonal.

7.12 Verify (7.5.8).

7.13 Verify (7.5.10).

7.14 Verify (7.5.16).

7.15 Verify (7.5.25).

7.16 Obtain $S^*_{n+1,n}$ for the quadratic and cubic cases and verify that the diagonal elements for degrees zero through three are as given on p. 371 with n replaced by L.

7.17 Verify that $S^*_{n+h,n}$ of (7.5.27) reduces to (7.5.28).

7.18 Using (7.5.28) obtain the 2×2 matrix $S^*_{n+1,n}$ from (7.5.25).

7.19 Using the matrices $S^*_{n+1,n}$ from Ex. 7.16, verify the assertions made in Case A and Case B on pp. 250 and 251.

7.20 Set up the filter

$$B = WY_{(n)}$$

where B is the vector of β's of (7.3.8). Show that if $R_{(n)} = I$, then the covariance matrix of B is diagonal.

7.21 a) Plot the functions $\left[\varphi_k(L + h)\right]^2 \Big|_{L=4}$ for $k = 0,1,2$, and h in the range -10 to $+6$.

b) By adding the above plots, obtain $\left[S^*_{n+h,n}\right]_{0,0}$ of (7.7.8) for $m = 0,1$ and 2.

c) Verify that in each case the above functions are strictly increasing for $h > 0$ and $h < -4$.

7.22 a) For the discrete Legendre polynomials $\varphi_0(r)$, $\varphi_1(r)$, $\varphi_2(r)$ set $r = \rho/L$ and let $L \to \infty$, thereby obtaining

$$\varphi_0(\rho) = \left(\frac{1}{L}\right)^{1/2}$$

$$\varphi_1(\rho) = \left(\frac{3}{L}\right)^{1/2} (1 - 2\rho)$$

$$\varphi_2(\rho) = \left(\frac{5}{L}\right)^{1/2} (1 - 6\rho + 6\rho^2)$$

b) Plot $L\sum_{k=0}^{2}\left[\varphi_k(\rho)\right]^2$ for $0 < \rho < 1$ and verify that the curve in Figure 7.5 is obtained. Thus given L, we can use that curve to obtain $\left[S^*_{n+h,n}\right]_{0,0}$ for any value of h. In a similar manner we can obtain the curves for any of the diagonal elements of $S^*_{n+h,n}$ for any degree.

c) Reconcile the end-point and center values of Figure 7.5 with the entries $m = 2$, $i = 0$ of Tables 7.2 and 7.3.

7.23 a) Using Table 7.2, verify that for L large, the first-derivative updated estimate based on a quadratic has a $VRF = 192/\tau^2 L^3$, and hence that its variance is $\sigma_\nu^2\,(192/\tau^2 L^3)$.

b) Using Table 7.2, obtain the variance of a 1-step prediction of the second derivative, based on a cubic, for L large.

$$\text{Ans: Var} = \sigma_\nu^2 \frac{25920}{\tau^4 L^5}$$

c) Repeat both of the above assuming smoothing to the center.

$$\text{Ans: a) Var} = \sigma_\nu^2 \frac{12}{\tau^2 L^3}, \quad \text{b) Var} = \sigma_\nu^2 \frac{720}{\tau^4 L^5}.$$

Note how much smaller the variances are when smoothing is to the center.

7.24 Let the input to a zeroth-degree filter be $\pi(t) = t$ with $\tau = 1$. Then by (7.10.8), $\tilde{y}_{n-L+r} = n - L + r$.

a) Using (7.10.11) verify that

$$\left(\tilde{\gamma}_0\right)_n = n - \frac{L}{2} \qquad \left(\tilde{\gamma}_1\right)_n = -\frac{L}{2}$$

and hence show that (7.10.9) sums to

$$\sum_{j=0}^{1} \left(\tilde{\gamma}_j\right)_n p_j(r) = n - L + r$$

b) Verify that (7.10.17) gives the bias error as

$$\left[b^*(r)\right]_n = (-L/2)\,[1 - (2r/L)]$$

Hence infer that the bias error in the updated estimate is

$[b^*(L)]_n = L/2$, and that the bias error at the center of the observation interval is zero. Note how $[b^*(L)]_n$ increases with L.

c) Verify that the zeroth-degree Fixed-Memory Polynomial Filter is

$$x_0^* = \frac{1}{L+1} \sum_{k=0}^{L} y_{n-L+k}$$

Show that for the observations $y_{n-L+k} = n - L + k$, this gives $x_0^* = n - (L/2)$. Verify that this implies a bias error equal to $L/2$ if smoothing is to the end point and a bias error of zero if smoothing is to the center. Reconcile this with b) above.

7.25 Assuming a filter which is first degree and that the input is $\pi(t) = t^2$, with $\tau = 1$.

a) Show that (7.10.11) leads to

$$\left(\tilde{\gamma}_0\right)_n = \left[(n-L)^2 + \frac{L}{2}(2n-L) - \frac{L(L-1)}{6}\right]$$

$$\left(\tilde{\gamma}_1\right)_n = -\frac{L}{2}(2n-L)$$

$$\left(\tilde{\gamma}_2\right)_n = \frac{L(L-1)}{6}$$

and verify that

$$\sum_{j=0}^{2} \left(\tilde{\gamma}_j\right)_n p_j(r) = (n-L+r)^2$$

b) Verify that the bias error is

$$\left[b^*(r)\right]_n = \frac{L(L-1)}{6}\left[1 - 6\frac{r}{L} + 6\frac{r(r-1)}{L(L-1)}\right]$$

and hence that $\left[b^*(L)\right]_n = [L(L-1)]/6$, and

$$\left[b^*\left(\frac{L}{2}\right)\right]_n = -\frac{L(L+2)}{12}$$

Note that the bias errors increase like L^2.

c) Verify that (7.13.5) gives the same answer as b) above.

d) Infer that if the filter were zeroth degree with $\pi(t) = t^2$, then the bias error would be

$$\left[b^*(r)\right]_n = -\frac{L}{2}(2n - L)\left(1 - 2\frac{r}{L}\right) + \frac{L(L-1)}{6}\left[1 - 6\frac{r}{L} + 6\frac{r(r-1)}{L(L-1)}\right]$$

and that

$$\left[b^*(L)\right]_n = \frac{L(6n - 2L - 1)}{6}$$

Note that $\left[b^*(L)\right]_n$ is now a first-degree polynomial in n, whereas in b) it was a zeroth-degree polynomial in n.

e) Starting from the zeroth-degree estimator [see Ex. 7.24c)]

$$\left(x_0^*\right)_{n,n} = \frac{1}{L+1}\sum_{k=0}^{L} y_{n-L+k}$$

verify directly that when the input is $\pi(t) = t^2$, the bias error $\left[b^*(L)\right]_n$ is as given in d) above.

7.26 a) Verify that for a first-degree Fixed-Memory Polynomial Filter the *updated* position estimate has

$$\sigma_{pos} = \sqrt{\frac{2(2L+3)}{L(L+1)}}\,\sigma_\nu$$

Let $\sigma_\nu = 200$ ft. and compute σ_{pos} for $L = 5, 10, 15, 20$.

b) Assume that

$$R(t) = R_0 + t\dot{R}_0 + \frac{t^2}{2}\ddot{R}_0$$

and verify that the systematic error in position is bounded by

$$\left|\frac{\ddot{R}}{2}\right| \frac{\tau^2 L(L-1)}{6}$$

For $\ddot{R} = 32$ ft/sec^2, $\tau = 1$, compute $\left[b^*(L)\right]_n$ for $L = 5, 10, 15, 20$.

c) Verify that the systematic error in position equals the 1σ random error in position when $L \approx 8$.

REFERENCES

1. Szego, G., "Orthogonal Polynomials," Amer. Math. Soc. Solloq. Pub. 23, rev. ed., 1959.
2. Blackman, R. B., "Data Smoothing and Prediction," Addison-Wesley, 1965.
3. Hastings, C., "Approximations for Digital Computers," Princeton University Press, 1955.

See Also

4. Joksch, H. C., "Random Errors of Derivatives Obtained from Least-Squares Approximations to Empirical Functions," SIAM Review, Jan. 1966, Vol. 8, No. 1, pp. 47 et seq.
5. Joksch, H. C., "Reduction of the Variance," SIAM Review, April 1966, Vol. 8, No. 2, pp. 211 et seq.

8

GENERALIZED FIXED-MEMORY FILTERING

8.1 INTRODUCTION

In this chapter we consider a scheme which is the generalization of the Fixed-Memory Polynomial Filter considered in the previous chapter. We call the present technique the *Generalized Fixed-Memory Filter,* and we shall show how any or all of the following extensions can be incorporated. (The reader should compare the following enumeration to the conditions implicit in Chapter 7, and he will readily observe how very much more flexible the present scheme is by comparison with the previous one.)

A. The Observation Method

1. Observation instants need not be equally spaced.
2. A vector of observations of mixed dimensions can be made at each instant, and the entire vector can be incorporated into the formation of the estimate.
3. Two or more instruments observing the same quantity can be employed simultaneously.

4. The vector of observations made at each instant can be either linearly or nonlinearly related to the process state-vector.

B. The Process Model

The model on which the process is assumed to be based can be either a constant coefficient linear differential equation, which of course includes polynomial models, a time-varying linear differential equation, or a nonlinear differential equation with possibly time-varying parameters.

C. Incorporation of the Data

Observations can be incorporated into the estimate with differing emphasis, i.e. observations which are known to be high-quality can be stressed more heavily than observations which are known to be of low-quality. The latter need not be discarded, but can be incorporated precisely according to their merit.

It is easy to see that the implementation of all of these possibilities into a fixed-memory scheme constitutes a significant generalization over the fixed-memory filters of the preceding chapter. However, nothing so luxurious comes without a price, and in the present case the price will be three-fold.

First, the filters will not emerge in the *explicit* form to which we became accustomed in the previous chapter, where the weight-matrix of numbers, W, could be computed off-line, once and for all, and used whenever required. The generalized filters will consist, rather, of an *implicit* algorithm, a sequence of mathematical operations, which will have to be executed repeatedly by the computer, on-line, once each time the filter is cycled.

The second aspect of the price we pay for the extreme flexibility of the filters under development is the difficulty of performing a precise analysis on their properties. In Chapter 7 it was relatively easy, using the discrete Legendre polynomials, to analyze the Fixed-Memory Polynomial Filters. Such is not the case here, and we shall have to content ourselves with far fewer precise results, and perhaps a few more heuristic arguments about the nature of these filters.

Finally, as will be readily noted, the generalized filters are computationally much more expensive than their simpler predecessors which we presented in Chapter 7.

The generalized filters are based directly on the ideas developed in Chapter 6, namely the minimum-variance and the least-squares theorems. We shall review these briefly in the next section. When either the process model or the observation relation is nonlinear, the filtering scheme which results is known as *differential-correction,* and has been used by astronomers since the time of Gauss, who introduced the method, in virtually its present form, very early in the nineteenth century [8.1]. We will also discuss this technique in the present chapter.

8.2 THE BASIC SCHEME

Consider first the simplified case, in which the process satisfies a *linear* differential equation and in which the observations are *linearly* related to the state-vector. Thus, let $X(t)$ be a state-vector for the model and suppose that X satisfies the differential equation

$$\frac{d}{dt} X(t) = A(t) X(t) \tag{8.2.1}$$

where $A(t)$ is a square, possibly time-dependent, matrix of numbers, independent of $X(t)$. Also, let Y_n be the vector of observations made on the process at time t_n, let N_n be the vector of errors in those observations, and assume that the observations are related to the model by

$$Y_n = M_n X_n + N_n \tag{8.2.2}$$

where M_n is a matrix of possibly time-dependent numbers. The components of Y_n can typically be quantities such as position, velocity, acceleration, range, range-rate, azimuth, elevation, etc., and Y_n at any time can be formed from varying mixtures of these.

In Section 6.3 we considered the situation where observations were made at times $t_n, t_{n-1}, \ldots, t_{n-L}$, and we showed that the total observation vector $Y_{(n)}$ could be related to X_n by an equation of the form

$$Y_{(n)} = T_n X_n + N_{(n)} \tag{8.2.3}$$

in which $Y_{(n)}$ and $N_{(n)}$ were defined in (6.3.4), and where, as in (6.3.9), we define

$$T_n \equiv \begin{pmatrix} M_n \\ \hline M_{n-1} \Phi(t_{n-1}, t_n) \\ \hline \vdots \\ \hline M_{n-L} \Phi(t_{n-L}, t_n) \end{pmatrix} \tag{8.2.4}$$

Assuming that the elements of $N_{(n)}$ are zero-mean random variables, with the known covariance matrix $R_{(n)}$, and considering all estimates of X_n derivable as linear transformations on $Y_{(n)}$, i.e.

$$X^*_{n,n} = W_n Y_{(n)} \tag{8.2.5}$$

then the covariance matrix of the random errors in the estimate will be

$$S^*_{n,n} = W_n R_{(n)} W_n^T \tag{8.2.6}$$

In Chapter 6, we studied the problem of selecting W_n subject to the exactness constraint (6.3.16), so that the $S^*_{n,n}$ induced by (8.2.6) will have the smallest possible diagonal elements. This criterion gave rise to what we called the minimum-variance filter, i.e.

$$\mathring{X}^*_{n,n} = \mathring{W}_n Y_{(n)} \tag{8.2.7}$$

where

$$\mathring{W}_n \equiv (T_n^T R_{(n)}^{-1} T_n)^{-1} T_n^T R_{(n)}^{-1} \tag{8.2.8}$$

The induced covariance matrix $S^*_{n,n}$ was shown to be equal to

$$\mathring{S}^*_{n,n} = (T_n^T R_{(n)}^{-1} T_n)^{-1} \tag{8.2.9}$$

These are the equations of the *Minimum-Variance Generalized Fixed-Memory Filter* assuming complete linearity in both the model and the observation relations.

In case we wish to use the least-squares criterion, rather than minimum-variance, we showed in Chapter 6 that this was accomplished by setting

$$\hat{X}^*_{n,n} = \hat{W}_n Y_{(n)} \tag{8.2.10}$$

where

$$\hat{W}_n \equiv (T_n^T T_n)^{-1} T_n^T \tag{8.2.11}$$

The induced covariance matrix of the errors in the estimate is then

$$\hat{S}^*_{n,n} = \hat{W}_n R_{(n)} \hat{W}_n^T \tag{8.2.12}$$

However, if $R_{(n)}$ is of the form

$$R_{(n)} = \sigma_\nu^2 I \tag{8.2.13}$$

then the induced estimate covariance matrix simplifies to

$$\hat{S}^*_{n,n} = \sigma_\nu^2 (T_n^T T_n)^{-1} \tag{8.2.14}$$

which is also the result obtained from the minimum-variance approach when (8.2.13) holds.

The actual implementation of either of the above two algorithms is straightforward, and consists essentially of setting up the matrix T_n of (8.2.4), followed by the computation of $\mathring{W}_n$ or $\hat{W}_n$ as the case may be. In obtaining T_n, the transition matrices shown in (8.2.4) must be computed using the techniques developed in Chapter 4.

8.3 GENERALIZED FIXED-MEMORY POLYNOMIAL FILTERS

The filters developed in Chapter 7 were based on the assumption that the observations were equally spaced in time. When this is not the case, then the techniques of that chapter become invalid, and we now discuss briefly the alternate procedure which we are forced to adopt.

Assume that a sequence of scalar observations $\ldots y_{n-1}, y_n \ldots$ is being obtained from a process, and that we wish to fit a polynomial of given degree to a fixed length record of them. The observation instants are assumed to be unequally spaced and we store both these as well as the observations themselves in push-down tables of adequate length.

As the state-vector for the polynomial model we take

$$Z(t) \equiv \begin{pmatrix} x(t) \\ \dfrac{d}{dt} x(t) \\ \dfrac{1}{2!} \dfrac{d^2}{dt^2} x(t) \\ \vdots \\ \dfrac{1}{m!} \dfrac{d^m}{dt^m} x(t) \end{pmatrix} \tag{8.3.1}$$

where the factorial scale-factors are introduced for convenience. Assume specifically that we wish to fit a quadratic to the observations. Then the transition relation for $Z(t)$ would be

$$\begin{pmatrix} x \\ \dot{x} \\ \frac{1}{2!}\ddot{x} \end{pmatrix}_{t_n+\zeta} = \begin{pmatrix} 1 & \zeta & \zeta^2 \\ & 1 & 2\zeta \\ & & 1 \end{pmatrix} \begin{pmatrix} x \\ \dot{x} \\ \frac{1}{2!}\ddot{x} \end{pmatrix}_{t_n} \tag{8.3.2}$$

as is easily verified by the use of Taylor's theorem. The general i, j^{th} term of the transition matrix is [see (4.2.15)]

$$\left[\Phi(\zeta)\right]_{ij} \equiv \binom{j}{i}\zeta^{j-i} \qquad 0 \le i, j \le m \tag{8.3.3}$$

For the quadratic case, each of the observations... y_{n-1}, y_n ... is related to $X(t)$ by an equation of the form

$$y_n = (1,\ 0,\ 0)\begin{pmatrix} x \\ \dot{x} \\ \frac{1}{2!}\ddot{x} \end{pmatrix}_n + \nu_n \tag{8.3.4}$$

and so M_n of (8.2.2) has the constant form

$$M = (1,\ 0,\ 0) \tag{8.3.5}$$

for all n.

Combining (8.3.5) with (8.3.3), it thus follows from (8.2.4) that, for the quadratic case,

$$T_n = \begin{pmatrix} 1 & 0 & 0 \\ 1 & -(t_n - t_{n-1}) & (t_n - t_{n-1})^2 \\ 1 & -(t_n - t_{n-2}) & (t_n - t_{n-2})^2 \\ \vdots & \vdots & \vdots \\ 1 & -(t_n - t_{n-L}) & (t_n - t_{n-L})^2 \end{pmatrix} \tag{8.3.6}$$

Note that in the formation of (8.3.6), we have assumed that $L + 1$ observations are to be used to obtain the polynomial estimate. Note also, that by including the scale-factors $1/m!$ in the definition of $Z(t)$ in (8.3.1), we have avoided their occurrence in (8.3.6). This was purely a matter of personal preference.

It now follows immediately from (8.2.11) that the least-squares polynomial approximation to the sequence of observations

$$\mathsf{Y}_{(n)} \equiv (y_n, y_{n-1}, \ldots, y_{n-L})^T \tag{8.3.7}$$

will be

$$\hat{Z}{}^*_{n,n} = (\mathsf{T}_n^T \mathsf{T}_n)^{-1} \mathsf{T}_n^T \mathsf{Y}_{(n)} \tag{8.3.8}$$

This is the state-vector of the estimating polynomial, and the validity instant is at the leading edge of the observation interval. It is readily verified that for a quadratic, (8.3.6) gives us the 3×3 matrix

$$\mathsf{T}_n^T \mathsf{T}_n = \begin{pmatrix} \sum_{k=0}^{L} 1 & -\sum_{k=0}^{L} (t_n - t_{n-k}) & \sum_{k=0}^{L} (t_n - t_{n-k})^2 \\ -\sum_{k=0}^{L} (t_n - t_{n-k}) & \sum_{k=0}^{L} (t_n - t_{n-k})^2 & -\sum_{k-0}^{L} (t_n - t_{n-k})^3 \\ \sum_{k=0}^{L} (t_n - t_{n-k})^2 & -\sum_{k=0}^{L} (t_n - t_{n-k})^3 & \sum_{k=0}^{L} (t_n - t_{n-k})^4 \end{pmatrix} \tag{8.3.9}$$

Obtaining the inverse of this matrix is the major computational problem in the derivation of (8.3.8). It is also easily verified that for the quadratic case

$$\mathsf{T}_n^T \mathsf{Y}_n = \begin{pmatrix} \sum_{k=0}^{L} y_{n-k} \\ -\sum_{k=0}^{L} (t_n - t_{n-k}) y_{n-k} \\ \sum_{k=0}^{L} (t_n - t_{n-k})^2 y_{n-k} \end{pmatrix} \tag{8.3.10}$$

and so the least-squares estimate of (8.3.8) can now be obtained. Extension to higher degree polynomials follows readily.

In the event that $R_{(n)}$, the covariance matrix of the errors in $Y_{(n)}$, were available, we could, if we prefer, set up the *minimum-variance* polynomial estimator

$$\overset{\circ}{Z}{}^{*}_{n,n} = (T_n^{\,T} R_{(n)}^{-1} T_n)^{-1} T_n^{\,T} R_{(n)}^{-1} Y_{(n)} \tag{8.3.11}$$

where T_n was given above. This, however, calls for additional computation in that $R_{(n)}$ must be inverted. But if this matrix happens to be diagonal, as is often the case, then the additional amount of computation is extremely slight.

It is sometimes the case that we know the true differential equations of the process to which polynomial filtering is being applied. *In that event the method of trend-removal which we developed in* Section 7.14 *is strongly recommended.* As we pointed out there, for a slight increase in the amount of computation we can achieve a marked improvement in the properties of the filter. The systematic errors are reduced, and this in turn permits us to use greater memory lengths, thereby smoothing the random errors more heavily.

We suggest then that the reader review the material of Section 7.14 and that he thereafter work Example 8.6 in which trend-removal is applied.

8.4 NONLINEAR SYSTEMS

Consider the following differential equation

$$\ddot{x}(t) + \omega^2 x(t) = 0 \tag{8.4.1}$$

If ω is a given constant then this is a *linear* differential equation, and of course defines the family of sines and cosines whose angular frequency is ω. We now examine the situation where ω is *unknown,* and is to be estimated along with $x(t)$ and $\dot{x}(t)$ by making observations on the process.

We assume first, that from prior considerations, it is known that ω *remains constant* with respect to time. What we then seek is the value of that constant. Under these circumstances, we now think of ω as the time function which satisfies the differential equation

$$\frac{d}{dt}\omega(t) = 0 \tag{8.4.2}$$

and combining this with (8.4.1), gives us the *extended differential equation*

$$\frac{d}{dt}\begin{pmatrix} x(t) \\ \dot{x}(t) \\ \omega(t) \end{pmatrix} = \begin{pmatrix} \dot{x}(t) \\ -\omega^2 x(t) \\ 0 \end{pmatrix} \tag{8.4.3}$$

This is a *nonlinear* differential equation, of the form

$$\frac{d}{dt}[X(t)] = F[X(t)] \tag{8.4.4}$$

in which the state-vector, $X(t)$, has been augmented to include the parameter ω. *The latter has thus been converted into one of the state-variables.* If we were now able to devise a method of applying estimation techniques to nonlinear differential equations, then clearly ω could be estimated from observations.

It is clear that the above approach is open to complete generalization. For example, on p. 105 we discussed the equations of motion of a body moving through the atmosphere, and we simplified the problem by assuming that the parameter α was a known constant. Now a very important class of problems arises in precisely the case where this parameter is *unknown,* and it is desired that its value be estimated by making successive observations on the position of the body. Studies on viscous motion provide us with some prior knowledge on the dynamics of α and its dependence on the state variables of position and possibly velocity. This gives us a differential equation, of the form say,

$$\frac{d}{dt}\begin{pmatrix} \alpha(t) \\ \dot{\alpha}(t) \end{pmatrix} = F\left[x_0, x_1, x_2, \dot{x}_0, \dot{x}_1, \dot{x}_2, \alpha(t), \dot{\alpha}(t)\right] \tag{8.4.5}$$

which can now be combined with the 6-vector of (4.8.11). The result is an 8-vector of state-variables, which is then the candidate for our estimation procedures. Again the model differential equation is nonlinear.

Nonlinearities can also enter through the observation scheme, and in fact it is more common in practice to encounter nonlinear, rather than linear, observation schemes. As a common example which occurs very frequently, the reader is referred to p. 168 where in (6.2.11) the equations are given which relate polar to Cartesian coordinates. They are clearly nonlinear.

A brief examination of the material of Section 8.2 shows that nonlinearities in either the differential equations or in the observation relation will invalidate all of the estimation techniques thus far developed. In the former case the transition matrix is undefined *per se,* and in the latter the matrix M_n cannot be set up. Either or both will then rule out the formation of T_n, and so we are then unable to set up either the least-squares or the minimum-variance filter algorithms.

The remainder of this chapter is, accordingly, devoted to a discussion of how we estimate in the presence of nonlinearities. The procedure to be outlined is called *iterative differential-correction* and will be seen to be a fusion of our already developed methods together with numerical iteration.

8.5 OBTAINING A NOMINAL TRAJECTORY

Assume that a nonlinearity exists in *either* or *both* of the process differential equations and the observation relation. This invalidates our existing estimation techniques which were developed on the assumption of complete linearity.

However, in Sections 4.8 and 6.2, we demonstrated that if a so-called *nominal trajectory* exists, i.e. one which is reasonably close to the true one, then by first order linearization techniques we can set up an associated differential equation and observation relation, *both of which are linear.* This then offers us the possibility of being able to apply our existing linear estimation techniques. In the present section we examine, precisely, how that nominal trajectory can be arrived at. In the next section we examine the details of the associated estimation procedures.

As a first method of obtaining a nominal trajectory, it is possible that an approximation to the true process may already be available from *prior knowledge.* A common example of this situation occurs in satellite work, where the insertion parameters existing at the time of separation are known. These can be used to construct an estimate of the value of the state-vector describing the orbit in question, and this then constitutes the required nominal trajectory.

When such prior knowledge does not exist, we must resort to constructing a nominal trajectory by the actual use of observations. *Accordingly we now fall back, temporarily, on the method of polynomial approximation,* as discussed in Section 8.3. This is admittedly a compromise, since, as we know, polynomial estimates contain systematic errors, depending upon the degree of the polynomial selected. *However the indisputable advantage of polynomial estimation is that our ability to apply it is completely unaffected by the presence of nonlinearities in the system equations.* Polynomial smoothing bypasses these completely, and very conveniently enables us to obtain the sought-after nominal trajectory.

Specifically, assume that we are observing the range, ρ, azimuth, ψ, and elevation, θ, of a body in motion. These, together with the observation instants give rise to the sequences of numbers

$$\begin{array}{l} \ldots\ \rho_{n-2},\ \rho_{n-1},\ \rho_n,\ \ldots \\ \ldots\ \psi_{n-2},\ \psi_{n-1},\ \psi_n,\ \ldots \\ \ldots\ \theta_{n-2},\ \theta_{n-1},\ \theta_n,\ \ldots \\ \ldots\ t_{n-2},\ t_{n-1},\ t_n,\ \ldots \end{array}$$

which we store in four push-down tables, each of length $L + 1$. It is now decided that a quadratic shall be fitted to each of the three sets of observations† and so we form the matrix T_n, as given in (8.3.6). This then, by the use of either the least-squares or the minimum-variance algorithms given in (8.3.8) or (8.3.11), enables us to obtain the three quadratic estimate state-vectors

$$\mathrm{P}^*_{n,n} \equiv \begin{pmatrix} \rho^* \\ \dot{\rho}^* \\ \ddot{\rho}^* \end{pmatrix}_{n,n} \qquad \Psi^*_{n,n} \equiv \begin{pmatrix} \psi^* \\ \dot{\psi}^* \\ \ddot{\psi}^* \end{pmatrix}_{n,n} \qquad \Theta^*_{n,n} \equiv \begin{pmatrix} \theta^* \\ \dot{\theta}^* \\ \ddot{\theta}^* \end{pmatrix}_{n,n} \tag{8.5.1}$$

We accordingly assume that we now have the quadratic fixed-memory polynomial estimates for each of the selected data sequences.

We direct our attention next to the true observation relations, and show how the results of the polynomial estimation procedure discussed above can be applied. Specifically, let the state-vector be made up of the three position and three velocity coordinates of a body in motion, i.e.

$$X(t) = (x_0,\ x_1,\ x_2,\ \dot{x}_0,\ \dot{x}_1,\ \dot{x}_2)^T \tag{8.5.2}$$

Then the simulated observation equations, relating the quantities under observation, namely ρ, ψ, and θ, to $X(t)$, are (see (6.2.11))

†The decision as to what degree polynomial shall be fitted, is based on the considerations outlined in Chapter 7.

$$
\begin{aligned}
\rho &= \left(x_0^{\,2} + x_1^{\,2} + x_2^{\,2}\right)^{1/2} \\
\psi &= \tan^{-1}\left(\frac{x_1}{x_0}\right) \\
\theta &= \tan^{-1}\left[\frac{x_2}{\left(x_0^{\,2} + x_1^{\,2}\right)^{1/2}}\right]
\end{aligned}
\tag{8.5.3}
$$

Suppose that we were given a set of values for ρ_n, ψ_n, and θ_n. Then, by inverting the above equations, we could solve for corresponding estimates of x_0, x_1, and x_2, valid at $t = t_n$. However, these three quantities do not, as yet, constitute a nominal trajectory, since from (8.5.2) we see that we also require values of $\dot{x}_0$, $\dot{x}_1$, and $\dot{x}_2$. *Moreover the latter three quantities are not present in* (8.5.3). This last fact is equivalent to the statement that the true observation equations of (8.5.3), when written in terms of the entire state-vector $X(t)$ in the form

$$
\begin{pmatrix} \rho \\ \psi \\ \theta \end{pmatrix}_t = G[X(t)] \tag{8.5.4}
$$

are *singular,* i.e. they cannot be inverted to give

$$
X(t) = G^{-1}(\rho, \psi, \theta)_t \tag{8.5.5}
$$

The singular nature of the set (8.5.3) is very common in practice, and arises out of the fact that, either by choice or by necessity, *we are not observing a sufficient number of independent quantities.* However the application of polynomial filtering to the ρ, ψ, and θ sequences has provided us with estimates of the *additional independent quantities* $\dot{\rho}$, $\dot{\psi}$, and $\dot{\theta}$, quantities which we are not actually observing. These latter estimates can now be utilized to eliminate the abovementioned singular situation.

First we invert (8.5.3). This gives us the three equations (see Ex. 8.13)

$$
\begin{aligned}
x_0 &= \rho \cos\theta \cos\psi \\
x_1 &= \rho \cos\theta \sin\psi \\
x_2 &= \rho \sin\theta
\end{aligned}
\tag{8.5.6}
$$

Then, by differentiating these, *three further independent relationships come about,* namely,

$$\begin{aligned} \dot{x}_0 &= \dot{\rho}\cos\theta\cos\psi - \dot{\theta}\rho\sin\theta\cos\psi - \dot{\psi}\rho\cos\theta\sin\psi \\ \dot{x}_1 &= \dot{\rho}\cos\theta\sin\psi - \dot{\theta}\rho\sin\theta\sin\psi + \dot{\psi}\rho\cos\theta\cos\psi \\ \dot{x}_2 &= \dot{\rho}\sin\theta + \dot{\theta}\rho\cos\theta \end{aligned} \tag{8.5.7}$$

These six equations clearly give us the *entire* state-vector, $X(t)$, in terms of ρ, ψ, θ, $\dot{\rho}$, $\dot{\psi}$ and $\dot{\theta}$, *and since estimates of the latter six do exist, an estimate of* $X(t)$ *follows immediately.*

In this manner, by using the polynomial filters to give us

$$\rho^*_{n,n},\ \dot{\rho}^*_{n,n},\ \psi^*_{n,n},\ \dot{\psi}^*_{n,n},\ \theta^*_{n,n},\ \dot{\theta}^*_{n,n}$$

we have, in effect, *augmented the number of observables,* and this, in turn, has enabled us to remove the singularity in the basic observation equations. Once the augmentation has been effected, it then becomes a purely algebraic problem to obtain an estimate of $X(t)$. (We have of course only shown the augmentation and inversion procedure to be valid for the above specific case, but we assert that it is possible to do *in general* what we have done above, if the observation scheme is adequately constructed.)

The first of our tasks has thus been carried out, and the sought-after nominal trajectory has been obtained. Errors will exist in that nominal trajectory, both random and systematic. In what now follows we show how those errors can be further reduced by iteration.

8.6 ITERATIVE DIFFERENTIAL-CORRECTION

In the final section of Chapter 4 we examined the case of the nonlinear differential equation

$$\frac{d}{dt}X(t) = F[X(t)] \tag{8.6.1}$$

where F is a vector of nonlinear functions of the state-variables making up the vector $X(t)$. We showed that if there are two trajectories, $X(t)$ and $\overline{X}(t)$, both satisfying (8.6.1), with

$$X(t) = \overline{X}(t) + \delta X(t) \tag{8.6.2}$$

then if $\delta X(t)$ is a vector of sufficiently small functions relative to those in $\overline{X}(t)$, that *to first order* $\delta X(t)$ satisfies the *linear* differential equation

$$\frac{d}{dt}\delta X(t) = A\left[\overline{X}(t)\right]\delta X(t) \tag{8.6.3}$$

The matrix $A\left[\overline{X}(t)\right]$ is defined by

$$\left[A\left(\overline{X}(t)\right)\right]_{ij} = \left.\frac{\partial f_i(X)}{\partial x_j}\right|_{X = \overline{X}(t)} \tag{8.6.4}$$

It then also follows that there is a matrix Φ, known as the transition matrix, such that

$$\delta X(t_n + \zeta) = \Phi\left(t_n + \zeta, t_n; \overline{X}\right)\delta X(t_n) \tag{8.6.5}$$

We showed in Section 4.7 that Φ satisfies the differential equation

$$\frac{\partial}{\partial\zeta}\Phi\left(t_n + \zeta, t_n; \overline{X}\right) = A\left[\overline{X}(t_n + \zeta)\right]\Phi\left(t_n + \zeta, t_n; \overline{X}\right) \tag{8.6.6}$$

with initial conditions

$$\Phi\left(t_n, t_n; \overline{X}\right) = I \tag{8.6.7}$$

[see e.g. (4.8.25)].

In a like manner, in the second section of Chapter 6, we considered the case of nonlinear observation schemes. Thus let $X(t)$ be a trajectory giving rise, at $t = t_n$, to a vector of observations Y_n, where

$$Y_n = G(X_n) + N_n \tag{8.6.8}$$

and where G is nonlinear. Let $\overline{X}(t)$ be a trajectory which is close to $X(t)$, in the sense that

$$\delta X(t) \equiv X(t) - \overline{X}(t) \tag{8.6.9}$$

has small elements. Define the simulated observations based on $\overline{X}_n$, by $\overline{Y}_n$, i.e.

$$\overline{Y}_n \equiv G\left(\overline{X}_n\right) \tag{8.6.10}$$

and let

$$\delta Y_n \equiv Y_n - \overline{Y}_n \tag{8.6.11}$$

Then, to *first order*, (8.6.8) gives rise to the linear relation

$$\delta Y_n = M\left(\overline{X}_n\right)\delta X_n + N_n \tag{8.6.12}$$

where the matrix M is defined by

$$\left[M\left(\overline{X}_n\right)\right]_{ij} = \left.\frac{\partial g_i(X)}{\delta x_j}\right|_{X = \overline{X}(t)} \tag{8.6.13}$$

Finally, consider the *truly linear* cases, namely

$$\frac{d}{dt}X(t) = A(t)X(t) \tag{8.6.14}$$

and

$$Y_n = M_n X_n + N_n \tag{8.6.15}$$

and as before, assume that

$$X(t) = \overline{X}(t) + \delta X(t) \tag{8.6.16}$$

where by assumption $\overline{X}(t)$ also satisfies (8.6.14). Then by (8.6.14)

$$\frac{d}{dt}\overline{X}(t) + \frac{d}{dt}\delta X(t) = A(t)\overline{X}(t) + A(t)\delta X(t) \tag{8.6.17}$$

and so we have, *precisely*

$$\frac{d}{dt}\delta X(t) = A(t)\delta X(t) \tag{8.6.18}$$

Likewise (8.6.15) gives, *precisely*

$$\delta Y_n = M_n \delta X_n + N_n \tag{8.6.19}$$

where δY_n was defined in (8.6.11). Thus truly linear systems can be replaced, *without errors,* by equivalent systems involving the differential vectors $\delta X(t)$ and δY_n.

In summary then, the systems which we can expect to encounter can be replaced by *entirely linear systems* a) which are first-order approximations, if they replace nonlinear equations, and b) are exact, if they replace linear equations. The replacement systems are equations in the vectors of *differentials,* $\delta X(t)$ and δY_n, rather than in the entire quantities, $X(t)$ and Y_n.

Since it was shown in Section 8.5 that a nominal trajectory $\overline{X}(t)$ *can in fact be found,* we are thus able to carry out any of the linearizations described above, and to set up the entirely linear system of equations in the vector differentials,

$$\begin{aligned} \frac{d}{dt}\delta X(t) &= A\left[\overline{X}(t)\right]\delta X(t) \\ \delta Y_n &= M\left(\overline{X}_n\right)\delta X_n + N_n \end{aligned} \tag{8.6.20}$$

In order to estimate the true trajectory $X(t)$, *we now need only estimate the perturbation vector* $\delta X(t)$, and we can then compute $X(t)$ by simple addition, using

$$X(t) = \overline{X}(t) + \delta X(t) \tag{8.6.21}$$

Now, (8.6.20) is of precisely the same form as (8.2.1) and (8.2.2), and so we see that $\delta X(t)$ and hence $X(t)$ *can in fact be estimated,* by applying the methods developed in Section 8.2. The precise sequence of steps which must be followed is now given.

Fixed-Memory Minimum-Variance Differential-Correction

Assume that a nominal trajectory state-vector $\overline{X}_{c,n}$ has been obtained,† either by the method outlined in Section 8.5, or some other means. Assume also that the differential equation of the model is

†The vector $\overline{X}_{c,n}$ is a nominal trajectory state-vector valid at some instant inside the interval $t_{n-L} \le t \le t_n$, rather than at the leading edge, in order to keep the systematic errors down. One convenient point is the *average* of the instants $t_{n-L}, t_{n-L+1}, \ldots, t_{n-1}, t_n$ in the current observation window, which we designate as $t_{c,n}$.

$$\frac{d}{dt} X(t) = F[X(t)] \tag{8.6.22}$$

For definiteness consider the case where X is a 3-vector. Then (8.6.22) will be

$$\begin{aligned}
\frac{d}{dt} x_0(t) &= f_0\left[x_0(t), x_1(t), x_2(t)\right] \\
\frac{d}{dt} x_1(t) &= f_1\left[x_0(t), x_1(t), x_2(t)\right] \\
\frac{d}{dt} x_2(t) &= f_2\left[x_0(t), x_1(t), x_2(t)\right]
\end{aligned} \tag{8.6.23}$$

Also assume that the observation vectors are related to the true state-vector by

$$Y_{n-k} = G(X_{n-k}) + N_{n-k} \qquad 0 \le k \le L \tag{8.6.24}$$

For definiteness let Y be a 2-vector, i.e.

$$\begin{aligned}
(y_0)_n &= g_0\left[x_0(t_n), x_1(t_n), x_2(t_n)\right] \\
(y_1)_n &= g_1\left[x_0(t_n), x_1(t_n), x_2(t_n)\right]
\end{aligned} \tag{8.6.25}$$

(I) Differentiate each of the above functions with respect to each of the independent variables, and set up the following matrices, in functional form:

$$\left[A(X)\right]_{ij} = \frac{\partial f_i(X)}{\partial x_j} \tag{8.6.26}$$

$$\left[M(X)\right]_{ij} = \frac{\partial g_i(X)}{\partial x_j} \tag{8.6.27}$$

(II) Using a sufficiently accurate numerical integration scheme, with appropriately chosen time intervals, obtain the nominal trajectory state-vector,

$\overline{X}(t)$, for each of the observation instants $t_{n-L}, \ldots, t_n$. Do this by integrating the true process differential equations both forwards and backwards, using as initial conditions the nominal trajectory state-vector $\overline{X}_{c,n}$.

(III) Evaluate the matrix $M(X)$ of (8.6.27) at each of the observation instants $t_{n-L}, \ldots, t_n$ using the numerical values of the nominal trajectory $\overline{X}(t)$, as obtained from II above.

(IV) Integrate the differential equations of the transition matrix, starting from $t = t_{c,n}$ and working both forwards and backwards to each end of the observation interval. Thus, integrate (over $0 \leq \zeta \leq t_n - t_{c,n}$ and $0 \geq \zeta \geq t_{n-L} - t_{c,n}$) the matrix differential equation

$$\frac{\partial}{\partial \zeta} \Phi\left(t_{c,n} + \zeta, t_{c,n}; \overline{X}\right) = A\left[\overline{X}(t_{c,n} + \zeta)\right] \Phi\left(t_{c,n} + \zeta, t_{c,n}; \overline{X}\right) \tag{8.6.28}$$

using as initial conditions

$$\Phi\left(t_{c,n}, t_{c,n}; \overline{X}\right) = I \tag{8.6.29}$$

The matrix $A\left[\overline{X}(t_{c,n} + \zeta)\right]$ is obtained by evaluating the function-matrix $A(X)$ of (8.6.26), using the numerical values of the trajectory $\overline{X}(t)$ obtained from II above.

(V) Evaluate the simulated observation vectors at each of the observation instants, i.e. compute the vectors

$$\begin{aligned} \overline{Y}_n &\equiv G\left[\overline{X}(t_n)\right] \\ \overline{Y}_{n-1} &\equiv G\left[\overline{X}(t_{n-1})\right] \\ &\vdots \\ \overline{Y}_{n-L} &\equiv G\left[\overline{X}(t_{n-L})\right] \end{aligned} \tag{8.6.30}$$

where $\overline{X}(t_n) \ldots \overline{X}(t_{n-L})$ are obtained in II above and G is defined in (8.6.24). We call the above vector $G\left(\overline{X}_{c,n}\right)$.

(VI) Assume that the actual observation vectors are

$$Y_{n-L}, Y_{n-L+1}, \ldots, Y_n$$

Compute

$$\begin{aligned} \delta Y_n &\equiv Y_n - \overline{Y}_n \\ \delta Y_{n-1} &\equiv Y_{n-1} - \overline{Y}_{n-1} \\ &\vdots \\ \delta Y_{n-L} &\equiv Y_{n-L} - \overline{Y}_{n-L} \end{aligned} \tag{8.6.31}$$

Then by (8.6.12),

$$\begin{aligned} \delta Y_n &= M_n \delta X_n + N_n \\ \delta Y_{n-1} &= M_{n-1} \delta X_{n-1} + N_{n-1} \\ &\vdots \\ \delta Y_{n-L} &= M_{n-L} \delta X_{n-L} + N_{n-L} \end{aligned} \tag{8.6.32}$$

where the matrices $M_n, \ldots, M_{n-L}$ were obtained in III above. Using the transition matrices obtained in IV above, (8.6.32) can be written

$$\begin{aligned} \delta Y_n &= M_n \Phi\left(t_n, t_{c,n}; \overline{X}\right) \delta X_{c,n} + N_n \\ \delta Y_{n-1} &= M_{n-1} \Phi\left(t_{n-1}, t_{c,n}; \overline{X}\right) \delta X_{c,n} + N_{n-1} \\ &\vdots \\ \delta Y_{n-L} &= M_{n-L} \Phi\left(t_{n-L}, t_{c,n}; \overline{X}\right) \delta X_{c,n} + N_{n-L} \end{aligned} \tag{8.6.33}$$

We write this as

$$\delta Y_{(n)} = T_{c,n} \delta X_{c,n} + N_{(n)} \tag{8.6.34}$$

in which we have defined

$$T_{c,n} \equiv \begin{pmatrix} M_n \Phi\left(t_n, t_{c,n}; \overline{X}\right) \\ \hline M_{n-1} \Phi\left(t_{n-1}, t_{c,n}; \overline{X}\right) \\ \hline \vdots \\ \hline M_{n-L} \Phi\left(t_{n-L}, t_{c,n}; \overline{X}\right) \end{pmatrix} \tag{8.6.35}$$

(VII) Compute the matrix $T_{c,n}$ defined by (8.6.35) above.

(VIII) Given that the covariance matrix of the vector $N_{(n)}$ is $R_{(n)}$, then to first order, the minimum-variance estimate of $\delta X_{c,n}$ of (8.6.34), is

$$\delta X^*_{c,n} \equiv W_n \delta Y_{(n)} \tag{8.6.36}$$

where

$$W_n \equiv \left(T^T_{c,n} R^{-1}_{(n)} T_{c,n}\right)^{-1} T^T_{c,n} R^{-1}_{(n)} \tag{8.6.37}$$

and, to second order, the covariance matrix of $\delta X^*_{c,n}$ is

$$S^*_{c,n} \equiv \left(T^T_{c,n} R^{-1}_{(n)} T_{c,n}\right)^{-1} \tag{8.6.38}$$

(IX) Compute the estimate of $X(t)$ at $t = t_{c,n}$, from

$$X^*_{c,n} = \overline{X}_{c,n} + \delta X^*_{c,n} \tag{8.6.39}$$

Combining this with (8.6.36) thus gives us

$$X^*_{c,n} = \overline{X}_{c,n} + \left(T^T_{c,n} R^{-1}_{(n)} T_{c,n}\right)^{-1} T^T_{c,n} R^{-1}_{(n)} \left[Y_{(n)} - G\left(\overline{X}_{c,n}\right)\right] \tag{8.6.40}$$

in which $G\left(\overline{X}_{c,n}\right)$ is the vector shown in (8.6.30). The covariance matrix of $X^*_{c,n}$ is the same as the covariance matrix of $\delta X^*_{c,n}$. Thus to second order the covariance matrix of the estimate $X^*_{c,n}$ is $S^*_{c,n}$ computed in VIII above.

The above method was based on the linearization procedures applied to the nonlinear functions $F(X)$ and $G(X)$ in the process and observation equations. Only first order terms of the Taylor's expansions were retained, and it was assumed that the differential vector $\delta X(t)$ was made up of elements small enough so that the higher order Taylor's expansion terms could be neglected. If $\overline{X}(t)$ *is* in fact close enough to $X(t)$, then the method will be found to operate successfully, and $X^*_{c,n}$ of (8.6.40) *will* be a better estimate of $X(t_{c,n})$ than $\overline{X}_{c,n}$ was. We will in fact have reduced the systematic errors in the estimate. The key to success thus hinges on ensuring that $\overline{X}(t)$ is close enough to $X(t)$, so that the higher order terms in the Taylor expansions can, in fact, be neglected. If not, $X^*_{c,n}$ will be a *worse* estimate of $X(t_{c,n})$ than $\overline{X}_{c,n}$ was, and so $\overline{X}_{c,n}$ must be estimated as carefully as possible to ensure success. Assuming that $X^*_{c,n}$ is an improvement of $\overline{X}_{c,n}$, *we then proceed to iterate on the same set of observations.*

Prior to discussing that iteration procedure, we point out that we could also perform *least-squares* differential correction if we so choose.† The only change in the above scheme would be steps VIII and IX where we would omit the matrix $R_{(n)}^{-1}$.

8.7 ITERATION PROCEDURE

The basic differential-correction scheme has been outlined in the preceding two sections. Starting from an *initial estimate* $\overline{X}_{c,n}$, we obtained an *improved estimate* $X^*_{c,n}$. The thought now immediately occurs to us – why not next use $X^*_{c,n}$ as the initial estimate in place of $\overline{X}_{c,n}$, repeat the differential-correction procedure using the same observations and obtain a *further* improvement, and so on? Specifically we do the following (refer to Figure 8.1).

Iteration Procedure (Differential-Correction)

(I) Obtain $\overline{X}_{c,n}$ (see Section 8.5).

(II) Using the method of differential-correction as outlined in Section 8.6, improve $\overline{X}_{c,n}$ thereby obtaining $X^*_{c,n}$.

(III) Now use $X^*_{c,n}$ as the nominal trajectory in place of $\overline{X}_{c,n}$, i.e. set $X^*_{c,n}$ into $\overline{X}_{c,n}$ and return to step II above. Repeat the differential-correction process *using the same set of observations* to obtain a further improvement. We thus iterate the following algorithm [c/f (8.6.40)]

$$\left(X^*_{c,n}\right)_{r+1} = \left(X^*_{c,n}\right)_r + \left[T\left(\overline{X}_{c,n}\right)^T R_{(n)}^{-1} T\left(\overline{X}_{c,n}\right)\right]^{-1} T\left(\overline{X}_{c,n}\right)^T R_{(n)}^{-1}\left[Y_{(n)} - G\left(\overline{X}_{c,n}\right)\right]\Bigg|_{\overline{X}_{c,n} = \left(X^*_{c,n}\right)_r} \tag{8.7.1}$$

in which T is shown to be a function of $\overline{X}_{c,n}$.

(IV) Iterate as often as time permits, or include a test to terminate the process, e.g.

$$\left[\left(X^*_{c,n}\right)_{r+1} - \left(X^*_{c,n}\right)_r\right]^T\left[\left(X^*_{c,n}\right)_{r+1} - \left(X^*_{c,n}\right)_r\right] < \epsilon\ ?$$

where ϵ is an appropriately small positive quantity, and r is the iteration count-number.

†Subject to the comments of Section 6.8 (see p. 200).

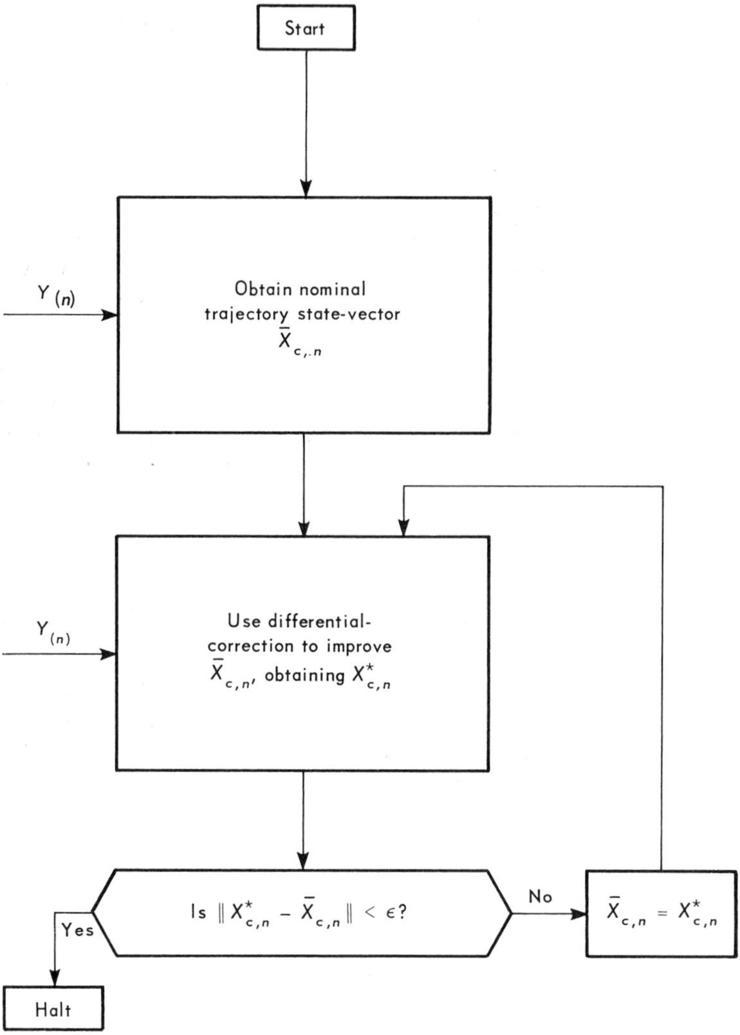

Fig. 8.1 *Flowchart for iterative differential-correction procedure.*

The rationale behind the iteration procedure is that each member of the sequence of estimates

$$\overline{X}_{c,n},\ \left(X^*_{c,n}\right)_1,\ \left(X^*_{c,n}\right)_2,\ \ldots,\ \left(X^*_{c,n}\right)_r,\ \ldots$$

should be a better estimate than its predecessor. This will be true if the initial vector $\overline{X}_{c,n}$ is close enough to the truth. Thereafter the errors arising out of neglecting terms of order higher than the first in the linearization process, *become rapidly smaller and smaller.* The linearization procedure thus becomes more and more error-free with each cycle of the iteration, and, as time proceeds, the equations (8.6.20) on which we are estimating $\delta X(t)$, become closer and closer to being exact. In the limit, the vector $\delta X(t)$ tends to the null vector (in theory at any rate), and so the differential-correction added to $\overline{X}_{c,n}$, after each pass, eventually disappears.

Of course the limiting value of the estimate $\left(X^*_{c,n}\right)_\infty$ after an arbitrarily large number of iteration cycles, will *not* be exactly $X(t_{c,n})$. This is so because we have only a finite number of observations on which that limiting value depends, and *those observations contain observational errors.* However that limiting value will certainly be the *best* that this particular set of observations can give. The iteration scheme serves to reduce the systematic errors as far as possible (but not to zero), and we are left with predominantly random errors in the estimate. Their covariance matrix is computed from (8.6.38) when the iteration scheme is terminated.

The reader might well wonder if the limiting value of $\left(X^*_{c,n}\right)_\infty$ from the iteration scheme forms an *unbiased estimate* of $X(t_{c,n})$ if $N_{(n)}$ is a vector of zero-mean errors, i.e. we might ask whether

$$E\left\{\left(X^*_{c,n}\right)_\infty - X(t_{c,n})\right\} = \emptyset \tag{8.7.2}$$

The answer is unfortunately negative, and this is so because the process whereby $\left(X^*_{c,n}\right)_\infty$ was obtained is *nonlinearly* related to the starting value $\overline{X}_{c,n}$, and so the errors in the limiting value $\left(X^*_{c,n}\right)_\infty$ are nonlinearly related to $N_{(n)}$. The limiting value $\left(X^*_{c,n}\right)_\infty$ thus *may not* form an unbiased estimate of $X(t_{c,n})$, even though the observation errors $N_{(n)}$ have zero mean.

However, this is really of no consequence. From a practical standpoint, iterative differential-correction can be made to give good estimates and that is all that matters. The method was introduced by Gauss, more or less in the above form in 1795; he limited himself to uncorrelated errors, which is the case most frequently encountered in practice. The procedure is still in use today in almost any situation where estimation is being performed on a nonlinear process, or where the observations are nonlinearly related to the state or both.

In Section 11.4 we discuss in detail what vector the iterative differential-correction procedure converges to, assuming that convergence occurs. Not unexpectedly we will see that it is the *nonlinear* counterpart of the *weighted least-squares criterion* [see (6.9.5)] which is minimized.

The scheme depicted in the flowchart on p. 312 shows how an estimate $X^*_{c,n}$ is obtained by iterative differential-correction on the data base given by $Y_{(n)}$. That estimate defines a trajectory when considered as initial conditions for the model differential equation. Hence, if at a later time a new observation vector, say $Y_{(n+1)}$, is obtained, then the *previous* estimate, namely $X^*_{c,n}$, can serve as the basis of a *new* nominal trajectory. From $X^*_{c,n}$ and $Y_{(n+1)}$ we would then obtain a new estimate $X^*_{c,n+1}$ and so on.

8.8 COMMENTS ON COMPUTATIONAL PROBLEMS

The Generalized Fixed-Memory Filters discussed in the preceding sections depend essentially on the equation

$$X^*_{n,n} = \left(T_n^T R_{(n)}^{-1} T_n\right)^{-1} T_n^T R_{(n)}^{-1} Y_{(n)} \tag{8.8.1}$$

We now consider some of the problems involved in actually carrying out this computation.

As a start, it is necessary that we invert the matrix $R_{(n)}$. By assumption the observation errors are linearly independent, and so $R_{(n)}$ is positive definite. If it is a diagonal matrix, then inverting it is a trivial problem; we need merely find the reciprocals of each of those diagonal elements. However when $R_{(n)}$ is not diagonal, then recognizing that it might be a matrix of large order, its inversion ceases to be completely trivial. While extensive work has been done on the problem of numerical errors in matrix inversion we nevertheless make a few elementary comments here.†

Inversion of positive definite matrices of large order is possible either by iterative methods or else by Gaussian elimination (see Ex. 5.11), and the

†For a detailed analysis of the problem of numerical inversion of matrices the reader is referred to [8.2] which is only one possible source in a very extensive literature.

amount of computation involved in most cases in practice, barring unexpected problems due to the matrix being nearly singular, is within reason on modern computing equipment. Of course in extreme cases the smoothing scheme under consideration may no longer be usable in real-time, but this would have to be determined on the basis of the details of the situation. The success of either the Gaussian-elimination or the iterative methods depends on the *conditioning* of the matrix, a concept which we now examine briefly.

Consider the effect which finite-length arithmetic calculations can have on matrix inversion when the matrix is nearly singular. Thus, suppose that we form the sum

$$s = 1 + \epsilon \tag{8.8.2}$$

on a computer which has 6 decimal digits in its arithmetic capability. If ϵ is less than .00001, say .000008, then the computer simply ignores it, and instead of obtaining the correct sum, namely 1.000008, we obtain instead 1.00000. In a less severe case, assume that ϵ is .000015. Then instead of obtaining 1.000015, we obtain 1.00001. Although the machine has in effect caused a 33-1/3% relative change in $\epsilon \left(\text{i.e., } \dfrac{.000005}{.000015}\right)$, the relative error in the sum is very small, being only about 5 parts in $10^6 \left(\text{i.e., } \dfrac{.000005}{1.000015}\right)$. At first sight this does not seem to be at all serious.

Suppose, however, that the term $s \equiv 1 + \epsilon$ is actually an element in a *nearly singular* matrix, e.g.

$$A = \begin{pmatrix} s & 1 \\ 1 & 1 \end{pmatrix} \tag{8.8.3}$$

and suppose that A is to be inverted. Algebraically we have

$$A^{-1} = \frac{1}{s-1}\begin{pmatrix} 1 & -1 \\ -1 & s \end{pmatrix} \tag{8.8.4}$$

If there are no errors in s, then the matrix A^{-1} is obtained without errors. Thus if $\epsilon = .000015$ and there is no truncation in the formation of s, we obtain the correct inverse, namely

$$A^{-1} = \frac{\begin{pmatrix} 1 & -1 \\ -1 & 1.000015 \end{pmatrix}}{0.000015} \approx 10^4 \begin{pmatrix} 6.66 & -6.66 \\ -6.66 & 6.66 \end{pmatrix} \tag{8.8.5}$$

However if ϵ was first truncated to .00001 in the formation of s, then we obtain

$$A^{-1} = \frac{\begin{pmatrix} 1 & -1 \\ -1 & 1.00001 \end{pmatrix}}{0.00001} \approx 10^4 \begin{pmatrix} 10 & -10 \\ -10 & 10 \end{pmatrix} \tag{8.8.6}$$

We now see that the 5 parts in 10^6 truncation error in s, which we thought was completely insignificant, shows up as a 50% error in the entire matrix A^{-1}. ◆◆

What we have attempted to do was to show, by a rather trivial example, how the errors of a finite-precision arithmetic computation can show up in the inversion of nearly singular matrices. The trouble can of course be cured by extending the precision of the arithmetic, but this is usually time-consuming and we do not wish to do it unless it is really warranted. We now discuss one possible test which can be applied to a matrix to suggest, ahead of time, if numerical troubles should be anticipated during its inversion.

Suppose that the matrix R, which is assumed to be positive-definite, is to be inverted. As a first step we form the *diagonal* matrix P comprised of the diagonal elements of R. The latter are all positive, and so their square roots are real. We are thus also able to form the real matrix $P^{-1/2}$ whose i, jth term is

$$\left[P^{-1/2}\right]_{ij} \equiv \frac{1}{\left([R]_{ii}\right)^{1/2}} \delta_{ij} \tag{8.8.7}$$

We then form the product

$$R' \equiv P^{-1/2} R P^{-1/2} \tag{8.8.8}$$

and it is easily seen (see Ex. 8.14) that R' is positive definite with *ones on its diagonal.* Moreover it can be proved that the determinant of R' lies between zero and one. This is done by the use of Lagrange's method of Undetermined Multipliers in Ex. 8.16, and as is also shown in that example, when

$$\det R' = 1 \tag{8.8.9}$$

then R is a diagonal matrix and so it is perfectly conditioned for inversion. Also, when

$$\det R' = 0 \tag{8.8.10}$$

then R is singular and cannot be inverted. Thus the value of det R' gives us an indication of how well R is conditioned for inversion, *the closer to unity* det R' *is, the more easy the inversion of R will be; the closer to zero, the more trouble we can expect with round-off errors due to the finite word-length of the arithmetic.*

Our remarks have of necessity been extremely superficial.† The intent, however, was to alert the reader to the possibility of troubles and it is to be assumed that he will make it his business to ensure that the inversion of R (or any other matrix for that matter) is satisfactorily carried out, making recourse as necessary to the large body of knowledge in existence relative to this problem. Without further ado, we now assume that R has been satisfactorily inverted *and that the product of R and the computed R^{-1} been obtained, and is equal to the identity matrix to within acceptable errors.* We now turn our attention to possible further sources of trouble.

The matrix $T^T R^{-1} T$ will be positive definite if R^{-1} is positive definite and if T has full column rank (see Ex. 6.4). *The question that we now address ourselves to is the rank of T.*

In theory, if the observation scheme gives us

$$Y = TX + N \tag{8.8.11}$$

then we saw in Chapter 6 that T must have full column-rank if we hope to estimate all of the elements of X. Suppose, for definiteness that X is a 6-vector and Y a 30-vector. Then T will be 6×30.

Consider first the question of *the length of the observation interval.* Thus suppose that the 30 observations in the above Y are all made on the same

†The reader is also referred to comments on ill-conditioned systems in [8.3, p. 100] and to Ex. 8.17 and 8.18.

quantity in the process $X(t)$, and are made within a span of time, designated Δt seconds. The matrix T relates all of those observations to the state-vector at time t_n, i.e.

$$\begin{pmatrix} y_{29} \\ y_{28} \\ \vdots \\ y_1 \\ y_0 \end{pmatrix} = \begin{pmatrix} T_{29} \\ \hline T_{28} \\ \hline \vdots \\ \hline T_1 \\ \hline T_0 \end{pmatrix} X_n + N_{(n)} \tag{8.8.12}$$

where $T_{29} \ldots T_0$ are the rows of T.

Now it is quite obvious that if Δt is reduced, the differences in the row-vectors of T will become smaller, and a stage will be reached where, even with *infinite* precision arithmetic, the full-rank of T begins to become marginal. In practice we of course have only *finite* precision and so the problem is further aggravated, for now as the differences in the rows of T diminish, the precision of the arithmetic becomes insufficient to keep track of those small differences. Truncation will then cause those rows to start becoming batched into groups of identical rows, and this of course will lead to appreciable errors in forming $(T^T R^{-1} T)^{-1}$. Eventually as Δt continues to be reduced, the matrix T will cease to have full rank, there being not enough distinct rows remaining to form a nonzero 6×6 determinant. The computational errors in inverting $(T^T R^{-1} T)^{-1}$ now give way to outright impossibility in obtaining the inverse. *Thus by taking the observations over too small a time span Δt, we can invariably expect trouble.*

Of course, we now ask, "How small is a small value of Δt?" Obviously the answer depends on the process under observation. From a heuristic standpoint however, it is clear that unless we can space our observations far enough apart in time so that the process has changed by a *meaningful amount* between the first and the last of those observations, then we really must admit that we have not gathered data on all of the state variables in an adequate way.

We recall that in the above discussion we assumed that all of the observations were on the same quantity. This led to a similarity in all of the rows of T. It is now obvious that the situation will be drastically improved if we use mixed observations whenever possible,† since this will have the immediate

†e.g., instead of observing *only* range, say, we observe range, azimuth and elevation.

effect of introducing basic differences into the rows of T. We see that with, say three different types of observations, there are then three essentially different types of rows in T, *and so T has at least rank* 3. This will be true regardless of how close the observations are spaced.

In summary then, observations should be well spaced, and should be made on as many independent quantities as possible in order to ensure that the full rank of T is strong and not weak. Only then will we be able to carry out the estimation satisfactorily.

Next we examine the question of what is commonly termed *unfavorable geometry*. Consider the situation depicted in Figure 8.2 in which a body is moving according to some law along the x-axis. Observations are being made from point A on the y-axis and the intention is to determine the state of the system, i.e. position, velocity, etc., of the body. *Suppose first that we make only range-measurements*, and assume that, for some reason beyond our control, we were forced to locate A at some distance up the y-axis. Then we see that as the body passes through the origin, the range-measurements (ρ) will change very little with time. Hence the situation will not be perceptibly different than if the body were completely stationary. When the body is close to the origin then, making only range-measurements will give very poor estimates of position, velocity, acceleration, etc., and we say that *the geometry is very unfavorable for range-measurements* in that

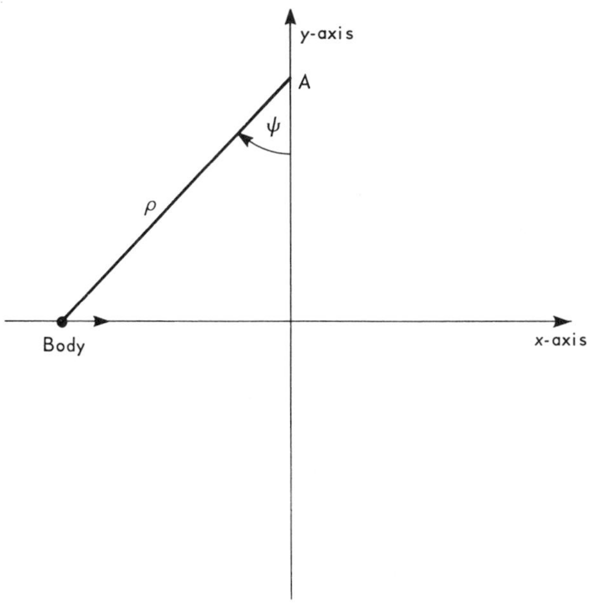

Fig. 8.2 *Geometrical considerations.*

region. As the body leaves the origin and moves further along the x-axis, the range-geometry consistently improves.

Suppose on the other hand *we were measuring only the azimuth angle* ψ. (See Figure 8.2.) Clearly if we were now forced to locate A close to the x-axis, the azimuth measurements would be less valuable when the body is far from the origin and would be of most value when it is passing through it.

The above two cases demonstrate how the geometry of a process can influence the value of some of the measurements and can make them almost worthless under certain circumstances. Note that the geometry we considered above took place in a simple two-dimensional space defined by the x and y axes. However spaces can be considerably more complicated in practice, being the hyper-spaces in which the state-vectors of the processes are defined. These involve variables which are not normally thought of as forming spaces, e.g. velocity, temperature, mass, etc., as well as the values of any parameters which we are attempting to estimate. The question of unfavorable geometry is thus seen to be nontrivial in these more complex situations and becomes much more difficult to envisage and hence to analyze.

The course of action needed to minimize the deleterious effects of poor geometry is clearly to make as many "orthogonal" measurements as possible. In the above example, *if we were simultaneously to measure both range and azimuth, then we see that in the very region where the geometry becomes bad for the one, it is best for the other.*

In constructing an observation scheme we must thus bear in mind that unfavorable geometry can and will arise in individual measurements. Thus redundancy of types of measurements must be provided in order to ensure that all of the measurements are never confronted by poor geometry at one and the same time.

We now close this chapter with a brief analysis of a fundamental relationship between M and Φ which ensures satisfactory operation of the filter.

8.9 RELATIONSHIP BETWEEN M AND Φ

In constructing an observation scheme the approach which is usually adopted is to observe as many different quantities as possible. Intuitively this is obviously the best course, and in the discussion given in the preceding section we attempted to show why this is the case. We now turn our attention to a brief study of the following strongly related question: *Given the model on which the filter is based, i.e. given the form of the differential equation, what are the constraints on the choice of the matrix* M *for satisfactory operation of the filter?*

We recall that the least-squares estimate is given by

$$\hat{X}^*_{n,n} = \left(T_n^T T_n\right)^{-1} T_n^T Y_{(n)} \tag{8.9.1}$$

and the minimum-variance estimate by

$$\mathring{X}^*_{n,n} = \left(T_n^T R_n^{-1} T_n\right)^{-1} T_n^T R_{(n)}^{-1} Y_{(n)} \tag{8.9.2}$$

These estimates exist if, respectively, $T_n^T T_n$ and $T_n^T R_{(n)}^{-1} T_n$ are nonsingular, and in Ex. 6.4 we saw that this is true in both cases, *if and only if* T_n *has full column-rank.* Our investigation in this section is accordingly directed towards developing the necessary and sufficient constraints on M, given Φ, so that T_n does have full column-rank.†

When we use the term *full column-rank,* what we mean is that the columns of T_n are linearly independent. Thus, letting U be a non-null vector, if T_n has full column-rank then this means that for *any* such U,

$$T_n U \neq \emptyset \tag{8.9.3}$$

We shall restrict our discussion to the case 1) where the observation matrix M does not vary with time, and 2) where the differential equation of the model is constant coefficient linear. Then [see e.g. (8.2.4)],

$$T = \begin{pmatrix} M\Phi(0) \\ \hline M\Phi(-1) \\ \hline \vdots \\ \hline M\Phi(-L) \end{pmatrix} \tag{8.9.4}$$

We now prove the following very useful result.

Let A be the matrix appearing in (4.6.27). Then:

Theorem 8.1

The matrix T *of* (8.9.4) *achieves full column-rank for* L *sufficiently large if and only if the matrix*

†The question is closely related to the concept of *observability* which appears in the Control Theory literature (see e.g. [8.4]).

$$G \equiv \begin{pmatrix} MA^{\circ} \\ --- \\ MA \\ --- \\ \vdots \\ --- \\ MA^{i} \end{pmatrix} \tag{8.9.5}$$

has full column-rank, where $i + 1$ *is the degree of the minimal polynomial*† *of* A.

Proof

By (4.6.27) we have

$$\begin{aligned} \Phi(k) &= \exp(kA) \\ &= \sum_{j=0}^{\infty} \frac{k^j}{j!} A^j \end{aligned} \tag{8.9.6}$$

But as is pointed out in [8.5, p. 609], this infinite series in powers of A collapses into a *finite* series in terms up to A^i, i.e. there exist coefficients $\alpha_j(k)$ such that (8.9.6) can be written

$$\Phi(k) = \sum_{j=0}^{i} \alpha_j(k) A^j \tag{8.9.7}$$

From this it follows that (8.9.4) becomes

$$T = \begin{pmatrix} \sum_{j=0}^{i} \alpha_j(0) MA^j \\ ---------- \\ \vdots \\ ---------- \\ \sum_{j=0}^{i} \alpha_j(-L) MA^j \end{pmatrix} \tag{8.9.8}$$

†See [8.6] for a definition of *minimal polynomial.*

Suppose for simplicity that $i = 1$. Then (8.9.8) can be written as

$$T = \begin{pmatrix} \alpha_0(0)I & \alpha_1(0)I \\ \alpha_0(-1)I & \alpha_1(-1)I \\ \vdots & \vdots \\ \alpha_0(-L)I & \alpha_1(-L)I \end{pmatrix} \begin{pmatrix} MA^0 \\ MA^1 \end{pmatrix} \tag{8.9.9}$$

Call the first matrix on the right J and the second K. We thus write (8.9.9) as

$$T = JK \tag{8.9.10}$$

which means that

$$T^T T = K^T J^T JK \tag{8.9.11}$$

Now, it is also shown in [8.5, p. 609] that the numbers $\alpha_0(k)$ and $\alpha_1(k)$ are linearly independent for k in any interval of nonzero length, and so it follows that the matrix J above has full column-rank for L sufficiently large. Hence $J^T J$ is positive definite. This then means (see Ex. 5.7) that $T^T T$ is positive definite if and only if K has full column-rank. Finally then T has full column-rank if and only if K has full column-rank. This completes the proof. ◆◆

The above theorem is very useful in that it provides us with a readily implemented test to establish whether or not the observation scheme and the model are properly matched so that a successful estimation algorithm can be set up.

Example 1

Suppose that $M = (1, 0, 0)$ and

$$A = \begin{pmatrix} 0 & 1 & 0 \\ 0 & 0 & 1 \\ 0 & 0 & 0 \end{pmatrix} \tag{8.9.12}$$

Then G of (8.9.5) becomes

$$G = \begin{pmatrix} 1 & 0 & 0 \\ 0 & 1 & 0 \\ 0 & 0 & 1 \end{pmatrix} \tag{8.9.13}$$

which clearly has full rank. (In this case $i = 2$.) This implies that T of (8.9.4) will have full column-rank if L is large enough.

In fact, from (8.9.12) and (4.6.28),

$$\Phi(k) = \begin{pmatrix} 1 & k & k^2/2 \\ & 1 & k \\ & & 1 \end{pmatrix} \tag{8.9.14}$$

and so

$$T = \begin{pmatrix} 1 & 0 & 0 \\ 1 & -1 & 1/2 \\ 1 & -2 & 4/2 \\ \vdots & \vdots & \\ 1 & -L & L^2/2 \end{pmatrix} \tag{8.9.15}$$

which has full column-rank for $L \geq 2$. ◆◆

Example 2

As an example where M and A are *not* properly matched, take $M = (0, 1, 0)$ and A as in (8.9.12). Then the reader can verify that in this case

$$G = \begin{pmatrix} 0 & 1 & 0 \\ 0 & 0 & 1 \\ 0 & 0 & 0 \end{pmatrix} \tag{8.9.16}$$

which only has rank 2, implying that T cannot attain full column-rank. In fact, for this M and $\Phi(k)$ of (8.9.14), we obtain

$$T = \begin{pmatrix} 0 & 1 & 0 \\ 0 & 1 & -1 \\ \vdots & & \\ 0 & 1 & -L \end{pmatrix} \tag{8.9.17}$$

whose rank never exceeds 2, regardless of how large we take L to be. ◆◆

We now state an alternate set of necessary and sufficient conditions relating M and the model, so that T will have full column-rank for L large enough.

Theorem 8.2

The matrix T will attain full column-rank for L large enough if and only if M does not annihilate any of the eigenvectors of Φ. The proof is quite straightforward but is unfortunately too lengthy to be included here. We accordingly sketch out a proof in Examples 8.19 through 8.21. ◆◆

Since both theorems in this section state necessary and sufficient conditions for the same result, *it follows that those conditions are equivalent and can be used interchangeably.*

We conclude this chapter with two simple examples demonstrating Theorem 8.2.

Example 3

Let $\Phi(k)$ be as given in (8.9.14). Then Φ has a single eigenvector, namely

$$V \equiv \alpha \begin{pmatrix} 1 \\ 0 \\ 0 \end{pmatrix} \tag{8.9.18}$$

where α is nonzero but otherwise arbitrary. Assuming that $M = (1,\ 0,\ 0)$ we see that

$$MV = \alpha \neq 0 \tag{8.9.19}$$

which means, by Theorem 8.2, that T should attain full column rank. This is in fact the case, as we saw in (8.9.15).

In the event that $M = (0, 1, 0)$ then

$$MV = 0 \tag{8.9.20}$$

and again, by Theorem 8.2, we expect T to have less than full rank. This is the case as we saw in (8.9.17). ◆◆

Example 4

Let

$$\Phi(k) = \begin{pmatrix} \cos\omega k & \frac{1}{\omega}\sin\omega k \\ -\omega\sin\omega k & \cos\omega k \end{pmatrix} \tag{8.9.21}$$

This is the transition matrix associated with the system†

$$\frac{d^2}{dt^2}x(t) = -\omega^2 x(t) \tag{8.9.22}$$

$\Phi(k)$ has two eigenvalues† namely $e^{j\omega}$ and $e^{-j\omega}$, and in general $e^{j\omega} \neq e^{-j\omega}$. The matrix Φ has two eigenvectors, namely

$$V_0 = \alpha_0\begin{pmatrix} 1 \\ j\omega \end{pmatrix} \qquad V_1 = \alpha_1\begin{pmatrix} 1 \\ -j\omega \end{pmatrix} \tag{8.9.23}$$

Thus if we take *either* $M = (1, 0)$ *or* $M = (0, 1)$, then both $MV_0 \neq \emptyset$ and $MV_1 \neq \emptyset$ and so, by Theorem 8.2, T will have full column-rank. This means that it is sufficient if we observe either one of the state variables.

In the case where ω is a multiple of π however, then both eigenvalues of Φ become equal and, as is easily verified, every nontrivial 2-vector *is now an eigenvector.* By the statement of Theorem 8.2, *we then require that* M *shall not annihilate any vector in the entire* 2-space, and so this means that M must have rank 2. Thus for example

†See Ex. 8.2 and Ex. 8.22.

$$M = \begin{pmatrix} 1 & 0 \\ 0 & 1 \end{pmatrix} \tag{8.9.24}$$

will be satisfactory to ensure that T attain full column-rank. ◆◆

In closing we point out that precisely the conditions of Theorem 8.2 (and hence by implication Theorem 8.1) *also ensure that the estimate covariance matrix* $S^*_{n,n}$ *goes to a null matrix as L is increased without bound.* This will be proved in Chapter 14. The reader is referred to Examples 8.22 through 8.27 where we consider further applications of Theorems 8.1 and 8.2.

This concludes our brief discussion. We have only examined the case where M and A are constant matrices but the time-varying cases are substantially more difficult and beyond the space limitations of the present work. We now leave the topic of Fixed-Memory Filters and turn to a discussion of the Expanding-Memory schemes.

EXERCISES

8.1 A sinusoidal process (period T) is observed and the sequence of scalar observations

$$Y_{(n)} \equiv (y_n, y_{n-1}, \ldots, y_{n-L})^T$$

is obtained. The inter-sample spacing is not necessarily constant. We wish to fit the function

$$f(t) \equiv \alpha_0 \cos\omega t + \alpha_1 \sin\omega t \tag{I}$$

to the data, in a least-squares sense, where $\omega \equiv 2\pi/T$ is known.

a) Set up the least-squares error criterion

$$e_n = \sum_{k=0}^{L} \left[y_{n-k} - f(t_{n-k} - t_n)\right]^2$$

and minimize it over α_0 and α_1. Hence set up a linear algorithm for α_0 and α_1 in terms of the observations.

b) Set up the linear algorithm which gives the updated estimate of the process and its first derivative in terms of the data.

8.2 a) Verify that the differential equation

$$\frac{d^2}{dt^2} x(t) = -\omega^2 x(t) \tag{I}$$

has as its transition relation

$$\begin{pmatrix} x(t_n + \zeta) \\ \dot{x}(t_n + \zeta) \end{pmatrix} = \begin{pmatrix} \cos\omega\zeta & \frac{1}{\omega}\sin\omega\zeta \\ -\omega \sin\omega\zeta & \cos\omega\zeta \end{pmatrix} \begin{pmatrix} x(t_n) \\ \dot{x}(t_n) \end{pmatrix}$$

where $x(t_n) = \alpha_0 \cos\omega t_n + \alpha_1 \sin\omega t_n$.

b) Assume that we observe $x(t)$ at times $t_n, t_{n-1}, \ldots, t_{n-L}$ (not necessarily equidistant), thereby obtaining the sequence of numbers

$$Y_{(n)} \equiv (y_n, y_{n-1}, \ldots, y_{n-L})^T$$

Set up the matrix T_n such that

$$Y_{(n)} = T_n X(t_n) + N_{(n)} \tag{II}$$

c) Now set up the least-squares algorithm

$$X^*_{n,n} = W_n Y_{(n)} \tag{III}$$

based on (II) above.

d) Reconcile (III) with part b) of Ex. 8.1.

8.3 Repeat Ex. 8.1, but

a) Assume that the covariance matrix of $Y_{(n)}$, namely $R_{(n)}$, is given, and obtain the minimum-variance algorithms for the quantities f and $\dot{f}$. [$R_{(n)}$ diagonal.]

b) Can a) above be carried out for Ex. 8.1 if $R_{(n)}$ is not diagonal?

8.4 a) Repeat Ex. 8.2 assuming that $R_{(n)}$, the covariance matrix of $Y_{(n)}$, is given. [$R_{(n)}$ diagonal.] Obtain the minimum-variance estimator for $X(t_n)$ and reconcile the resulting algorithm with that of Ex. 8.3, part a).

b) Extend the above to the case where $R_{(n)}$ is not diagonal.

8.5 a) Write a computer program which generates a sinusoidal signal plus random errors whose standard deviation is about 1% of the peak value of the sine wave.

b) Obtain a sequence of 30 unequally spaced observations,† recording both their values and sampling times.

c) Program and run the filtering algorithms obtained in the preceding four examples and estimate the state-vector of the true process.

d) Predict forward using the appropriate transition matrix (see Ex. 8.2 part a) and compute the prediction errors.

8.6 a) Repeat Ex. 8.5, but use instead, a quadratic Fixed-Memory Polynomial Filter to estimate $x(t)$ and $\dot{x}(t)$. Use a memory length equal to about 1/4 period of the sine wave and about 30 observations.

b) Compare the polynomial estimate state-vector with that obtained in Ex. 8.5.

c) Predict forward using the polynomial transition matrix and compute the prediction errors. Compare these with the errors obtained in Ex. 8.5 part d). Observe that using a sinusoidal model rather than a polynomial one enables us to predict much more effectively.

d) Now apply trend removal to the data being fed to the filter (see Section 7.14). (Make use of the fact that the differential equation of the true process is given in (I) of Ex. 8.2.)

8.7 Assume that we are observing a sinusoidal process whose angular frequency, ω, is an *unknown constant.*

a) Verify that the state-equation is

$$\frac{d}{dt}\begin{pmatrix} x(t) \\ \dot{x}(t) \\ \omega(t) \end{pmatrix} = \begin{pmatrix} \dot{x}(t) \\ -\omega^2 x(t) \\ 0 \end{pmatrix} \tag{I}$$

b) Assuming that we fit a quadratic polynomial to the data, using a Fixed-Memory (least-squares) Polynomial Filter, show how the elements of that polynomial state-vector can be used to give an estimate of the state-vector $X(t)$ of (I).

c) Using the program of Ex. 8.6, carry out b) above on a computer and obtain a nominal trajectory $\overline{X}(t)$.

8.8 Assume that

$$\frac{d^2}{dt^2} x(t) = -[\dot{x}(t)]^2 \tag{I}$$

†About 120 observations per sine-wave period.

a) Show how a least-squares polynomial estimate can be used to obtain a nominal trajectory.

b) Using a computer, generate a sequence of observations using (I) plus a noise generator.

c) Fit a quadratic polynomial with the aid of a Fixed-Memory Polynomial Filter.

d) Make predictions by using the polynomial directly and then by first obtaining a nominal trajectory based on that polynomial, followed by integration of (I) above. Observe how much better the latter approach is.

8.9 a) Assume that

$$\frac{d^2}{dt^2} x(t) = -k[\dot{x}(t)]^2 \qquad (k > 0)$$

where k is unknown, but constant. Show that a quadratic least-squares polynomial estimate can serve as the basis of a nominal trajectory for the state-vector

$$X(t) = \begin{pmatrix} x(t) \\ \dot{x}(t) \\ k \end{pmatrix}$$

b) If k were a function satisfying

$$\frac{d}{dt} k(t) = \omega k(t) \qquad \omega > 0$$

where ω is known, verify that a quadratic polynomial can still be used to form a nominal trajectory. What must we do if ω is an unknown constant?

8.10 a) Carry out part b) of Ex. 8.9 above using a computer to generate *noise-free* data from

$$\frac{d^2}{dt^2} x(t) = -k[\dot{x}(t)]^2$$

$$\frac{d}{dt} k(t) = \omega k(t) \qquad (\omega > 0, \text{ constant but unknown})$$

Obtain a cubic polynomial estimate on the data.

b) Make predictions using the cubic polynomial obtained in a) above.

c) Make predictions using the *nominal trajectory* obtained from the polynomial in a) above. Note the improvement over b).

d) Add random errors to the data and repeat a), b) and c) above.

8.11 Program and run (on a computer) a complete iterative differential-correction algorithm for the system of Ex. 8.8. Generate first noise-free and then noisy observations and obtain a nominal trajectory. Correct the latter by iterative differential-correction. Vary the amounts of noise added and the number of iteration cycles used.

8.12 Repeat Ex. 8.11 but use the system of Ex. 8.9, part a).

8.13 Verify that (8.5.6) is the inverse of (8.5.3).

8.14 Verify that R' of (8.8.8) has ones on its diagonal.

8.15 Starting from the eigenvalue equation

$$AX = \lambda X$$

where A is a matrix, X a vector and λ a scalar, verify that

a) $$\mathrm{Tr}(A) = \sum_i \lambda_i$$

i.e., the *trace* of A equals the *sum* of its eigenvalues. Also verify that

b) $$\mathrm{Det}(A) = \prod_i \lambda_i$$

i.e. the *determinant* of A equals the *product* of its eigenvalues.

8.16 Considering the 3 × 3 positive definite matrix

$$R' = \begin{pmatrix} 1 & r_{01} & r_{02} \\ r_{01} & 1 & r_{12} \\ r_{02} & r_{12} & 1 \end{pmatrix}$$

we see by Ex. 8.15 that

$$\lambda_1 + \lambda_2 + \lambda_3 = 3 \tag{I}$$

i.e. the sum of the eigenvalues of R' equals 3. The determinant of R' is positive. *We are interested in finding out how large* det R' *can be subject to* (I). By Ex. 8.15,

$$\det(R') = \lambda_1 \lambda_2 \lambda_3 \tag{II}$$

and so the problem reduces to one of *maximizing* (II) *subject to the constraint* (I) *above.*

Using Lagrange's method of undetermined multipliers (see e.g. p. 187) show that the above problem leads to the maximum values

$$\lambda_1 = \lambda_2 = \lambda_3 = 1 \tag{III}$$

and hence that

$$0 < \det(R') \leq 1 \tag{IV}$$

In general then, if R' is positive definite with ones on the diagonal, we have proved that (IV) must hold.

Verify that $\det(R') = 1$ if and only if R' is an identity matrix.

8.17 a) Test the conditioning of the matrix

$$A(\alpha) = \begin{pmatrix} 5+\alpha & 4+\alpha & 3+2\alpha \\ 4+\alpha & 5+\alpha & 3+2\alpha \\ 3+2\alpha & 3+2\alpha & 2+4\alpha \end{pmatrix}$$

for the three cases $\alpha = 1$, $\alpha = 0.1$, $\alpha = 0.01$, by the method discussed in Section 8.8.

b) In each of the three cases, invert the matrix $A(\alpha)$ by Gaussian elimination (by desk calculator) *using two-digit truncated precision.* Multiply the "inverse" into the matrix A in each case and compare the result to the identity matrix.

8.18 a) Test the conditioning of the coefficient matrix of the linear algebraic system

$$10000x + 20000y = 30000$$
$$20000x + 39999y = 59999$$

b) Verify that the solutions are $x = 1,\ y = 1$.

c) Making the very slight modification

$$10000x + 20000y = 30000$$

$$20000x + 39998y = 59999$$

verify that the solution becomes $x = 2,\ y = 1/2$. Clearly then the solution is extremely sensitive to errors in the fifth place of significance, a consequence of the very poor conditioning of the system matrix.

8.19 Given the model $\dot{X}(t) = AX(t)$, assume without loss of generality that A is in Jordan normal form, i.e. A is the direct sum of matrices $J(\gamma_0)$, $J(\gamma_1), \ldots,$ where each of the blocks J has the form

$$J(\gamma_i) = \begin{pmatrix} \gamma_i & 1 & & \\ & \gamma_i & \ddots & \\ & & \ddots & 1 \\ & & & \gamma_i \end{pmatrix} \qquad \text{(I)}$$

Using $\Phi(\zeta) = \exp(\zeta A)$ prove that $\Phi(\zeta)$ is the direct sum of matrices of the form

$$\exp\left[\zeta J(\gamma_i)\right] = \lambda_i^{\zeta} \begin{pmatrix} 1 & \zeta & \frac{1}{2}\zeta^2 \\ & 1 & \zeta \\ & & 1 \end{pmatrix} \qquad \text{(II)}$$

where we have shown a 2×2 for definiteness and where $\lambda_i \equiv \exp(\gamma_i)$.

8.20 Assume that $\Phi(\zeta)$ is 5×5, consisting of the direct sum of a 3×3 and a 2×2 matrix of the form shown in (II) of Ex. 8.19 above. Verify that Φ has only two eigenvectors, $V_0 \equiv (1, 0, 0, \mid 0, 0)^T$ and $V_3 \equiv (0, 0, 0, \mid 1, 0)^T$. Show that the vectors $V_1 \equiv (0, 1, 0, \mid 0, 0)^T$ and $V_2 \equiv (0, 0, 1, \mid 0, 0)^T$ form a hierarchy with V_0, in the sense that

$$\Phi(\zeta) V_0 = \lambda_0{}^{\zeta} V_0$$

$$\Phi(\zeta) V_1 = \lambda_0{}^{\zeta}(V_1 + \zeta V_0)$$

$$\Phi(\zeta) V_2 = \lambda_0{}^{\zeta}\left(V_2 + \zeta V_1 + \frac{\zeta^2}{2} V_0\right)$$

Verify that a similar hierarchy exists between V_3 above and $V_4 \equiv (0, 0, 0, \mid 0, 1)^T$.

8.21 The matrix T of (8.9.4) has full column-rank if and only if for any vector U, $TU \neq \emptyset$.

a) Considering a typical block from T, namely $M\Phi(-k)$, and writing any vector U in the form

$$U = \sum_{i=0}^{4} u_i V_i$$

where the V_i were defined in Ex. 8.20, verify that

$$M\Phi(-k)U = \lambda_0{}^{-k}\left[\left(u_0 - ku_1 + \frac{k^2}{2}u_2\right)MV_0 + (u_1 - ku_2)MV_1 + u_2 MV_2\right] + \lambda_1{}^{-k}\left[(u_3 - ku_4)MV_3 + u_4 MV_4\right]$$

b) Infer first that if $\lambda_0 \neq \lambda_1$ then $M\Phi(-k)U \neq \emptyset$ if and only if $MV_0 \neq \emptyset$ and $MV_3 \neq \emptyset$ (except perhaps for a few values of k). Now infer that if $\lambda_0 = \lambda_1$ then $M\Phi(-k)U \neq \emptyset$ if and only if M does not annihilate any linear combination of V_0 and V_3. (Note that the latter is also an eigenvector of Φ.)

c) By comparing the above two results to the statement of Theorem 8.2, verify that the theorem is true.

8.22 a) Verify that (8.9.21) is the transition matrix of (8.9.22).

b) Verify that the eigenvalues of (8.9.21) are $e^{j\omega}$ and $e^{-j\omega}$, and that the eigenvectors are as given in (8.9.23).

c) Find a matrix M which annihilates V_0 of (8.9.23), and hence verify that for this M the resultant T does not have full column-rank.

d) From (8.9.22) find the matrix A for this system, and using the same M as in c) above, apply the test given by Theorem 8.1.

8.23 For the transition matrix

$$\Phi(\zeta) = \begin{pmatrix} 1 & \zeta & 0 \\ & 1 & 0 \\ & & 1 \end{pmatrix}$$

verify both by the use of Theorem 8.2 as well as directly that

a) $M \equiv (1, 0, 1)$ *will not* give a T with full column-rank, whereas $M \equiv \begin{pmatrix} 1 & 0 & 1 \\ 1 & 0 & 0 \end{pmatrix}$ *will* give a T with full column-rank.

b) Verify that the A which gives rise to Φ above is

$$A = \begin{pmatrix} 0 & 1 & 0 \\ 0 & 0 & 0 \\ 0 & 0 & 0 \end{pmatrix}$$

Apply the test of Theorem 8.1 for both M's in a).

c) Verify by the use of (8.9.4), that if in a) above we were to use instead the *time-varying* matrix $M_n \equiv (n, 0, 1)$ then T would have full column-rank.

8.24 Investigate the class of observation matrices which must be used in conjunction with

$$\Phi(\zeta) \equiv \begin{pmatrix} 1 - \zeta & \zeta \\ -\zeta & 1 + \zeta \end{pmatrix}$$

so that T has full column rank.

8.25 a) Prove that

$$\Phi(\zeta) \equiv \begin{pmatrix} 1 & \zeta & 0 \\ 0 & 1 & 0 \\ 0 & \zeta & 1 \end{pmatrix}$$

is a transition matrix.

b) Verify that it has eigenvalues all equal to unity.

c) Verify that it has two linearly independent eigenvectors. Hence infer from Theorem 8.2 that for T to have full column-rank, M must have column-rank of two or more.

d) Verify directly by the use of (8.9.4) that if $M = (1, 0, -1)$ then T has column-rank 1, and that if $M = (1, 0, 1)$ then T has column-rank 2. However if $M = \begin{pmatrix} 1 & 0 & -1 \\ 1 & 0 & 1 \end{pmatrix}$, then T will have column-rank 3, i.e. full column-rank. Does this agree with c) above?

8.26 a) Consider the constant coefficient linear differential equation

$$\frac{d}{dt} X(t) = AX(t) \tag{I}$$

Let the initial conditions be $X(0) = (\alpha_0, \alpha_1, \alpha_2)^T$ and assume that A has the form

$$A = \begin{pmatrix} \gamma & 1 & 0 \\ & \gamma & 1 \\ & & \gamma \end{pmatrix}$$

Prove that

$$X(t) = \begin{pmatrix} \alpha_0 + \alpha_1 t + \alpha_2 t^2/2 \\ \alpha_1 + \alpha_2 t \\ \alpha_2 \end{pmatrix} \lambda^t \tag{II}$$

where $\lambda \equiv e^\gamma$.

b) Infer that the solutions to (I) die out as $t \to \infty$ if and only if $\text{Re}\gamma < 0$. *Thus* (I) *is a stable*† *system if and only if* A's *eigenvalues are in the left half of the complex plane.*

c) Verify that the transition matrix for $X(t)$ of (II) is

†A system is said to be stable if its natural modes (i.e. solutions to the homogeneous part of its differential equation) die out as t or $n \to \infty$.

$$\Phi(\zeta) \equiv \lambda^{\zeta} \begin{pmatrix} 1 & \zeta & \zeta^2/2 \\ & 1 & \zeta \\ & & 1 \end{pmatrix} \qquad \text{(III)}$$

and hence infer that this system is stable *if and only if* Φ's *eigenvalues are within the unit circle.*

d) Verify that if all three of A's eigenvalues are distinct, i.e. A's Jordan form is

$$A = \begin{pmatrix} \gamma_0 & & \\ & \gamma_1 & \\ & & \gamma_2 \end{pmatrix}$$

then (II) above becomes

$$X(t) = \begin{pmatrix} \alpha_0 \lambda_0^{\,t} \\ \alpha_1 \lambda_1^{\,t} \\ \alpha_2 \lambda_2^{\,t} \end{pmatrix}$$

where $\lambda_0 = e^{\gamma_0}$, etc. Show also that (III) above becomes diagonal.

8.27 a) Examine part d) of Ex. 8.26 above if $\gamma_0 = \gamma_1 = \gamma_2$. What requirements must we place on M so that T has full column-rank in (8.9.4)?

b) What forms do X and Φ assume if $\gamma_0 = \gamma_1 = \gamma_2 = 0$, i.e., $A = \emptyset$?

8.28 a) Generalize Ex. 8.26 to the case where A is initially not in Jordan normal form and has more than one Jordan block. Verify that with an appropriate extension, the same results hold concerning stability.

8.29 a) Verify that if

$$\frac{d}{dt} X(t) = AX(t)$$

then $X(t)$ is a polynomial state-vector if and only if all of A's eigenvalues are zero.

Hint: See Ex. 8.26 part a).

b) Infer that

$$X(t_n + \zeta) = \Phi(\zeta)\, X(t_n)$$

is a polynomial transition relation, if and only if all of Φ's eigenvalues equal unity.

8.30 a) Let

$$X(t_n + \zeta) = \Phi(\zeta)\, X(t_n)$$

and suppose that Φ has all of its eigenvalues on the unit circle. Verify that $X(t_n)$ has, for its elements, sums of sines and cosines, possibly multiplied by polynomials in n.

b) Under what conditions will the polynomial multipliers be present or absent?

REFERENCES

1. Blackman, R. B., "Methods of Orbit Refinement," Bell Syst. Tech. Jour., Vol. 43, May 1964, Part II, pp. 886 to 890.
2. Wilkinson, J. H., "Rounding Errors in Algebraic Processes," Prentice-Hall, New Jersey, 1963.
3. Deutsch, R., "Estimation Theory," Prentice-Hall, New Jersey, 1965.
4. Kalman, R. E., "On the General Theory of Control Systems," Proceedings of the First International Conference of Automatic Control (IFAC), Moscow, 1960.
5. Zadeh, L. A., and Desoer, C. A., "Linear System Theory," McGraw-Hill Book Company, 1963.
6. Mirsky, L., "An Introduction to Linear Algebra," Oxford, 1961, p. 203.

See also

7. Whittaker, E. T., and Robinson, G., "The Calculus of Observations," Blackie, London, 1924.
8. Brouwer, D., and Clemence, G. M., "Methods of Celestial Mechanics," Academic Press, New York, 1961, Chapters 8 and 9.

PART 3

EXPANDING–MEMORY FILTERING

The schemes that we have considered up to now have been characterized by the fact that estimates were repeatedly obtained on the basis of a *fixed-length* record of observations.

Consider a function such as $\sin \omega t$. If we select a fixed-length record out of it, then it is easy to see that if that length is small enough, we can fit a parabola to it very satisfactorily. We imagine a chain that is essentially flexible, subject to the constraint that its form is always parabolic. We now drag this chain along the curve $\sin \omega t$, and imagine that it fits itself to the curve in the least-squares sense. Heuristically, this is the concept of the Fixed-Memory Polynomial Filter discussed in Chapters 7 and 8.

As we lengthen the chain, the systematic errors worsen. However the smoothing effect of the filter depends only on the length of the chain without regard to the form of the signal – a direct consequence of the linearity of the algorithms which we developed – and so as we lengthen the chain, the estimates become smoother. This trade-off between smoothness and systematic errors was what, in the final analysis, fixed the length of the smoothing interval.

Fixed-memory filtering is ideally suited to a number of situations. As a first example, suppose that the signal entering such a filter changes abruptly from time to time, and so when one of these changes occurs, the filter output suddenly acquires errors. However because of the *fixed length* of the filter memory, these errors persist *for at most one memory-length in time,* and thereafter they are completely eradicated since the prevailing estimates are then based on memory-lengths containing completely new data. In fact, we might say that "transient" phenomena in fixed-memory filters disappear *completely* in a finite amount of time equal to the memory-span.

A second case to which they are well suited is the situation where the true process model is either unknown or is excessively complex. It is generally possible, *over short enough record-lengths,* to find simpler functions which adequately describe the true process. Increasing that length causes the fit to become unacceptably bad, while decreasing it causes the estimate to become excessively sensitive to the random fluctuations caused by the observational errors.

These facts that

a) Transient phenomena caused by abrupt changes in the input die out completely in a fixed time,

b) Simple functions can effectively describe complex ones over short enough record-lengths,

are where the true efficacy of the fixed-memory schemes reside.

However, those schemes also possess two very serious drawbacks. First, the observations made over the entire memory-span must always be retained, and we can only delete data when they are staler than the memory-length. It is easy to see that this could lead to excessive memory requirements, particularly if the observation interval is very long or if the number of filters being operated in the computer becomes large. The second drawback is the amount of computation involved, since the entire memory-length of data must be reprocessed every time an estimate is derived. In the case of differential-correction this amount of computation can be large.

If these factors do not constitute drawbacks in a given situation, then the fixed-memory filters should definitely be used. However when these drawbacks become serious we begin to look about for methods of reducing the memory and computation requirements, and it is in this quest that we consider the possibilities offered by filters with an *expanding* memory-length.

9

THE EXPANDING-MEMORY POLYNOMIAL FILTER

9.1 INTRODUCTION

Consider the situation where it has been decided to obtain an estimate by using a least-squares polynomial fit to, say, 100 equally spaced observations. Initially none of these observations has been obtained, and at some time, $t = 0$, the observation period commences.

As the data begin to arrive the question arises as to whether anything can be done to process them immediately, rather than to wait until they have all been received, and doing no processing in the interim. It is conceivable that an appropriate algorithm might be found, so that as soon as a datum has been received it could be immediately processed and then discarded. This would necessitate only *one* data storage location as against *one hundred,* called for by the alternate method of doing nothing until all hundred are in.

The question of whether or not something can be done immediately the data begin to arrive is particularly important in *tracking systems.* In such schemes the observation instrument has a limited field of view (e.g. a high-powered telescope or a tracking radar), and unless a method is devised whereby the instrument can be pointed approximately at the object being observed, the object disappears from the field of view and is lost. Further observations thereupon cease until the system reacquires the object.

The scheme which we now describe, known as the Expanding-Memory Recursive Polynomial Filter, is the first of a series of algorithms intended specifically to cope with such situations.

Essentially we will be performing a least-squares polynomial fit to the data in a manner very similar to that used in the Fixed-Memory Polynomial Filter. One major difference however is that the memory-length of the present filter will be steadily increasing, so that as each new observation arrives it will be incorporated, *together with all of its predecessors,* into the formation of a new estimate. This estimate is then immediately usable, if need be, to assist the observation instrument in its tracking function, in order that further observations can be made.

The second major difference will be that the algorithms will be recursive; the new estimate will be a linear combination of its predecessor and the latest observation. As a result only one datum need be retained, and when the recursion is cycled, that datum can be discarded and room made for its successor. Only one memory location is thus needed for the data.

After we have synthesized the algorithms we shall analyze their properties. Their covariance matrices and systematic errors will be seen to be closely related to those of the Fixed-Memory Polynomial Filters of Chapter 7. However, because of the steadily increasing data-base, we will find that the estimates produced by the present filters *become steadily smoother as time passes.* In this sense the filters of this chapter differ markedly from those of Chapter 7.

The present chapter is founded on the discrete Legendre polynomials discussed in Section 3.2, and we will assume complete familiarity with that material.

9.2 THE APPROXIMATING POLYNOMIAL

Assume that scalar observations are made, all on the *same* quantity in a process, at *equally* spaced instants, τ seconds apart. We call the first of them y_0 and at time $t = n\tau$, measured from time of starting, our data record consists of the set of numbers

$$y_0, y_1, y_2, \ldots, y_{n-1}, y_n$$

In Figure 9.1 we depict the situation showing the time axis quantized every τ seconds.

The $n + 1$ observations of the above sequence are to be used to form an estimate of the process, and we choose as the model a polynomial in the continuous variable r. The r-axis is shown in the figure, its origin located at $n = 0$, and r increasing positively with time. The estimating polynomial, by assumption of degree m, is designated $\left[p^*(r)\right]_n$, the subscript of course being intended to show that this is the polynomial computed on the basis of the sequence $y_0, y_1, \ldots, y_{n-1}, y_n$. *A completely new polynomial will be computed when* y_{n+1} *arrives,* based on the sequence $y_0, y_1, \ldots, y_n, y_{n+1}$, and it will be designated $\left[p^*(r)\right]_{n+1}$.

Just as we let p^* be the estimate of the process function, so we decide that the time-derivatives of p^* will serve as estimates of its derivatives. Reference to Figure 9.1 shows that

$$t = r\tau \tag{9.2.1}$$

and so

$$\frac{d}{dt} = \frac{1}{\tau}\frac{d}{dr} \tag{9.2.2}$$

Hence we see that

$$\frac{d}{dt}\left[p^*(r)\right]_n = \frac{1}{\tau}\frac{d}{dr}\left[p^*(r)\right]_n \tag{9.2.3}$$

We write the state-vector of successive time-derivatives of $\left[p^*(r)\right]_n$ as $X^*_{r,n}$. Thus

$$\begin{pmatrix} x^* \\ \dot{x}^* \\ \vdots \\ D^m x^* \end{pmatrix}_{r,n} \equiv \begin{pmatrix} p^*(r) \\ \dfrac{1}{\tau}\dfrac{d}{dr}\,p^*(r) \\ \vdots \\ \dfrac{1}{\tau^m}\dfrac{d^m}{dr^m}\,p^*(r) \end{pmatrix}_n \tag{9.2.4}$$

and we now show how the state-vector $X^*_{r,n}$ defined above can be obtained by the use of the *least-squares* criterion.

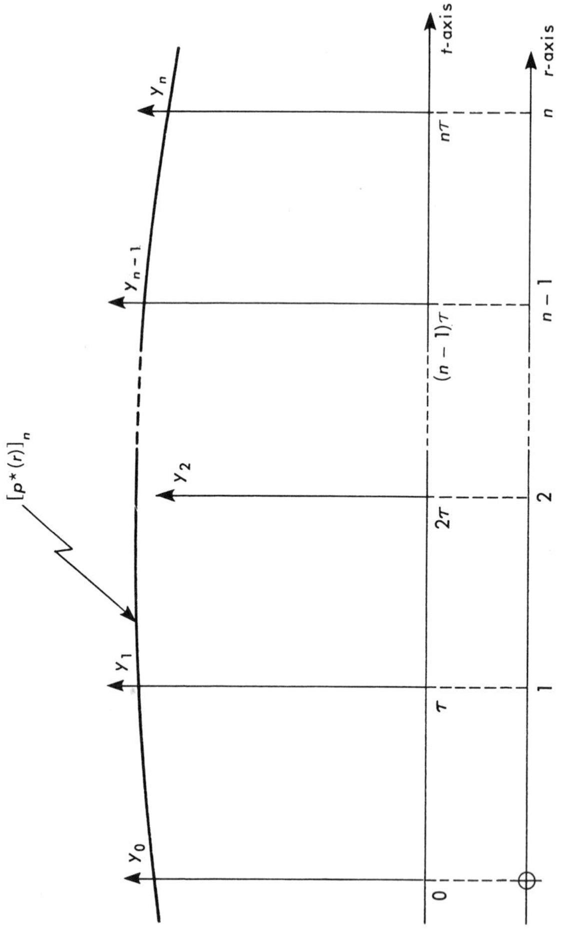

Fig. 9.1 *The t and r-axes.*

In Section 3.2 we derived the *discrete Legendre polynomials* which satisfy the orthogonality condition

$$\sum_{r=0}^{L} p(r; i,L)\, p(r; j,L) = 0, \qquad i \neq j \tag{9.2.5}$$

Assume that time is frozen just after the nth sampling instant and so, for the present, the quantity n is a *fixed integer.* This freeze is maintained *throughout this section.* We select as our interval of orthogonality the range $0 \leq r \leq n$, i.e. we set L in (9.2.5) equal to n, where n is the temporarily frozen time-index shown in Figure 9.1. Equation (9.2.5) then becomes

$$\sum_{r=0}^{n} p(r; i, n)\, p(r; j, n) = 0 \qquad i \neq j \tag{9.2.6}$$

From (3.2.20) the polynomials satisfying this condition are given explicitly by

$$p(r; j, n) \equiv \sum_{\nu=0}^{j} (-1)^{\nu} \binom{j}{\nu} \binom{j+\nu}{\nu} \frac{r^{(\nu)}}{n^{(\nu)}} \tag{9.2.7}$$

Also, the quantity $c(j, n)$, defined by

$$[c(j, n)]^2 \equiv \sum_{r=0}^{j} [p(r; j, n)]^2 \tag{9.2.8}$$

is given in (3.2.30) as

$$[c(j, n)]^2 = \frac{(n + j + 1)^{(j+1)}}{(2j + 1)\, n^{(j)}} \tag{9.2.9}$$

Equations (9.2.7) *and* (9.2.9) *define a separate set of orthogonal polynomials for each given value of* n.

We commence the estimation process by writing the estimating polynomial $\left[p^*(r)\right]_n$ as a linear combination of the first $m + 1$ of the above orthogonal polynomials. Thus, letting $\beta_0, \beta_1, \ldots, \beta_m$ be a set of (as yet unknown) constants, we write

$$\left[p^*(r)\right]_n = \sum_{j=0}^{m} \left(\beta_j\right)_n p(r;\, j, n) \tag{9.2.10}$$

Note that the β's in the above sum have been subscripted with an n to show their association with $\left[p^*(r)\right]_n$.

From Figure 9.1 it is seen that the difference between the kth observation and the estimating polynomial is $y_k - \left[p^*(k)\right]_n$. This is the *residual* as defined in Section 6.5, and so the sum of the squared residuals is

$$e_n = \sum_{k=0}^{n} \left\{y_k - \left[p^*(k)\right]_n\right\}^2 \tag{9.2.11}$$

We have decided to use least-squares and it thus follows that $\left[p^*(r)\right]_n$ must be chosen so as to minimize this scalar.

First we set (9.2.10) into (9.2.11) thereby obtaining

$$e_n = \sum_{k=0}^{n} \left[y_k - \sum_{j=0}^{m} \left(\beta_j\right)_n p(k;\, j, n)\right]^2 \tag{9.2.12}$$

which is now seen to be a function of the β's. We differentiate e_n with respect to each of those parameters, setting the result to zero, i.e. we set

$$\frac{\partial e_n}{\partial \left(\beta_i\right)_n} = 0 \qquad 0 \le i \le m \tag{9.2.13}$$

and the solution of these $m+1$ equations then provides us with the β's which minimize (9.2.12).

Performing the indicated differentiations gives

$$\sum_{k=0}^{n} \left[y_k - \sum_{j=0}^{m} \left(\beta_j\right)_n p(k;\, j, n)\right] p(k;\, i, n) = 0 \qquad 0 \le i \le m \tag{9.2.14}$$

which can be written

$$\sum_{k=0}^{n} \sum_{j=0}^{m} \left(\beta_j\right)_n p(k;\, i, n)\, p(k;\, j, n) = \sum_{k=0}^{n} p(k;\, i, n)\, y_k \qquad 0 \le i \le m \tag{9.2.15}$$

We interchange the order of summation on the left, obtaining

$$\sum_{j=0}^{m} \left(\beta_j\right)_n \sum_{k=0}^{n} p(k;\, i, n)\, p(k;\, j, n) = \sum_{k=0}^{n} p(k;\, i, n)\, y_k \tag{9.2.16}$$

and then, by invoking (9.2.6) and (9.2.8), this gives us

$$\left(\beta_i\right)_n [c(i, n)]^2 = \sum_{k=0}^{n} p(k;\, i, n)\, y_k \tag{9.2.17}$$

Now dividing through by $[c(i,n)]^2$, we obtain an explicit form for the computation of the ith constant β_i, used in (9.2.10). Thus,

$$\left(\beta_j\right)_n = \frac{\sum_{k=0}^{n} p(k;\, j, n)\, y_k}{[c(j, n)]^2} \qquad 0 \le j \le m \tag{9.2.18}$$

Finally, using this result in (9.2.10), the estimating polynomial is given explicitly in terms of the observations as

$$\left[p^*(r)\right]_n = \sum_{j=0}^{m} \left\{ \frac{\sum_{k=0}^{n} p(k;\, j, n)\, y_k}{[c(j, n)]^2} \right\} p(r;\, j, n) \tag{9.2.19}$$

Comparison with (7.3.14) shows that we have obtained precisely the fixed-memory polynomial estimate, with the exception that the L in (7.3.14) has been replaced by n to give us (9.2.19).

This usage of n is not a case of mistaken identity of symbols. We have deliberately used n *both as the counting symbol for the data as well as the memory-length designator.* Our intention was to develop an algorithm which gives us the fixed-memory estimate over $n+1$ observations, with n steadily increasing. That is precisely what (9.2.19) does in fact provide.

Returning to (9.2.4), we are now able to write

$$\left(D^i x^*\right)_{r,n} = \frac{1}{\tau^i} \sum_{j=0}^{m} \left\{ \frac{\sum_{k=0}^{n} p(k; j, n)\, y_k}{[c(j, n)]^2} \right\} \frac{d^i}{dr^i}\, p(r; j, n) \tag{9.2.20}$$

which, by an interchange in the order of summation, gives us

$$\left(D^i x^*\right)_{r,n} = \frac{1}{\tau^i} \sum_{k=0}^{n} \left\{ \sum_{j=0}^{m} \frac{p(k; j, n)\,(d^i/dr^i)\, p(r; j, n)}{[c(j, n)]^2} \right\} y_k \tag{9.2.21}$$

From this last result it is clear that any estimates obtained from this scheme will be linear combinations of the observations. This linearity property will be present throughout our discussion and will form the basis for some of our proofs.

9.3 RECURSIVE FORMULATION

The expression for $\left(D^i x^*\right)_{r,n}$ given in (9.2.21) requires that all of the data $y_0 \ldots y_n$ be stored and available. Moreover the $n+1$ weights multiplying the data in (9.2.21) must either be stored or else computed on-line. Finally $n+1$ multiplies are required to form $\left(D^i x^*\right)_{r,n}$. All of these objections can be largely overcome by developing a *recursive formulation* for $\left[p^*(r)\right]_n$ which relates it to its predecessors $\left[p^*(r)\right]_{n-1}$, $\left[p^*(r)\right]_{n-2}$, etc.

We shall see in Chapter 10 that the carrying out of this task, for the algorithms under discussion, could be treated as a special case of the very powerful procedures to be developed in that chapter. However, unless we go from the special to the general we will fail to acquire an intuitive appreciation of what it is that we are really doing. Accordingly we shall proceed, step by step, at times perhaps a trifle heuristically, and obtain a set of results which we will have the satisfaction of generalizing when we reach Chapter 10.

The first of our heuristic decisions is that we will consider only the 1-step predictors, i.e. we shall limit ourselves to the case

$$r = n + 1 \tag{9.3.1}$$

We note however that this is in reality no restriction, since if we have the 1-step prediction of the estimate state-vector of (9.2.4), then we can readily obtain its value at any other validity instant by the use of the appropriate polynomial transition matrix, i.e., by the use of

$$X^*_{n+h,n} = \Phi(h-1)X^*_{n+1,n} \tag{9.3.2}$$

where the matrix Φ is given in (4.2.5).

By (9.2.19) we have

$$\left[p^*(r)\right]_n = \sum_{j=0}^{m}\left[\sum_{k=0}^{n} p(k;j,n)\,y_k\right]\frac{p(r;j,n)}{[c(j,n)]^2} \tag{9.3.3}$$

and so, setting $r = n + 1$ gives us

$$\left[p^*(n+1)\right]_n = \sum_{j=0}^{m}\left[\sum_{k=0}^{n} p(k;j,n)\,y_k\right]\frac{p(n+1;j,n)}{[c(j,n)]^2} \tag{9.3.4}$$

Now, it is possible to show either by direct algebraic means or else by the use of the hypergeometric series of Gauss (see [9.1]) that

$$\frac{p(n+1;j,n)}{[c(j,n)]^2} = \frac{(-1)^j(2j+1)}{n+1} \tag{9.3.5}$$

Making use of this result, (9.3.4) becomes

$$\left[p^*(n+1)\right]_n = \sum_{j=0}^{m} \left[\sum_{k=0}^{n} p(k; j, n)\, y_k\right] \frac{(-1)^j(2j+1)}{n+1} \tag{9.3.6}$$

We write this out for $m = 0, 1, 2$. Thus:

$m = 0$:

$$\left[p^*(n+1)\right]_n = \frac{1}{n+1} \sum_{k=0}^{n} y_k \tag{9.3.7}$$

$m = 1$:

$$\left[p^*(n+1)\right]_n = \frac{1}{n+1} \left[\sum_{k=0}^{n} y_k - 3 \sum_{k=0}^{n} \left(1 - 2\frac{k}{n}\right) y_k\right] \tag{9.3.8}$$

$m = 2$:

$$\left[p^*(n+1)\right]_n = \frac{1}{n+1} \left\{\sum_{k=0}^{n} y_k - 3 \sum_{k=0}^{n} \left(1 - 2\frac{k}{n}\right) y_k + 5 \sum_{k=0}^{n} \left[1 - 6\frac{k}{n} + 6\frac{k(k-1)}{n(n-1)}\right] y_k\right\} \tag{9.3.9}$$

and so forth.

Multiplying both sides of the above three equations by $(n+1)^{(m+1)}$ we now obtain

$m = 0$:

$$(n+1)\left[p^*(n+1)\right]_n = \sum_{k=0}^{n} y_k \tag{9.3.10}$$

$m = 1$:

$$(n+1)\, n\left[p^*(n+1)\right]_n = n \sum_{k=0}^{n} y_k - 3 \sum_{k=0}^{n} (n - 2k)\, y_k \tag{9.3.11}$$

$m = 2:$

$$(n+1)n(n-1)\left[p^*(n+1)\right]_n = n(n-1)\sum_{k=0}^{n} y_k - 3(n-1)\sum_{k=0}^{n}(n-2k)y_k$$

$$+ 5\sum_{k=0}^{n}[n(n-1) - 6(n-1)k + 6k(k-1)]y_k \tag{9.3.12}$$

and in general

$$(n+1)^{(m+1)}\left[p^*(n+1)\right]_n = L\text{ (data)} \tag{9.3.13}$$

where L(data) is an appropriate linear combination of the data.

If we examine the right-hand sides of (9.3.10) through (9.3.12) then we see that in each case they are *polynomials of degree* m *in the discrete variable* n. In Section 2.5 we showed that if f_n is a polynomial of degree m or less in n, then

$$\nabla^{m+1} f_n = 0 \tag{9.3.14}$$

(see (2.5.20)). We apply this result as follows.

Specifically, consider (9.3.11) where $m = 1$. By forming the second backward difference of both sides with respect to n, we obtain

$$\nabla^2\left\{(n+1)n\left[p^*(n+1)\right]_n\right\}$$

$$= \nabla^2\left(n\sum_{k=0}^{n} y_k\right) - \nabla^2\left[3\sum_{k=0}^{n}(n-2k)y_k\right]$$

$$= \left[n\sum_{k=0}^{n} y_k - 2(n-1)\sum_{k=0}^{n-1} y_k + (n-2)\sum_{k=0}^{n-2} y_k\right]$$

$$-3\left[\sum_{k=0}^{n}(n-2k)y_k - 2\sum_{k=0}^{n-1}(n-1-2k)y_k + \sum_{k=0}^{n-2}(n-2-2k)y_k\right]$$

$$= n(y_n + y_{n-1}) - 2(n-1)y_{n-1} + (\nabla^2 n)\sum_{k=0}^{n-2} y_k$$

$$- 3\Big[-ny_n + (-n+2)y_{n-1} + 2(n-1)y_{n-1}\Big] - 3\sum_{k=0}^{n-2}\Big[\nabla^2(n-2k)\Big]y_k$$

$$= 4ny_n - 2(2n-1)y_{n-1} + (\nabla^2 n)\sum_{k=0}^{n-2} y_k - 3\sum_{k=0}^{n-2}\Big[\nabla^2(n-2k)\Big]y_k \tag{9.3.15}$$

But now, by virtue of (9.3.14), this reduces to

$$\nabla^2\Big\{(n+1)n\Big[p^*(n+1)\Big]_n\Big\} = 4ny_n - 2(2n-1)y_{n-1} \tag{9.3.16}$$

This is a recursion in $\Big[p^*(n+1)\Big]_n$, a fact which is more easily seen if we expand the left-hand side, i.e.

$$\begin{aligned}(n+1)n\Big[p^*(n+1)\Big]_n &= 2n(n-1)\Big[p^*(n)\Big]_{n-1} \\ &\quad - (n-1)(n-2)\Big[p^*(n-1)\Big]_{n-2} \\ &\quad + 4ny_n - 2(2n-1)y_{n-1}\end{aligned} \tag{9.3.17}$$

It is now evident that $\Big[p^*(n+1)\Big]_n$ can be obtained *from a linear combination of its two predecessors and the two most recent observations.* Regardless of the value of n, we need thus retain only the three numbers

$$\Big[p^*(n)\Big]_{n-1}, \quad \Big[p^*(n-1)\Big]_{n-2}, \quad y_{n-1}$$

and as soon as y_n arrives, the new value of the 1-step prediction $\Big[p^*(n+1)\Big]_n$ can be computed by the use of (9.3.17).

The same argument that took us from (9.3.11) to the recursive form (9.3.16) can, naturally, also be applied to (9.3.10) or (9.3.12). Corresponding recursive algorithms would result. It is thus easily seen that the general form (9.3.13) would give rise to the recursive form

$$\nabla^{m+1}\left[(n+1)^{(m+1)}p^*(n+1)_n\right] = L\ (\text{data}) \tag{9.3.18}$$

where L is also a linear combination of the data.

We return to (9.3.17). Define

$$z^*_{n+1,n} \equiv \left[p^*(n+1)\right]_n \qquad \text{(See Note)} \tag{9.3.19}$$

Then (9.3.17) becomes

$$z^*_{n+1,n} = \frac{1}{(n+1)n}\left[2n(n-1)z^*_{n,n-1} - (n-1)(n-2)z^*_{n-1,n-2} + 4ny_n - 2(2n-1)y_{n-1}\right] \tag{9.3.20}$$

At this stage we introduce the *prediction error* ϵ_n, defined by

$$y_n = z^*_{n,n-1} + \epsilon_n \tag{9.3.21}$$

Thus ϵ_n is the difference between what is actually observed at time n (namely y_n), and what was predicted for time n based on observations up to time $n-1$ (namely $z^*_{n,n-1}$). By substituting (9.3.21) into (9.3.20) it is now easily verified that the latter gives

$$z^*_{n+1,n} = 2z^*_{n,n-1} - z^*_{n-1,n-2} + \frac{4}{n+1}\epsilon_n - \frac{2(2n-1)}{(n+1)n}\epsilon_{n-1} \tag{9.3.22}$$

This form is particularly well-suited to a tracking situation where it is precisely the prediction error that is measured by the observation instrument.

Equation (9.3.22) is a linear difference equation with homogeneous part

$$\nabla^2 z^*_{n+1,n} = 0 \tag{9.3.23}$$

and a forcing function of the form

$$f_n \equiv a_n\epsilon_n + b_n\epsilon_{n-1} \tag{9.3.24}$$

Note: Up to now $\left[p^*(n+1)\right]_n$ has been called $x^*_{n+1,n}$ (see e.g. (9.2.4)). It will soon become clear why we have introduced the symbol z. (See (9.4.1).)

This is consistent with what we expect *when the prediction errors are zero for all n.* For example, suppose that the input was a series of error-free samples of a first-degree polynomial. We are then consistently able to predict the inputs without error, and so those predictions would also constitute a sampled first-degree polynomial in n. That is precisely what (9.3.23) implies. When prediction errors are present however, then the forcing function of (9.3.24) is nonzero, and this serves to modify the prediction as soon as the latest observation is obtained. We shall consider this further in the next section. However, prior to proceeding with that discussion we list the algorithms corresponding to (9.3.22), obtained by applying the same approach, for each of the cases $m = 0, 1, 2, 3$ (see Ex. 9.4, 9.5 and 9.6).

$m = 0$:

$$z^*_{n+1,n} = z^*_{n,n-1} + \frac{1}{n+1}\epsilon_n \tag{9.3.25}$$

$m = 1$:

$$z^*_{n+1,n} = 2z^*_{n,n-1} - z^*_{n-1,n-2} + \frac{4}{n+1}\epsilon_n - \frac{2(2n-1)}{(n+1)n}\epsilon_{n-1} \tag{9.3.26}$$

$m = 2$:

$$\begin{aligned} z^*_{n+1,n} = {} & 3z^*_{n,n-1} - 3z^*_{n-1,n-2} + z^*_{n-2,n-3} \\ & + \frac{9}{n+1}\epsilon_n - \frac{18(n-1)}{(n+1)n}\epsilon_{n-1} + \frac{3(3n^2-9n+8)}{(n+1)n(n-1)}\epsilon_{n-2} \end{aligned} \tag{9.3.27}$$

$m = 3$:

$$\begin{aligned} z^*_{n+1,n} = {} & 4z^*_{n,n-1} - 6z^*_{n-1,n-2} + 4z^*_{n-2,n-3} - z^*_{n-3,n-4} \\ & + \frac{16}{n+1}\epsilon_n - \frac{24(2n-3)}{(n+1)n}\epsilon_{n-1} + \frac{48(n^2-4n+5)}{(n+1)n(n-1)}\epsilon_{n-2} \\ & - \frac{8(2n^3-15n^2+43n-45)}{(n+1)n(n-1)(n-2)}\epsilon_{n-3} \end{aligned} \tag{9.3.28}$$

9.4 COUPLING PROCEDURE

The algorithms given in (9.3.25) through (9.3.28) provide us with recursion formulae for the 1-step predictions of the *zeroth derivative.* We now show that very compact recursive algorithms can be obtained which compute the *entire state-vector* of the estimating polynomial, rather than just its zeroth derivative.

First, define the scaled-derivative estimate vector

$$Z^*_{n,n} \equiv \begin{pmatrix} z^*_0 \\ z^*_1 \\ \vdots \\ z^*_m \end{pmatrix}_{n,n} \equiv \begin{pmatrix} x^* \\ \tau\dot{x}^* \\ \vdots \\ \dfrac{\tau^m}{m!} D^m x^* \end{pmatrix}_{n,n} \tag{9.4.1}$$

Suppose that the process giving rise to the observations is in fact a quadratic in t, and assume initially that we are able to observe it *without* errors. Then the sequence of estimating polynomials

$$\ldots \left[p^*(r)\right]_{n-3}, \left[p^*(r)\right]_{n-2}, \left[p^*(r)\right]_{n-1} \ldots$$

will be one and the same polynomial, and we would expect that

$$\left[p^*(n)\right]_n = \left[p^*(n)\right]_{n-1} \tag{9.4.2}$$

i.e. the updated estimate at time t_n would equal the 1-step prediction based on observations up to t_{n-1}. Clearly, if (9.4.2) holds, then so will

$$Z^*_{n,n} = Z^*_{n,n-1} \tag{9.4.3}$$

Using the transition matrix for $Z^*_{n,n}$ as defined in (4.2.15) we have

$$Z^*_{n,n} = \Phi(-1)\, Z^*_{n+1,n} \tag{9.4.4}$$

which we combine with (9.4.3) to give

$$\Phi(-1)\, Z^*_{n+1,n} = Z^*_{n,n-1} \tag{9.4.5}$$

Consider next the observation errors. The polynomial $\left[p^*(r)\right]_{n-1}$ is based on the set of observations

$$y_0, y_1, \ldots, y_{n-2}, y_{n-1}$$

and $\left[p^*(r)\right]_n$ is based on

$$y_0, y_1, \ldots, y_{n-2}, y_{n-1}, y_n$$

They share all data in common, except for y_n. Thus if y_n is precisely what is predicted on the basis of $\left[p^*(r)\right]_{n-1}$, i.e. if

$$y_n = \left[p^*(n)\right]_{n-1} \tag{9.4.6}$$

then the two polynomials $\left[p^*(r)\right]_n$ and $\left[p^*(r)\right]_{n-1}$ will in fact be the same. However if (9.4.6) does not hold, i.e. if the prediction error

$$\epsilon_n \equiv y_n - \left[p^*(n)\right]_{n-1} \tag{9.4.7}$$

is not zero, then $\left[p^*(r)\right]_n$ will differ from $\left[p^*(r)\right]_{n-1}$. Moreover all derivatives of $\left[p^*(r)\right]_n$ and $\left[p^*(r)\right]_{n-1}$ will differ when ϵ_n is nonzero, and so (9.4.3) will not hold true.

We now postulate that the vectors $Z^*_{n,n}$ and $Z^*_{n,n-1}$ *will differ by a vector which depends linearly on the prediction error* ϵ_n. Thus we assume (c/f (9.4.5)) that

$$\Phi(-1)\, Z^*_{n+1,n} = Z^*_{n,n-1} + H_n \epsilon_n \tag{9.4.8}$$

where H_n is a *vector of weights,* as yet unknown, which possibly depend on n, and ϵ_n is of course the scalar prediction error defined in (9.4.7) which we rewrite as

$$\epsilon_n \equiv y_n - \left(z^*_0\right)_{n,n-1} \tag{9.4.9}$$

The vector of weights H_n *will now be derived by making* (9.4.8) *consistent with the algorithms on* p. 354 *for each chosen degree* m.

First we write (9.4.8) as

$$[\Phi(-1) - qI]\, Z^*_{n+1,n} = H_n \epsilon_n \tag{9.4.10}$$

where q is the backward-shifting operator. Assume that $m = 1$. Then (9.4.10) can be written out as

$$\begin{pmatrix} 1-q & -1 \\ 0 & 1-q \end{pmatrix} \begin{pmatrix} z_0^* \\ z_1^* \end{pmatrix}_{n+1,n} = \begin{pmatrix} h_0 \\ h_1 \end{pmatrix}_n \epsilon_n \tag{9.4.11}$$

Solving for the vector on the left results in

$$\begin{pmatrix} z_0^* \\ z_1^* \end{pmatrix}_{n+1,n} = \frac{1}{(1-q)^2} \begin{pmatrix} 1-q & 1 \\ 0 & 1-q \end{pmatrix} \begin{pmatrix} h_0 \\ h_1 \end{pmatrix}_n \epsilon_n \tag{9.4.12}$$

and the first line of this equation then gives

$$(1-q)^2 \left(z_0^*\right)_{n+1,n} = (1-q)\left[\left(h_0\right)_n \epsilon_n\right] + \left(h_1\right)_n \epsilon_n \tag{9.4.13}$$

i.e.

$$\nabla^2 \left(z_0^*\right)_{n+1,n} = \left[\left(h_0\right)_n + \left(h_1\right)_n\right] \epsilon_n - \left(h_0\right)_{n-1} \epsilon_{n-1} \tag{9.4.14}$$

We now compare (9.4.14) to (9.3.26), and we choose $\left(h_0\right)_n$ and $\left(h_1\right)_n$ so as to make both equations identical. To do this we set

$$\left[\left(h_0\right)_n + \left(h_1\right)_n\right] \epsilon_n - \left(h_0\right)_{n-1} \epsilon_{n-1} = \frac{4}{n+1} \epsilon_n - \frac{2(2n-1)}{(n+1)n} \epsilon_{n-1} \tag{9.4.15}$$

and by equating coefficients of ϵ_n and ϵ_{n-1}, we then obtain the two equations

$$\left(h_0\right)_n + \left(h_1\right)_n = \frac{4}{n+1} \tag{9.4.16}$$

and

$$\left(h_0\right)_{n-1} = \frac{2(2n-1)}{(n+1)n} \tag{9.4.17}$$

Setting n to $n + 1$ in (9.4.17) now gives us the result

$$\left(h_0\right)_n = \frac{2(2n+1)}{(n+2)^{(2)}} \tag{9.4.18}$$

and finally, using (9.4.18) in (9.4.16) results in

$$\left(h_1\right)_n = \frac{6}{(n+2)^{(2)}} \tag{9.4.19}$$

These are the required expressions for the vector H_n when $m = 1$. We set them into (9.4.11), thereby obtaining

$$\begin{pmatrix} 1 & -1 \\ 0 & 1 \end{pmatrix} \begin{pmatrix} z_0^* \\ z_1^* \end{pmatrix}_{n+1,n} = \begin{pmatrix} z_0^* \\ z_1^* \end{pmatrix}_{n,n-1} + \begin{pmatrix} \dfrac{2(2n+1)}{(n+2)^{(2)}} \\ \dfrac{6}{(n+2)^{(2)}} \end{pmatrix} \epsilon_n \tag{9.4.20}$$

which is the required recursion for $Z^*_{n+1,n}$ in terms of $Z^*_{n,n-1}$, assuming least-squares polynomial estimation.

As the next step we rearrange (9.4.8) into a form more suited to efficient computation. Thus, let $m = 2$. Then (9.4.8) becomes (c/f (9.4.20))

$$\begin{pmatrix} 1 & -1 & 1 \\ & 1 & -2 \\ & & 1 \end{pmatrix} \begin{pmatrix} z_0^* \\ z_1^* \\ z_2^* \end{pmatrix}_{n+1,n} = \begin{pmatrix} z_0^* \\ z_1^* \\ z_2^* \end{pmatrix}_{n,n-1} + \begin{pmatrix} h_0 \\ h_1 \\ h_2 \end{pmatrix}_n \epsilon_n \tag{9.4.21}$$

where the vector H_n can be derived by a method analogous to the one which yielded the weights in (9.4.20). (See Ex. 9.7.) Now, *starting from the bottom line* in (9.4.21), we obtain, by simple transpositions:

$$\left(z_2^*\right)_{n+1,n} = \left(z_2^*\right)_{n,n-1} + \left(h_2\right)_n \epsilon_n \tag{9.4.22}$$

$$\left(z_1^*\right)_{n+1,n} = \left(z_1^*\right)_{n,n-1} + 2\left(z_2^*\right)_{n+1,n} + \left(h_1\right)_n \epsilon_n \tag{9.4.23}$$

$$\left(z_0^*\right)_{n+1,n} = \left(z_0^*\right)_{n,n-1} + \left(z_1^*\right)_{n+1,n} - \left(z_2^*\right)_{n+1,n} + \left(h_0\right)_n \epsilon_n \tag{9.4.24}$$

which is the form in which the actual computation is best carried out.

Observe that the computation starts with the three quantities

$$\left(z_0^*,\ z_1^*,\ z_2^*\right)_{n,\,n-1}$$

already in memory. The observation y_n is then received, and we immediately form the further quantity

$$\epsilon_n \equiv y_n - \left(z_0^*\right)_{n,\,n-1} \tag{9.4.25}$$

Four memory locations are required to hold these four numbers. We then compute $\left(z_2^*\right)_{n+1,n}$ using (9.4.22) and we store it into the same location that $\left(z_2^*\right)_{n,\,n-1}$ was contained in, i.e. (9.4.22) is a *replacement* operation. Next, we execute (9.4.23) and this is *again* seen to be a replacement operation, since $\left(z_1^*\right)_{n,\,n-1}$ can be discarded once this computation is completed. Finally, the same is true for (9.4.24) and *we see that we never require more than four memory locations* to execute the quadratic algorithm, three permanent ones to hold the state-vector Z^* and one temporary storage location to hold ϵ_n.

For completeness, the coupled algorithms for $m = 0, 1, 2, 3$ are displayed in Table 9.1 (see over).

Observe that by defining Z^* as we did in (9.4.1), the resultant algorithms are independent of the intersample time τ. Hence we are able to give their forms in Table 9.1 in which τ does not appear. Of course, τ must eventually be taken into account, and this takes place when we form the unscaled-derivative state-vector

$$X^*_{n+1,n} \equiv \begin{pmatrix} x^* \\ \dot{x}^* \\ \vdots \\ D^m x^* \end{pmatrix}_{n+1,n} \tag{9.4.26}$$

This is related to $Z^*_{n+1,n}$ by the use of (4.2.18), i.e.

$$X^*_{n+1,n} = D(\tau)\, Z^*_{n+1,n} \tag{9.4.27}$$

at which time τ must be specified.

Table 9.1 The Expanding-Memory Polynomial Filter

Define:

$$\begin{pmatrix} z_0^* \\ z_1^* \\ z_2^* \\ z_3^* \end{pmatrix}_{n+1,n} \equiv \begin{pmatrix} x^* \\ \tau \dot{x}^* \\ \frac{\tau^2}{2!} \ddot{x}^* \\ \frac{\tau^3}{3!} \dddot{x}^* \end{pmatrix}_{n+1,n}$$

$$\epsilon_n \equiv y_n - \left(z_0^*\right)_{n,n-1}$$

Degree 0†

$$\left(z_0^*\right)_{n+1,n} = \left(z_0^*\right)_{n,n-1} + \frac{1}{n+1}\epsilon_n$$

Degree 1†

$$\left(z_1^*\right)_{n+1,n} = \left(z_1^*\right)_{n,n-1} + \frac{6}{(n+2)(n+1)}\epsilon_n$$

$$\left(z_0^*\right)_{n+1,n} = \left(z_0^*\right)_{n,n-1} + \left(z_1^*\right)_{n+1,n} + \frac{2(2n+1)}{(n+2)(n+1)}\epsilon_n$$

Degree 2†

$$\left(z_2^*\right)_{n+1,n} = \left(z_2^*\right)_{n,n-1} + \frac{30}{(n+3)(n+2)(n+1)}\epsilon_n$$

$$\left(z_1^*\right)_{n+1,n} = \left(z_1^*\right)_{n,n-1} + 2\left(z_2^*\right)_{n+1,n} + \frac{18(2n+1)}{(n+3)(n+2)(n+1)}\epsilon_n$$

$$\left(z_0^*\right)_{n+1,n} = \left(z_0^*\right)_{n,n-1} + \left(z_1^*\right)_{n+1,n} - \left(z_2^*\right)_{n+1,n} + \frac{3(3n^2+3n+2)}{(n+3)(n+2)(n+1)}\epsilon_n$$

†In all cases n starts at zero.

Table 9.1 The Expanding-Memory Polynomial Filter *(Continued)*

Degree 3†

$$\left(z_3^*\right)_{n+1,n} = \left(z_3^*\right)_{n,n-1} + \frac{140}{(n+4)(n+3)(n+2)(n+1)}\epsilon_n$$

$$\left(z_2^*\right)_{n+1,n} = \left(z_2^*\right)_{n,n-1} + 3\left(z_3^*\right)_{n+1,n} + \frac{120(2n+1)}{(n+4)(n+3)(n+2)(n+1)}\epsilon_n$$

$$\left(z_1^*\right)_{n+1,n} = \left(z_1^*\right)_{n,n-1} + 2\left(z_2^*\right)_{n+1,n} - 3\left(z_3^*\right)_{n+1,n} + \frac{20(6n^2+6n+5)}{(n+4)(n+3)(n+2)(n+1)}\epsilon_n$$

$$\left(z_0^*\right)_{n+1,n} = \left(z_0^*\right)_{n,n-1} + \left(z_1^*\right)_{n+1,n} - \left(z_2^*\right)_{n+1,n} + \left(z_3^*\right)_{n+1,n} + \frac{8(2n^3+3n^2+7n+3)}{(n+4)(n+3)(n+2)(n+1)}\epsilon_n$$

†In all cases n starts at zero.

The weight-vectors H_n of the algorithms given in Table 9.1, if derived by the method used to obtain (9.4.18) and (9.4.19), would cause us to expend a considerable amount of effort. This amount of effort increases rapidly with each increase in the degree of the estimating polynomial. We now give the *general form* for those weight-vectors, derivation being deferred.‡

Define the square $(m+1) \times (m+1)$ matrix $P(n)$, whose i, j^{th} term is

$$\left[P(n)\right]_{ij} \equiv \frac{1}{i!}\frac{d^i}{dr^i}p(r; j, n)\bigg|_{r=n} \qquad 0 \le i, j \le m \tag{9.4.28}$$

Also define the $(m+1)$-vector K, whose j^{th} term is

$$\left[K(n)\right]_j \equiv \frac{(-1)^j}{[c(j,n)]^2} \qquad 0 \le j \le m \tag{9.4.29}$$

where of course $p(r; j, n)$ and $[c(j,n)]^2$ are as defined in (9.2.7) and (9.2.9). Then the weight-vector H_n of (9.4.8) is given by the $(m+1)$-vector

$$H_n \equiv P(n)K(n) \tag{9.4.30}$$

‡The derivation of this general form is given in the examples for Chapter 12. (See Ex. 12.3)

As an example, let $m = 1$. Then by (9.4.28),

$$P(n) = \begin{pmatrix} 1 & -1 \\ 0 & -\frac{2}{n} \end{pmatrix} \tag{9.4.31}$$

and by (9.4.29)

$$K(n) = \begin{pmatrix} \dfrac{1}{n+1} \\ -\dfrac{3n}{(n+2)(n+1)} \end{pmatrix} \tag{9.4.32}$$

Hence (9.4.30) gives

$$H_n = \begin{pmatrix} 1 & -1 \\ 0 & -\frac{2}{n} \end{pmatrix} \begin{pmatrix} \dfrac{1}{n+1} \\ -\dfrac{3n}{(n+2)(n+1)} \end{pmatrix} = \begin{pmatrix} \dfrac{2(2n+1)}{(n+2)(n+1)} \\ \dfrac{6}{(n+2)(n+1)} \end{pmatrix} \tag{9.4.33}$$

which is in precise agreement with (9.4.18) and (9.4.19). ◆◆

The attention of the reader is directed to the fact that forming the matrix P, defined by (9.4.28), will be difficult when m exceeds 3, since the polynomials $p(r; j, n)$ are not differentiated easily. This problem occurred in Chapter 7 and was solved by the use of (7.4.15).

9.5 STABILITY

Having shown that recursive algorithms can be set up which perform least-squares polynomial estimation on an expanding data base, we now turn our attention to a very fundamental question concerning recursive forms, namely their *stability*. We shall define the concept of stability in a precise manner presently, but for now we give an intuitive account of what it is we have in mind.

The recursion algorithms which we have created, in themselves constitute self-contained *systems*. They are in each case a difference equation, which, like a differential equation, possesses a characteristic behavior and natural modes. This behavior is unrelated to the data being processed, and unless it is kept under adequate control it can easily make the algorithm's outputs completely useless. What we thus require is that the natural modes, if ever excited, should of their own accord die out in time. A recursive algorithm whose natural modes eventually die out is then said to be *stable* and if they fail to die out or if they build up the algorithm is called as *unstable.*

We now present, as a simple demonstration, a recursive algorithm whose natural modes never die out, thereby making it unstable. As we show, this also makes it of questionable value as a practical filter.

Example: Consider the Fixed-Memory Polynomial Filter of degree zero, discussed in Chapter 7, namely

$$x^*_{n,n} = \frac{1}{L+1} \sum_{k=0}^{L} y_{n-L+k} \tag{9.5.1}$$

The predecessor of this estimate was

$$x^*_{n-1,n-1} = \frac{1}{L+1} \sum_{k=0}^{L} y_{n-L+k-1} \tag{9.5.2}$$

and so, by subtraction, we obtain

$$x^*_{n,n} = x^*_{n-1,n-1} + \frac{1}{L+1}(y_n - y_{n-L-1}) \tag{9.5.3}$$

This is a recursive form, and in theory, gives the same answers as (9.5.1). Quite evidently, (9.5.3) is computationally far more efficient than (9.5.1).

However, suppose that an error enters, at some time, into $x^*_{n,n}$ in (9.5.3). Then it is readily verified *that such an error will never die out.* However, in (9.5.1) it disappears the very next time the algorithm is cycled. Thus (9.5.3), despite its being computationally more compact than the non-recursive form given in (9.5.1), suffers from this very serious defect *of being entirely unable to eliminate any past errors in its output.* ◆◆

The property to which we are alluding is evidently completely contained in the homogeneous part of the algorithm, which, in the case of (9.5.3), is

$$x^*_{n,n} - x^*_{n-1,n-1} = 0 \tag{9.5.4}$$

Suppose we assume this system to be completely relaxed, and then at some instant assume it to be excited by a Kronecker delta. We thus have

$$x^*_{n,n} - x^*_{n-1,\,n-1} = \delta_{n,k} \tag{9.5.5}$$

with

$$x^*_{n,n} = 0 \qquad n < k \tag{9.5.6}$$

The Kronecker delta is of course defined by

$$\delta_{n,k} \equiv \begin{cases} 0 & n \neq k \\ 1 & n = k \end{cases} \tag{9.5.7}$$

Then the solution sequence would be

$$x^*_{n,n} = \{0, 0, 0, 0, \ldots, 0, 1, 1, 1, \ldots\} \tag{9.5.8}$$

Clearly the sequence of 1's initiated by the Kronecker delta never dies out. Since (9.5.8) is a time-sequence of a natural mode of the algorithm, it is clear that stability (or instability) is thus a property which resides in the homogeneous part.

We now present a formal definition of the concept of stability.

Definition

An algorithm is said to be stable if the solution sequence, generated by the completely relaxed homogeneous part being stimulated by a Kronecker delta at any time, tends to zero as $n \to \infty$. If it does not tend to zero the algorithm is said to be unstable.†

As a further example, it is shown in Ex. 9.9 that the counterpart of (9.5.3) for a *first-degree* polynomial, has as its homogeneous part

$$x^*_{n,n} - 2x^*_{n-1,\,n-1} + x^*_{n-2,\,n-2} = 0 \tag{9.5.9}$$

Subjecting this to a Kronecker delta at some instant gives the solution sequence

$$x^*_{n,n} = \{0, 0, \ldots, 0, 1, 2, 3, 4, \ldots\} \tag{9.5.10}$$

†The reader is referred to Section 2.8 in which we examined the *impulse response* of a linear difference equation.

and in this case, as $n \to \infty$, *the solution becomes unbounded.* The algorithm is thus unstable.

An unstable algorithm is obviously of marginal value as a practical procedure, for if once errors arise in the output, as they invariably will sooner or later, then the effects of those errors fail to die out. In most cases the effects build up (see (9.5.10)) and soon completely swamp the legitimate output. *For this reason we restrict ourselves to only stable recursion formulae, and the onus is on the analyst to verify that any recursive form which he creates is stable, in addition to being correct.* We accordingly examine the algorithms given in Table 9.1 and verify that they are, in fact, stable.

The algorithms in question are algebraic restatements of the recursive formulation first obtained in Section 9.3. The homogeneous part of the general case is seen, from (9.3.18), to be

$$\nabla^{m+1}\left[(n+1)^{(m+1)}\left(x_0^*\right)_{n+1,n}\right] = 0 \qquad \text{(See Note)} \qquad (9.5.11)$$

If this is stable then so are all of the algorithms of Table 9.1, for their stability properties are one and the same as those of (9.5.11).

We accordingly excite (9.5.11) by a Kronecker delta at time $n = k$, and obtain

$$\nabla^{m+1}\left[(n+1)^{(m+1)}\left(x_0^*\right)_{n+1,n}\right] = \delta_{n,k} \qquad (9.5.12)$$

To solve this, we set

$$\zeta_n \equiv (n+1)^{(m+1)}\left(x_0^*\right)_{n+1,n} \qquad (9.5.13)$$

and then (9.5.12) becomes

$$\nabla^{m+1}\zeta_n = \delta_{n,k} \qquad (9.5.14)$$

Note: Observe that (9.5.11) is a linear difference equation with *time-varying* coefficients. The elementary approach to the stability of *constant-coefficient* systems is based on the location of their eigenvalues. If these are within the unit circle, then the natural modes will die out in time and so the system is stable. However, we have deliberately defined stability on the basis of the natural modes themselves, and have thereby also included time-varying systems such as (9.5.11), which do not possess eigenvalues.

Now ζ_n is identically zero for $n < k$, since, by assumption, the system is initially completely relaxed. For simplicity, let $m = 0$. Then (9.5.14) is

$$\zeta_n - \zeta_{n-1} = \delta_{n,k} \tag{9.5.15}$$

whose solution is

$$\zeta_n = \begin{cases} 0 & n < k \\ 1 & n \geq k \end{cases} \tag{9.5.16}$$

Suppose next that $m = 1$. Then (9.5.14) is

$$\zeta_n - 2\zeta_{n-1} + \zeta_{n-2} = \delta_{n,k} \tag{9.5.17}$$

whose solution is

$$\zeta_n = \begin{cases} 0 & n < k \\ n - k + 1 & n \geq k \end{cases} \tag{9.5.18}$$

Finally, for the general case, it is easily verified (see Ex. 9.10) that the solution to (9.5.14) is

$$\zeta_n = \begin{cases} 0 & n < k \\ \dbinom{n-k+m}{m} & n \geq k \end{cases} \tag{9.5.19}$$

It then follows by (9.5.13) that the solution to (9.5.12) is

$$\left(x_0^*\right)_{n+1,n} = \begin{cases} 0 & n < k \\ \dfrac{\dbinom{n-k+m}{m}}{(n+1)^{(m+1)}} & n \geq k \end{cases} \tag{9.5.20}$$

But $\dbinom{n-k+m}{m}$ is a polynomial in n of degree m, whereas $(n+1)^{(m+1)}$ is a polynomial in n of degree $m + 1$. Hence

$$\lim_{n\to\infty} \frac{\binom{n-k+m}{m}}{(n+1)^{(m+1)}} = 0 \tag{9.5.21}$$

and so all algorithms developed in this chapter are stable. (See Ex. 9.11.)

9.6 INITIALIZATION

A recursive algorithm must be initialized. Thus, if the degree-1 algorithm of Table 9.1 is being employed, initial values must be specified for the position and scaled-velocity variables $\left(z_0^*\right)_{n,n-1}$ and $\left(z_1^*\right)_{n,n-1}$ when $n = 0$.

One approach to this problem might be to perform an interpolation on the first few observations using the Lagrange interpolation method of Section 4.5. The resultant state-vector could then be used to initialize our recursive algorithms. However, a much more direct approach is possible. Thus, *it is completely immaterial what initial values are chosen for* $Z^*_{n,n-1}$ *at* $n = 0$, *in the case of the Expanding-Memory Recursive Polynomial Filters, for after precisely* $m + 1$ *cycles of the* m^{th} *degree algorithm, those initial values will have been completely dropped, and the algorithm will have automatically performed a least-squares fit on the* $m + 1$ *data points presented to it. Thereafter the output continues to be precisely the least-squares polynomial fit, based on the input data.*

Prior to proving this assertion, we give an example. For $m = 0$, the algorithm is (see Table 9.1)

$$\left(x_0^*\right)_{n+1,n} = \left(x_0^*\right)_{n,n-1} + \frac{1}{n+1}\left[y_n - \left(x_0^*\right)_{n,n-1}\right] \tag{9.6.1}$$

When $n = 0$, suppose we set

$$\left(x_0^*\right)_{n,n-1} = \alpha \tag{9.6.2}$$

where α is some arbitrarily chosen number. Then on the first cycle of (9.6.1) we obtain

$$\begin{aligned}\left(x_0^*\right)_{1,0} &= \alpha + (y_0 - \alpha)\\ &= y_0\end{aligned} \tag{9.6.3}$$

Thus the initial value used for $\left(x_0^*\right)_{n,\, n-1}$ has been completely deleted and the correct zeroth-degree 1-step prediction, based on least-squares, has been made. The reader should now cycle the degree-1 algorithm of Table 9.1 using pencil and paper (see Ex. 9.12), and convince himself that a similar process occurs, i.e. that the initial values assigned to the vector $Z^*_{n,\, n-1}$ on the right-hand side will also disappear completely after precisely *two* cycles. *We now prove that our assertion is also true in general for degree m.*

From (9.3.18) we recall that

$$\nabla^{m+1}\left[(n+1)^{(m+1)}\left(x_0^*\right)_{n+1,n}\right] = L\ (\text{data}) \tag{9.6.4}$$

where the term $L(\text{data})$ means a linear combination of the data. But then, by (2.5.14) we can restate this as

$$\sum_{k=0}^{m+1} (-1)^k \binom{m+1}{k} (n+1-k)^{(m+1)} \left(x_0^*\right)_{n+1-k,\, n-k} = L\ (\text{data}) \tag{9.6.5}$$

i.e.

$$(n+1)^{(m+1)}\left(x_0^*\right)_{n+1,n} = \sum_{k=1}^{m+1} (-1)^k \binom{m+1}{k} (n+1-k)^{(m+1)} \left(x_0^*\right)_{n+1-k,\, n-k} + L\ (\text{data}) \tag{9.6.6}$$

Then when $n = m$, each of the terms in the sum on the right is nulled out by the factor $(n - m)$ contained in $(n + 1 - k)^{(m+1)}$ for each of $k = 1, 2, \ldots, m + 1$. Thus for $n = m$, $\left(x_0^*\right)_{n+1,n}$ will correctly depend only on the data and not on the initial-value terms. Thereafter, the correctness of the output is maintained by the remaining factors in $(n + 1 - k)^{(m+1)}$ which become zero at appropriate times. (See Ex. 9.13.)

What is true for x_0^* must also be true for the remaining elements of the polynomial state-vector $X^*_{n+1,n}$, since these other elements are, after all, the derivatives of the estimating polynomial on which x_0^* is based.

Thus after precisely $m + 1$ cycles the initial values will have been completely discarded, and the estimating polynomial will be based solely on the data by least-squares. We are thus free to set whatever values we please into the predecessor terms of the algorithms given in Table 9.1. From a practical standpoint these initial values may then as well be zeros, *and the schemes are thus all completely self-starting.* ◆◆

9.7 VARIANCE REDUCTION

The covariance matrix of the estimate vector $Z^*_{n+1,n}$, obtained by the algorithms of Table 9.1, is of interest in determining how well the algorithm is performing its smoothing function. However, the Expanding-Memory Polynomial Filter output, for each value of n, is identical to the output of the 1-step predictor Fixed-Memory Polynomial Filter, *for an equivalent length data-base.* Thus everything we need to know about the output covariance matrix for the filters of the present chapter is already available from Chapter 7.

As a start, let $R_{(n)}$ be the covariance matrix of the input errors. Then by (7.5.4), the covariance matrix of the output errors for the Expanding-Memory algorithms will be

$$S^*_{n+1,n} = W(1) R_{(n)} W(1)^T \tag{9.7.1}$$

where $W(1)$ is defined in (7.4.10), and where, in computing the matrices P, C and B in that formula for use in (9.7.1), an n must now be substituted for L wherever it occurs. *Thus* $S^*_{n+1,n}$ *above is time-varying since its elements are all functions of* n.

When the covariance matrix of the input errors has the form

$$R_{(n)} = \sigma_\nu^2 I \tag{9.7.2}$$

then the simplifications noted in Chapter 7 all apply here. By (7.5.22) we now have

$$S^*_{n+1,n} = \sigma_\nu^2 Q(n+1) Q(n+1)^T \tag{9.7.3}$$

where the matrix $Q(n+1)$ is obtained using (7.5.11) with $r = n + 1$. The reader can now use (9.7.3) to set up the covariance matrix $S^*_{n+1,n}$ for any degree. Recall that $S^*_{n+1,n}$ is the covariance matrix of the *scaled*-derivative state-vector $Z^*_{n+1,n}$, used in Table 9.1. If it is the covariance matrix of the *unscaled* state-vector which is needed, then the matrix $D(\tau)$ of (4.2.19) must be used, giving

$$S^*_{n+1,n} = D(\tau) S^*_{n+1,n} D(\tau)^T \tag{9.7.4}$$

As an example, by the direct use of (7.5.25) we can write, for the first-degree Expanding-Memory Polynomial Filter,

$$S^*_{n+1,n} = \sigma_\nu^2 \begin{pmatrix} \dfrac{2(2n+3)}{(n+1)n} & \dfrac{6}{(n+1)n} \\ \dfrac{6}{(n+1)n} & \dfrac{12}{(n+2)(n+1)n} \end{pmatrix} \tag{9.7.5}$$

This is the covariance matrix of the scaled-derivative state-vector $Z^*_{n+1,n}$ defined in (9.4.1) as

$$Z^*_{n+1,n} \equiv \begin{pmatrix} x^* \\ \tau\dot{x}^* \end{pmatrix}_{n+1,n} \tag{9.7.6}$$

The covariance matrix of the unscaled estimate vector, namely

$$X^*_{n+1,n} \equiv \begin{pmatrix} x^* \\ \dot{x}^* \end{pmatrix}_{n+1,n} \tag{9.7.7}$$

is then obtained, using (9.7.4). This gives us

$$S^*_{n+1,n} = \sigma_\nu^2 \begin{pmatrix} \dfrac{2(2n+3)}{(n+1)n} & \dfrac{6}{\tau(n+1)n} \\ \dfrac{6}{\tau(n+1)n} & \dfrac{12}{\tau^2(n+2)(n+1)n} \end{pmatrix} \tag{9.7.8}$$

◆◆

In Table 9.2 we give the *diagonal* elements of the covariance matrices for the unscaled estimate vector, up to degree 3. *In each case the reader will observe that the terms go to zero as n increases.* This was proved, in general, in Section 7.6 where we showed that the entire covariance matrix goes to a null matrix as L, or in this case, as n goes to infinity. In Table 9.3 we give the same terms as in Table 9.2, but on the assumption that n is very large. These are the asymptotic forms to which the diagonal terms of the covariance matrix $S^*_{n+1,n}$ tend as $n \to \infty$. The rows of Table 9.3 should be compared with the first four columns of Table 7.2 on p. 258.

Table 9.2 VRF for Expanding-Memory 1-step Predictors† (Diagonal elements of $S^*_{n+1,n}$)

Degree (m)	Output	VRF
0	$x^*_{n+1,n}$	$\dfrac{1}{(n+1)^{(1)}}$
1	$\dot{x}^*_{n+1,n}$	$\dfrac{12}{\tau^2(n+2)^{(3)}}$
	$x^*_{n+1,n}$	$\dfrac{2(2n+3)}{(n+1)^{(2)}}$
2	$\ddot{x}^*_{n+1,n}$	$\dfrac{720}{\tau^4(n+3)^{(5)}}$
	$\dot{x}^*_{n+1,n}$	$\dfrac{192n^2+744n+684}{\tau^2(n+3)^{(5)}}$
	$x^*_{n+1,n}$	$\dfrac{9n^2+27n+24}{(n+1)^{(3)}}$
3	$\dddot{x}^*_{n+1,n}$	$\dfrac{100800}{\tau^6(n+4)^{(7)}}$
	$\ddot{x}^*_{n+1,n}$	$\dfrac{25920n^2+102240n+95040}{\tau^4(n+4)^{(7)}}$
	$\dot{x}^*_{n+1,n}$	$\dfrac{1200n^4+10200n^3+31800n^2+43800n+23200}{\tau^2(n+4)^{(7)}}$
	$x^*_{n+1,n}$	$\dfrac{16n^3+72n^2+152n+120}{(n+1)^{(4)}}$

†See p. 257 for a definition of Variance Reduction Factor (VRF).

The behavior of the output covariance matrices as we vary the degree (m), the inter-sample time (τ), or even the validity instant, using a polynomial transition matrix (see (9.3.2)), was discussed in depth in Chapter 7. We leave it to the reader to obtain the required information by re-examining that material, replacing the parameter L by an n wherever appropriate.

Table 9.3 Asymptotic VRF for Expanding-Memory 1-step Predictors (n large) (Diagonal elements of $S^*_{n+1,n}$ as $n \to \infty$)

Degree (m)	$x^*_{n+1,n}$	$\dot{x}^*_{n+1,n}$	$\ddot{x}^*_{n+1,n}$	$\dddot{x}^*_{n+1,n}$
0	$\frac{1}{n}$			
1	$\frac{4}{n}$	$\frac{1}{\tau^2}\frac{12}{n^3}$		
2	$\frac{9}{n}$	$\frac{1}{\tau^2}\frac{192}{n^3}$	$\frac{1}{\tau^4}\frac{720}{n^5}$	
3	$\frac{16}{n}$	$\frac{1}{\tau^2}\frac{1200}{n^3}$	$\frac{1}{\tau^4}\frac{25920}{n^5}$	$\frac{1}{\tau^6}\frac{100800}{n^7}$

9.8 SYSTEMATIC ERRORS

As with variance reduction, discussed in Section 9.7, the problem of the systematic errors of the Expanding-Memory Polynomial Filter has essentially been covered in detail in Chapter 7. Little more need be said, and the reader can review that material, setting n in place of L and recalling that the algorithms of the present chapter are nominally 1-step predictors.

Of particular interest is the discussion given in Section 7.13, where we examined a method by which the systematic errors can be balanced against the random errors by the choice of L. In the present instance, this means that some value of n will be reached beyond which the systematic errors will make the filter outputs unacceptable. *This then shows that the Expanding-Memory Polynomial Filters should not be cycled indefinitely.*† They are intended essentially *only for short-term use,* and where a long-term recursive polynomial filter is called for, the Fading-Memory Polynomial Filters of Chapter 13 should be used. As we point out there, the Expanding-Memory Polynomial Filters of the present chapter are ideally suited for initialization of the algorithms of Chapter 13.

EXERCISES

9.1 Following the method by which (9.3.17) was obtained for the first-degree case, obtain the equivalent results for the zeroth-degree and second-degree cases.

†Unless of course the true process really is a polynomial of degree no greater than that of the filter model.

9.2 Verify that (9.3.21) in (9.3.20) gives (9.3.22).

9.3 Using the results of Ex. 9.1 together with (9.3.21) obtain (9.3.25) and (9.3.27).

9.4 Obtain (9.3.26) as follows. Start with $\nabla^2 z^*_{n+1,n} = a_n \epsilon_n + b_n \epsilon_{n-1}$. Then write ϵ_n as $y_n - z^*_{n,n-1}$ (see (9.3.21)). Now collect terms in $z^*_{n+1,n}, \ldots, z^*_{n-1,n-2}$ and solve for a_n and b_n so that (9.3.13) is satisfied.

9.5 Repeat Ex. 9.4 to obtain (9.3.27).

9.6 Repeat Ex. 9.4 to obtain (9.3.28).

9.7 Obtain the vector H_n of (9.4.21) by the method with which (9.4.18) and (9.4.19) were derived. Compare the result to the degree-2 algorithm of Table 9.1.

9.8 By making use of (9.4.30) verify the results of Ex. 9.7 above.

9.9 a) Starting from (7.3.20), verify that for first-degree

$$\left(x_0^*\right)_{n,n} = \sum_{k=0}^{L} \sum_{j=0}^{1} \varphi_j(k)\varphi_j(L)\, y_{n-L+k} \tag{I}$$

$$\left(x_0^*\right)_{n-1,n-1} = \sum_{k=-1}^{L-1} \sum_{j=0}^{1} \varphi_j(k+1)\varphi_j(L)\, y_{n-L+k}$$

$$\left(x_0^*\right)_{n-2,n-2} = \sum_{k=-2}^{L-2} \sum_{j=0}^{1} \varphi_j(k+2)\varphi_j(L)\, y_{n-L+k}$$

b) Now form $\nabla^2 \left(x_0^*\right)_{n,n}$ and verify that the sum over k on the right disappears, leaving an algorithm of the form

$$\nabla^2 \left(x_0^*\right)_{n,n} = \alpha_0 y_n + \alpha_1 y_{n-1} + \alpha_2 y_{n-L-1} + \alpha_3 y_{n-L-2}$$

c) Verify that this recursive algorithm for $\left(x_0^*\right)_{n,n}$ is computationally far more efficient than (I) above, but that it is unstable. (See p. 364.)

9.10 a) Verify that $\nabla_n \binom{n+m}{m} = \binom{n+m-1}{m-1}$. (By ∇_n we mean the backward difference with respect to the variable n.) Hence infer that $\nabla_n^m \binom{n+m}{m} = \binom{n}{0}$.

b) Prove that $\nabla\binom{n}{0} = \delta_{n,0}$ and so verify that $\nabla_n^{m+1}\binom{n+m}{m} = \delta_{n,0}$.

(Note: $\binom{n}{0}$ is 0 for $n < 0$ and 1 for $n \geq 0$.)

c) Infer that $\nabla_n^{m+1}\binom{n-k+m}{m} = \delta_{n,k}$ thus verifying (9.5.19).

9.11 a) Starting from (9.3.18) verify that the homogeneous portions of the zeroth and first-degree algorithms are

$m = 0$:

$$x^*_{n,n} = \frac{n}{n+1} x^*_{n-1,n-1} \tag{I}$$

$m = 1$:

$$x^*_{n,n} = 2\frac{n-1}{n+1} x^*_{n-1,n-1} - \frac{(n-1)(n-2)}{(n+1)n} x^*_{n-2,n-2} \tag{II}$$

b) Starting from $x^*_{0,0} = 1$, verify that (I) above gives the sequence

$$x^*_{n,n} = \left\{1, \frac{1}{2}, \frac{1}{3}, \frac{1}{4}, \right\} \ldots$$

c) Starting from $x^*_{1,1} = 1/2$, verify that (II) gives the sequence

$$x^*_{n,n} = \left\{\frac{1}{2}, \frac{1}{3}, \frac{1}{4}, \ldots\right\}$$

d) Reconcile b) and c) above with (9.5.20).

e) Infer from the above that the zeroth and first-degree algorithms are stable.

9.12 a) Let $x_0 = \alpha$, $x_1 = \beta$. Verify that the 1-step predictions based on interpolation are

$$x^*_{2,1} = 2\beta - \alpha$$

$$\tau\dot{x}^*_{2,1} = \beta - \alpha$$

b) Letting $y_0 = \alpha$, $y_1 = \beta$, cycle the degree-1 algorithm on p. 360 and verify that, *for any choice of initial values,* the output is

$$\begin{pmatrix} z^*_0 \\ z^*_1 \end{pmatrix}_{2,1} = \begin{pmatrix} \beta - \alpha \\ 2\beta - \alpha \end{pmatrix}$$

Thus the correct state-vector is obtained, regardless of the initializing vector.

9.13 a) Consider the term $(n + 1 - k)^{(m+1)}$ in (9.6.6) for $m = 3$. Since k goes from 1 to $m + 1$ in the summation, we thus have the terms

$$n^{(4)},\ (n - 1)^{(4)},\ (n - 2)^{(4)},\ (n - 3)^{(4)} \qquad \text{(I)}$$

appearing in (9.6.6). Verify that the first of these is zero at $n = 0,1,2,3$ and the second at $n = 1,2,3,4$, etc. Hence infer that when $n = 3$, the entire summation term in (9.6.6) is zero, making $\left(x_0^*\right)_{4,3}$ on the left depend only on the data and independent of the choice of initial values.

b) Verify that when $n = 4$, the summation term contains only $\left(x_0^*\right)_{4,3}$ and that the values chosen for $\left(x_0^*\right)_{3,2} \cdots \left(x_0^*\right)_{0,-1}$ are still nulled out by the zeros contained in the factors in (I) above. Infer that this nulling out continues appropriately, so that the chosen initial values *never* enter the estimate which thus depends only on the data.

9.14 Let the input sequence to a first-degree Expanding-Memory Polynomial Filter be a Kronecker delta at $n = 0$, i.e.

$$y_n = \delta_{n,0} \qquad \text{(I)}$$

a) Starting from (9.3.6) prove that

$$\left(x_0^*\right)_{n+1,n} = -\frac{2}{n + 1} \qquad \text{(II)}$$

b) Cycle (9.3.20) using (I) above as the input and verify that the output is (II) above, regardless of the choice of initial conditions.

c) Repeat b) above on (9.3.26).

d) Cycle the first-degree algorithm of Table 9.1 using (I) as the input, and verify that the output is (II) above, regardless of initial conditions.

9.15 Repeat Ex. 9.14 but use instead a second-degree filter.

9.16 Verify that for very large n, Table 9.3 approximates Table 9.2.

9.17 a) Plot the square-root of the degree-1 VRF's of Table 9.2, multiplied by 3, for $\tau = 1$, as a function of n.

b) Use a computer to generate a sequence of zero-mean uncorrelated Gaussian random numbers whose variance is constant. Feed these numbers into the degree-1 algorithm of Table 9.1 and plot the absolute value of the output. Verify that this plot is satisfactorily bounded by the plot of part a) above.

9.18 a) Assuming that (9.7.2) holds true, obtain the expression for the covariance matrix of a filter made up of two Expanding-Memory Polynomial Filters arranged so that the position output of the first is the input to the second.

b) Verify your results by the use of a computer, making a number of runs and estimating the variances of the outputs for successive values of n.

9.19 a) Generate the sequence of numbers

$$y_n = \sin n \frac{2\pi}{100}$$

using a computer, and feed it into the second- and third-degree algorithms of Table 9.1. Observe that the systematic errors are quite small initially but eventually become quite large.

b) Assume that random noise is present in the observations above, whose covariance matrix has the form $R_{(n)} = \alpha I$. Using the method outlined in Section 7.13, calculate the permissible smoothing interval so that the bias errors never exceed the 0.1σ value of the output errors. Retain α as a parameter.

REFERENCES

1. Joksch, H. C., "Reduction of the Variance," SIAM Review, April 1966, Vol. 8, No. 2, pp. 217 et. seq.

See also

2. Levine, N., "Increasing the Flexibility of Recursive Least-Squares Data Smoothing," Bell System Tech. Journal, 40 (1961), pp. 821-840.

10

GENERALIZED EXPANDING-MEMORY FILTERS — THE BAYES FORMULATION

10.1 INTRODUCTION

We now turn our attention to a filtering technique which is the generalization of the recursive scheme considered in the previous chapter.

We saw there that the recursive structure of the Expanding-Memory Polynomial Filter gives rise to algorithms which are extremely desirable from both a storage and a computation standpoint. Moreover the expanding memory inherent in those algorithms leads to a steady improvement in the smoothness of the estimates. However if the true process is not really a polynomial, then the systematic errors in those estimates increase steadily, and after a while, even though the estimates are extremely quiet, they are in all likelihood in serious error due to bias.

The scheme which we consider here has most of the advantages of the filters developed in Chapter 9 with few of the drawbacks. What we now consider is a set of recursive expanding-memory algorithms which use an arbitrary linear differential equation as the definition of the model. If the true process equations are known then they can be used in the filter, and the systematic errors, formerly associated with lengthening memory times, are entirely avoided.

Just as the filters of Chapter 9 were derived by recasting those of Chapter 7 into a recursive form, so the present ones are essentially obtained by recasting the Generalized Fixed-Memory Filters of Chapter 8 into a recursive form.

We direct the attention of the reader to a *distinctly new concept* which arises in the present filters and some of those to follow, and that is the idea of the *a priori estimate.* Briefly, the algorithms to be developed will have the predictor-corrector structure

$$X^*_{n+1,n} = \Phi(n+1,n)\,X^*_{n,n}$$

$$X^*_{n+1,n+1} = X^*_{n+1,n} + H_n(Y_n - MX^*_{n+1,n})$$

When the first observation vector Y_1 is obtained, we see that in order to cycle the second of these equations an initializing vector $X^*_{1,0}$ must be provided. In the preceding chapter a similar situation arose, but we saw there that the initializing vector could be chosen in a completely arbitrary manner since it had no effect on the resultant estimation process. Hence very little attention had to be given to the problem of how to obtain that vector.

In the present case, however, we will have the choice of *whether or not* we wish to have the initializing vector affect the subsequent estimate. Thus suppose that nothing whatever is known about the true state of the process at the time of initialization. *Then the initial vector $X^*_{1,0}$ will have to be chosen completely arbitrarily, and we would not want it to affect the estimate thereafter. On the other hand if $X^*_{1,0}$ can be chosen as a result of some definite a priori knowledge of the true state, then we might wish to have that knowledge incorporated into the subsequent estimate.* The filtering scheme which we now develop gives us the option of having either choice.

In addition to examining the case where the differential equation describing the evolution of the process is assumed to be linear, we also examine† the *nonlinear case.* We show how the results of the linear case can again be incorporated into an *iterative differential-correction scheme*, thereby

†In Chapter 11.

permitting us to handle the situations most frequently encountered in practice, namely processes defined by nonlinear differential equations or systems involving nonlinear observation relations.

10.2 THE BAYES FILTER—LINEAR CASE

Suppose that a process under observation is modelled by a *linear differential equation* of the form

$$\frac{d}{dt} X(t) = A(t) X(t) \tag{10.2.1}$$

and assume that a number of vectors of observations are made. Let these be called

$$Y_n, \; Y_\alpha, \; \ldots, \; Y_\gamma$$

assumed to have been made at times $t_n, t_\alpha, \ldots, t_\gamma$, and let them be related to the state-vector $X(t)$ by the *linear observation relations*

$$\begin{aligned} Y_n &= M_n X_n + N_n \\ Y_\alpha &= M_\alpha X_\alpha + N_\alpha \\ &\vdots \\ Y_\gamma &= M_\gamma X_\gamma + N_\gamma \end{aligned} \tag{10.2.2}$$

Just after time t_n we assemble $Y_n, Y_\alpha, \ldots, Y_\gamma$ into the single vector

$$Y_{(n)} \equiv \begin{pmatrix} Y_n \\ -- \\ Y_\alpha \\ -- \\ \vdots \\ -- \\ Y_\gamma \end{pmatrix} \tag{10.2.3}$$

Then, letting $\Phi(m, n)$ be the transition matrix derived from (10.2.1), we see that $Y_{(n)}$ is related to X_n by

$$Y_{(n)} = \begin{pmatrix} M_n \\ \hline M_\alpha \Phi(\alpha, n) \\ \hline \vdots \\ \hline M_\gamma \Phi(\gamma, n) \end{pmatrix} X_n + N_{(n)} \tag{10.2.4}$$

Defining the matrix on the right of (10.2.4) as T_n, enables us to write that equation as

$$Y_{(n)} = T_n X_n + N_{(n)} \tag{10.2.5}$$

Note that at no stage have we stated that the instants $t_n, t_\alpha, \ldots, t_\gamma$ are distinct, and in fact this is not required. They may be assumed to lie on the time-axis in a completely arbitrary fashion, but as a matter of convenience, we regard t_n to be the most recent in the set.

Assume that all of the observation errors have zero mean with covariance matrix $R_{(n)}$. Then we know, from Chapter 6, that the minimum-variance estimate of X_n, based on (10.2.5) is given by

$$\overset{\circ}{X}{}^*_{n,n} = \overset{\circ}{W}_n Y_{(n)} \tag{10.2.6}$$

where

$$\overset{\circ}{W}_n \equiv \left(T_n^T R_{(n)}^{-1} T_n\right)^{-1} T_n^T R_{(n)}^{-1} \tag{10.2.7}$$

Moreover, the covariance matrix of the random errors in $\overset{\circ}{X}{}^*_{n,n}$ is

$$\overset{\circ}{S}{}^*_{n,n} \equiv \left(T_n^T R_{(n)}^{-1} T_n\right)^{-1} \tag{10.2.8}$$

This formed the basis of the Generalized Fixed-Memory Filter algorithms which were discussed in Chapter 8.

We now pose the following problem. Suppose that the above estimate $\overset{\circ}{X}{}^*_{n,n}$ and its covariance matrix $\overset{\circ}{S}{}^*_{n,n}$ have been obtained, based on $Y_{(n)}$. At some time later a further vector of observations, which we call $Y_{(n+1)}$, becomes available. By assumption the observation errors in $Y_{(n+1)}$ are *statistically uncorrelated* with those in $Y_{(n)}$. We wish to incorporate $Y_{(n+1)}$ into our existing estimate of the process. *What is the most efficient way in which to proceed?*

As a first course of action, we might consider repeating (10.2.6), replacing (10.2.5) by the *augmented* relation

$$\begin{pmatrix} Y_{(n+1)} \\ \hline Y_{(n)} \end{pmatrix} = \begin{pmatrix} T_{n+1} \\ \hline T_n \Phi(n, n+1) \end{pmatrix} X_{n+1} + \begin{pmatrix} N_{(n+1)} \\ \hline N_{(n)} \end{pmatrix} \tag{10.2.9}$$

This is of the form

$$Y = TX_{n+1} + N \tag{10.2.10}$$

where we have defined

$$Y \equiv \begin{pmatrix} Y_{(n+1)} \\ \hline Y_{(n)} \end{pmatrix} \qquad N \equiv \begin{pmatrix} N_{(n+1)} \\ \hline N_{(n)} \end{pmatrix} \tag{10.2.11}$$

and

$$T \equiv \begin{pmatrix} T_{n+1} \\ \hline T_n \Phi(n, n+1) \end{pmatrix} \tag{10.2.12}$$

Since $N_{(n+1)}$ and $N_{(n)}$ are by assumption uncorrelated, we also see that

$$E\left\{ \begin{pmatrix} N_{(n+1)} \\ \hline N_{(n)} \end{pmatrix} \begin{pmatrix} N^T_{(n+1)} & \vert & N^T_{(n)} \end{pmatrix} \right\} = \left(\begin{array}{c|c} R_{(n+1)} & \emptyset \\ \hline \emptyset & R_{(n)} \end{array} \right) \tag{10.2.13}$$

and so, by virtue of (10.2.10), we are able to write, as the minimum-variance estimate of X_{n+1},

$$\overset{\circ}{X}{}^*_{n+1,n+1} = (T^T R^{-1} T)^{-1} T^T R^{-1} Y \tag{10.2.14}$$

where Y and T are as defined and where, by (10.2.13),

$$R \equiv \left(\begin{array}{c|c} R_{(n+1)} & \emptyset \\ \hline \emptyset & R_{(n)} \end{array} \right) \tag{10.2.15}$$

While the above scheme would certainly give the correct value for $\overset{\circ}{X}{}^*_{n+1,n+1}$, we see that it is costly in memory space since both $Y_{(n+1)}$ and $Y_{(n)}$ are required for its execution. Moreover the effort that went into obtaining $\overset{\circ}{X}{}^*_{n,n}$, prior to the arrival of $Y_{(n+1)}$, is seen to have been completely wasted. Equation (10.2.14) makes no use of it whatever.

The information in the vector $Y_{(n)}$ was reduced, so to speak, into the much more compact vector $\overset{\circ}{X}{}^*_{n,n}$ by (10.2.6). *What we now show is how* $Y_{(n)}$ *can thereafter be dropped from memory, and that* $\overset{\circ}{X}{}^*_{n,n}$ *can be used in its place.* Considerable memory and computational savings will be accomplished.

First we examine the errors in $\overset{\circ}{X}{}^*_{n,n}$, $Y_{(n)}$ and $Y_{(n+1)}$. These are respectively $\overset{\circ}{N}{}^*_{n,n}$, $N_{(n)}$ and $N_{(n+1)}$. By assumption $N_{(n)}$ and $N_{(n+1)}$ are uncorrelated. Hence

$$\begin{aligned} E\left\{\overset{\circ}{N}{}^*_{n,n} N^T_{(n+1)}\right\} &= E\left\{\overset{\circ}{W}_n N_{(n)} N^T_{(n+1)}\right\} \\ &= \overset{\circ}{W}_n E\left\{N^T_{(n)} N^T_{(n+1)}\right\} \\ &= \emptyset \end{aligned} \qquad (10.2.16)$$

Thus $\overset{\circ}{N}{}^*_{n,n}$ and $N_{(n+1)}$ are also uncorrelated.

Next, consider the unbiased minimum-variance linear prediction of X_{n+1} based on observations up to t_n. By (6.7.21) we compute it from the transition relation

$$\overset{\circ}{X}{}^*_{n+1,n} = \Phi(n+1,n)\,\overset{\circ}{X}{}^*_{n,n} \qquad (10.2.17)$$

Associated with this is the error equation

$$\overset{\circ}{N}{}^*_{n+1,n} = \Phi(n+1,n)\,\overset{\circ}{N}{}^*_{n,n} \qquad (10.2.18)$$

and so the covariance matrix of $\overset{\circ}{N}{}^*_{n+1,n}$ is given by

$$\overset{\circ}{S}{}^*_{n+1,n} = \Phi(n+1,n)\,\overset{\circ}{S}{}^*_{n,n}\,\Phi(n+1,n)^T \qquad (10.2.19)$$

Since the vector $\overset{\circ}{N}{}^*_{n,n}$ is uncorrelated with $N_{(n+1)}$, it follows from (10.2.18) that $\overset{\circ}{N}{}^*_{n+1,n}$ is also. This in turn means that the covariance matrix of the vector

$$\begin{pmatrix} \overset{\circ}{N}{}^*_{n+1,n} \\ \hline N_{(n+1)} \end{pmatrix}$$

will be

$$E\left\{\begin{pmatrix} \overset{\circ}{N}{}^*_{n+1,n} \\ \hline N_{(n+1)} \end{pmatrix} \begin{pmatrix} \overset{\circ}{N}{}^{*T}_{n+1,n} & \vline & N^T_{(n+1)} \end{pmatrix}\right\} = \begin{pmatrix} \overset{\circ}{S}{}^*_{n+1,n} & \vline & \emptyset \\ \hline \emptyset & \vline & R_{(n+1)} \end{pmatrix} \tag{10.2.20}$$

Returning now to the main problem in hand, we know that $\overset{\circ}{X}{}^*_{n+1,n}$ and $Y_{(n+1)}$ are related to the model state-vector X_{n+1} as follows:

$$\overset{\circ}{X}{}^*_{n+1,n} = X_{n+1} + \overset{\circ}{N}{}^*_{n+1,n} \tag{10.2.21}$$

$$Y_{(n+1)} = T_{n+1}X_{n+1} + N_{(n+1)} \tag{10.2.22}$$

These now provide the basis of our required procedure. Define

$$Y \equiv \begin{pmatrix} \overset{\circ}{X}{}^*_{n+1,n} \\ \hline Y_{(n+1)} \end{pmatrix} \qquad T \equiv \begin{pmatrix} I \\ \hline T_{n+1} \end{pmatrix} \tag{10.2.23}$$

and

$$N \equiv \begin{pmatrix} \overset{\circ}{N}{}^*_{n+1,n} \\ \hline N_{(n+1)} \end{pmatrix} \qquad R \equiv \begin{pmatrix} \overset{\circ}{S}{}^*_{n+1,n} & \vline & \emptyset \\ \hline \emptyset & \vline & R_{(n+1)} \end{pmatrix} \tag{10.2.24}$$

Then using Y, T and N so defined, enables us to combine (10.2.21) and (10.2.22) into the *single* equation

$$Y = TX_{n+1} + N \tag{10.2.25}$$

where we note from (10.2.20) that the covariance matrix of N is the block-diagonal matrix R of (10.2.24).

From (10.2.25) *it follows immediately* that the minimum-variance estimate of X_{n+1} will be

$$\overset{\circ}{X}{}^*_{n+1,n+1} = (T^T R^{-1} T)^{-1} T^T R^{-1} Y \tag{10.2.26}$$

and its covariance matrix will be

$$\overset{\circ}{S}{}^*_{n+1,n+1} = (T^T R^{-1} T)^{-1} \tag{10.2.27}$$

We expand (10.2.26) and (10.2.27).

Consider first, by (10.2.24)

$$R^{-1} = \left(\begin{array}{c|c} \left(\overset{\circ}{S}{}^*_{n+1,n}\right)^{-1} & \emptyset \\ \hline \emptyset & R^{-1}_{(n+1)} \end{array}\right) \tag{10.2.28}$$

Hence

$$\begin{aligned} T^T R^{-1} &= \left(I \;\middle|\; T^T_{n+1}\right) \left(\begin{array}{c|c} \left(\overset{\circ}{S}{}^*_{n+1,n}\right)^{-1} & \emptyset \\ \hline \emptyset & R^{-1}_{(n+1)} \end{array}\right) \\ &= \left(\left(\overset{\circ}{S}{}^*_{n+1,n}\right)^{-1} \;\middle|\; T^T_{n+1} R^{-1}_{(n+1)}\right) \end{aligned} \tag{10.2.29}$$

and so

$$\begin{aligned} T^T R^{-1} T &= \left(\left(\overset{\circ}{S}{}^*_{n+1,n}\right)^{-1} \;\middle|\; T^T_{n+1} R^{-1}_{(n+1)}\right) \left(\begin{array}{c} I \\ \hline T_{n+1} \end{array}\right) \\ &= \left(\overset{\circ}{S}{}^*_{n+1,n}\right)^{-1} + T^T_{n+1} R^{-1}_{(n+1)} T_{n+1} \end{aligned} \tag{10.2.30}$$

Hence by (10.2.27)

$$\overset{\circ}{S}{}^*_{n+1,n+1} = \left[\left(\overset{\circ}{S}{}^*_{n+1,n}\right)^{-1} + T^T_{n+1} R^{-1}_{(n+1)} T_{n+1}\right]^{-1} \tag{10.2.31}$$

This will be the covariance matrix of the random errors in $\overset{\circ}{X}{}^*_{n+1,n+1}$.

Next, using (10.2.29) and (10.2.23),

$$\begin{aligned} T^T R^{-1} Y &= \left(\left(\overset{\circ}{S}{}^*_{n+1,n}\right)^{-1} \;\middle|\; T^T_{n+1} R^{-1}_{(n+1)}\right) \left(\begin{array}{c} \overset{\circ}{X}{}^*_{n+1,n} \\ \hline Y_{(n+1)} \end{array}\right) \\ &= \left(\overset{\circ}{S}{}^*_{n+1,n}\right)^{-1} \overset{\circ}{X}{}^*_{n+1,n} + T^T_{n+1} R^{-1}_{(n+1)} Y_{(n+1)} \end{aligned} \tag{10.2.32}$$

and so, inserting (10.2.32) and (10.2.30) in (10.2.26) gives us

$$\overset{\circ}{X}{}^*_{n+1,n+1}$$
$$= \left[\left(\overset{\circ}{S}{}^*_{n+1,n}\right)^{-1} + T^T_{n+1} R^{-1}_{(n+1)} T_{n+1}\right]^{-1} \left[\left(\overset{\circ}{S}{}^*_{n+1,n}\right)^{-1} \overset{\circ}{X}{}^*_{n+1,n} + T^T_{n+1} R^{-1}_{(n+1)} Y_{(n+1)}\right] \tag{10.2.33}$$

This can be simplified as follows.

First, we observe, by the use of (10.2.31), that (10.2.33) is just

$$\overset{\circ}{X}{}^*_{n+1,n+1} = \overset{\circ}{S}{}^*_{n+1,n+1} \left[\left(\overset{\circ}{S}{}^*_{n+1,n}\right)^{-1} \overset{\circ}{X}{}^*_{n+1,n} + T^T_{n+1} R^{-1}_{(n+1)} Y_{(n+1)}\right] \tag{10.2.34}$$

We now add and subtract the quantity $T^T_{n+1} R^{-1}_{(n+1)} T_{n+1} \overset{\circ}{X}{}^*_{n+1,n}$ inside the final bracket on the right of the above expression. Thus

$$\begin{aligned}
\overset{\circ}{X}{}^*_{n+1,n+1} &= \overset{\circ}{S}{}^*_{n+1,n+1} \Big[\left(\overset{\circ}{S}{}^*_{n+1,n}\right)^{-1} \overset{\circ}{X}{}^*_{n+1,n} + T^T_{n+1} R^{-1}_{(n+1)} T_{n+1} \overset{\circ}{X}{}^*_{n+1,n} \\
&\quad + T^T_{n+1} R^{-1}_{(n+1)} Y_{(n+1)} - T^T_{n+1} R^{-1}_{(n+1)} T_{n+1} \overset{\circ}{X}{}^*_{n+1,n}\Big] \\
&= \overset{\circ}{S}{}^*_{n+1,n+1} \left[\left(\overset{\circ}{S}{}^*_{n+1,n}\right)^{-1} + T^T_{n+1} R^{-1}_{(n+1)} T_{n+1}\right] \overset{\circ}{X}{}^*_{n+1,n} \\
&\quad + \overset{\circ}{S}{}^*_{n+1,n+1} T^T_{n+1} R^{-1}_{(n+1)} \left(Y_{(n+1)} - T_{n+1} \overset{\circ}{X}{}^*_{n+1,n}\right)
\end{aligned} \tag{10.2.35}$$

Then by the use of (10.2.31) this becomes

$$\overset{\circ}{X}{}^*_{n+1,n+1} = \overset{\circ}{X}{}^*_{n+1,n} + \overset{\circ}{S}{}^*_{n+1,n+1} T^T_{n+1} R^{-1}_{(n+1)} \left(Y_{(n+1)} - T_{n+1} \overset{\circ}{X}{}^*_{n+1,n}\right) \tag{10.2.36}$$

which gives us the new estimate $\overset{\circ}{X}{}^*_{n+1,n+1}$ as a linear combination of the prediction $\overset{\circ}{X}{}^*_{n+1,n}$, and the vector of observations $Y_{(n+1)}$.

Using (10.2.17), we note that $\overset{\circ}{X}{}^*_{n+1,n}$ is easily computed from $\overset{\circ}{X}{}^*_{n,n}$ and so we have thus obtained a predictor-corrector type recursion algorithm, namely:

$$\overset{\circ}{X}{}^*_{n+1,n} = \Phi(n+1,n) \overset{\circ}{X}{}^*_{n,n} \tag{10.2.37}$$

$$\overset{\circ}{X}{}^*_{n+1,n+1} = \overset{\circ}{X}{}^*_{n+1,n} + \overset{\circ}{H}_{n+1} \left(Y_{(n+1)} - T_{n+1} \overset{\circ}{X}{}^*_{n+1,n}\right) \tag{10.2.38}$$

where $\overset{\circ}{H}_{n+1}$ is seen from (10.2.36) to be defined by

$$\overset{\circ}{H}_{n+1} \equiv \overset{\circ}{S}{}^{*}_{n+1,n+1} T^{T}_{n+1} R^{-1}_{(n+1)} \tag{10.2.39}$$

This is the sought-after method of combining $\overset{\circ}{X}{}^{*}_{n,n}$ with $Y_{(n+1)}$ in order to obtain $\overset{\circ}{X}{}^{*}_{n+1,n+1}$.

The above recursive algorithm provides us with a set of equations known in the literature as the *Bayes Filter*, since it can also be derived using Bayes theorem on conditional probabilities [10.1, p. 63]. We now give an explicit itemization of the steps needed to perform one pass of the Bayes recursion algorithm.

Assume that $\overset{\circ}{X}{}^{*}_{n,n}$ and $\overset{\circ}{S}{}^{*}_{n,n}$ have been obtained and that a new observation vector $Y_{(n+1)}$ with covariance matrix $R_{(n+1)}$ are then received. The observation errors in $Y_{(n+1)}$ are assumed to be uncorrelated with the errors in all of the observations on which $\overset{\circ}{X}{}^{*}_{n,n}$ was based. Then the following algorithm serves to incorporate the new observations recursively into the estimate:

Bayes Filter (Batched Observations)

$$\overset{\circ}{X}{}^{*}_{n+1,n} = \Phi(n+1,n)\,\overset{\circ}{X}{}^{*}_{n,n} \tag{10.2.40}$$

$$\overset{\circ}{S}{}^{*}_{n+1,n} = \Phi(n+1,n)\,\overset{\circ}{S}{}^{*}_{n,n}\,\Phi(n+1,n)^{T} \tag{10.2.41}$$

$$\overset{\circ}{S}{}^{*}_{n+1,n+1} = \left[\left(\overset{\circ}{S}{}^{*}_{n+1,n}\right)^{-1} + T^{T}_{n+1} R^{-1}_{(n+1)} T_{n+1}\right]^{-1} \tag{10.2.42}$$

$$\overset{\circ}{H}_{n+1} = \overset{\circ}{S}{}^{*}_{n+1,n+1} T^{T}_{n+1} R^{-1}_{(n+1)} \tag{10.2.43}$$

$$\overset{\circ}{X}{}^{*}_{n+1,n+1} = \overset{\circ}{X}{}^{*}_{n+1,n} + \overset{\circ}{H}_{n+1}\left(Y_{(n+1)} - T_{n+1}\overset{\circ}{X}{}^{*}_{n+1,n}\right) \tag{10.2.44}$$

Using this set of equations we are able to process the batch of observations $Y_{(n+1)}$, with only $\overset{\circ}{X}{}^{*}_{n,n}$ and $\overset{\circ}{S}{}^{*}_{n,n}$ being retained from the preceding computations. Once $Y_{(n+1)}$ has been processed, it too can be discarded and we then need retain only $\overset{\circ}{X}{}^{*}_{n+1,n+1}$ and $\overset{\circ}{S}{}^{*}_{n+1,n+1}$. We see that we will be in a position to cycle the algorithm once more, should a further batch of observations become available.

It is now a simple matter to extend our thinking as follows. Assume that successive batches of observations

$$Y_{(1)},\ Y_{(2)},\ \ldots,\ Y_{(n)},\ \ldots$$

are arriving and assume that their error vectors

$$N_{(1)}, N_{(2)}, \ldots, N_{(n)}, \ldots$$

are all uncorrelated with each other, i.e. that their *overall* error covariance matrix is of the *block-diagonal* form

$$R = \begin{pmatrix} R_{(n)} & & & \\ & R_{(n-1)} & & \\ & & \ddots & \\ & & & R_{(1)} \end{pmatrix} \tag{10.2.45}$$

This characterizes the observations as being *batch-wise uncorrelated.*

Suppose, first, that we process $Y_{(1)}$ using a Generalized Fixed-Memory Filter, thereby obtaining

$$\mathring{X}^*_{1,1} \equiv (T_1{}^T R_{(1)}^{-1} T_1)^{-1} T_1{}^T R_{(1)}^{-1} Y_{(1)} \tag{10.2.46}$$

and

$$\mathring{S}^*_{1,1} \equiv (T_1{}^T R_{(1)}^{-1} T_1)^{-1} \tag{10.2.47}$$

Then by (10.2.16), $\mathring{N}^*_{1,1}$ and $N_{(2)}$ are uncorrelated and so we can use the Bayes algorithm to combine $\mathring{X}^*_{1,1}$, $\mathring{S}^*_{1,1}$, $Y_{(2)}$ and $R_{(2)}$ to obtain $\mathring{X}^*_{2,2}$ and $\mathring{S}^*_{2,2}$. Clearly, since the data are batch-wise uncorrelated, we can continue to cycle the Bayes algorithm every time the next batch of observations arrives. *The resultant estimate $\mathring{X}^*_{n,n}$ will be precisely the minimum-variance estimate which could also have been obtained by using a single pass of the minimum-variance filter of* (6.6.29), i.e.

$$\mathring{X}^*_{n,n} = (T^T R^{-1} T)^{-1} T^T R^{-1} Y \tag{10.2.48}$$

where

$$Y \equiv \begin{pmatrix} Y_{(n)} \\ \hline Y_{(n-1)} \\ \hline \vdots \\ \hline Y_{(2)} \\ \hline Y_{(1)} \end{pmatrix} \qquad T \equiv \begin{pmatrix} T_n \\ \hline T_{n-1}\Phi(n-1, n) \\ \hline \vdots \\ \hline T_2\Phi(2, n) \\ \hline T_1\Phi(1, n) \end{pmatrix} \tag{10.2.49}$$

and R is the matrix of (10.2.45). (See also Ex. 10.1.) However, using the Bayes scheme, the amount of memory space required to operate the filter is substantially less and the computational load considerably lighter than that required for (10.2.48). This becomes of greater and greater significance if n is increasing and (10.2.48) must be recycled over and over, with an ever increasing amount of data and a rapidly expanding computational load. Thus, as the memory expands and the data-record lengthens, the Bayes scheme is seen to require essentially a *fixed amount of storage* and a *fixed amount of computation* for each cycle.

If the data are not batch-wise uncorrelated, then in general the only way to obtain the truly minimum-variance estimate is to use the method of equation (10.2.48), repeating it with more and more effort as n increases. If the correlation between batches is only slight, then of course we can, if we so choose, treat them as though they were uncorrelated, and process them using the Bayes scheme. The estimates would not be truly minimum-variance, but at least we would have the satisfaction of having a workable scheme. Subject to certain conditions, even *with* batch-wise correlation, it is possible to obtain the minimum-variance estimate using a modified Bayes scheme.† However, since space is limited, we will not detail that procedure. In practice it is very frequently the case that the data *are* in fact batch-wise uncorrelated and then, clearly, the Bayes scheme is of tremendous advantage. It will serve as the basis for a considerable number of extensions which we present in the work to follow.

Finally, we note that if all of the observation instants in (10.2.2) are concurrent, then $Y_{(n)}$ of (10.2.3) can be written simply as Y_n and so (10.2.4) becomes

$$Y_n = M_n X_n + N_n \tag{10.2.50}$$

†A. J. Claus, Unpublished Memorandum, Bell Telephone Laboratories, Whippany, New Jersey.

Then, *assuming as before that we have stage-wise uncorrelated errors,* the Bayes Filter on p. 386 reduces to the following:

Bayes Filter (Concurrent Observations)

$$\overset{\circ}{X}^*_{n+1,n} = \Phi(n+1,n)\,\overset{\circ}{X}^*_{n,n} \tag{10.2.51}$$

$$\overset{\circ}{S}^*_{n+1,n} = \Phi(n+1,n)\,\overset{\circ}{S}^*_{n,n}\,\Phi(n+1,n)^T \tag{10.2.52}$$

$$\overset{\circ}{S}^*_{n+1,n+1} = \left[\left(\overset{\circ}{S}^*_{n+1,n}\right)^{-1} + M^T_{n+1}R^{-1}_{n+1}M_{n+1}\right]^{-1} \tag{10.2.53}$$

$$\overset{\circ}{H}_{n+1} = \overset{\circ}{S}^*_{n+1,n+1}M^T_{n+1}R^{-1}_{n+1} \tag{10.2.54}$$

$$\overset{\circ}{X}^*_{n+1,n+1} = \overset{\circ}{X}^*_{n+1,n} + \overset{\circ}{H}_{n+1}\left(Y_{n+1} - M_{n+1}\overset{\circ}{X}^*_{n+1,n}\right) \tag{10.2.55}$$

In the remainder of this chapter, we analyze the filter on p. 386, but almost without exception every comment we make also applies to the filter given above.

The reader is referred to the final paragraphs of Chapter 5 where we discussed some applications of the *Chi-squared distribution function.* Two tests were described there, both for off-line use, to be applied during the computer program debugging phase. The first was for verifying whether or not an estimate state-vector was consistent with its error covariance matrix. *This can be applied to testing* (10.2.44) *in relation to* (10.2.42). The second test concerned the validity of a prediction and its covariance matrix. *This can be used to test* (10.2.40) *relative to* (10.2.41). It will be shown in Chapter 12 that the Kalman Filter has a structure which is very similar to the Bayes, and so these tests should also be used there.

The Chi-squared tests are perhaps *most* useful when applied to the debugging of the *iterative differential-correction schemes* based on the Bayes and Kalman filters, to be described in Chapters 11 and 12. As will be seen, the nonlinearities are handled by the use of first order approximation techniques, and so a method of verifying the validity of these approximations will be sorely needed. *The Chi-squared tests are ideally suited to that task, and we cannot stress their value sufficiently.*

10.3 THE *A PRIORI* ESTIMATE

The set of equations on p. 386 which defines the Bayes Filter, assumes the prior existence of $\overset{\circ}{X}^*_{n,n}$ and $\overset{\circ}{S}^*_{n,n}$. When $Y_{(n+1)}$ and $R_{(n+1)}$ become available, a new estimate $\overset{\circ}{X}^*_{n+1,n+1}$ and covariance matrix $\overset{\circ}{S}^*_{n+1,n+1}$ are obtained. As long as new data (whose errors are assumed to be uncorrelated with the errors in *all* the preceding data in the estimate) keep arriving, the filter equations can be recycled. *The algorithm is, in fact, self-sustaining.*

In order to use the recursive procedure when $Y_{(1)}$ and $R_{(1)}$ first arrive, it is necessary that we have a set of values for $\mathring{X}^*_{1,0}$ and $\mathring{S}^*_{1,0}$. Suppose that *nothing whatever* is known initially about the state of the process. Clearly then, any vector of numbers which we choose for $\mathring{X}^*_{1,0}$ will have to be completely arbitrary. Thus, for example, we might simply choose to set all zeros or any other set of values into $\mathring{X}^*_{1,0}$. Obviously we cannot have any confidence whatever in this choice, and we would not want it to affect the subsequent estimate. This zero-confidence can be conveyed to the filter equations as follows.

The covariance matrix of the *a priori* estimate, namely $\mathring{S}^*_{1,0}$, is inverted in order to commence computation of (10.2.42). Suppose we arbitrarily assign to $\left(\mathring{S}^*_{1,0}\right)^{-1}$ the *null matrix*. Then (10.2.42) becomes

$$\begin{aligned} \mathring{S}^*_{1,1} &= (\emptyset + T_1^T R_{(1)}^{-1} T_1)^{-1} \\ &= (T_1^T R_{(1)}^{-1} T_1)^{-1} \end{aligned} \tag{10.3.1}$$

The matrix of weights $\mathring{H}_1$ of (10.2.43) now becomes

$$\begin{aligned} \mathring{H}_1 &= \mathring{S}^*_{1,1} T_1^T R_{(1)}^{-1} \\ &= (T_1^T R_{(1)}^{-1} T_1)^{-1} T_1^T R_{(1)}^{-1} \end{aligned} \tag{10.3.2}$$

and so we obtain finally,

$$\begin{aligned} \mathring{X}^*_{1,1} &= \mathring{X}^*_{1,0} + \mathring{H}_1 \left(Y_{(1)} - T_1 \mathring{X}^*_{1,0}\right) \\ &= \mathring{X}^*_{1,0} + (T_1^T R_{(1)}^{-1} T_1)^{-1} T_1^T R_{(1)}^{-1} \left(Y_{(1)} - T_1 \mathring{X}^*_{1,0}\right) \end{aligned} \tag{10.3.3}$$

which reduces to

$$\mathring{X}^*_{1,1} = (T_1^T R_{(1)}^{-1} T_1)^{-1} T_1^T R_{(1)}^{-1} Y_{(1)} \tag{10.3.4}$$

We will have thus obtained the minimum-variance estimate $\mathring{X}^*_{1,1}$ based solely on $Y_{(1)}$, and the a priori estimate $\mathring{X}^*_{1,0}$ is seen to have been completely *ignored.*† Also, from (10.3.1), we see that the covariance matrix of the errors in $\mathring{X}^*_{1,1}$ is independent of the covariance matrix $\mathring{S}^*_{1,0}$. Clearly $\mathring{X}^*_{1,1}$ and $\mathring{S}^*_{1,1}$ are the correct minimum-variance values based on $Y_{(1)}$ and $R_{(1)}$, and so we see that we are able, if we so choose, *to start the recursion process with an arbitrarily chosen vector* $\mathring{X}^*_{1,0}$ *which does not, in any way, affect subsequent estimation.*

Heuristically the notion of making $\left(\mathring{S}^*_{1,0}\right)^{-1}$ the null matrix makes good sense. Thus, suppose that we were able to estimate a number perfectly. The estimates would have zero variance. The less perfect our ability to estimate a number, the greater would be the variance of those estimates and in the limit, *complete inability* to estimate the number would lead to an *infinite* variance in the estimates. Reversing the argument, we see that if we *assign* a variance to an estimate, then we are expressing a certain degree of confidence in the accuracy of that estimate. Assigning *zero* variance implies that we believe our estimate to be perfect, and the larger the variance we assign, the less faith we have in that estimate. In the limit, to convey the idea of *zero confidence,* we assign *infinite variance.*

In the case of the initializing estimate $\mathring{X}^*_{1,0}$, in which we supposedly have zero confidence, we assign a covariance matrix which is say diagonal, with "infinities" on each of the diagonal elements, i.e.

$$\mathring{S}^*_{1,0} = \infty I \tag{10.3.5}$$

Then the algebra of the filter takes over and correctly gives us (10.3.1) and (10.3.4). Thus the indicated technique of conveying zero confidence in the initializing vector makes sense heuristically as well as giving us the required answer.

Suppose, next, *that something meaningful is known about the process state-vector at time zero.* As an example, let the state-vector be the six numbers

$$X = (x, y, z, \dot{x}, \dot{y}, \dot{z})^T \tag{10.3.6}$$

defining position and velocity in cartesian coordinates. Then the knowledge about these quantities must also, in some way, be accompanied by information which enables us to assign variances to those estimates, *for if not, then in*

†Note that the existence of $\mathring{X}^*_{1,1}$ in (10.3.4) depends on whether or not $T_{(1)}$ has full column-rank. If it does not, then (10.3.4) cannot be executed. Under these conditions we could not start the algorithm in this way. This will be discussed again later.

reality we cannot have any confidence in the estimates, and infinite variances must be assigned.

Suppose that our *a priori* knowledge permits us to assign variance σ_p^2 to each of the position estimates x_0, y_0 and z_0, and the quantity σ_v^2 as the variance of the velocity components, $\dot{x}_0$, $\dot{y}_0$ and $\dot{z}_0$. We thus have

$$\overset{\circ}{S}{}^*_{1,0} = \left(\begin{array}{c|c} \sigma_p^2 I & \emptyset \\ \hline \emptyset & \sigma_v^2 I \end{array}\right) \tag{10.3.7}$$

where each of the indicated submatrices is 3×3. This matrix, together with the vector chosen for $\overset{\circ}{X}{}^*_{1,0}$ are now sent to the filter and the first estimate, after $Y_{(1)}$ has been processed, will be (see (10.2.33)),

$$\overset{\circ}{X}{}^*_{1,1} = \left[\left(\overset{\circ}{S}{}^*_{1,0}\right)^{-1} + T_1{}^T R_{(1)}^{-1} T_1\right]^{-1} \left[T_1{}^T R_{(1)}^{-1} Y_{(1)} + \left(\overset{\circ}{S}{}^*_{1,0}\right)^{-1} \overset{\circ}{X}{}^*_{1,0}\right] \tag{10.3.8}$$

This is seen to be a weighted combination of $Y_{(1)}$ and $\overset{\circ}{X}{}^*_{1,0}$, with the weighting being essentially "inversely proportional" to the covariance matrices $R_{(1)}$ and $\overset{\circ}{S}{}^*_{1,0}$. Depending on the matrices $T_1{}^T R_{(1)}^{-1}$ and $\left(\overset{\circ}{S}{}^*_{1,0}\right)^{-1}$, the vectors $Y_{(1)}$ and $\overset{\circ}{X}{}^*_{1,0}$ will be stressed lightly or heavily in the formation of $\overset{\circ}{X}{}^*_{1,1}$. If for example $T_1{}^T R_{(1)}^{-1}$ has very large terms in relation to $\left(\overset{\circ}{S}{}^*_{1,0}\right)^{-1}$, then $\overset{\circ}{X}{}^*_{1,1}$ will be essentially based on $Y_{(1)}$. On the other hand if $T_1{}^T R_{(1)}^{-1}$ is comparable to $\left(\overset{\circ}{S}{}^*_{1,0}\right)^{-1}$, then $\overset{\circ}{X}{}^*_{1,1}$ will depend equally on $Y_{(1)}$ and $\overset{\circ}{X}{}^*_{1,0}$ and so forth. We are thus able, *by the choice of the matrix* $\overset{\circ}{S}{}^*_{1,0}$, *to influence the subsequent estimate to a greater or lesser extent, and to cause it to depend heavily, lightly or not at all on* $\overset{\circ}{X}{}^*_{1,0}$.

It is clear then that the Bayes algorithm provides us with a mechanism for making use of any prior knowledge which we might have about the process, at the time we start observing it. However the reader should recognize that unless some effort is put into the selection of $\overset{\circ}{X}{}^*_{1,0}$ and $\overset{\circ}{S}{}^*_{1,0}$, difficulties could be encountered in operating the filter.

As a few examples in which troubles might occur if we are not sufficiently careful, we consider first the case where the user hopes to start the filter with *no prior knowledge.* Reference to (10.3.1) shows that if we set $\left(\overset{\circ}{S}{}^*_{1,0}\right)^{-1} = \emptyset$ then, *for an estimate to exist,* T *must have full column-rank.* Only then will we be able to start up in this way. (See Examples 10.5 and 10.6.)

Consider next the case *where prior knowledge exists about some of the state-variables and little or nothing about the remainder.* For example let σ_p^2 in (10.3.7) be relatively small and σ_v^2 very large. Again troubles could be encountered *because we are now essentially asking the filter to estimate velocity from only the first observation vector* $Y_{(1)}$ and this may or may not be possible. The problem will show up as a difficulty in inverting $\left(\mathring{S}^*_{1,0}\right)^{-1} + T_1 R_{(1)}^{-1} T_1$ in the execution of (10.2.42).

In conclusion we point out that unless there are strong reasons to the contrary, the reader should use only diagonal matrices for $\mathring{S}^*_{1,0}$. If off-diagonal terms are included, then they convey that the errors in the estimate $\mathring{X}^*_{1,0}$ are correlated, and it is difficult enough to estimate $\mathring{X}^*_{1,0}$, let alone estimate the correlation coefficients of its errors. This last comment should be viewed in the light of the comments made on the Chi-squared density function in Section 5.6, where we demonstrated how strongly the likelihood of an error vector can depend on the off-diagonal terms of its covariance matrix.

10.4 BASIC STRUCTURE OF THE BAYES FILTER

The Bayes Filter equations which we derived in Section 10.2, serve to combine the vectors $\mathring{X}^*_{n,n}$ and $Y_{(n+1)}$, in order to form the minimum-variance *composite* estimate $\mathring{X}^*_{n+1,n+1}$. The covariance matrices $\mathring{S}^*_{n,n}$ and $R_{(n+1)}$ were also used as inputs, and the matrix $\mathring{S}^*_{n+1,n+1}$ was obtained along with $\mathring{X}^*_{n+1,n+1}$. We now show that those equations are only a special application of a more general method, and that an understanding of the latter will provide us with a very flexible technique which can be modified in many ways to suit a large range of situations.

The present chapter, like Chapter 8, is an extension and application of the very basic concept of *minimum-variance estimation* as developed in Chapter 6. In Chapter 6 we also showed how the inverse covariance matrix serves as a very natural weighting function, by which we can stress high quality data more heavily than low grade data. We continue to utilize the same basic ideas, but now, by a slight extension in our thinking, we will be able to derive a much more widely usable set of algorithms, the Bayes filter of Section 10.2 being just one of them.

The basis of the derivation in Section 10.2 was the pair of equations (10.2.21) and (10.2.22). We rewrite them here:

$$\mathring{X}^*_{n+1,n} = X_{n+1} + \mathring{N}^*_{n+1,n} \tag{10.4.1}$$

$$Y_{(n+1)} = T_{n+1} X_{n+1} + N_{(n+1)} \tag{10.4.2}$$

These were then combined into the single relation

$$\begin{pmatrix} \overset{\circ}{X}{}^*_{n+1,n} \\ \hline Y_{(n+1)} \end{pmatrix} = \begin{pmatrix} I \\ \hline T_{n+1} \end{pmatrix} X_{n+1} + \begin{pmatrix} \overset{\circ}{N}{}^*_{n+1,n} \\ \hline N_{(n+1)} \end{pmatrix} \tag{10.4.3}$$

The two equations (10.4.1) and (10.4.2) show that the vectors $\overset{\circ}{X}{}^*_{n+1,n}$ and $Y_{(n+1)}$ are related to the state-vector X_{n+1} by the linear transformations I and T_{n+1} respectively, to within their respective error vectors $\overset{\circ}{N}{}^*_{n+1,n}$ and $N_{(n+1)}$.

Suppose we now consider *any two linear transformations,* say T_A and T_B and let

$$V_A = T_A X_n + N_A \tag{10.4.4}$$

$$V_B = T_B X_n + N_B \tag{10.4.5}$$

These can initially be thought of as two linear observation relations, where V_A and V_B are the observation vectors, and N_A and N_B are the corresponding error vectors. We shall show additionally, however, that V_A and V_B may also be estimates of the state-vector X_n, and it is this generalization which forms the basis of the present section. Combining (10.4.4) and (10.4.5) gives

$$\begin{pmatrix} V_A \\ \hline V_B \end{pmatrix} = \begin{pmatrix} T_A \\ \hline T_B \end{pmatrix} X_n + \begin{pmatrix} N_A \\ \hline N_B \end{pmatrix} \tag{10.4.6}$$

We now make the basic assumption that N_A and N_B are uncorrelated. Let the covariance matrices of N_A and N_B be S_A and S_B respectively. Then the covariance matrix of the vector

$$\begin{pmatrix} N_A \\ \hline N_B \end{pmatrix}$$

is clearly

$$R \equiv \begin{pmatrix} S_A & \emptyset \\ \emptyset & S_B \end{pmatrix} \tag{10.4.7}$$

Now, (10.4.6) is of the form

$$Y = TX_n + N \tag{10.4.8}$$

We thus see that the minimum-variance estimate of X_n, based on the vectors V_A and V_B, is

$$\mathring{X}^* = (T^T R^{-1} T)^{-1} T^T R^{-1} Y \tag{10.4.9}$$

where

$$Y = \begin{pmatrix} V_A \\ V_B \end{pmatrix} \qquad T \equiv \begin{pmatrix} T_A \\ T_B \end{pmatrix} \tag{10.4.10}$$

and R is defined in (10.4.7). The covariance matrix of the estimate will be

$$\mathring{S}^* = (T^T R^{-1} T)^{-1} \tag{10.4.11}$$

If we substitute (10.4.7) and (10.4.10) into (10.4.9) and (10.4.11), then it is easily verified (see Ex. 10.12) that the result is the pair of equations

$$\mathring{S}^* = (T_A{}^T S_A{}^{-1} T_A + T_B{}^T S_B{}^{-1} T_B)^{-1} \tag{10.4.12}$$

and

$$\mathring{X}^* = \mathring{S}^* (T_A{}^T S_A{}^{-1} V_A + T_B{}^T S_B{}^{-1} V_B) \tag{10.4.13}$$

The Bayes Filter derivation in Section 10.2 was just a special application of this result, obtained for the case where

$$\begin{aligned} V_A &\equiv \overset{\circ}{X}{}^*_{n+1,n} & V_B &\equiv Y_{(n+1)} \\ T_A &\equiv I & T_B &\equiv T_{n+1} \\ N_A &\equiv \overset{\circ}{N}{}^*_{n+1,n} & N_B &\equiv N_{(n+1)} \\ S_A &\equiv \overset{\circ}{S}{}^*_{n+1,n} & S_B &\equiv R_{(n+1)} \end{aligned} \tag{10.4.14}$$

but it is clear now that *any vectors* V_A and V_B related to the state-vector by relations of the form of (10.4.4) and (10.4.5) can be combined by the use of (10.4.12) and (10.4.13) *to yield a composite minimum-variance estimate.* We now demonstrate the above technique by applying it to a few cases of practical interest.

Example 1

We wish to combine a minimum-variance prediction of X_n with an observation vector, to form the composite minimum-variance estimate. Thus we have, as the "observation" relations

$$\overset{\circ}{X}{}^*_{n,n-1} = X_n + \overset{\circ}{N}{}^*_{n,n-1} \tag{10.4.15}$$

and

$$Y_{(n)} = T_n X_n + N_{(n)} \tag{10.4.16}$$

where $\overset{\circ}{N}{}^*_{n,n-1}$ and $N_{(n)}$ are, by assumption, uncorrelated with each other and where $\overset{\circ}{S}{}^*_{n,n-1}$ and $R_{(n)}$ are their respective covariance matrices. Then the composite minimum-variance estimate is, by virtue of (10.4.13),

$$\overset{\circ}{X}{}^*_{n,n} = \overset{\circ}{S}{}^*_{n,n}\left[\left(\overset{\circ}{S}{}^*_{n,n-1}\right)^{-1} \overset{\circ}{X}{}^*_{n,n-1} + T_n{}^T R_{(n)}^{-1} Y_{(n)}\right] \tag{10.4.17}$$

where by (10.4.12),

$$\overset{\circ}{S}{}^*_{n,n} = \left[\left(\overset{\circ}{S}{}^*_{n,n-1}\right)^{-1} + T_n{}^T R_{(n)}^{-1} T_n\right]^{-1} \tag{10.4.18}$$

This last pair is seen to be precisely the same as (10.2.34) and (10.2.31) respectively, and so they are the basis of the Bayes Filter as obtained in Section 11.2. ◆◆

Example 2

Let the two "observation" relations be

$$X_A^* = X_n + N_A^* \tag{10.4.19}$$

$$X_B^* = X_n + N_B^* \tag{10.4.20}$$

i.e. we have *two estimates* of X_n which we wish to combine, and assume that their errors are statistically uncorrelated. Then the composite minimum-variance estimate based on these two relations is

$$\mathring{X}^*_{(A,B)} \equiv \mathring{S}^*_{(A,B)} \left[(S_A^*)^{-1} X_A^* + (S_B^*)^{-1} X_B^* \right] \tag{10.4.21}$$

where

$$\mathring{S}^*_{(A,B)} = \left[(S_A^*)^{-1} + (S_B^*)^{-1} \right]^{-1} \tag{10.4.22}$$

Now let a further estimate become available, i.e.,

$$X_C^* = X_n + N_C^* \tag{10.4.23}$$

whose errors are assumed to be uncorrelated with those in X_A^* and X_B^* (and hence also with those in $\mathring{X}^*_{(A,B)}$). Then the overall composite estimate, including X_C^*, will be

$$\mathring{X}^*_{(A,B,C)} = \mathring{S}^*_{(A,B,C)} \left[(\mathring{S}^*_{(A,B)})^{-1} \mathring{X}^*_{(A,B)} + (S_C^*)^{-1} X_C^* \right] \tag{10.4.24}$$

where

$$\mathring{S}^*_{(A,B,C)} = \left[(\mathring{S}^*_{(A,B)})^{-1} + (S_C^*)^{-1} \right]^{-1} \tag{10.4.25}$$

(It is shown in Ex. 10.13 that these two results can be further reduced. See also Ex. 10.3.) ◆◆

Basically two types of vectors have been considered in the above examples, namely *estimates,* characterized by a relation of the form

$$X^* = X_n + N^* \tag{10.4.26}$$

and *observations* characterized by

$$Y_{(n)} = T_n X_n + N_{(n)} \tag{10.4.27}$$

The former have been vectors which are defined in the state-space of the process, and the latter were vectors defined in an observation-space related to the process state-space by the linear transformation T_n. We have seen that the method of combination can be generalized at will, and among those possibilities, the Bayes Filter of Section 10.2 was only one example, characterized by the fact that precisely one process state-space vector, namely $\overset{\circ}{X}{}^*_{n,n-1}$, was combined with one observation-space vector, $Y_{(n)}$. It was this particular choice which permitted us to go from (10.2.34) to the very simple recursive form of (10.2.36). However, for any particular problem in hand the reader should always bear in mind that that choice was just one of many possible, and that some other form may well be better suited to his particular need.†

In the final analysis, nothing really new has been added, other than a few additional ideas on manipulation. The basic concept is still that of minimum-variance estimation as discussed in Chapter 6, and all that we have done is to draw the attention of the reader to simplifications which occur when the total observational error vector can be partitioned into statistically uncorrelated sub-vectors. We have also pointed out that *estimates of the state-vector* can themselves be treated in the same way as observation vectors, in the combination process of obtaining a composite minimum-variance estimate.

10.5 PROPERTIES OF THE COVARIANCE MATRIX

For the Bayes Filter, the covariance matrix of the composite estimate is given by (10.2.42) as

$$\overset{\circ}{S}{}^*_{n,n} = \left[\left(\overset{\circ}{S}{}^*_{n,n-1}\right)^{-1} + T_n^T R_{(n)}^{-1} T_n\right]^{-1} \tag{10.5.1}$$

We consider briefly some of the properties of this equation.

As a start we note that, *just prior* to incorporation of the data, the covariance matrix of the estimate is $\overset{\circ}{S}{}^*_{n,n-1}$, and *just after* incorporation it is $\overset{\circ}{S}{}^*_{n,n}$. If something meaningful has been added to our knowledge of X_n by the most recent observations, then in some sense $\overset{\circ}{S}{}^*_{n,n}$ must be "smaller"

†For an application of these techniques, the reader is referred to [10.2].

than $\overset{\circ}{S}{}^*_{n,n-1}$. By (10.5.1) we see that

$$\left(\overset{\circ}{S}{}^*_{n,n}\right)^{-1} = \left(\overset{\circ}{S}{}^*_{n,n-1}\right)^{-1} + T_n{}^T R_{(n)}^{-1} T_n \tag{10.5.2}$$

and so because of the addition on the right, the inverse of $\overset{\circ}{S}{}^*_{n,n}$ is, in some way, "larger" than the inverse of $\overset{\circ}{S}{}^*_{n,n-1}$. If the above matrices were scalars, then clearly we could state outright, by virtue of (10.5.2), that

$$\overset{\circ}{S}{}^*_{n,n} < \overset{\circ}{S}{}^*_{n,n-1} \tag{10.5.3}$$

However, being matrices, we cannot write anything as trivial as this, and a more detailed study of the relationships between the covariance matrices of the estimate before and after data-incorporation is called for. We accordingly analyze (10.5.1) in some depth and we will prove the following:

Theorem 10.1

At least one diagonal element of $\overset{\circ}{S}{}^*_{n,n}$ is strictly less than its counterpart in $\overset{\circ}{S}{}^*_{n,n-1}$ with the remainder of the diagonal elements in the former being less than or equal to their counterparts in the latter.

To carry out this task a tool known as the inversion lemma† is required.

Lemma

Let S and R be positive definite matrices and let R be possibly of different order than S. Let T be a matrix such that $T^T R^{-1} T$ is of the same order as S. Then

$$(S^{-1} + T^T R^{-1} T)^{-1} = S - ST^T (R + TST^T)^{-1} TS \tag{10.5.4}$$

Proof

Form the product

$$(S^{-1} + T^T R^{-1} T)\left[S - ST^T (R + TST^T)^{-1} TS\right]$$

Then by direct expansion and a small manipulation (see Ex. 12.1), it becomes a simple matter to verify that the above is precisely equal to the identity matrix. This proves the lemma. ◆◆

†The origins of this lemma are somewhat obscure. However, the left side of (10.5.4) emerges naturally in the Bayes derivation (c/f (10.2.31)) whereas the right side emerges in the derivation of the Kalman Filter (see Ex. 12.6), and so this is probably the way in which the lemma originated.

We now return to the proof of Theorem 10.1. First we apply (10.5.4) to (10.2.42). This enables us to write the latter (after setting n to $n-1$) as

$$\mathring{S}^*_{n,n} = \mathring{S}^*_{n,n-1} - \mathring{S}^*_{n,n-1} T_n^T \left(R_{(n)} + T_n \mathring{S}^*_{n,n-1} T_n^T\right)^{-1} T_n \mathring{S}^*_{n,n-1} \tag{10.5.5}$$

which in turn means that

$$\mathring{S}^*_{n,n-1} - \mathring{S}^*_{n,n} = \mathring{S}^*_{n,n-1} T_n^T \left(R_{(n)} + T_n \mathring{S}^*_{n,n-1} T_n^T\right)^{-1} T_n \mathring{S}^*_{n,n-1} \tag{10.5.6}$$

Now both $R_{(n)}$ and $\mathring{S}^*_{n,n-1}$ are positive definite. Hence so is $\left(R_n + T_n \mathring{S}^*_{n,n-1} T_n^T\right)^{-1}$ and hence the matrix on the right of (10.5.6) is nonnegative definite. (See Ex. 10.26.) Moreover it is not a null matrix unless T_n is null, a possibility which can be ruled out.

The above argument thus proves that the difference $\mathring{S}^*_{n,n-1} - \mathring{S}^*_{n,n}$ is nonnegative definite. But this then means that its diagonal elements are nonnegative, i.e.

$$\left[\mathring{S}^*_{n,n-1} - \mathring{S}^*_{n,n}\right]_{i,i} \geq 0 \tag{10.5.7}$$

and so we must have

$$\left[\mathring{S}^*_{n,n}\right]_{i,i} \leq \left[\mathring{S}^*_{n,n-1}\right]_{i,i} \tag{10.5.8}$$

Moreover, since $\mathring{S}^*_{n,n-1} - \mathring{S}^*_{n,n}$ is not a null matrix, then, being nonnegative definite, at least one of its diagonal elements must be strictly positive (see Ex. 10.17). Thus in (10.5.8), strict inequality holds for at least one value of i and so the theorem is proved. ◆◆

It follows then that the variance of at least one, and possibly more than one, of the elements of the estimate vector is definitely reduced by incorporation of the observations, with the variances of the remainder at worst remaining unchanged. A definite improvement thus always occurs when (10.5.1) is cycled.

Having considered the relationship between $\mathring{S}^*_{n,n}$ and $\mathring{S}^*_{n,n-1}$, we now examine the somewhat larger problem of how $\mathring{S}^*_{n,n}$ is related to $\mathring{S}^*_{n-1,n-1}$.

There are two stages in the computation of $\mathring{S}^*_{n,n}$ from $\mathring{S}^*_{n-1,n-1}$. The first is the *predictor* equation, namely

$$\mathring{S}^*_{n,n-1} = \Phi(n, n-1) \mathring{S}^*_{n-1,n-1} \Phi(n, n-1)^T \tag{10.5.9}$$

which is then followed by the *corrector* equation

$$\mathring{S}^*_{n,n} = \left[\left(\mathring{S}^*_{n,n-1}\right)^{-1} + T_n^T R_{(n)}^{-1} T_n\right]^{-1} \tag{10.5.10}$$

Now it is a fact that with most practical systems, *when we make predictions out of the observation interval on which the estimate is based, one or more of the diagonal elements of* $\mathring{S}^*$ *usually increases as the errors are propagated.* This is readily understood when we recall that the estimate was obtained by a curve-fitting process in which a model trajectory was selected *that best fits the data over the observation interval.* Outside of that interval we placed no constraints on the fit, and so it is natural to expect divergence between the model and true trajectories, increasing in size the further we move our prediction point out of the observation interval. It should be pointed out that the divergence we refer to here is not in the nature of a bias, for we assume that the chosen model matches the true process precisely. It is present because of the zero-mean random errors in the observations and it too is a vector of zero-mean random variables, satisfying the propagation equation

$$N^*_{t_{n-1}+\zeta,\, t_{n-1}} = \Phi(t_{n-1} + \zeta,\, t_{n-1}) N^*_{t_{n-1},\, t_{n-1}} \tag{10.5.11}$$

Then as ζ is increased and these random variables propagate forward, *it is usually the case that we get an increase in uncertainty,* manifested by a growth in some or all of the diagonal elements of the covariance matrix of (10.5.9).

We are of course not claiming that the uncertainty *always* increases for *all* systems as we predict forward, and it is in fact quite easy to demonstrate systems in which $\mathring{S}^*_{t_{n-1}+\zeta,\, t_{n-1}} \to \emptyset$ as $\zeta \to \infty$ (see Ex. 10.19). We are merely pointing out *that in general prediction leads to an increase in uncertainty although in very special cases it may reduce it.*

Returning now to (10.5.9) and (10.5.10) we thus recognize that the former usually causes an expansion in at least some of the diagonal elements of $\mathring{S}^*$ (although it is possible that a contraction occurs), whereas by (10.5.8) we know that (10.5.10) causes a definite contraction in some of the elements and leaves the remainder unchanged. *Clearly then, only if the net contraction on the average exceeds the net expansion in every one of the diagonal elements of* $\mathring{S}^*_{n,n}$ *will this matrix approach a null matrix as the filter is cycled,* and only then will our knowledge of the true state ultimately become perfect. However, it is certainly possible that the degradation caused by prediction consistently exceeds the improvement wrought by the incorporation of the observations. In this case one or more of the diagonal elements of $\overset{\circ}{S}{}^{*}_{n,n}$ will increase without bound with n. The operation of a predictor/corrector pair such as (10.2.40) and (10.2.44) is thus usually a see-saw battle between the losses due to prediction and the gains due to observation.

The ultimate objective in expanding-memory filtering is to obtain an estimate of the true process to within errors that go to zero as n goes to infinity. This is equivalent to having the covariance matrix $\overset{\circ}{S}{}^{*}_{n,n}$ shrink to a null matrix as n is increased. In practice this is not easily accomplished, but it serves as an ideal towards which we strive. In the light of the comments given above on the gains and losses made by correction and prediction we now make the following further statements.

As a start, it is clear that we should try to avoid having the intervals between observations excessively large in order to curtail the degradation caused by prediction. Secondly, the quality of the observations must be as high as possible so that the improvement induced by their incorporation is large enough to more than overcome the losses caused by prediction. However, even assuming that we have met both of these requirements a further and perhaps dominant aspect of the problem must be properly handled.

In Section 8.9 of Chapter 8 we examined the conditions needed to ensure that the fixed-memory algorithms could be executed. This reduced to developing the necessary and sufficient conditions for the matrix T of (8.2.4) to have full column-rank and led to Theorems 8.1 and 8.2. The discussion centered around the concept of *observability,* taken from the Control Theory literature.

In the present chapter we have developed recursive methods for obtaining precisely the same results as the algorithms of Chapter 8 provide. It would be upsetting, to say the least, *if our new techniques were able to succeed in situations where those of* Chapter 8 *fail.* However this is not the case, and complete consistency prevails between the two approaches, as we now show.

The general case of time-varying or nonlinear models is beyond our scope and we shall have to satisfy ourselves with an analysis of the constant coefficient linear model and observation scheme, as we did in Section 8.9. From this it is hoped that we can gain some insight into the more general problems which, in practice, we are usually forced to handle.

The recursive method differs from the fixed-memory method in that the former includes the possibility of the incorporation of *a priori* information. We now consider a simple example in order to contrast the two procedures against each other. As our model we take a first-degree polynomial, exemplified by the state-vector

$$X(t) \equiv \begin{pmatrix} x(t) \\ \dot{x}(t) \end{pmatrix} \tag{10.5.12}$$

and differential equation

$$\dot{X}(t) = AX(t) \tag{10.5.13}$$

where

$$A = \begin{pmatrix} 0 & 1 \\ 0 & 0 \end{pmatrix} \tag{10.5.14}$$

Then the transition matrix is readily shown to be (see (4.6.28))

$$\Phi(\zeta) = \begin{pmatrix} 1 & \zeta \\ 0 & 1 \end{pmatrix} \tag{10.5.15}$$

As our observation scheme we take the scalar equation

$$y_n = MX_n + \nu_n \tag{10.5.16}$$

in which

$$M \equiv (0, \ 1) \tag{10.5.17}$$

Then either by Theorem 8.1 or 8.2 the reader can verify that the *fixed-memory* scheme is inoperable since T will not have full column-rank. In fact by (8.2.4), assuming three observation instants, we get

$$T^T = \begin{pmatrix} 0 & 0 & 0 \\ 1 & 1 & 1 \end{pmatrix} \tag{10.5.18}$$

If we now take as our covariance matrix for the three observation errors

$$R = \begin{pmatrix} 1 & 0 & 0 \\ 0 & 1 & 0 \\ 0 & 0 & 1 \end{pmatrix} \tag{10.5.19}$$

then we see that

$$T^T R^{-1} T = \begin{pmatrix} 0 & 0 \\ 0 & 1 \end{pmatrix} \tag{10.5.20}$$

which is singular. Thus the fixed-memory algorithm cannot be executed.

Consider now the *recursive* method and assume that we have, as the covariance matrix of an *a priori* estimate, the matrix

$$\overset{\circ}{S}{}^*_{1,0} = \begin{pmatrix} 1 & 0 \\ 0 & 1 \end{pmatrix} \tag{10.5.21}$$

Then (10.2.53) gives us

$$\begin{aligned} \overset{\circ}{S}{}^*_{1,1} &= \left[\begin{pmatrix} 1 & 0 \\ 0 & 1 \end{pmatrix}^{-1} + \begin{pmatrix} 0 \\ 1 \end{pmatrix} 1 \, (0, \; 1) \right]^{-1} \\ &= \begin{pmatrix} 1 & 0 \\ 0 & \frac{1}{2} \end{pmatrix} \end{aligned} \tag{10.5.22}$$

in which we have used the 0,0 element of R in (10.5.19) as the variance of the first measurement error. This now means that the entire Bayes algorithm can be executed, giving us the estimate $\overset{\circ}{X}{}^*_{1,1}$. We thus begin to wonder whether the recursive method with an *a priori* estimate can succeed despite the fact that the fixed-memory method is inoperable.

If the above situation really were the case, then we would have grave cause for concern, but fortunately the contradiction we seem to have arrived at is only illusory. *In fact even though the recursive method is capable of putting out estimates, we find that it soon begins to degenerate because its covariance matrix develops diagonal elements which become unbounded.* This means that the confidence which we can have in those estimates must soon diminish to zero.

In Chapter 14 we will study the behavior of $\overset{\circ}{S}{}^*_{n,n}$ of the recursive algorithms of this chapter. We will show that in general, in the case of the linear constant-coefficient model and observation scheme with stationary error statistics, *precisely the same conditions as were developed in Theorems* 8.1 *and* 8.2 *are also necessary and sufficient for* $\overset{\circ}{S}{}^*_{n,n}$ *to go to a null matrix as*

$n \to \infty$.† For the present however, we must merely content ourselves with demonstrating that this is true in the case of the above simple example.

The reader can readily cycle (10.2.52) and (10.2.53) by hand,‡ starting from (10.5.22). He will find that the 0,0 element of $\mathring{S}^*_{n,n}$ rapidly begins to grow and continues to do so without bound. In Figure 10.1 we show the result of a computer run in which such a procedure was carried out, and in it we see how the 0,0 element of $\mathring{S}^*_{n,n}$ tends to infinity as the algorithm is cycled repeatedly. (By contrast in Figure 10.2 we show a run of the same situation with the exception that M of (10.5.17) is taken to be $(1, 0)$. In this

†We will also show at that time, that this behavior of $\mathring{S}^*_{n,n}$ constitutes proof that the algorithm is stable.

‡Note that these two equations can be cycled independently of the rest of the Bayes algorithm.

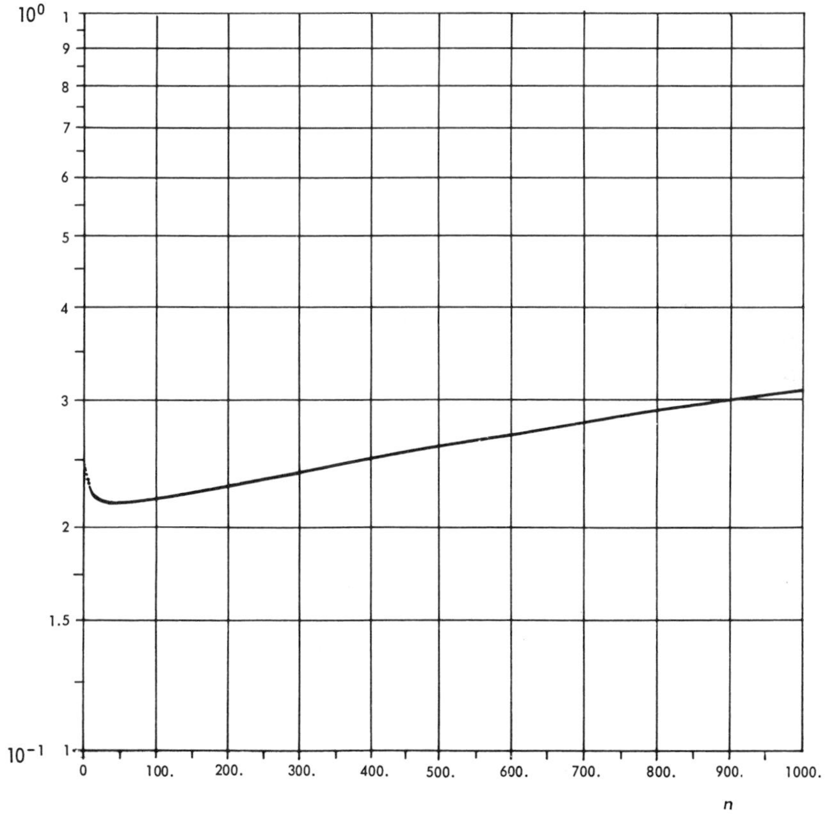

Fig. 10.1 $\left[\mathring{S}^*_{n,n}\right]_{0,0}$ *vs.* n, using $M = (0,1)$.

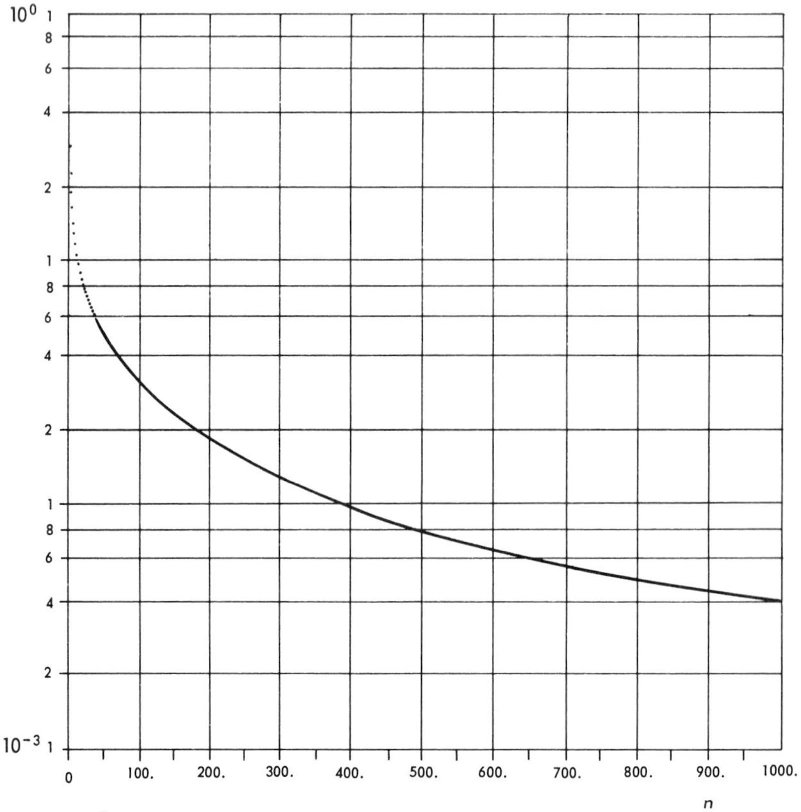

Fig. 10.2 $\left[\overset{\circ}{S}{}^*_{n,n}\right]_{0,0}$ *vs.* n, using $M = (1,0)$.

case we see that $\left[\overset{\circ}{S}{}^*_{n,n}\right]_{0,0}$ goes to zero. For both runs, the 1,1 element of $\overset{\circ}{S}{}^*_{n,n}$ went to zero.)

In summary then, if the fixed-memory method is inoperable because Theorem 8.1 or 8.2 is not satisfied, then the recursive method, even though formally operable, eventually becomes meaningless as an estimation procedure.

The treatment here has, of necessity, been very brief, and once again, as in Chapter 8, the concept of *observability* in the sense used in Control Theory, emerges as the key to a satisfactory match between the observation scheme and the model. (See [10.3], [10.4], [10.5].) If the conditions of observability *are* met, then the recursive filter operates properly, *and if not* then it fails. Thus complete consistency prevails between the fixed-memory and recursive methods.

Returning briefly to the time-varying or nonlinear cases, we recognize that these are much more difficult to analyze. Intuitively however, in order to ensure that in all cases $\overset{\circ}{S}{}^*_{n,n} \to \emptyset$ as $n \to \infty$, we see that attention must be given to the following three problem areas. a) First and foremost, the observation scheme must be so arranged that the process is being "properly observed." This is a subtle question that will have to be tackled by a mixture of intuition (gained from the constant-coefficient linear case) and simulation studies. b) Second, the observations should be made sufficiently frequently so that the losses due to prediction do not become excessive. c) Finally the quality of the observations must be good enough so that a large enough improvement does indeed result every time the algorithm is cycled.

The stability properties of the Bayes recursive algorithm will be investigated in Section 7 of Chapter 14.

10.6 COMPUTATIONAL ASPECTS

From a computational standpoint, the only equation in the Bayes Filter algorithm on p. 386 which offers any problems is (10.2.42), namely

$$\overset{\circ}{S}{}^*_{n,n} = \left[\left(\overset{\circ}{S}{}^*_{n,n-1}\right)^{-1} + T_n^T R_{(n)}^{-1} T_n\right]^{-1} \tag{10.6.1}$$

This equation calls for the inversion of three matrices, one of the order of the observation vector and two of the order of the state-vector. As a result it generally constitutes a large part of the total computational load of the algorithm. In a later section† we will show how, under certain circumstances, we can reduce the number of matrix inversions to two, thereby effecting a large saving, but for now we discuss (10.6.1) as given.

The first stage of the computation of (10.6.1) requires that we evaluate $R_{(n)}^{-1}$. In Chapter 8 we made some comments on possible matrix inversion problems, and naturally they apply here. The matrix $R_{(n)}$ may be of large or small order and may be easy or hard to invert, depending on how well it is conditioned. We do not dwell further on this problem here and turn our attention to other aspects of (10.6.1).

Unlike the filters of Chapter 8, the matrix T_n above need not necessarily have full column rank. However, whether it does or not, the product $T_n^T R_{(n)}^{-1} T_n$ is nonnegative definite. Then since $\left(\overset{\circ}{S}{}^*_{n,n-1}\right)^{-1}$ is positive definite, we see that the sum

$$\left(\overset{\circ}{S}{}^*_{n,n-1}\right)^{-1} + T_n^T R_{(n)}^{-1} T_n$$

†Section 11.3.

will be positive definite, since the sum of a positive definite and a non-negative definite matrix is again positive definite. (See Ex. 10.26.) Hence in the normal course of events, the matrix $\overset{\circ}{S}{}^*_{n,n}$ will be well defined and, subject to the usual problems of matrix inversion, fairly easily obtained. We consider now a number of *extreme situations* which could give rise to computational difficulties.

Suppose first that a set of *very precise uncorrelated observations* is presented to the filter. In this case $R_{(n)}$ will be a diagonal matrix of small numbers, and we can write it as

$$R_{(n)} = \epsilon Q_{(n)} \tag{10.6.2}$$

say, where ϵ is a small scalar and $Q_{(n)}$ is diagonal. For simplicity we assume for the present that $Q_{(n)}$ is the identity matrix. We thus have

$$T_n^T R_{(n)}^{-1} T_n = \frac{1}{\epsilon} T_n^T T_n \tag{10.6.3}$$

which is a matrix whose elements become larger as ϵ becomes smaller.

Suppose now *that T does not have full column-rank,* as it may well not. Then the above matrix which we add to $\left(\overset{\circ}{S}{}^*_{n,n-1}\right)^{-1}$, is a singular matrix, and even though the latter might be a well-conditioned positive definite matrix, *their sum will be badly behaved for ϵ very small.* To see this we have

$$\left(\overset{\circ}{S}{}^*_{n,n-1}\right)^{-1} + \frac{1}{\epsilon} T_n^T T_n = \frac{1}{\epsilon}\left[\epsilon\left(\overset{\circ}{S}{}^*_{n,n-1}\right)^{-1} + T_n^T T_n\right] \tag{10.6.4}$$

and so without loss of generality, we in fact need only examine a matrix sum of the form

$$\epsilon\left(\overset{\circ}{S}{}^*_{n,n-1}\right)^{-1} + T_n^T T_n$$

where ϵ is small. Now as ϵ diminishes this sum approaches $T_n^T T_n$, and so, since the latter was assumed to be singular, *the sum becomes increasingly difficult to invert.* This means that, for sufficiently small ϵ in relation to the precision of the arithmetic, the sum on the left of (10.6.4) will be badly distorted by the machine, the terms of $\left(\overset{\circ}{S}{}^*_{n,n-1}\right)^{-1}$ being lost either partially or totally in the addition operation, and its inverse, namely the matrix $\overset{\circ}{S}{}^*_{n,n}$, containing errors which become larger as ϵ decreases. (See Ex. 10.27.)

Clearly then extremely precise uncorrelated measurements could, under the above circumstances, give rise to trouble in operating the Bayes algorithm. This is frustrating, to say the least, since one would think that very precise uncorrelated measurements would be just the ones which are most to be desired.

The difficulty of course arises, not because the elements of $R_{(n)}$ are small, but because T_n is of inadequate rank. *The algorithm attempts to discard past data and to form an estimate based almost entirely on the most recent very precise set,* and unless T_n has full rank, we know that this is impossible. One way to avoid the problem would be to ensure that T_n is in fact of full rank by constructing our observation scheme accordingly.†

In practice however this may not always be possible, and what we are then forced to do is to multiply $R_{(n)}$ of (10.6.2) by some factor greater than unity, thereby artificially offsetting the smallness of ϵ and so downgrading those measurements. The net result is to continue to rely, to a larger extent than theoretically necessary, on the previous estimate, and so the estimation process is intentionally degraded. However since this avoids the serious pitfalls which we might encounter by using the true $R_{(n)}$, it is a small price to pay. A poor estimate is certainly preferable to no estimate at all.

We can generalize the above arguments slightly and consider precise observations which are not entirely uncorrelated. In this case $R_{(n)}$ is again made up of small elements but is no longer diagonal. We write it as

$$R_{(n)} = \epsilon Q_{(n)} \tag{10.6.5}$$

and then, clearly, its inverse, namely

$$R_{(n)}^{-1} = \frac{1}{\epsilon} Q_{(n)}^{-1} \tag{10.6.6}$$

will be considerably larger, term by term, than the inverse of a matrix such as $Q_{(n)}$, which was supposed to be typical of the covariance matrices which preceded $R_{(n)}$. Again the $1/\epsilon$ term will cause numerical problems and the arguments given above will continue to apply then T_n is of insufficient column rank.

In addition to problems arising when the observations are extremely accurate, difficulties will also arise *if the observational errors are highly correlated,* for, when that is the case, then the matrix $R_{(n)}$ will be nearly

†The reader is referred to Sections 8.8 and 8.9 where we examined some of the factors controlling the rank of T_n.

singular. This as we know means that the matrix is badly conditioned, and so numerical difficulties will be encountered when we attempt to invert it.

Thus the Bayes Filter algorithm could run into computational problems if a batch of either *very accurate* data or *highly correlated* data are presented as an input.

If the accuracy of the estimate is steadily improving i.e. if the diagonal elements of $\mathring{S}^*_{n,n}$ are diminishing monotonically with n (see Section 10.5), then the attention given to successive batches of observations will diminish steadily, assuming more or less constant accuracy of the data. Thus, referring to (10.2.43) and (10.2.44) we see that as $\mathring{S}^*_{n,n}$ shrinks, so the matrix $\mathring{H}_n$ shrinks, and the estimate begins more and more to follow one particular trajectory. This, after all, is precisely what we would want from our filter, and only if a vector of very accurate observations is obtained, would $Y_{(n)}$ again enter significantly into the choice of $\mathring{X}^*_{n,n}$.

This shrinking of the weight matrix $\mathring{H}_n$ of (10.2.43) can, in practice, cause serious troubles *if the differential equation used as the model is not a close representation of the differential equation really governing the process.* This could easily be the case a) if the true equations are not well understood, and we have been forced to assume a set of approximate differential equations in order to construct the filter, or b) if we have deliberately elected to use a simplified model in order to reduce the amount of computation in cycling the filter.

In such a situation, the estimate will develop a bias which corresponds precisely to the systematic errors discussed in Chapter 7. As time passes, and the weight matrix $\mathring{H}_n$ shrinks further and further, the term $Y_{(n)} - T_n \mathring{X}^*_{n,n-1}$ in (10.2.44), which is actually the *prediction error,* and which should serve to tell the filter that it has settled onto an erroneous trajectory, becomes ignored to a greater and greater extent. The filter, so to speak, adopts the attitude: "Don't confuse me with the facts – my mind is already made up!," and while the output is becoming smoother and smoother as $\mathring{S}^*$ shrinks, it will also become more and more biased by systematic errors.

The problem of course lies in the decision to use an expanding memory. *Only if the true process equations are very closely duplicated by the model can we continue to expand the observation interval.* In such a case the true trajectory will be approximated better and better, free of bias errors, with an ever diminishing error covariance matrix. *However, if those differential equations are not well known, or are too complex for accurate implementation, then the use of an expanding memory beyond a certain point is clearly undesirable.* There are two alternatives.

First, we can resort to a Generalized Fixed-Memory Filter as discussed in Chapter 8, in which case the bias errors can be balanced against the random errors in the estimate by the choice of memory length. The price paid in going to a fixed-memory filter is high however, since the computations expended in obtaining one estimate are not used when the next estimate is derived. Moreover the memory space required for storing the data is generally large.

The second alternative, which has all of the computational and memory-saving advantages of expanding-memory filtering, is to use what we call a *Fading-Memory Filter*. This class of filters will be discussed in the final three chapters of this book where we show that, in their case, the diagonal elements of $\mathring{S}^*_{n,n}$ asymptotically approach a *plateau above zero* which keeps $\mathring{H}_n$ non-null, and thereby permits the prediction error $Y_{(n)} - T_n \mathring{X}^*_{n,n-1}$ of (10.2.44) *always* to influence the estimate.

EXERCISES

10.1 Generalize the weighted least-squares criterion (6.9.5) using the residual vector

$$E(X^*_{n,n}) \equiv \begin{pmatrix} \mathring{X}^*_{n,n-1} - X^*_{n,n} \\ \hline Y_{(n)} - T_n X^*_{n,n} \end{pmatrix} \tag{I}$$

and covariance matrix

$$\Gamma_{(n)} \equiv \left(\begin{array}{c|c} \mathring{S}^*_{n,n-1} & \\ \hline & R_n \end{array}\right) \tag{II}$$

where $\mathring{S}^*_{n,n-1}$ and $R_{(n)}$ are, respectively, the covariance matrices of $\mathring{X}^*_{n,n-1}$ and $Y_{(n)}$ in (I). (Note that we have assumed the errors in these vectors to be uncorrelated with each other.) Minimize the resultant scalar

$$e(X^*_{n,n}) \equiv \left[E(X^*_{n,n})\right]^T \Gamma^{-1}_{(n)} E(X^*_{n,n})$$

and show that we obtain the Bayes Filter.

10.2 a) Let $Y_{(n)}$ of (10.2.3) be made up of *two* vectors, i.e.

$$Y_{(n)} = \begin{pmatrix} Y_n \\ \hline Y_{n-1} \end{pmatrix}$$

and let the errors in Y_n and Y_{n-1} be uncorrelated, i.e.

$$R_{(n)} = \left(\begin{array}{c|c} R_n & \\ \hline & R_{n-1} \end{array}\right) \tag{I}$$

Verify that (10.2.42) gives

$$\mathring{S}^*_{n,n} = \Big[\left(\mathring{S}^*_{n,n-2}\right)^{-1} + M_n^{\ T} R_n^{\ -1} M_n + \Phi(n-1,n)^T M^T_{n-1} R^{-1}_{n-1} M_{n-1} \Phi(n-1,n)\Big]^{-1} \tag{II}$$

b) We now cycle the vectors Y_n and Y_{n-1} *separately* through the Bayes Filter on p. 389. Verify that (10.2.52) and (10.2.53), after being cycled twice, give exactly the same as (II) above.

c) If $R_{(n)}$ in (I) above is *not* block diagonal can we perform (b) above?

10.3 Starting from the two "observation" relations

$$\mathring{X}^*_A = X_n + \mathring{N}^*_A$$

$$\mathring{X}^*_B = X_n + \mathring{N}^*_B$$

with errors which are uncorrelated with each other, verify that the weighted least-squares error criterion (c/f (6.9.5)) becomes

$$e(X^*_{n,n}) = E(X^*_{n,n})\,\Gamma^{-1}_{(n)}\,E(X^*_{n,n})$$

where

$$E(X^*_{n,n}) \equiv \begin{pmatrix} \overset{\circ}{X}{}^*_A - X^*_{n,n} \\ \hline \overset{\circ}{X}{}^*_B - X^*_{n,n} \end{pmatrix} \qquad \Gamma_{(n)} = \left(\begin{array}{c|c} \overset{\circ}{S}{}^*_A & \\ \hline & \overset{\circ}{S}{}^*_B \end{array}\right)$$

Minimize e over $X^*_{n,n}$ and verify that (10.4.21) results.

10.4 Show that if we multiply the covariance matrices of all inputs to a Bayes Filter by a scalar λ^2 and also multiply the *a priori* covariance matrix $\overset{\circ}{S}{}^*_{1,0}$ by λ^2, then the estimate $\overset{\circ}{X}{}^*_{n,n}$ is unchanged. Infer then that we need only know all error statistics *to within a constant* in order to obtain the minimum-variance estimate (c/f (6.10.7)).

10.5 Suppose that we initialize the Bayes filter on p. 386 arbitrarily and that we accordingly set

$$\left(\overset{\circ}{S}{}^*_{1,0}\right)^{-1} = \emptyset \tag{I}$$

to signify zero confidence in the initializing vector. Suppose, however, that $T_1^T R_{(1)}^{-1} T_1$ is singular and that (10.2.42) cannot be executed. What does this indicate? How shall we proceed if we insist on using (I) above?

10.6 a) The minimum-variance estimate

$$\overset{\circ}{X}{}^*_{n,n} = \left(T_n^T R_{(n)}^{-1} T_n\right)^{-1} T_n^T R_{(n)}^{-1} Y_{(n)}$$

requires that $T_n^T R_{(n)}^{-1} T_n$ be nonsingular. Suppose that this matrix is singular but that $R_{(n)}$ is block-diagonal. We are thus able to break $Y_{(n)}$ into subvectors (corresponding to the blocks in $R_{(n)}$) which we process successively through a Bayes Filter. How is it possible that we now obtain an estimate?

b) If the Bayes Filter in a) above is initialized with $\overset{\circ}{S}{}^*_{1,0} = \emptyset$ (i.e. completely ignoring the *a priori* estimate) can we ever obtain an estimate?

10.7 A computer is cycling the Bayes algorithm on p. 389 with the observation instants equally spaced. Y_{n-1} has been received and is incorporated, giving $\overset{\circ}{X}{}^*_{n-1,n-1}$ and $\overset{\circ}{S}{}^*_{n-1,n-1}$. At time t_n however Y_n fails to arrive. There are two courses of action.

a) Predict forward to t_{n+1} by a double operation of (10.2.51).

b) Insert arbitrarily chosen numbers for Y_n (e.g. zeros) and cycle the filter at time t_n.

Since the second method does not disrupt the operation of the filter we may prefer to use it. Can the filter be informed that we have zero confidence in Y_n? Verify that b) can in fact be made to give the identical answer to a) by appropriate action on the inputs to the algorithm.

10.8 Let $\overset{\circ}{S}{}^*_{n,n}$ of (10.2.52) be equal to $\epsilon Q_{n,n}$ where ϵ is a very small positive scalar. Prove that for ϵ sufficiently small, (10.2.53) gives

$$\overset{\circ}{S}{}^*_{n+1,n+1} = \overset{\circ}{S}{}^*_{n+1,n}$$

Interpret this in physical terms.

10.9 a) Verify that the time-varying linear differential equation

$$\frac{d}{dt}\begin{pmatrix} x(t) \\ \dot{x}(t) \end{pmatrix} = A(t)\begin{pmatrix} x(t) \\ \dot{x}(t) \end{pmatrix} \tag{I}$$

where

$$A(t) \equiv \frac{1}{1 + \sin t + \cos t}\begin{pmatrix} 0 & 1 + \sin t + \cos t \\ -1 & \cos t - \sin t \end{pmatrix} \tag{II}$$

has as its solution

$$\begin{pmatrix} x(t) \\ \dot{x}(t) \end{pmatrix} = P(t)\begin{pmatrix} x(0) \\ \dot{x}(0) \end{pmatrix} \tag{III}$$

in which

$$P(t) \equiv \frac{1}{2}\begin{pmatrix} 1 + \cos t & 1 + 2\sin t - \cos t \\ -\sin t & 2\cos t + \sin t \end{pmatrix} \tag{IV}$$

b) Infer that the transition matrix of the system is

$$\Phi(t_n + \zeta, t_n) = P(t_n + \zeta)\left[P(t_n)\right]^{-1} \tag{V}$$

10.10 a) Write a computer program which numerically integrates

$$\frac{\partial}{\partial \zeta} \Phi(t_n + \zeta, t_n) = A(t_n + \zeta)\Phi(t_n + \zeta, t)$$

$$\Phi(t_n, t_n) = I$$

where Φ and A are 2×2 and where $A(t)$ is defined by (II) of Ex. 10.9. Integrate by the use of Heun's method (see Ex. 4.20 part b) on p. 122), or by any other convenient method.

b) Compare the results of a) above to the exact value of $\Phi(t_n + \zeta, t_n)$ obtained in part b) of Ex. 10.9.

10.11 a) Generate a trajectory by integrating (I) of Ex. 10.9 using a computer.

b) Generate zero-mean random numbers which are uncorrelated with one another and of known constant variance. Generate "observations" by adding these numbers to the state-vectors obtained in a) above.

c) Program and operate the Bayes Filter given on p. 386. For the observation vector $Y_{(n)}$ use "observations" obtained in b) above on three successive instants. For the first run, observe only $x(t)$, i.e. $M = (1, 0)$. For the second observe only $\dot{x}(t)$, i.e. $M = (0, 1)$ and on a third run use $M = (1, 1)$. Initialize the filter in the first run with a state-vector obtained by least-squares fixed-memory polynomial estimation on $x(t)$. Initialize the filter in the second and third runs with arbitrarily chosen numbers and indicate zero confidence in those numbers by choosing the initial covariance matrix appropriately. Determine in which of the three runs $\overset{\circ}{S}{}^*_{n,n} \to \emptyset$ as $n \to \infty$.

10.12 Verify that (10.4.12) and (10.4.13) follow directly from (10.4.7) and (10.4.9) through (10.4.11).

10.13 Verify that (10.4.25) is equal to

$$\overset{\circ}{S}{}^*_{(A,B,C)} = \left[(S^*_A)^{-1} + (S^*_B)^{-1} + (S^*_C)^{-1}\right]^{-1}$$

and that (10.4.24) is equal to

$$\overset{\circ}{X}{}^*_{(A,B,C)} = \overset{\circ}{S}{}^*_{(A,B,C)}\left[(S^*_A)^{-1} X^*_A + (S^*_B)^{-1} X^*_B + (S^*_C)^{-1} X^*_C\right]$$

10.14 The "observation" vectors $X^*_A, X^*_B, \ldots, X^*_Z$ are all related to the true state X_n by relations of the form

$$X^*_J = X_n + N^*_J$$

The error-vectors $N_A^*, N_B^*, \ldots, N_Z^*$ are all uncorrelated with one another, and have covariance matrices $S_A^*, S_B^*, \ldots$ etc.

a) Process the "observations" by feeding them individually into the Bayes Filter on p. 389 and verify that the composite estimate is

$$\overset{\circ}{X}{}^*_{(A,\ldots,Z)} = \overset{\circ}{S}{}^*_{(A,\ldots,Z)} \left[\sum_{i=A}^{Z} (S_i^*)^{-1} X_i^* \right]$$

where

$$\overset{\circ}{S}{}^*_{(A,\ldots,Z)} = \left[\sum_{i=A}^{Z} (S_i^*)^{-1} \right]^{-1}$$

b) Process the vectors $X_A^*, X_B^*, \ldots$ *in two batches* and cycle the Bayes Filter on p. 386 twice. Verify that (I) and (II) above are obtained.

10.15 Let x_1^* and x_2^* be two statistically independent unbiased estimates of the scalar process x_n, with variances σ_1^2 and σ_2^2 respectively. We form the *convex linear combination*

$$x^*_{(1,2)} \equiv \frac{\alpha_1 x_1^* + \alpha_2 x_2^*}{\alpha_1 + \alpha_2}$$

where $\alpha_1, \alpha_2 > 0$.

a) Verify that $x^*_{(1,2)}$ so defined is an unbiased estimate of x_n.

b) Verify that the variance of $x^*_{(1,2)}$ is

$$s^*_{(1,2)} \equiv \frac{\alpha_1^2\sigma_1^2 + \alpha_2^2\sigma_2^2}{(\alpha_1 + \alpha_2)^2}.$$

c) Minimize $s^*_{(1,2)}$ over α_1 and α_2 and hence prove that the minimum-variance linear unbiased estimate occurs when we choose $\alpha_1 = 1/\sigma_1^2$, $\alpha_2 = 1/\sigma_2^2$.

d) Reconcile the results of c) with (10.4.12) and (10.4.13).

10.16 a) Let X_A^* and X_B^* of (10.4.19) and (10.4.20) be scalars. Verify that the minimum-variance composite estimate is given by

$$\mathring{x}^*_{(A,B)} = \frac{x_A^*/\sigma_A^2 + x_B^*/\sigma_B^2}{1/\sigma_A^2 + 1/\sigma_B^2} \tag{I}$$

and that

$$\sigma^2_{(A,B)} = \frac{1}{1/\sigma_A^2 + 1/\sigma_B^2} \tag{II}$$

b) Now incorporate x_C^* with variance σ_C^2 into $\mathring{x}^*_{(A,B)}$ and verify that

$$\mathring{x}^*_{(A,B,C)} = \frac{x_A^*/\sigma_A^2 + x_B^*/\sigma_B^2 + x_C^*/\sigma_C^2}{1/\sigma_A^2 + 1/\sigma_B^2 + 1/\sigma_C^2}$$

$$\sigma^2_{(A,B,C)} = \frac{1}{1/\sigma_A^2 + 1/\sigma_B^2 + 1/\sigma_C^2}$$

10.17 Prove that if a nonnegative definite matrix has only zeros on the diagonal, then it is a null matrix.

10.18 Assume that a definite improvement takes place in a particular diagonal element of $\mathring{S}^*_{n,n-1}$ when (10.5.1) is cycled. Prove that we can make that improvement as small as we please if we multiply R_n by a sufficiently large scalar. Hence infer that if the observations are of sufficiently poor quality, the behavior of the diagonal elements of $\mathring{S}^*_{n,n}$ can be made to be as close as we please to the behavior under pure prediction *without* data incorporation.

10.19 a) Starting from the transition matrix

$$\Phi(\zeta) \equiv (1/2)^{\zeta}\begin{pmatrix} 1 & \zeta \\ 0 & 1 \end{pmatrix} \tag{I}$$

verify that the diagonal elements of $\mathring{S}^*_{t_n+\zeta,t_n} \to \emptyset$ as $\zeta \to \infty$. Thus we can construct examples in which the uncertainty eventually disappears completely under prediction.

b) Verify that the state-vectors for which (I) is the transition matrix all go to null-vectors as $\zeta \to \infty$.

10.20 a) Starting from the transition matrix

$$\Phi(\zeta) = \begin{pmatrix} 1-\zeta & \zeta \\ -\zeta & 1+\zeta \end{pmatrix}$$

and the covariance matrix

$$S^*_{n,n} = I$$

verify that

$$S^*_{t_n+\zeta-k,t_n} = \begin{pmatrix} 2(\zeta-k-1/2)^2 + 1/2 & 2(\zeta-k)^2 \\ 2(\zeta-k)^2 & 2(\zeta-k+1/2)^2+1/2 \end{pmatrix} \quad \text{(I)}$$

b) Show that both diagonal elements of (I) above diminish monotonically as ζ increases in the interval $-\infty < \zeta < k - 1/2$.

c) Verify that when $\zeta > k + 1/2$, both diagonal elements of (I) above *increase* monotonically without bound as ζ is increased. Since we can make k as large as we please, *this means that we can construct examples in which the uncertainty first diminishes with prediction for as long as we please and thereafter expands without bound.*

10.21 Consider the system discussed in Example 8.2 whose transition matrix is

$$\Phi(\zeta) = \begin{pmatrix} \cos\omega\zeta & \frac{1}{\omega}\sin\omega\zeta \\ -\omega\sin\omega\zeta & \cos\omega\zeta \end{pmatrix}$$

Show that the diagonal elements of its covariance matrices $S^*_{t_n+\zeta,t_n}$ are periodic functions of ζ. Infer that for this system the uncertainty increases and decreases in a periodic fashion under prediction.

10.22 a) Consider the system

$$\frac{d}{dt}\begin{pmatrix} x_0 \\ x_1 \end{pmatrix} = \begin{pmatrix} 0 \\ 0 \end{pmatrix}$$

Verify that

$$\Phi(\zeta) = I$$

Let $M = (1,\ 0)$, i.e. we observe only x_0, and assume that all observations are uncorrelated and of variance σ^2. Starting from the *a priori* covariance matrix

$$\overset{\circ}{S}{}^*_{0,0} \equiv \begin{pmatrix} \alpha^2 & 0 \\ 0 & \beta^2 \end{pmatrix}$$

show that repeatedly cycling (10.2.52) and (10.2.53) gives us

$$\overset{\circ}{S}{}^*_{n,n} = \begin{pmatrix} \dfrac{1}{1/\alpha^2 + n/\sigma^2} & 0 \\ 0 & \beta^2 \end{pmatrix}$$

Hence infer that as $n \to \infty$, the uncertainty in x_0 disappears completely but that no improvement occurs in our *a priori* knowledge of x_1. The chosen observation scheme is thus unsatisfactory.

b) Repeat a) above but assume instead that $M = (1,\ 1)$. Verify that

$$\left(\overset{\circ}{S}{}^*_{n,n}\right)^{-1} = \begin{pmatrix} \dfrac{1}{\alpha^2} + \dfrac{n}{\sigma_\nu^2} & \dfrac{n}{\sigma_\nu^2} \\ \dfrac{n}{\sigma_\nu^2} & \dfrac{1}{\beta^2} + \dfrac{n}{\sigma_\nu^2} \end{pmatrix}$$

and hence infer that as $n \to \infty$

$$\overset{\circ}{S}{}^*_{n,n} \to \frac{\begin{pmatrix} 1 & -1 \\ -1 & 1 \end{pmatrix}}{1/\alpha^2 + 1/\beta^2}$$

showing that the uncertainty in x_0 and x_1 never disappears. Again we see that we have chosen an unsatisfactory M.

c) Repeat a) above but now assume that $M = \begin{pmatrix} 1 & 0 \\ 0 & 1 \end{pmatrix}$. Verify that $\mathring{S}^*_{n,n} \to \emptyset$ as $n \to \infty$. Finally then we have selected a satisfactory M.

d) Reconcile the results of this exercise with Theorem 8.2.

10.23 a) For the system defined by the equation

$$\frac{d}{dt}(x, y, z, \dot{x}, \dot{y}, \dot{z})^T = (\dot{x}, \dot{y}, \dot{z}, 0, 0, 0)^T$$

assume that the observation matrix is $M = (0, 0, 1, 0, 0, 0)$. Starting with a *diagonal* covariance matrix $\mathring{S}^*_{1,0} = I$, cycle the Bayes Filter by hand and verify that the knowledge we have of z and $\dot{z}$ ultimately becomes perfect, whereas the knowledge of x and y deteriorates without bound.

b) Repeat but now use

$$M = \begin{pmatrix} 1 & 0 & 0 & 0 & 0 & 0 \\ 0 & 1 & 0 & 0 & 0 & 0 \\ 0 & 0 & 1 & 0 & 0 & 0 \end{pmatrix}$$

and show that $\mathring{S}^*_{n,n} \to \emptyset$ as $n \to \infty$.

10.24 Consider the system discussed in Example 8.2, and assume that $\omega = \pi/4$, $\zeta = 1$ and $\mathring{S}^*_{1,0} = I$.

a) Assume that we observe it using $M = (1, 0)$. Write a computer program to cycle (10.2.52) and (10.2.53) and hence verify that $\mathring{S}^*_{n,n} \to \emptyset$ as $n \to \infty$.

b) Repeat using $M = (0, 1)$ and verify the same result.

c) Let $\omega = \pi$ and rerun a) and b) again. Now note that $\mathring{S}_{n,n}$ does not go to a null matrix as $n \to \infty$.

d) Rerun c) but use

$$M = \begin{pmatrix} 1 & 0 \\ 0 & 1 \end{pmatrix}$$

and verify that $\mathring{S}^*_{n,n} \to \emptyset$ as $n \to \infty$.

e) Reconcile these results with Theorem 8.2.

10.25 a) Assume that

$$\overset{\circ}{S}{}^*_{1,0} = \begin{pmatrix} 1 & 0 \\ 0 & 1 \end{pmatrix}$$

and let $M_n = (1,\ 0)$ with $R_n = 1$ (i.e. a scalar). Verify that (10.2.53) gives

$$\overset{\circ}{S}{}^*_{1,1} = \begin{pmatrix} 0.5 & 0 \\ 0 & 1 \end{pmatrix}$$

Thus only the state-variable we are observing is improved on this first pass.

b) Repeat a) above but start with

$$\overset{\circ}{S}{}^*_{1,0} = \begin{pmatrix} 1 & 0.99 \\ 0.99 & 1 \end{pmatrix} \tag{I}$$

and verify that (10.2.53) now gives

$$\overset{\circ}{S}{}^*_{1,1} = \begin{pmatrix} 0.500 & 0.495 \\ 0.495 & 0.510 \end{pmatrix}$$

We see then that even though we are observing only one of the state-variables, when strong correlation exists between its errors and those of another, *then improvement takes place in both.* Explain this in physical terms.

c) Let the transition matrix for the above system be $\Phi(h) = I$. Cycle (10.2.52) and (10.2.53) on a computer starting from (I) in part b) above as the initial conditions, and using M and R as given in part a). Verify that only the 0,0 element of $\overset{\circ}{S}{}^*_{n,n}$ goes to zero. Infer then that in spite of the *initial* improvement in the other element of $\overset{\circ}{S}{}^*_{n,n}$ owing to the high correlation, this choice of observation scheme is unsatisfactory for the system whose transition matrix is as given.

d) Reconcile this with Theorem 8.2.

10.26 Prove that the sum of a positive definite and a nonnegative definite matrix is positive definite.

10.27 Assume that

$$\overset{\circ}{S}{}^*_{1,0} = \begin{pmatrix} 1 & -0.5 \\ -0.5 & 0.75 \end{pmatrix} \qquad R = \sigma^2 I \quad (I \text{ is } 4 \times 4)$$

and

$$T^T = \begin{pmatrix} 1 & 1 & 1 & 1 \\ 1 & 1 & 1 & 1 \end{pmatrix}$$

a) Verify that (10.6.1) gives

$$\overset{\circ}{S}{}^*_{1,1} = \left[\begin{pmatrix} 1.5 & 1 \\ 1 & 2 \end{pmatrix} + \frac{1}{\sigma^2}\begin{pmatrix} 4 & 4 \\ 4 & 4 \end{pmatrix}\right]^{-1} \tag{I}$$

b) Set $\sigma^2 = 0.01$ and verify that (I) gives us

$$\overset{\circ}{S}{}^*_{1,1} \approx \begin{pmatrix} 0.66778 & -0.66611 \\ -0.66611 & 0.66694 \end{pmatrix}$$

c) Now assume that we have only 3-digit capability in our arithmetic. Show that for $\sigma^2 = 0.01$, and $\overset{\circ}{S}{}^*_{1,0}$ as above, we obtain

$$\overset{\circ}{S}{}^*_{1,1} \approx \begin{pmatrix} 1.003 & -1.00 \\ -1.00 & 1.00 \end{pmatrix}$$

This shows the drastic errors which can occur in the above circumstances† when the arithmetic is of insufficient precision.

d) Repeat a), b) and c) above but assume instead that

$$T = \begin{pmatrix} 1 & 1 \\ 1 & 1 \\ 1 & 1 \\ 1 & 2 \end{pmatrix}$$

†i.e. T of less than full rank and a set of extremely precise observations. See pp. 408 and 409.

Verify now that

$$\overset{\circ}{S}{}^{*}_{1,1} = \begin{pmatrix} 0.02275 & -0.01624 \\ -0.01624 & 0.01301 \end{pmatrix}$$

and that with 3-digit precision we get

$$\overset{\circ}{S}{}^{*}_{1,1} = \begin{pmatrix} 0.0230 & -0.0164 \\ -0.0164 & 0.0131 \end{pmatrix}$$

Thus when T is of full rank, even if σ^2 is very small, the precision of the arithmetic has only a small effect.

REFERENCES

1. Lee, R. C. K., "Optimal Estimation, Identification and Control," Research Mono. 28, Mass. Inst. of Tech. Press, Cambridge, Mass., Chapter 3.
2. Claus, A. J., Blackman, R. B., Halline, E. G., and Ridgeway, W. C. III, "Orbit Determination and Prediction, and Computer Programs, (Telstar)," Bell Systems Tech. Jour., Vol. 42, July 1963, pp. 1357-1382.
3. Kalman, R. E., "On the General Theory of Control Systems," Proceedings of the First International Conference of Automatic Control (IFAC), Moscow, 1960.
4. Deutsch, R., "Estimation Theory," Prentice-Hall, 1965.
5. Zadeh, L. A., and Desoer, C. A., "Linear System Theory," McGraw-Hill Book Company, 1963, p. 502 et seq.

See also reference list at the end of Chapter 11.

11

BAYES ALGORITHM WITH ITERATIVE DIFFERENTIAL-CORRECTION

11.1 INTRODUCTION

In the preceding chapter we discussed the Bayes Filter as applied to the estimation of a process assumed to be governed by a linear differential equation and assuming a linear observation relation. We showed that recursive algorithms could be derived which accept as inputs $Y_{(n)}$ and $R_{(n)}$, combine these with the previous outputs $\overset{\circ}{X}{}^*_{n-1,n-1}$ and $\overset{\circ}{S}{}^*_{n-1,n-1}$, and by appropriate computation then obtain the new outputs $\overset{\circ}{X}{}^*_{n,n}$ and $\overset{\circ}{S}{}^*_{n,n}$. Such algorithms were displayed on pages 386 and 389, and are depicted in block diagram form in Figure 11.1 on the following page. We examine that figure briefly.

The outputs are computed by the algorithm as shown, and are then placed in storage and retrieved at a later time to be fed back into the input side. We thus discern a *loop* in the structure which we call Loop A. Such loops are characteristic of *all* recursive algorithms, i.e. ones in which the current output is computed from a combination of its predecessors and the newest observations. Loops are entirely absent in fixed-memory schemes since for them the current output is computed *solely* from the observations, and previous outputs are never again used in subsequent computation cycles.

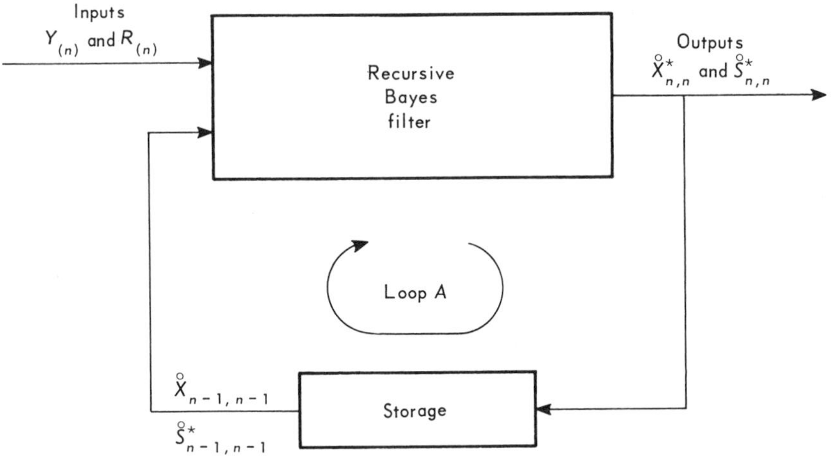

Fig. 11.1 *Block diagram of Bayes Filter.*

We now turn our attention to the expanding-memory estimation of processes which are governed by nonlinear differential equations, or are being observed by the use of nonlinear observation relations, or both. The Bayes Filter, as defined on either p. 386 or on p. 389, cannot thus be used directly, and some modification of those algorithms must accordingly be devised.

Our course of action will be *to use the previously derived estimate as a nominal trajectory about which we linearize the differential equations and the observation relation.* This replaces the nonlinear process equations in $X(t)$ by a set of *linear* equations in $\delta X(t)$, which we call the differential vector, defined as the difference between the true state-vector $X(t)$ and the nominal state-vector $\overline{X}(t)$. The Bayes Filter is then applied to the estimation of $\delta X(t)$, using as inputs $Y_{(n)}$, $R_{(n)}$ and the previous estimate† which we call $X^*_{k,k}$ and its covariance matrix $S^*_{k,k}$. Once that estimate of $\delta X(t)$ is derived, it is added to $\overline{X}(t)$ and a better approximation of $X(t)$ is thus obtained. This new improved estimate will be called $\left(X^*_{n,n}\right)_1$.

†In our discussion here we have in mind the algorithm for *batched* observations (see p. 386). Where necessary we shall comment on the special case of *concurrent* observations (p. 389).

Since errors were committed during the linearization process by the deletion of second and higher order terms, we expect $\left(X^*_{n,n}\right)_1$ to be in error even though it is a better estimate than $\overline{X}_n$ was. But precisely because it is closer to X_n than $\overline{X}_n$ was, if we were now to linearize about $\left(X^*_{n,n}\right)_1$ then the linearization errors would be smaller than formerly. We thus commence an *iteration cycle* as follows.

$\left(X^*_{n,n}\right)_1$ is used as a new nominal trajectory in place of $\overline{X}_n$ and the equations are again linearized. Using the *same* inputs as before, namely $Y_{(n)}$, $R_{(n)}$, $X^*_{k,k}$ and $S^*_{k,k}$, a new estimate of the differential vector is again obtained by the Bayes Filter. This is now added to $\left(X^*_{n,n}\right)_1$ to give us $\left(X^*_{n,n}\right)_2$, a further improvement on $\left(X^*_{n,n}\right)_1$. $\left(X^*_{n,n}\right)_2$ is now used as the nominal trajectory, and with the same inputs $Y_{(n)}$, $R_{(n)}$, $X^*_{k,k}$ and $S^*_{k,k}$ we obtain $\left(X^*_{n,n}\right)_3$, and so on.

At each cycling of the iteration a convergence test is performed, and when this is satisfied the iteration terminates. The output is then the current $X^*_{n,n}$ and its associated covariance matrix $S^*_{n,n}$, and computation ceases until a new observation vector arrives. When that occurs a new iteration procedure commences.

The attention of the reader is directed to the two distinct concepts involved – *recursion* and *iteration*. Their roles in the scheme under discussion are best seen from Figure 11.2, which depicts the Bayes Filter with iterative differential-correction in block diagram form. We examine the structure of that figure and contrast it to Figure 11.1.

The structure is still basically the same as that of the previous figure, with the outputs being computed, placed in storage, retrieved and used at a later time, together with the inputs $Y_{(n)}$ and $R_{(n)}$, to obtain the new estimate. This is the *recursive* part of the process, and is characterized by Loop A.

However, in addition to loop A, we observe a second loop, namely Loop B, which is the *iterative loop*. Conceptually, this loop is cycled many times for each time we cycle loop A. When the differential-correction procedure commences, $X^*_{k,k}$ is retrieved from storage and, by the use of numerical integration, it forms the basis of the first nominal trajectory, $\overline{X}_n$. The model and observation equations are then linearized and a Bayes estimation of the differential-vector is carried out, thereby providing a first value of the iterate, $X^*_{n,n}$. The output is then used repeatedly as the nominal trajectory until the convergence test is satisfied, whereupon iteration ceases. Recursion recommences when the next vector of observations arrives.

For the most part the techniques to be employed are very similar to those used in Sections 8.5 through 8.7. We assume familiarity with that material

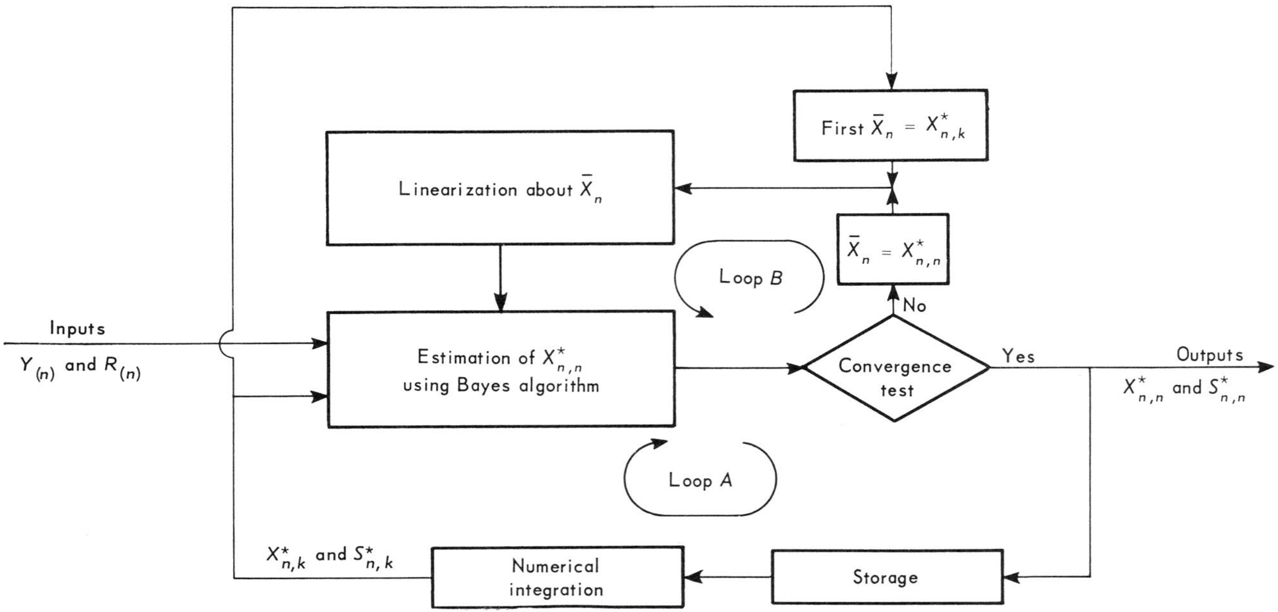

Fig. 11.2 *Bayes Filter with iterative differential-correction.*

as well as with the final section of Chapter 4 and the second section of Chapter 6.

11.2 ITERATIVE DIFFERENTIAL-CORRECTION—COMPUTATIONAL PROCEDURE

Assume, for generality, that a process under observation is modelled by the nonlinear differential equation

$$\frac{d}{dt} X(t) = F[X(t)] \tag{11.2.1}$$

and that the observations at t_n are related to X_n by the nonlinear relation

$$Y_n = G(X_n) + N_n \tag{11.2.2}$$

We propose to estimate X_n by combining an observation vector $\mathsf{Y}_{(n)}$, comprised of one or more subvectors of the form of (11.2.2), together with an already existent estimate. The latter, designated $X^*_{k,k}$, *is assumed to have been obtained at some earlier time,* based on observations whose errors are uncorrelated with those in $\mathsf{Y}_{(n)}$.

For simplicity in the discussion let $\mathsf{Y}_{(n)}$ be made up of, say, *three* subvectors which were obtained at times t_n, t_α and t_β (not necessarily distinct). Thus

$$\mathsf{Y}_{(n)} \equiv \begin{pmatrix} Y_n \\ -- \\ Y_\alpha \\ -- \\ Y_\beta \end{pmatrix} = \begin{pmatrix} G(X_n) \\ ---- \\ G(X_\alpha) \\ ---- \\ G(X_\beta) \end{pmatrix} + \mathsf{N}_{(n)} \tag{11.2.3}$$

where we have used the same function-vector G in each case, although this is not essential to the argument. Our first step is to linearize (11.2.3).

Suppose that we have a nominal trajectory, close to the true trajectory $X(t)$, which we call $\overline{X}(t)$. Then, given $\overline{X}(t)$ at some time, say t_n, we can use that vector $\overline{X}_n$ as a set of initial conditions, and by integrating (11.2.1), we can generate $\overline{X}(t)$ for any other t. In particular, we can form $\overline{X}_\alpha$ and $\overline{X}_\beta$, and using the function-vector G of (11.2.2), we compute the *simulated observation vector* $\mathsf{G}(\overline{X}_n)$, defined by

$$G(\overline{X}_n) \equiv \begin{pmatrix} G(\overline{X}_n) \\ ---- \\ G(\overline{X}_\alpha) \\ ---- \\ G(\overline{X}_\beta) \end{pmatrix} \tag{11.2.4}$$

This is the vector of observations that would have been made on $\overline{X}(t)$ if there were no observation errors. Note that we have shown $G(\overline{X}_n)$ as being a vector function of $\overline{X}_n$.

The assumption that $\overline{X}(t)$ is close to $X(t)$ means that the differential-vector $\delta X(t)$, defined by

$$\delta X(t) \equiv X(t) - \overline{X}(t) \tag{11.2.5}$$

has small elements relative to $\overline{X}(t)$. We subtract (11.2.4) from (11.2.3), obtaining

$$Y_{(n)} - G(\overline{X}_n) = \begin{pmatrix} G(X_n) - G(\overline{X}_n) \\ ---------- \\ G(X_\alpha) - G(\overline{X}_\alpha) \\ ---------- \\ G(X_\beta) - G(\overline{X}_\beta) \end{pmatrix} + N_{(n)} \tag{11.2.6}$$

and we recall from (6.2.16), that when δX_n is small, to first order

$$G(X_n) - G(\overline{X}_n) = M(\overline{X}_n)\,\delta X_n \tag{11.2.7}$$

where

$$\left[M(\overline{X}_n)\right]_{ij} = \left.\frac{\partial g_i(X)}{\partial x_j}\right|_{X = \overline{X}_n} \tag{11.2.8}$$

We thus see that (11.2.6) can be written

$$Y_{(n)} - G(\overline{X}_n) = \begin{pmatrix} M(\overline{X}_n)\,\delta X_n \\ ------ \\ M(\overline{X}_\alpha)\,\delta X_\alpha \\ ------ \\ M(\overline{X}_\beta)\,\delta X_\beta \end{pmatrix} + N_{(n)} \tag{11.2.9}$$

The next step is to linearize the model differential equations about $\overline{X}(t)$. From (4.8.21), we know that (11.2.1) yields, to first order

$$\frac{d}{dt}\delta X(t) = A\left[\overline{X}(t)\right]\delta X(t) \tag{11.2.10}$$

where

$$\left[A\left[\overline{X}(t)\right]\right]_{ij} \equiv \left.\frac{\partial f_i(X)}{\partial x_j}\right|_{X=\overline{X}(t)} \tag{11.2.11}$$

It now follows that since (11.2.10) is linear, there exists a transition matrix Φ, such that (see e.g. (4.8.24))

$$\delta X(t_p) = \Phi(t_p, t_q; \overline{X})\,\delta X(t_q) \tag{11.2.12}$$

(The question of how to obtain Φ in practice was discussed in detail in Chapter 4.)

Applying (11.2.12) to (11.2.9) we now obtain

$$Y_{(n)} - G(\overline{X}_n) = \begin{pmatrix} M(\overline{X}_n) \\ \hline M(\overline{X}_\alpha)\,\Phi(t_\alpha, t_n; \overline{X}) \\ \hline M(\overline{X}_\beta)\,\Phi(t_\beta, t_n; \overline{X}) \end{pmatrix} \delta X_n + N_{(n)} \tag{11.2.13}$$

and then by defining

$$\delta Y_{(n)} \equiv Y_{(n)} - G(\overline{X}_n) \tag{11.2.14}$$

and

$$T(\overline{X}_n) \equiv \begin{pmatrix} M(\overline{X}_n) \\ \hline M(\overline{X}_\alpha)\,\Phi(t_\alpha, t_n; \overline{X}) \\ \hline M(\overline{X}_\beta)\,\Phi(t_\beta, t_n; \overline{X}) \end{pmatrix} \tag{11.2.15}$$

we are able to write (11.2.13) as

$$\delta Y_{(n)} = T(\overline{X}_n)\,\delta X_n + N_{(n)} \tag{11.2.16}$$

This constitutes a *linearized observation relation* on δX_n, and will serve as the basis of a scheme for estimating that vector.

It was assumed at the beginning of our discussion that an estimate of $X(t)$ had been made at an earlier time. We called it $X^*_{k,k}$ and we designated its covariance matrix, also assumed available, as $S^*_{k,k}$. By the integration of (11.2.1), using $X^*_{k,k}$ as initial conditions, we now obtain an estimate of X_n which we call $X^*_{n,k}$, related to X_n by

$$X^*_{n,k} = X_n + N^*_{n,k} \tag{11.2.17}$$

The covariance matrix of the error vector $N^*_{n,k}$ in (11.2.17) is $S^*_{n,k}$, related to $S^*_{k,k}$ by the propagation equation†

$$S^*_{n,k} = \Phi(t_n, t_k; \overline{X}) S^*_{k,k} \Phi(t_n, t_k; \overline{X})^T \tag{11.2.18}$$

We now define

$$\delta X^*_{n,k} \equiv X^*_{n,k} - \overline{X}_n \tag{11.2.19}$$

Then by subtracting $\overline{X}_n$ from both sides of (11.2.17), we obtain

$$\delta X^*_{n,k} = \delta X_n + N^*_{n,k} \tag{11.2.20}$$

Next, we combine (11.2.20) and (11.2.16) into the single equation

$$\begin{pmatrix} \delta X^*_{n,k} \\ --- \\ \delta Y_{(n)} \end{pmatrix} = \begin{pmatrix} I \\ --- \\ T(\overline{X}_n) \end{pmatrix} \delta X_n + \begin{pmatrix} N^*_{n,k} \\ --- \\ N_{(n)} \end{pmatrix} \tag{11.2.21}$$

and recalling that $N_{(n)}$ and $N^*_{n,k}$ are by assumption uncorrelated, we see that the above equation is of the same form as (10.2.25). This means that we can immediately write, as the *composite minimum-variance estimate* of δX_n based on (11.2.21), the Bayes Filter equations (c/f p. 386)

$$\delta X^*_{n,n} = \delta X^*_{n,k} + H(\overline{X}_n)\left[\delta Y_{(n)} - T(\overline{X}_n)\,\delta X^*_{n,k}\right] \tag{11.2.22}$$

†See e.g., (5.5.33). Equation (11.2.18) is accurate only to second order terms.

where

$$H(\overline{X}_n) \equiv S^*_{n,n}\left[T(\overline{X}_n)\right]^T R^{-1}_{(n)} \tag{11.2.23}$$

and where

$$S^*_{n,n} \equiv \left\{(S^*_{n,k})^{-1} + \left[T(\overline{X}_n)\right]^T R^{-1}_{(n)} T(\overline{X}_n)\right\}^{-1} \tag{11.2.24}$$

The matrix $S^*_{n,n}$ as given by (11.2.24) is, to second order accuracy, the covariance matrix of the random errors in $\delta X^*_{n,n}$ of (11.2.22). Finally we form the vector $\left(X^*_{n,n}\right)_1$, defined by

$$\left(X^*_{n,n}\right)_1 \equiv \overline{X}_n + \delta X^*_{n,n} \tag{11.2.25}$$

and then *since* $\delta X^*_{n,n}$ *is an estimate of* δX_n *it follows from* (11.2.5) *that* $\left(X^*_{n,n}\right)_1$ *will be an improvement on* $\overline{X}_n$ *as an estimate of* X_n.

The obvious path to follow is now to re-use the improved estimate $\left(X^*_{n,n}\right)_1$ as the nominal trajectory in place of $\overline{X}_n$ in (11.2.22). Linearization about $\left(X^*_{n,n}\right)_1$ should result in smaller errors than was the case for linearization about $\overline{X}_n$, because the former is closer to the true trajectory than the latter was. *Hence if we now re-estimate* δX_n, *we will obtain a vector* $\left(X^*_{n,n}\right)_2$ *which should be a further improvement yet.* This suggests the following iterative procedure.

The algorithm (11.2.22) is first rearranged as follows. We add $\overline{X}_n$ to both sides, obtaining

$$X^*_{n,n} = X^*_{n,k} + H(\overline{X}_n)\left[\delta Y_{(n)} - T(\overline{X}_n)\,\delta X^*_{n,k}\right] \tag{11.2.26}$$

Now, using (11.2.14) and (11.2.19) to replace $\delta Y_{(n)}$ and $\delta X^*_{n,k}$ respectively, this becomes

$$X^*_{n,n} = X^*_{n,k} + H(\overline{X}_n)\left[Y_{(n)} - G(\overline{X}_n) - T(\overline{X}_n)(X^*_{n,k} - \overline{X}_n)\right] \tag{11.2.27}$$

which is the form in which the iterative computations are carried out.

To start the iteration, we use $X^*_{n,k}$ as the nominal trajectory, i.e. we set

$$\overline{X}_n = X^*_{n,k} \tag{11.2.28}$$

Making this substitution, (11.2.27) becomes

$$\left(X^*_{n,n}\right)_1 = X^*_{n,k} + H(X^*_{n,k})\left[Y_{(n)} - G(X^*_{n,k})\right] \tag{11.2.29}$$

which is the first cycle of the iteration.

The second cycle then uses $\left(X^*_{n,n}\right)_1$ as the nominal trajectory, i.e. we now set

$$\overline{X}_n = \left(X^*_{n,n}\right)_1 \tag{11.2.30}$$

and so (11.2.27) becomes

$$\begin{aligned}\left(X^*_{n,n}\right)_2 &= X^*_{n,k} \\ &\quad + H\left[\left(X^*_{n,n}\right)_1\right]\left\{Y_{(n)} - G\left[\left(X^*_{n,n}\right)_1\right] - T\left[\left(X^*_{n,n}\right)_1\right]\left[X^*_{n,k} - \left(X^*_{n,n}\right)_1\right]\right\}\end{aligned} \tag{11.2.31}$$

and then in general, for the rth cycle,

$$\left(X^*_{n,n}\right)_{r+1} = X^*_{n,k} + H(\overline{X}_n)\left[Y_{(n)} - G(\overline{X}_n) - T(\overline{X}_n)(X^*_{n,k} - \overline{X}_n)\right]\Big|_{\overline{X}_n = (X^*_{n,n})_r} \tag{11.2.32}$$

Observe that *throughout the iteration process,* $X^*_{n,k}$, $S^*_{n,k}$, $Y_{(n)}$ *and* $R_{(n)}$ *are left completely unchanged,* and only the terms involving $\overline{X}_n$ in (11.2.27) are modified from cycle to cycle.

Upon the completion of each of those cycles, a comparison is performed to see whether the estimate $\left(X^*_{n,n}\right)_{r+1}$ has changed by a meaningful amount from its predecessor. For this purpose, we use the convergence test

$$\text{“Is } \left[\left(X^*_{n,n}\right)_{r+1} - \left(X^*_{n,n}\right)_r\right]^T \left[\left(X^*_{n,n}\right)_{r+1} - \left(X^*_{n,n}\right)_r\right] < \epsilon\text{?”} \tag{11.2.33}$$

(or some other equivalent criterion). The quantity ϵ is a small positive number which is chosen to reflect the precision of the arithmetic and the amount of refinement the user wishes to accomplish. Once convergence has occurred, the resultant vector $\left(X^*_{n,n}\right)_{r+1}$ is then taken as the new estimate of X_n and the matrix $S^*_{n,n}$, which is evaluated during the final iteration cycle (see (11.2.24)), is taken as its covariance matrix.

In practice it is often the case that convergence occurs after the first pass of the iteration. This is particularly true when n is large and the initializing estimate $X^*_{n,k}$ is already a very good approximation to X_n, and so the linearization errors are extremely small. Under these conditions (11.2.29) constitutes the main algorithm rather than (11.2.32).

In Section 11.4 we investigate, more precisely, what vector the iteration algorithm converges to in the limit, assuming infinite precision arithmetic. We prove there *that it is the correct estimate in the weighted least-squares sense.* Owing to the inherent nonlinearities however, the matrix $S^*_{n,n}$, even in the limit, is only an approximation to the true covariance matrix of $X^*_{n,n}$.

Prior to that analysis we consider a small reorganization of the above scheme which materially reduces the possibility of round-off errors from corrupting the estimate during the recursion cycles.

11.3 CONTROL OF ERROR-PROPAGATION

In the iteration scheme outlined in the preceding section it was assumed that an estimate of X_n, obtained from prior observations, exists. We called it $X^*_{n,k}$. Clearly there must be a beginning to the process and at that time no such vector based on previous data would be available.

In order to get started, some form of initial estimation would have to be performed, and the most suitable method would probably be to operate a fixed-memory estimator, based on the first vector of observations, $Y_{(1)}$.† This could either be done using the polynomial approach discussed in Section 8.3, or could be the result of a fixed-memory iterative differential-correction computation applied to $Y_{(1)}$, as discussed in Sections 8.5, 8.6 and 8.7. Once this has been carried out, the vector $X^*_{n,k}$ discussed in the preceding section will then be available, and as new data vectors $Y_{(2)}, Y_{(3)}, \ldots, Y_{(n)}$ arrive, these can be incorporated recursively by performing an iterative differential-correction on each. The memory length is thus steadily increased, and if the filter is operating properly then the quality of the estimate will

†Assuming that no *a priori* estimate is available. Naturally if a dependable *a priori* estimate exists we would use it and commence directly with the recursive filter.

steadily improve. We now consider an approach† for controlling possible round-off-error accumulation. For simplicity we shall present our arguments in relation to the Bayes Filter on p. 389 which relates to the entirely linear situation with concurrent observations. The reader will easily be able to apply these ideas to the more general cases.

Consider the two equations (10.2.52) and (10.2.53), namely

$$\mathring{S}^*_{n,n-1} = \Phi(n, n-1)\mathring{S}^*_{n-1,n-1}\Phi(n, n-1)^T \tag{11.3.1}$$

$$\mathring{S}^*_{n,n} = \left[\left(\mathring{S}^*_{n,n-1}\right)^{-1} + M_n^T R_n^{-1} M_n\right]^{-1} \tag{11.3.2}$$

Observe that the sequence of operations by which $\mathring{S}^*_{n,n}$ is obtained from $\mathring{S}^*_{n-1,n-1}$ contains two inversions of the covariance matrix, both appearing in (11.3.2). If the algorithm is cycled repeatedly then a very strong danger exists that $\mathring{S}^*_{n,n}$ will degenerate because of the accumulation of round-off errors incurred in these repeated matrix inversions. Once this takes place the entire algorithm degenerates as well. However by a slight rearrangement we now show that this situation can be completely avoided.

Instead of working with $\mathring{S}^*_{n,n}$ *suppose we decide to work with its inverse.* We accordingly write (11.3.1) and (11.3.2) as

$$\left(\mathring{S}^*_{n,n-1}\right)^{-1} = \Psi(n, n-1)^T\left(\mathring{S}^*_{n-1,n-1}\right)^{-1}\Psi(n, n-1) \tag{11.3.3}$$

$$\left(\mathring{S}^*_{n,n}\right)^{-1} = \left(\mathring{S}^*_{n,n-1}\right)^{-1} + M_n^T R_n^{-1} M_n \tag{11.3.4}$$

in which $\Psi(n, n-1)$ is the inverse of $\Phi(n, n-1)$. *In this modified form we now see that the repeated inversions of* $\mathring{S}^*$ *are eliminated,* and so of course the problem of round-off accumulation disappears. Note that $\Psi(n, n-1)$ can be obtained directly from the model differential equations without actually inverting $\Phi(n, n-1)$ (see (4.7.31)) and moreover, if the prediction operation in (10.2.51) is performed by integrating the model equations, then the matrix Φ is never required. Of course $\mathring{S}^*_{n,n}$ itself is required for later use in (10.2.54), and so an inversion *will* have to be performed. However this inversion is now an ancillary operation *and does not lie in the repetitive loop by which* $\left(\mathring{S}^*_{n,n}\right)^{-1}$ *is computed,* namely (11.3.3) and (11.3.4). As such, any errors perpetrated in obtaining it from $\left(\mathring{S}^*_{n,n-1}\right)^{-1}$ will not propagate into subsequent computations as they would if (11.3.1) and (11.3.2) were being

†Due to A. J. Claus of Bell Telephone Laboratories.

cycled. We are thus able to avoid completely the difficulty of maintaining numerical accuracy in the $\overset{\circ}{S}{}^*$ computations by this very simple rearrangement. The reader is also referred to Ex. 11.16 where we show how the execution of (11.3.3) can be simplified.

11.4 CONVERGENCE OF ITERATIVE DIFFERENTIAL-CORRECTION

We have utilized the method of iterative differential-correction, both in Chapter 8 as well as in the present one, in order to obtain an estimate from a vector of observations when the observation scheme or the process differential equations are nonlinear. We now analyze more precisely the vectors to which those iterative schemes converge.

Consider first the Generalized Fixed-Memory case. We recall from (8.7.1) that the crucial equation in that scheme was

$$\left(X^*_{n,n}\right)_{r+1} = \left(X^*_{n,n}\right)_r$$
$$+ \left\{T\left[\left(X^*_{n,n}\right)_r\right]^T R^{-1}_{(n)} T\left[\left(X^*_{n,n}\right)_r\right]\right\}^{-1} T\left[\left(X^*_{n,n}\right)_r\right]^T R^{-1}_{(n)} \left\{Y_{(n)} - G\left[\left(X^*_{n,n}\right)_r\right]\right\} \tag{11.4.1}$$

where for simplicity in the present discussion we are now estimating to the *end* of the observation interval rather than to some interior point t_c as we did in (8.7.1). This was then iterated until convergence takes place as defined essentially by the condition

$$\left(X^*_{n,n}\right)_{r+1} = \left(X^*_{n,n}\right)_r \tag{11.4.2}$$

If we designate the vector prevailing *at the time of convergence* of (11.4.1) as X^*_∞, then we see, by combining the above two equations, that it satisfies

$$X^*_\infty = X^*_\infty + \left[T(X^*_\infty)^T R^{-1}_{(n)} T(X^*_\infty)\right]^{-1} T(X^*_\infty)^T R^{-1}_{(n)} \left[Y_{(n)} - G(X^*_\infty)\right] \tag{11.4.3}$$

i.e. that

$$\left[T(X^*_\infty)^T R^{-1}_{(n)} T(X^*_\infty)\right]^{-1} T(X^*_\infty)^T R^{-1}_{(n)} \left[Y_{(n)} - G(X^*_\infty)\right] = \emptyset \tag{11.4.4}$$

However, since $\left[T(X_{\infty}^{*})^{T}R_{(n)}^{-1}T(X_{\infty}^{*})\right]^{-1}$ is positive definite, (11.4.4) can only be true if

$$T(X_{\infty}^{*})^{T}R_{(n)}^{-1}\left[Y_{(n)} - G(X_{\infty}^{*})\right] = \emptyset \tag{11.4.5}$$

We have thus proved that if the iteration converges, then it converges to the vector X_{∞}^{*} *which satisfies* (11.4.5). We now demonstrate that this is precisely what we desire.

In Chapter 6, we defined the *residuals* as the difference between the actual observations and the simulated observations based on the selected estimate. In the iterative scheme considered above, it was assumed (see (8.6.24)) that the observations were related to the true state-vector by relations of the form:

$$\begin{aligned} Y_{n} &= G(X_{n}) + N_{n} \\ Y_{n-1} &= G(X_{n-1}) + N_{n-1} \\ &\vdots \\ Y_{n-k} &= G(X_{n-k}) + N_{n-k} \end{aligned} \tag{11.4.6}$$

These can be written as

$$Y_{(n)} = G(X_{n}) + N_{(n)} \tag{11.4.7}$$

and so if we let $X_{n,n}^{*}$ be an estimate of X_{n}, then from this last equation we see that the residual vector will be

$$E(X_{n,n}^{*}) \equiv Y_{(n)} - G(X_{n,n}^{*}) \tag{11.4.8}$$

It is now decided *that* $X_{n,n}^{*}$ *shall be chosen so as to minimize the weighted least-squares functional* (c/f (6.9.5))

$$e(X_{n,n}^{*}) \equiv \left[E(X_{n,n}^{*})\right]^{T}R_{(n)}^{-1}E(X_{n,n}^{*}) \tag{11.4.9}$$

As we already know from Section 6.9, this criterion leads to the *minimum-variance* filter if $E(X_{n,n}^{*})$ is a *linear* function of $X_{n,n}^{*}$.

In order to minimize $e(X^*_{n,n})$ of (11.4.9), we differentiate it with respect to each of the components of $X^*_{n,n}$, setting each of those derivatives to zero. This, as we have pointed out previously, is only a necessary condition for a minimum. However, since $e(X^*_{n,n})$ is a quadratic form on a positive definite matrix, it can easily be shown (see Ex. 6.7) that it is also a sufficient one. Thus we solve for the $X^*_{n,n}$ which minimizes e, by setting

$$\frac{\partial}{\partial\left(x_j^*\right)_{n,n}}\left[Y_{(n)} - G(X^*_{n,n})\right]^T R^{-1}_{(n)}\left[Y_{(n)} - G(X^*_{n,n})\right] = 0 \tag{11.4.10}$$

for each component $\left(x_j^*\right)_{n,n}$ of $X^*_{n,n}$.

Carrying out the above operation, we first obtain

$$\left\{\frac{\partial}{\partial\left(x_j^*\right)_{n,n}}\left[Y_{(n)} - G(X^*_{n,n})\right]^T\right\} R^{-1}_{(n)}\left[Y_{(n)} - G(X^*_{n,n})\right]$$
$$+\left[Y_{(n)} - G(X^*_{n,n})\right]^T R^{-1}_{(n)}\left\{\frac{\partial}{\partial\left(x_j^*\right)_{n,n}}\left[Y_{(n)} - G(X^*_{n,n})\right]\right\} = 0 \tag{11.4.11}$$

and since $R^{-1}_{(n)}$ is symmetric this reduces simply to

$$\left[\frac{\partial G(X^*_{n,n})}{\partial\left(x_j^*\right)_{n,n}}\right]^T R^{-1}_{(n)}\left[Y_{(n)} - G(X^*_{n,n})\right] = 0 \tag{11.4.12}$$

Define the matrix $B(X^*_{n,n})$ whose i, j^{th} term is

$$\left[B(X^*_{n,n})\right]_{ij} \equiv \frac{\partial g_i(X^*_{n,n})}{\partial\left(x_j^*\right)_{n,n}} \tag{11.4.13}$$

where $g_i(X^*_{n,n})$ is the ith function of the vector $G(X^*_{n,n})$. Then, combining each of the scalar equations obtained from (11.4.12), as j goes from 0 to m, we get the vector equation

$$B(X^*_{n,n})^T R^{-1}_{(n)}\left[Y_{(n)} - G(X^*_{n,n})\right] = \emptyset \tag{11.4.14}$$

We now investigate the function-matrix B, and show, quite simply, that it is of exactly the same form as the function-matrix T appearing in (11.4.5).

Thus, consider the difference $G(X_n) - G(\overline{X}_n)$. We have shown that this can be linearized to the form

$$G(X_n) - G(\overline{X}_n) = T(\overline{X}_n)\,\delta X_n + \text{higher order terms} \tag{11.4.15}$$

where

$$T(\overline{X}_n) = \begin{pmatrix} M(\overline{X}_n) \\ \hline M(\overline{X}_{n-1})\,\Phi(t_{n-1}, t_n; \overline{X}) \\ \vdots \\ \hline M(\overline{X}_{n-L})\,\Phi(t_{n-L}, t_n; \overline{X}) \end{pmatrix} \tag{11.4.16}$$

and where the matrix M is defined by

$$\left[M(\overline{X}_n)\right]_{ij} = \left.\frac{\partial g_i(X)}{\partial x_j}\right|_{X=\overline{X}_n} \tag{11.4.17}$$

(c/f (11.2.8)).

We now expand $G(X_n)$ about $\overline{X}_n$, using a *vector Taylor series.* This gives

$$\begin{pmatrix} g_0(X_n) \\ g_1(X_n) \\ \vdots \\ g_s(X_n) \end{pmatrix} = \begin{pmatrix} g_0(X) \\ g_1(X) \\ \vdots \\ g_s(X) \end{pmatrix}_{X=\overline{X}_n} +$$

$$\begin{pmatrix} \dfrac{\partial g_0(X)}{\partial x_0} & \dfrac{\partial g_0(X)}{\partial x_1} & \cdots & \dfrac{\partial g_0(X)}{\partial x_m} \\ \dfrac{\partial g_1(X)}{\partial x_0} & \dfrac{\partial g_1(X)}{\partial x_1} & \cdots & \dfrac{\partial g_1(X)}{\partial x_m} \\ \vdots & \vdots & & \vdots \\ \dfrac{\partial g_s(X)}{\partial x_0} & \dfrac{\partial g_s(X)}{\partial x_1} & \cdots & \dfrac{\partial g_s(X)}{\partial x_m} \end{pmatrix}_{X=\overline{X}_n} \begin{pmatrix} \delta x_0 \\ \delta x_1 \\ \vdots \\ \delta x_m \end{pmatrix} + \text{higher order terms} \tag{11.4.18}$$

But this is the same equation as (11.4.15) and so the matrix of partial derivatives on the right of (11.4.18) is precisely $T(\bar{X}_n)$. Thus, finally, comparing (11.4.13) with the above matrix of partial derivatives, we see that

$$B(X^*_{n,n}) = T(X^*_{n,n}) \tag{11.4.19}$$

We now return to (11.4.14). By (11.4.19) it becomes

$$T(X^*_{n,n})^T R^{-1}_{(n)} \left[Y_{(n)} - G(X^*_{n,n})\right] = \emptyset \tag{11.4.20}$$

and so the vector $X^*_{n,n}$ which satisfies this equation will minimize the weighted least-squares functional e_n of (11.4.9). But (11.4.20) is identical, in form, to (11.4.5). We have thus in fact proved *that the vector to which the differential-correction iteration scheme of Chapter 8 converges, is also the vector which satisfies the nonlinear weighted least-squares criterion,* and so if convergence takes place, then in this sense it is to the correct vector.

It is now a relatively simple matter to verify, by an approach similar to the above, that if (11.2.32) converges, then it also converges to the correct value. We merely outline the procedure, leaving the details for the reader.

In this case when convergence takes place, (11.2.32) becomes

$$X^*_\infty = X^*_{n,k} + H(X^*_\infty)\left[Y_{(n)} - G(X^*_\infty) - T(X^*_\infty)(X^*_{n,k} - X^*_\infty)\right] \tag{11.4.21}$$

which can be regrouped as

$$\left[I - H(X^*_\infty)T(X^*_\infty)\right](X^*_{n,k} - X^*_\infty) + H(X^*_\infty)\left[Y_{(n)} - G(X^*_\infty)\right] = \emptyset \tag{11.4.22}$$

We now examine the definition of H in (11.2.23) and obtain the result (see Ex. 11.18) that

$$I - H(X^*_\infty)T(X^*_\infty) = S(X^*_\infty)(S^*_{n,k})^{-1} \tag{11.4.23}$$

in which we have defined (c/f (11.2.24))

$$S(X^*_\infty) \equiv \left[(S^*_{n,k})^{-1} + T(X^*_\infty)^T R^{-1}_{(n)} T(X^*_\infty)\right]^{-1} \tag{11.4.24}$$

By the use of (11.4.23) and (11.2.23), equation (11.4.22) now reduces to

(see Ex. 11.19)

$$S(X^*_\infty)\left\{(S^*_{n,k})^{-1}(X^*_{n,k} - X^*_\infty) + T(X^*_\infty)^T R^{-1}_{(n)}\left[Y_{(n)} - G(X^*_\infty)\right]\right\} = \emptyset \quad (11.4.25)$$

and since $S(X^*_\infty)$ is positive definite, this can only be true if

$$(S^*_{n,k})^{-1}(X^*_{n,k} - X^*_\infty) + T(X^*_\infty)^T R^{-1}_{(n)}\left[Y_{(n)} - G(X^*_\infty)\right] = \emptyset \quad (11.4.26)$$

The vector X^*_∞ to which the Bayes version of the differential-correction scheme converges, will thus satisfy (11.4.26). We now show that this is precisely as required.

Consider the two relations

$$X^*_{n,k} = X_n + N^*_{n,k} \quad (11.4.27)$$

$$Y_{(n)} = G(X_n) + N_{(n)} \quad (11.4.28)$$

The former occurs as (11.2.17) and the latter as (11.2.3). We combine them, obtaining

$$\begin{pmatrix} X^*_{n,k} \\ --- \\ Y_{(n)} \end{pmatrix} = \begin{pmatrix} X_n \\ ---- \\ G(X_n) \end{pmatrix} + \begin{pmatrix} N^*_{n,k} \\ --- \\ N_{(n)} \end{pmatrix} \quad (11.4.29)$$

and letting $X^*_{n,n}$ be an estimate of X_n, the *residual vector* is seen to be (c/f (11.4.8))

$$E(X^*_{n,n}) = \begin{pmatrix} X^*_{n,k} - X^*_{n,n} \\ --------- \\ Y_{(n)} - G(X^*_{n,n}) \end{pmatrix} \quad (11.4.30)$$

We choose $X^*_{n,n}$ *so as to minimize the weighted least-squares functional*

$$e(X^*_{n,n}) \equiv \left[E(X^*_{n,n})\right]^T R^{-1} E(X^*_{n,n}) \quad (11.4.31)$$

where R is the covariance matrix of the error vector of (11.4.29), i.e.

$$R = \begin{pmatrix} S^*_{n,k} & \emptyset \\ \emptyset & R_{(n)} \end{pmatrix} \tag{11.4.32}$$

It is now easily verified (see Ex. 11.20) that if we use (11.4.30) and (11.4.32) in (11.4.31), we obtain

$$\begin{aligned} e(X^*_{n,n}) &= (X^*_{n,k} - X^*_{n,n})^T (S^*_{n,k})^{-1} (X^*_{n,k} - X^*_{n,n}) \\ &\quad + \left[Y_{(n)} - G(X^*_{n,n})\right]^T R^{-1}_{(n)} \left[Y_{(n)} - G(X^*_{n,n})\right] \end{aligned} \tag{11.4.33}$$

We minimize this by setting

$$\frac{\partial e_n(X^*_{n,n})}{\partial \left(x^*_j\right)_{n,n}} = 0 \qquad 0 \le j \le m \tag{11.4.34}$$

and this then gives us (see Ex. 11.21)

$$(S^*_{n,k})^{-1}(X^*_{n,k} - X^*_{n,n}) + \left[T(X^*_{n,n})\right]^T R^{-1}_{(n)} \left[Y_{(n)} - G(X^*_{n,n})\right] = \emptyset \tag{11.4.35}$$

This last result is identical in form to (11.4.26) *and so we have proved that the vector* X^*_∞ *which satisfies* (11.4.26) *also satisfies* (11.4.35). The iteration scheme of (11.2.32) thus converges to the same vector which minimizes (11.4.31) and is thus the required result, in the sense of weighted least-squares.

NOTES

Gauss essentially concerned himself only with situations where relatively small numbers of observations were to be used in forming an estimate. Thus his approach was basically that of Chapter 8. We may speculate, however, that he almost certainly encountered cases *where an estimate had previously been derived, and then at a later time it was desired that further data be incorporated.* His refinement procedure would then probably have consisted of using the previous estimate as a basis for a nominal trajectory for the incorporation, by differential-correction, of the new data.

He does not appear, from [11.1], to have developed a clear-cut *recursive* procedure whereby a steadily expanding data-base could be accepted. This is simply because he was not concerned with this problem, which did not really arise until the present times with the advent of artificial satellites, radar equipment and high-speed computers. There is very little doubt, however, that he would have developed the necessary algorithms had he needed them, since he was only one small step away from them as we see from the pair of equations in [11.1, p. 251].

Nevertheless, in all fairness, Gauss *did not explicitly derive a recursive differential-correction algorithm,* and it remained until 1958 when P. Swerling appears to have been the first to clearly propose one [11.2, 11.3]. Swerling's method is essentially the iterative differential-correction scheme we develop in this chapter, with the exception that no iteration is used; he terminates at our equation (11.2.29), thereafter waiting for new data.

In the early stages of an estimation operation, when each new batch of data is a significant contribution to the total data-base thus far received, iteration is definitely advantageous, i.e. the use of (11.2.32) and (11.2.33) is recommended for n small. However when n becomes large, the newest data-addition is a small contribution to the already existent data-base, and at this time iteration is of little help, most of the modification to the estimate by the newest data taking place on the *first* cycle of the iteration (see (11.2.29)). Thus once the estimation is well under way, Swerling's algorithm will probably result, even though repeated iteration is called for. Equation (11.2.33) will under normal circumstances terminate the iteration when $r = 0$ for large n.

In [11.4] Blackman presents a comparative study of the various schemes we have considered up to now, and it is on his opinion (i.e. that Swerling was the originator of the recursive schemes) that we base ours. Blackman's paper is an excellent supplement to our discussion.

We also recommend [11.5, 11.6, 11.7] in which the estimation scheme that was used in the highly successful Telstar project is discussed in detail.

For an excellent survey of Bayes filtering and the material to come, the reader is referred to [11.8]. The reader may also wish to consult [11.9], in which an extensive list of further references is presented.

EXERCISES

11.1 A stationary body is located at a point distant x from the origin, positively along the ξ-axis. (See Figure 11.3.)

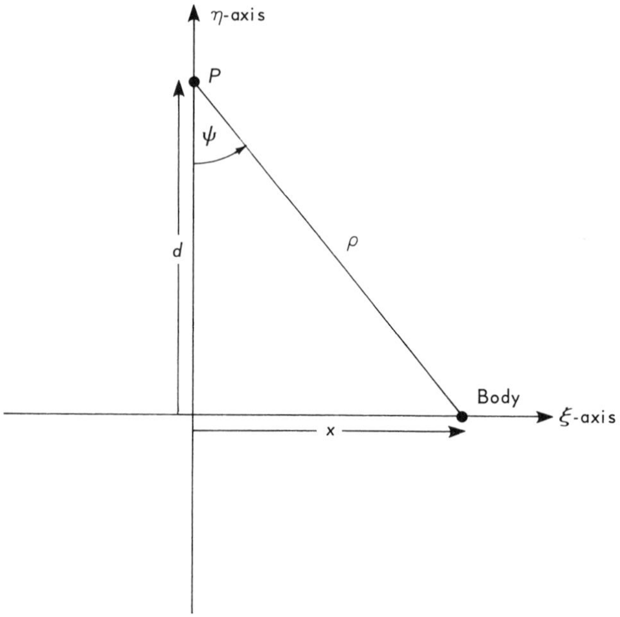

Fig. 11.3 *Location of body on ξ-axis.*

Measurements are made of the range ρ from a point P on the η-axis, distant d from the origin, where d is a known quantity. The measurements are all uncorrelated with each other and all have variance σ_ρ^2. The intention is to determine the value of x.

a) We take as our state-vector the scalar x, and so the differential equation for the process is $\dot{x} = 0$. Verify that the transition matrix is given by the scalar

$$\Phi(\zeta) = 1$$

Also verify that the observation relation is

$$\rho_n = (d^2 + x^2)^{1/2} + \nu_n \qquad \text{(I)}$$

Linearize this about a nominal value of x, namely $\bar{x}$, and obtain the relation

$$\delta\rho_n = m(\bar{x})\,\delta x + \nu_n$$

where

$$m(\bar{x}) \equiv \frac{\bar{x}}{(d^2 + \bar{x}^2)^{1/2}}$$

b) Assume that we have an *a priori* estimate of x which we call $x^*_{1,0}$ with variance $s^*_{1,0} = \alpha^2$. Verify that (11.2.24) gives

$$s^*_{1,1} = \frac{\alpha^2 \sigma_\rho^2}{\sigma_\rho^2 + \dfrac{\alpha^2 \bar{x}^2}{d^2 + \bar{x}^2}} \tag{II}$$

and show that this is a definite improvement over the variance of the *a priori* value, $s^*_{1,0}$.

c) Verify that (11.2.23) becomes

$$h(\bar{x}) = \frac{\alpha^2 \bar{x}}{(d^2 + \bar{x}^2)^{1/2}\left(\sigma_\rho^2 + \dfrac{\alpha^2 \bar{x}^2}{d^2 + \bar{x}^2}\right)} \tag{III}$$

d) Show that (11.2.27) gives

$$x^*_{1,1} = x^*_{1,0} + h(\bar{x})\left[\rho_1 - (d^2 + \bar{x}^2)^{1/2} - \frac{\bar{x}}{(d^2 + \bar{x}^2)^{1/2}}(x^*_{1,0} - \bar{x})\right] \tag{IV}$$

e) We decide to iterate (IV) using the *a priori* estimate $x^*_{1,0}$ as the first value of $\bar{x}$. Verify that on the first cycle of the iteration (IV) becomes

$$\left(x^*_{1,1}\right)_1 = x^*_{1,0} + h\left(x^*_{1,0}\right)\left[\rho_1 - \left(d^2 + x^{*2}_{1,0}\right)^{1/2}\right] \tag{V}$$

Now use the output of (IV) repeatedly in place of $\bar{x}$ in that equation and verify that we obtain the iteration algorithm

$$\left(x^*_{1,1}\right)_{r+1} = x^*_{1,0} + h(\bar{x})\left[\rho_1 - (d^2 + \bar{x}^2)^{1/2} - \frac{\bar{x}}{(d^2 + \bar{x}^2)^{1/2}}(x^*_{1,0} - \bar{x})\right]\Bigg|_{\bar{x} = \left(x^*_{1,1}\right)_r} \tag{VI}$$

11.2 Suppose that $d = 0$ in Ex. 11.1.

a) Verify that the observation relation is now linear and that (II) of Ex. 11.1 reduces to

$$s^*_{1,1} = \frac{\alpha^2 \sigma_\rho^2}{\alpha^2 + \sigma_\rho^2}$$

b) Show that (III) of Ex. 11.1 becomes

$$h = \frac{\alpha^2}{\sigma_\rho^2 + \alpha^2}$$

and that (IV) of that example reduces to

$$x^*_{1,1} = \frac{x^*_{1,0}/\alpha^2 + \rho_1/\sigma_\rho^2}{1/\alpha^2 + 1/\sigma_\rho^2} \qquad \text{(I)}$$

Reconcile the above three results with (10.4.12) and (10.4.13).

11.3 a) Write a computer program which generates "observations" according to (I) of Ex. 11.1.

b) Obtain an *a priori* estimate of x from the equation

$$x^*_{1,0} = \left(\rho_0^2 - d^2\right)^{1/2}$$

where ρ_0 is the first observation.

c) Program and operate the iteration algorithm (VI) of Ex. 11.1 and incorporate ρ_1 into the estimate. When the iteration converges obtain a new measurement and cycle the iteration once more to convergence, using the previous estimate of x to get started. In this way incorporate observations successively and obtain an estimate of x based on an expanding data base.

d) Estimate x by the following alternate method. Apply a zeroth-degree Expanding-Memory Polynomial Filter (see Table 9.1, p. 360) to the sequence of observations $\rho_0, \rho_1, \rho_2, \ldots \rho_n$ thereby obtaining $\rho^*_{n,n}$. Now estimate x using

$$x^*_{n,n} = \left[\left(\rho^*_{n,n}\right)^2 - d^2\right]^{1/2}$$

e) Compare the amount of effort in d) against that in c). Could we use the method in d) if the point was *moving* along the ξ-axis?

11.4 Referring to (VI) of Ex. 11.1 assume that convergence has taken place, as indicated by the fact that the output remains substantially unchanged from its previous value. Call this x^*_∞.

a) Verify that under these conditions

$$\left[1 - h(x^*_\infty)\frac{x^*_\infty}{(d^2 + x^{*2}_\infty)^{1/2}}\right](x^*_{1,0} - x^*_\infty) + h(x^*_\infty)\left[\rho_1 - (d^2 + x^{*2}_\infty)^{1/2}\right] = 0 \qquad \text{(I)}$$

b) Show also that (III) of Ex. 11.1 becomes

$$h(x^*_\infty) = \frac{\alpha^2 x^*_\infty}{(d^2 + x^{*2}_\infty)^{1/2}\left(\sigma_\rho^2 + \dfrac{\alpha^2 x^{*2}_\infty}{d^2 + x^{*2}_\infty}\right)} \qquad \text{(II)}$$

and that (II) of Ex. 11.1 becomes

$$s^*(x^*_\infty) = \frac{\alpha^2\sigma_\rho^2}{\sigma_\rho^2 + \dfrac{\alpha^2 x^{*2}_\infty}{d^2 + x^{*2}_\infty}} \qquad \text{(III)}$$

c) Now verify that (II) and (III) of this exercise give

$$1 - h(x^*_\infty)\frac{x^*_\infty}{(d^2 + x^{*2}_\infty)^{1/2}} = \frac{s^*(x^*_\infty)}{\alpha^2} \qquad \text{(IV)}$$

and reconcile this with (11.4.23).

d) Using (II) and (IV) in (I) above, obtain the result

$$\frac{1}{\alpha^2}(x^*_{1,0} - x^*_\infty) + \frac{x^*_\infty}{\sigma_\rho^2 (d^2 + x^{*2}_\infty)^{1/2}}\left[\rho_1 - (d^2 + x^{*2}_\infty)^{1/2}\right] = 0 \qquad \text{(V)}$$

e) Show that when $d = 0$, (V) above gives us

$$x^*_\infty = \frac{x^*_{1,0}/\alpha^2 + \rho_1/\sigma_\rho^2}{1/\alpha^2 + 1/\sigma_\rho^2}$$

and reconcile this with (I) of Ex. 11.2.

11.5 Considering the situation of Ex. 11.1, show that (11.4.33) gives

$$e(x^*_{1,1}) = \frac{1}{\alpha^2}(x^*_{1,0} - x^*_{1,1})^2 + \frac{1}{\sigma_\rho^2}\left[\rho_1 - (d^2 + x^{*2}_{1,1})^{1/2}\right]^2 \qquad \text{(I)}$$

Minimize this with respect to $x^*_{1,1}$ and verify that the resultant $x^*_{1,1}$ satisfies

$$\frac{1}{\alpha^2}(x^*_{1,0} - x^*_{1,1}) + x^*_{1,1}\frac{\rho_1 - (d^2 + x^{*2}_{1,1})^{1/2}}{\sigma_\rho^2 (d^2 + x^{*2}_{1,1})^{1/2}} = 0 \qquad \text{(II)}$$

Compare this to (V) of Ex. 11.4 and hence infer that the value of $x^*_{1,1}$ which minimizes (I) above is also the value to which the iteration in (VI) of Ex. 11.1 converges.

11.6 Consider the situation of Ex. 11.1, but assume now that the distance d is *unknown*. We hope to estimate both d and x.

a) Taking as our process state-vector

$$X \equiv \begin{pmatrix} x \\ d \end{pmatrix}$$

verify that the transition matrix is a 2×2 unit matrix. Verify also that the linearized observation relation is

$$\delta\rho_n = M(\overline{X})\delta X + \nu_n$$

where M is the 1×2 matrix

$$M(\overline{X}) \equiv \frac{1}{(\overline{d}^2 + \overline{x}^2)^{1/2}}(\overline{x}, \quad \overline{d})$$

b) Assume that we were to use a Generalized Fixed-Memory Filter to estimate X. Show that in this case the matrix T of (8.2.4) does *not* have full column rank and hence infer that we cannot execute (8.7.1). The observation scheme is thus unsatisfactory.

c) Reconcile the findings of b) above with Theorem 8.2 of Section 8.9 and show that they are predicted by that theorem.

11.7 Assume that in the previous example we also measure the azimuth angle ψ (see Figure 11.3).

a) Verify that in this case

$$M(\bar{X}) = \begin{pmatrix} \dfrac{\bar{x}}{\gamma} & \dfrac{\bar{d}}{\gamma} \\[2ex] \dfrac{\bar{d}}{\gamma^2} & -\dfrac{\bar{x}}{\gamma^2} \end{pmatrix}$$

where

$$\gamma \equiv (\bar{d}^2 + \bar{x}^2)^{1/2}$$

b) Now show that T has full column rank. Reconcile this with Theorem 8.2.

11.8 Consider the situation of Ex. 11.1 *but assume that the body is now moving along the* ξ-axis *with constant but unknown speed* $\dot{x}$. The same observations are made as in Ex. 11.1.

a) We choose as our process state-vector

$$X \equiv \begin{pmatrix} x \\ \dot{x} \end{pmatrix}$$

where x is the position at $t = 0$. Verify that the differential equation for X is

$$\frac{d}{dt} X(t) = \emptyset$$

and hence infer that

$$\Phi(\zeta) = I$$

Also verify that the observation relation is now

$$\rho_n = \left[d^2 + (x + t_n \dot{x})^2\right]^{1/2} + \nu_n \tag{I}$$

and that linearization about a nominal value of X, namely $\overline{X}$, leads to

$$\delta\rho_n = M_n(\overline{X})\, \delta X + \nu_n$$

where M is the 1×2 matrix

$$M_n(\overline{X}) \equiv \frac{\bar{x} + t_n \bar{\dot{x}}}{\left[d^2 + (\bar{x} + t_n \bar{\dot{x}})^2\right]^{1/2}}(1, \quad t_n)$$

b) Assume that we have an *a priori* estimate of X, namely $X^*_{1,0}$ with covariance matrix

$$S^*_{1,0} = \begin{pmatrix} \alpha^2 & 0 \\ 0 & \beta^2 \end{pmatrix} \tag{II}$$

Verify that (11.2.24) gives

$$S^*_{1,1} = \frac{1}{\lambda}\begin{pmatrix} \alpha^2\left(\sigma_\rho^2 + t_1^2 \beta^2\gamma^2\right) & -t_1\alpha^2\beta^2\gamma^2 \\ -t_1\alpha^2\beta^2\gamma^2 & \beta^2\left(\sigma_\rho^2 + \alpha^2\gamma^2\right) \end{pmatrix}$$

where $\lambda \equiv \sigma_\rho^2 + (\alpha^2 + t_1^2\beta^2)\gamma^2$ and where

$$\gamma \equiv \frac{\bar{x} + t_1\bar{\dot{x}}}{\left[d^2 + (\bar{x} + t_1 \bar{\dot{x}})^2\right]^{1/2}}$$

c) Verify that as $t_1 \to 0$,

$$S^*_{1,1} \to \begin{pmatrix} \dfrac{\alpha^2\sigma_\rho^2}{\sigma_\rho^2 + \dfrac{\alpha^2\bar{x}^2}{d^2 + \bar{x}^2}} & 0 \\ 0 & \beta^2 \end{pmatrix}$$

Observe that the 0,0 element of this matrix is identical to (II) of Ex. 11.1, and that the 1,1 element is the same as that of (II) above. Explain both facts in physical terms.

d) Verify that as $t_1 \to \infty$

$$S^*_{1,1} \to \begin{pmatrix} \alpha^2 & 0 \\ 0 & 0 \end{pmatrix}$$

Infer then that, as t_1 increases, the variance of the errors in our estimate of $\dot{x}$ goes to zero, whereas the variance of the errors in our estimate of x tends to its *a priori* value. This seems to be a contradiction in which the measurement ρ_1 improves our knowledge of $\dot{x}$ but *not* our knowledge of x. Explain.

11.9 a) For the situation of Ex. 11.8, assume that the first six observations $\rho_0, \ldots, \rho_5$ where smoothed, using an Expanding-Memory Polynomial Filter to give the estimates $\rho^*_{0,5}$ and $\dot{\rho}^*_{0,5}$, valid at time t_0. Show that we can estimate x and $\dot{x}$ from

$$\begin{aligned} x^* &= \left((\rho^*_{0,5})^2 - d^2\right)^{1/2} \\ \dot{x}^* &= \frac{\rho^*_{0,5}\,\dot{\rho}^*_{0,5}}{\left((\rho^*_{0,5})^2 - d^2\right)^{1/2}} \end{aligned} \qquad \text{(I)}$$

Now verify that to first order

$$\begin{pmatrix} \delta x^* \\ \delta \dot{x}^* \end{pmatrix} = \Gamma(\rho^*, \dot{\rho}^*) \begin{pmatrix} \delta \rho^* \\ \delta \dot{\rho}^* \end{pmatrix} \qquad \text{(II)}$$

where Γ is the 2×2 matrix

$$\Gamma(\rho^*, \dot{\rho}^*) \equiv \begin{pmatrix} \dfrac{\rho^*}{((\rho^*)^2 - d^2)^{1/2}} & 0 \\ \dfrac{-d^2 \dot{\rho}^*}{((\rho^*)^2 - d^2)^{3/2}} & \dfrac{\rho^*}{((\rho^*)^2 - d^2)^{1/2}} \end{pmatrix} \qquad \text{(III)}$$

The covariance matrix of the vector $\left(\rho^*_{0,5},\ \dot{\rho}^*_{0,5}\right)^T$ is obtained from the polynomial filter. Let it be called $Z^*_{0,5}$. Infer from (II) and (III) above that the covariance matrix of $(x^*,\ \dot{x}^*)^T$ in (I) is given by

$$S^* = \Gamma(\rho^*_{0,5},\ \dot{\rho}^*_{0,5})\, Z^*_{0,5} \left(\Gamma(\rho^*_{0,5},\ \dot{\rho}^*_{0,5})\right)^T \tag{IV}$$

b) Simulate the situation of Ex. 11.8 on a computer by generating observations according to (I) of that example. Apply the Bayes Filter with iterative differential-correction on an expanding data-base. Initialize the filter using the estimate and covariance matrix obtained in part a) of this exercise. (See (I) and (IV) above.)

11.10 Repeat Ex. 11.8 but assume instead that we observe the azimuth angle ψ, shown in Figure 11.3. The observations are all uncorrelated and of variance $\sigma_\psi^{\,2}$.

a) Verify that

$$\psi_n = \arctan\left(\frac{x + t_n \dot{x}}{d}\right) + \nu_n \tag{I}$$

and hence that

$$\delta\psi = M_n(\overline{X})\delta X + \nu_n$$

where M is the 1×2 matrix

$$M_n(\overline{X}) = \frac{d}{d^2 + (\bar{x} + t_n \bar{\dot{x}})^2}(1,\ t_n)$$

b) Assuming an *a priori* estimate $X^*_{1,0}$ with covariance matrix

$$S^*_{1,0} = \begin{pmatrix} \alpha^2 & 0 \\ 0 & \beta^2 \end{pmatrix}$$

apply (11.2.24) to obtain $S^*_{1,1}$ and verify that

$$S^*_{1,1} = \frac{1}{\lambda}\begin{pmatrix} \alpha^2(\sigma_\psi^2 + t_1{}^2\beta^2\gamma^2) & -t_1\alpha^2\beta^2\gamma^2 \\ -t_1\alpha^2\beta^2\gamma^2 & \beta^2(\sigma_\psi^2 + \alpha^2\gamma^2) \end{pmatrix}$$

where

$$\lambda \equiv \sigma_\psi{}^2 + (\alpha^2 + t_1{}^2\beta^2)\gamma^2$$

and

$$\gamma \equiv d/[d^2 + (\bar{x} + t_1\dot{\bar{x}})^2]$$

c) Investigate what happens to $S^*_{1,1}$ if $d \to 0$ and also what happens as time passes and d shrinks relative to the distance of the body from the origin. Hence infer that the scheme works best when d is large relative to that distance.

d) Show that as $\sigma_\psi{}^2 \to \infty$ then $S^*_{1,1} \to S^*_{1,0}$. Interpret this.

e) Show that as $t_1 \to 0$,

$$S^*_{1,1} \to \begin{pmatrix} \dfrac{\alpha^2\sigma_\psi{}^2}{\sigma_\psi{}^2 + \dfrac{\alpha^2 d^2}{(d^2 + \bar{x}^2)^2}} & 0 \\ 0 & \beta^2 \end{pmatrix}$$

f) Show that as $t_1 \to \infty$

$$S^*_{1,1} \to \begin{pmatrix} \alpha^2 & 0 \\ 0 & 0 \end{pmatrix}$$

and explain this apparent contradiction.

11.11 a) For the situation of Ex. 11.10, filter the first six measurements of ψ in an Expanding-Memory Polynomial Filter. Following the approach of Ex. 11.9 a), use the resultant state-vector $\left(\psi^*_{0,5},\ \dot{\psi}^*_{0,5}\right)^T$ to obtain an estimate of $(x,\ \dot{x})^T$. Also obtain a covariance matrix for the latter from the covariance matrix derived by the polynomial filter.

b) Simulate the situation of Ex. 11.10 by generating "measurements" of ψ on a computer, according to (I) of Ex. 11.10. Now apply a Bayes Filter with iterative differential-correction, using the state-vector and covariance matrix obtained in part a) above for initialization.

11.12 Assume that both ρ and ψ are observed simultaneously (see Examples 11.8 and 11.10).

a) Infer that the observation relation now becomes

$$\begin{pmatrix} \rho \\ \psi \end{pmatrix}_n = \begin{pmatrix} \left[d^2 + (x + t_n \dot{x})^2\right]^{1/2} \\ \arctan\left(\dfrac{x + t_n \dot{x}}{d}\right) \end{pmatrix} + \begin{pmatrix} \nu_\rho \\ \nu_\psi \end{pmatrix} \tag{I}$$

and hence that, to first order

$$\begin{pmatrix} \delta\rho \\ \delta\psi \end{pmatrix}_n = M_n(\overline{X}) \begin{pmatrix} \delta x \\ \delta \dot{x} \end{pmatrix} + \begin{pmatrix} \nu_\rho \\ \nu_\psi \end{pmatrix}$$

where M is the 2×2 matrix

$$M_n(\overline{X}) \equiv \begin{pmatrix} \dfrac{\bar{x} + t_n \bar{\dot{x}}}{\lambda} & t_n \dfrac{\bar{x} + t_n \bar{\dot{x}}}{\lambda} \\ \dfrac{d}{\lambda^2} & \dfrac{t_n d}{\lambda^2} \end{pmatrix} \tag{II}$$

with

$$\lambda \equiv \left[d^2 + (\bar{x} + t_n \bar{\dot{x}})^2\right]^{1/2}$$

b) Generate observations according to (I) above on a computer. The errors in the ρ and ψ observations are all uncorrelated, the former having variance σ_ρ^2 and the latter σ_ψ^2. Operate a Bayes Filter with iterative differential-correction and compare the time-histories of $S^*_{n,n}$ in the three cases of Examples 11.9, 11.11 and the present one. Verify that with only ρ being observed the improvement is worst when the body is close to the origin and best when far removed, whereas when ψ is observed the situation is reversed. Thus the ρ and ψ observations complement each other very well in the face of "poor geometry" (see Section 8.8).

11.13 Considering the situation of Example 11.8, suppose we observe only range-rate, i.e. $d\rho/dt$.

a) Show that

$$\dot{\rho}_n = \frac{(x + t_n \dot{x})\dot{x}}{\left[d^2 + (x + t_n \dot{x})^2\right]^{1/2}} + \nu_n$$

Linearize this relation about a nominal trajectory to obtain

$$\delta\dot{\rho}_n = M_n(\bar{X})\delta X + \nu_n$$

where M is the 1×2 matrix

$$M_n(\bar{X}) \equiv \frac{1}{k^3}\left[d^2\bar{\dot{x}}, \quad d^2(\bar{x} + 2t_n \bar{\dot{x}}) + (\bar{x} + t_n \bar{\dot{x}})^3\right]$$

with

$$k \equiv \left[d^2 + (\bar{x} + t_n \bar{\dot{x}})^2\right]^{1/2}$$

b) Now show that if we set up the matrix T_n of (8.2.4), that it has full column-rank and hence infer that we can apply a Generalized Fixed-Memory Filter to estimate x and $\dot{x}$ using only range-rate measurements.

c) Consider the first-degree polynomial state-vector $(x, \ \dot{x})^T$ with transition matrix

$$\Phi(\zeta) = \begin{pmatrix} 1 & \zeta \\ 0 & 1 \end{pmatrix}$$

and assume we observe only velocity, i.e.

$$M = (0, \ 1)$$

Verify, either directly or else by the use of Theorem 8.2, that T in this case *does not* have full column-rank. Explain the apparent contradiction between this situation and that of part b) above.

d) Investigate the rank of T_n in b) above as $d \to 0$. Show that in the limit we can determine the entire state-vector if and only if we are observing the body as it passes through the origin. If not, the chosen observation scheme is unsatisfactory.

11.14 Referring to Figure 11.3, assume that the body is moving along the ξ-axis retarded by a drag force which is proportional to its speed. Its equation of motion is

$$\frac{d^2}{dt^2}x(t) = -k\left[\frac{d}{dt}x(t)\right]^2 \qquad k > 0 \tag{I}$$

where k is a known constant.

a) Verify that the solutions to this differential equation are

$$x(t) = \frac{1}{k}\ln\left[kt + \frac{1}{\dot{x}(0)}\right] + x(0) + \frac{1}{k}\ln\dot{x}(0) \tag{II}$$

b) Set (I) above in the form

$$\frac{d}{dt}\begin{pmatrix} x \\ \dot{x} \end{pmatrix}_t = \begin{pmatrix} \dot{x} \\ -k(\dot{x})^2 \end{pmatrix}_t \tag{III}$$

and linearize these equations about a nominal trajectory $\overline{X}(t)$ to obtain

$$\frac{d}{dt}\begin{pmatrix} \delta x \\ \delta\dot{x} \end{pmatrix}_t = A(\overline{X}(t))\begin{pmatrix} \delta x \\ \delta\dot{x} \end{pmatrix}_t$$

where

$$A(\overline{X}(t)) = \begin{pmatrix} 0 & 1 \\ 0 & -2k\dot{\bar{x}}(t) \end{pmatrix} \tag{IV}$$

Verify that the transition matrix for the linearized system is governed by the differential equation

$$\frac{\partial}{\partial \zeta} \Phi(t_n + \zeta, t_n; \overline{X}) = A\left[\overline{X}(t_n + \zeta)\right]\Phi(t_n + \zeta, t_n; \overline{X}) \tag{V}$$

$$\Phi(t_n, t_n; \overline{X}) = I$$

where A is given in (IV) above.

c) As an alternate method for obtaining the transition matrix, verify that (I) above gives us, neglecting terms in ζ^3 and higher,

$$x(t_n + \zeta) = x(t_n) + \zeta \dot{x}(t_n) - \frac{\zeta^2}{2} k\left[\dot{x}(t_n)\right]^2$$

$$\dot{x}(t_n + \zeta) = \dot{x}(t_n) - \zeta k\left[\dot{x}(t_n)\right]^2 + \zeta^2 k^2 \left[\dot{x}(t_n)\right]^3$$

Show that on this basis

$$\Phi(t_n + \zeta, t_n; \overline{X}) = \begin{pmatrix} 1 & \zeta - \zeta^2 k \bar{\dot{x}}(t_n) \\ 0 & 1 - 2\zeta k \bar{\dot{x}}(t_n) + 3\zeta^2 k^2 \left[\bar{\dot{x}}(t_n)\right]^2 \end{pmatrix} \tag{VI}$$

Now show that if ζ is sufficiently small so that terms involving ζ^2 and higher can be neglected, that

$$\Phi(t_n + \zeta, t_n; \overline{X}) = \begin{pmatrix} 1 & \zeta \\ 0 & 1 - 2\zeta k \bar{\dot{x}}(t_n) \end{pmatrix} \tag{VII}$$

d) Cycle (V) of part b) above once by hand, using (4.7.18), and reconcile the result with (VII).

11.15 Assume that the body in Ex. 11.14 is being observed from P (see Figure 11.3), and that both ρ and ψ are being observed. *The distance d is known.* The ρ and ψ observations are all uncorrelated, the former all having variance σ_ρ^2 and the latter σ_ψ^2.

a) Verify that the observation relation is

$$\begin{pmatrix} \rho_n \\ \psi_n \end{pmatrix} = \begin{pmatrix} (d^2 + x_n^2)^{1/2} \\ \arctan \dfrac{x_n}{d} \end{pmatrix} + \begin{pmatrix} \nu_\rho \\ \nu_\psi \end{pmatrix} \tag{I}$$

Linearize this about a nominal trajectory $\overline{X}_n$ and obtain

$$\begin{pmatrix} \delta\rho_n \\ \delta\psi_n \end{pmatrix} = M(\overline{X}_n)\begin{pmatrix} \delta x \\ \delta \dot{x} \end{pmatrix}_n + \begin{pmatrix} \nu_\rho \\ \nu_\psi \end{pmatrix}$$

where M is the 2×2 matrix

$$M(\overline{X}_n) \equiv \begin{pmatrix} \dfrac{x}{(d^2 + x^2)^{1/2}} & 0 \\ \dfrac{d}{d^2 + x^2} & 0 \end{pmatrix} \qquad \text{(II)}$$

b) Assume that we were to smooth the first k measurements of ρ in an Expanding-Memory Polynomial Filter, thereby obtaining the smoothed values $\rho^*_{0,k}$ and $\dot{\rho}^*_{0,k}$ together with their joint covariance matrix $S^*_{0,k}(\rho,\dot{\rho})$. Assume that we do the same with the ψ measurements, obtaining $\psi^*_{0,k}$ and $\dot{\psi}^*_{0,k}$ together with their covariance matrix $S^*_{0,k}(\psi,\dot{\psi})$. Verify that we can estimate x_0 and $\dot{x}_0$ from

$$\begin{aligned} x^*_{0,k} &= (\rho^* \sin\psi^*)_{0,k} \\ \dot{x}^*_{0,k} &= (\dot{\rho}^* \sin\psi^* + \rho^*\dot{\psi}^* \cos\psi^*)_{0,k} \end{aligned} \qquad \text{(III)}$$

c) Now verify that the above can be linearized about $\left(\rho^*,\ \dot{\rho}^*,\ \psi^*,\ \dot{\psi}^*\right)^T_{0,k}$ to give

$$\begin{pmatrix} \delta x^* \\ \delta \dot{x}^* \end{pmatrix}_{0,k} = \Gamma_{0,k}(\rho^*, \dot{\rho}^*, \psi^*, \dot{\psi}^*)\begin{pmatrix} \delta\rho^* \\ \delta\dot{\rho}^* \\ \delta\psi^* \\ \delta\dot{\psi}^* \end{pmatrix}_{0,k} \qquad \text{(IV)}$$

where Γ is the 2×4 matrix

$$\Gamma_{0,k} \equiv \begin{pmatrix} \sin\psi^* & 0 & \rho^* \cos\psi^* & 0 \\ \dot{\psi}^* \cos\psi^* & \sin\psi^* & \theta & \rho^* \cos\psi^* \end{pmatrix}_{0,k} \qquad \text{(V)}$$

in which $\theta \equiv \dot{\rho}^* \cos\psi^* - \rho^*\dot{\psi}^* \sin\psi^*.$

d) Verify that the covariance matrix of the vector $\left(\delta\rho^*,\ \delta\dot{\rho}^*,\ \delta\psi^*,\ \delta\dot{\psi}^*\right)^T_{0,k}$ is given by

$$S^*(\rho,\dot{\rho},\psi,\dot{\psi}) \equiv \left(\begin{array}{c|c} S^*(\rho,\dot{\rho}) & \emptyset \\ \hline \emptyset & S^*(\psi,\dot{\psi}) \end{array}\right) \tag{VI}$$

where the blocks on the right of this equation are obtained from the ρ and ψ polynomial filters respectively.

e) Finally infer from (IV) that the covariance matrix of $\left(x^*_{0,k},\ \dot{x}^*_{0,k}\right)^T$ in (III) can be approximated by

$$S^*_{0,k}(x,\dot{x}) \equiv \Gamma_{0,k} S^*_{0,k}(\rho,\dot{\rho},\psi,\dot{\psi})\Gamma^T_{0,k} \tag{VII}$$

where $\Gamma_{0,k}$ is defined in (V) and $S^*_{0,k}(\rho,\dot{\rho},\psi,\dot{\psi})$ in (VI). Thus (III) and (VII) constitute an estimate state-vector and a covariance matrix for x and $\dot{x}$.

11.16 a) Verify that the differential equation for $\mathring{S}^*(t_{n-1}+\zeta, t_{n-1})$ is

$$\frac{\partial}{\partial\zeta}\mathring{S}^*(t_{n-1}+\zeta, t_{n-1}) \tag{I}$$

$$= A(t_{n-1}+\zeta)\,\mathring{S}^*(t_{n-1}+\zeta, t_{n-1}) + \mathring{S}^*(t_{n-1}+\zeta, t_{n-1})\,A(t_{n-1}+\zeta)^T$$

with given initial conditions $\mathring{S}^*(t_{n-1}, t_{n-1})$.

b) Given that $\mathring{S}^*(t_{n-1}, t_{n-1}) = I\ (2\times 2)$, and that $A(t)$ is as given in Ex. 10.9, integrate the above differential equation (on a computer). Now compute $\mathring{S}^*(t_{n-1}+\zeta, t_{n-1})$ directly from (10.5.9) using $\Phi(t_n+\zeta, t_n)$ obtained in Ex. 10.10. Verify that the results are identical but that the integration of (I) above requires *less* effort than the direct method, if we include in the latter the derivation of Φ from its differential equation.

c) Repeat part a) above to obtain the differential equation satisfied by $\left(\mathring{S}^*\right)^{-1}$ and show how this can be used to replace (11.3.3).

11.17 Program and operate on a computer, a Bayes Filter with iterative differential-correction for the situation of Ex. 11.14 with the observation scheme of Ex. 11.15. Update the estimate state-vector by the integration of (III) in Ex. 11.14. Update the *inverse* of the covariance matrix by integrating the differential equation obtained in part c) of Ex. 11.16 and operate the filter by the method outlined in Section 11.3. To initialize the entire scheme use the polynomial pre-filter discussed in Ex. 11.15.

11.18 Verify that (11.2.23) in the left side of (11.4.23) leads to the right side of (11.4.23).

11.19 Verify that (11.4.23) and (11.2.23) in (11.4.22) results in (11.4.25).

11.20 Verify that (11.4.30) and (11.4.32) in (11.4.31) gives (11.4.33).

11.21 Show that (11.4.34) applied to (11.4.33) leads to (11.4.35).

REFERENCES

1. Gauss, K. F., "Theory of the Motion of the Heavenly Bodies Moving About the Sun in Conic Sections," 1809. English Trans., Dover, Inc., New York, 1963.
2. Swerling, P. A., "A Proposed Stagewise Differential Correction Procedure for Satellite Tracking and Prediction," Rand Corp., Santa Monica, California, Report P-1292, January 8, 1958.
3. Swerling, P., "First Order Error Propagation in a Stagewise Smoothing Procedure for Satellite Observations," Jour. Astronautical Sciences, Vol. 6, Autumn 1959, pp. 46-52.
4. Blackman, R. B., "Methods of Orbit Refinement," Bell Systems Tech. Jour., Vol. 43, May 1964, pp. 885-909.
5. Claus, A. J., Blackman, R. B., Halline, E. G., and Ridgeway, W. C. III, "Orbit Determination and Prediction, and Computer Programs," Telstar, Bell Systems Tech. Jour., Vol. 42, July 1963, pp. 1357-1382.
6. Claus, A. J., "On Systematic Errors in Trajectory Determination Problems," Proc. Internat. Fed. of Auto. Control (IFAC) Instrument Soc. of America, Fall 1965.
7. Claus, A. J., "Orbit Determination for Communications Satellite from Angular Data Only," American Rocket Society, 17th Annual Meeting, November 1962, Los Angeles.
8. Lee, R. C. K., "Optimal Estimation, Identification and Control," Research Mono. 28, Mass. Inst. of Tech. Press, Cambridge, Mass., Chapter 3.
9. Deutsch, R., "Estimation Theory," Prentice-Hall, New Jersey, 1965.

See also reference list at end of Chapter 10.

12

GENERALIZED EXPANDING-MEMORY FILTERS — THE KALMAN FORMULATION

12.1 INTRODUCTION

The Bayes Filter as derived in Chapter 10 provides us with a recursive algorithm for obtaining the linear minimum-variance unbiased estimate of a process, based on a steadily expanding data base. If so desired, available *a priori* knowledge can also be incorporated, according to its merit relative to the subsequent observations.

The only assumption on which the Bayes Filter derivation was based, was that the errors in each batch of observations should be uncorrelated with those in every other batch. Note that almost no restrictions were placed concerning probability density functions. The errors could possess any density function whatever, provided only that their second order statistics exist and that their means be zero.

The requirement that individual batches of observations be uncorrelated with each other is frequently the case in practice, and is almost certain to prevail in a situation such as where many observation instruments are observing a process. A typical example of this might be a global tracking network observing a satellite in orbit. Radars located around the globe each observe the body while it is in view, and then feed their respective batches of data to a central processor where the orbit refinement computations are undertaken.

We have also shown how the Bayes algorithm, derived on the assumption of complete linearity, can be used as the corner-stone of an iterative differential-correction scheme. This permits of estimation when nonlinear relationships exist, and in practice these are certainly the most likely ones to be encountered.

Regardless of the order of the observation vector, at least two matrices must always be inverted when cycling the Bayes Filter.† The first of these is $R_{(n)}$ and when this is diagonal, as is often the case in practice, its inversion is trivial. The other matrix to be inverted is $\left(\mathring{S}^*_{n,n}\right)^{-1}$ which is of the same order as the state-vector, and is seldom if ever diagonal.

We now show how the Bayes equations can be reformulated so that *only one* matrix must be inverted, *of the order of the observation vector.* For the situation where the state-vector of the estimate has, say, eight elements, and only three observations are to be incorporated, this means that only one 3×3 inversion is called for, a decided advantage over the corresponding Bayes case in which an 8×8 would have to be inverted.

The filter equations we are about to derive are well suited to the following type of situations. Suppose first that the observation instrument requires assistance in order to keep an object under observation in a narrow field of view, e.g. a radar operating in a tracking mode. Clearly as soon as one measurement vector is made (such as for example range and two angles) we would want to incorporate it into the estimate in order to be better able to predict where next to look for the object. As a second example we consider the case where observations are being made infrequently because they are time-consuming and very costly, e.g. astronauts in a space-ship making a position fix. We would certainly want to incorporate those numbers into the estimate as soon as we obtain them, primarily in order to be able to make immediate corrective action.

†See Section 11.3 where we discussed a modification to the Bayes Filter in which only $R_{(n)}$ and $\left(\mathring{S}^*_{n,n}\right)^{-1}$ had to be inverted.

The filter algorithms which we now derive are intended to be used under circumstances similar to the above. As with the Bayes Filter, we first obtain the equations assuming complete linearity. This provides us with the basic algorithm on which we construct a more widely applicable iterative differential-correction scheme for use in nonlinear cases. We also examine some of the computational and practical aspects of the resulting algorithms.

The present chapter is a direct extension of the previous two, and complete familiarity will be assumed with that material.

12.2 THE KALMAN FILTER – LINEAR CASE WITH NO DRIVING-NOISE†

The Bayes Filter shown on p. 386 was derived as the algorithm which provides the composite minimum-variance estimate of X_n, based on the two relations (see (10.2.21) and (10.2.22))

$$\overset{\circ}{X}{}^*_{n,n-1} = X_n + \overset{\circ}{N}{}^*_{n,n-1} \tag{12.2.1}$$

$$Y_{(n)} = T_n X_n + N_{(n)} \tag{12.2.2}$$

If all of the observations in $Y_{(n)}$ were to be made *simultaneously*, then we can write

$$T_n = M_n \tag{12.2.3}$$

and so (12.2.2) becomes simply

$$Y_n = M_n X_n + N_n \tag{12.2.4}$$

Let the covariance matrix of N_n be R_n and assume, as we have done up to now, that N_n is uncorrelated with the error vectors of all other observation vectors to be presented to the filter. Under these assumptions the Bayes Filter based on (12.2.1) and (12.2.4) takes on the form shown on p. 389, which we repeat on the next page for convenience.

A brief examination of those equations shows that, aside from R_n, two matrices of the order of the state-vector must be inverted, namely $\overset{\circ}{S}{}^*_{n,n-1}$ and $\left(\overset{\circ}{S}{}^*_{n,n}\right)^{-1}$. If we are prepared to replace (12.2.5) by a numerical integration of the process differential equations, then we know from Section 11.3, that the

†Refer to Chapter 15 where we discuss precisely what we mean by the term "driving-noise."

Bayes Filter – Concurrent Observations

$$\overset{\circ}{X}{}^*_{n,n-1} = \Phi(n, n-1)\overset{\circ}{X}{}^*_{n-1,n-1} \tag{12.2.5}$$

$$\overset{\circ}{S}{}^*_{n,n-1} = \Phi(n, n-1)\overset{\circ}{S}{}^*_{n-1,n-1}\Phi(n, n-1)^T \tag{12.2.6}$$

$$\overset{\circ}{S}{}^*_{n,n} = \left[\left(\overset{\circ}{S}{}^*_{n,n-1}\right)^{-1} + M_n{}^T R_n{}^{-1} M_n\right]^{-1} \tag{12.2.7}$$

$$\overset{\circ}{H}_n = \overset{\circ}{S}{}^*_{n,n} M_n{}^T R_n{}^{-1} \tag{12.2.8}$$

$$\overset{\circ}{X}{}^*_{n,n} = \overset{\circ}{X}{}^*_{n,n-1} + \overset{\circ}{H}_n\left(Y_n - M_n\overset{\circ}{X}{}^*_{n,n-1}\right) \tag{12.2.9}$$

---

need to invert $\overset{\circ}{S}{}^*_{n,n-1}$ in (12.2.7) can be entirely avoided. However even then, $\overset{\circ}{S}{}^*_{n,n}$ must still be computed, and since the order of $\left(\overset{\circ}{S}{}^*_{n,n}\right)^{-1}$ is the same as that of the process state-vector, its inversion could be computationally costly. Adding to this the cost of inverting R_n in case it were not diagonal we begin to wonder whether a more economical algorithm might not be found. As a first step in that direction we now show that (12.2.7) and (12.2.8) can be replaced by two equivalent expressions.

The reader will recall that in Section 10.5 we proved a result known as the *inversion lemma*. Thus, referring to (10.5.4), we see that if we replace T by M and R by R, then we can write

$$\begin{aligned}&\left(\overset{\circ}{S}{}^{*-1}_{n,n-1} + M_n{}^T R_n{}^{-1} M_n\right)^{-1} \\ &= \overset{\circ}{S}{}^*_{n,n-1} - \overset{\circ}{S}{}^*_{n,n-1} M_n{}^T\left(R_n + M_n\overset{\circ}{S}{}^*_{n,n-1} M_n{}^T\right)^{-1} M_n\overset{\circ}{S}{}^*_{n,n-1}\end{aligned} \tag{12.2.10}$$

We now post-multiply both sides by $M_n{}^T R_n{}^{-1}$ and obtain (after deleting subscripts and stars)

$$(S^{-1} + M^T R^{-1} M)^{-1} M^T R^{-1} = \left[S - SM^T(R + MSM^T)^{-1} MS\right] M^T R^{-1} \tag{12.2.11}$$

By direct expansion and a small manipulation on the right-hand side (see Ex. 12.2) this reduces to

$$(S^{-1} + M^T R^{-1} M)^{-1} M^T R^{-1} = SM^T(R + MSM^T)^{-1} \tag{12.2.12}$$

and if we now make use of (12.2.7) in (12.2.12), then we obtain

$$\mathring{S}^*_{n,n} M_n^T R_n^{-1} = \mathring{S}^*_{n,n-1} M_n^T \left(R_n + M_n \mathring{S}^*_{n,n-1} M_n^T\right)^{-1} \tag{12.2.13}$$

The right-hand side of this expression thus gives us an alternate expression for $\mathring{H}_n$ in (12.2.8). Using the symbol $\mathring{H}_n$ for the term on the right of (12.2.13) now means that (12.2.10) can be written as

$$\left(\mathring{S}^{*-1}_{n,n-1} + M_n^T R_n^{-1} M_n\right)^{-1} = \left(I - \mathring{H}_n M_n\right) \mathring{S}^*_{n,n-1} \tag{12.2.14}$$

which gives us an alternate expression for $\mathring{S}^*_{n,n}$ in (12.2.7). The Bayes Filter of p. 464 thus gives us the following *algebraically equivalent* algorithm:

Kalman Filter (No Driving-Noise)

$$\mathring{X}^*_{n,n-1} = \Phi(n, n-1)\mathring{X}^*_{n-1,n-1} \tag{12.2.15}$$

$$\mathring{S}^*_{n,n-1} = \Phi(n, n-1)\mathring{S}^*_{n-1,n-1}\Phi(n, n-1)^T \tag{12.2.16}$$

$$\mathring{H}_n = \mathring{S}^*_{n,n-1} M_n^T \left(R_n + M_n \mathring{S}^*_{n,n-1} M_n^T\right)^{-1} \tag{12.2.17}$$

$$\mathring{S}^*_{n,n} = \left(I - \mathring{H}_n M_n\right)\mathring{S}^*_{n,n-1} \tag{12.2.18}$$

$$\mathring{X}^*_{n,n} = \mathring{X}^*_{n,n-1} + \mathring{H}_n\left(Y_n - M_n \mathring{X}^*_{n,n-1}\right) \tag{12.2.19}$$

This set of equations can be shown to be equivalent to those obtained by Kalman in [12.1]† *when driving-noise in the process is omitted.* It is for this reason that they constitute what is generally termed the Kalman Filter without driving-noise. In Chapter 15 we show that when driving-noise *is* present then only (12.2.16) is affected. Our approach to the derivation of the Kalman Filter differs markedly from that used in [12.1] and is possible only when driving-noise is absent.

The algorithm developed above, will, in the case of a polynomial model and equally spaced observations on the zeroth derivative of the process, be equivalent to the Expanding-Memory Polynomial Filter developed in Chapter 9. This is discussed further in Ex. 12.3, where we show how the Bayes Filter on p. 464 can be used to derive the general form of the weight-vector H_n given in (9.4.30) for the filters of Chapter 9.

†See also [12.2], [12.3] and [12.4].

As a conclusion to this section, we note from (12.2.7) and (12.2.10) that, in the Kalman Filter, the matrix $\mathring{S}^*_{n,n}$ is effectively being obtained from the relation

$$\mathring{S}^*_{n,n} = \mathring{S}^*_{n,n-1} - \mathring{S}^*_{n,n-1} M_n^T \left(R_n + M_n \mathring{S}^*_{n,n-1} M_n^T\right)^{-1} M_n \mathring{S}^*_{n,n-1} \tag{12.2.20}$$

While this form is *algebraically* equivalent to (12.2.7) by the inversion lemma, we shall see presently that there are *strong computational differences* between them which serve to differentiate the Bayes and Kalman Filters very clearly from each other.

We now examine the properties of the Kalman Filter and, as a start, we re-examine some of the important criteria which both it and the Bayes Filter satisfy.

12.3 CRITERIA SATISFIED BY THE BAYES AND KALMAN FILTERS

The Kalman Filter, as defined on p. 465, was derived from the Bayes Filter by the application of the inversion lemma. That being the case, they must thus be algebraically equivalent.

We recall, first, that in Chapter 10 the Bayes Filter was shown to be simply a recursive reformulation of the minimum-variance rule derived in Chapter 6, assuming that the observation errors are stage-wise uncorrelated. *It thus follows that the Kalman Filter also gives the minimum-variance estimate.*

Consider next the predictor-corrector pair

$$X^*_{n,n-1} = \Phi(n, n-1) X^*_{n-1,n-1} \tag{12.3.1}$$

$$X^*_{n,n} = X^*_{n,n-1} + H_n (Y_n - M_n X^*_{n,n-1}) \tag{12.3.2}$$

where H_n is *any* matrix of the same dimensions as $\mathring{H}_n$ of (12.2.8) or (12.2.17). It is readily verified that the above pair yields an unbiased estimate $X^*_{n,n}$ for *any* such H_n (see Ex. 12.4). However only when H_n is the same as $\mathring{H}_n$, is the covariance matrix of the errors in $X^*_{n,n}$ minimum-variance.

Consider the form of the covariance matrix of the errors in $X^*_{n,n}$ of (12.3.2). Recalling that

$$X^*_{n,n} = X_n + N^*_{n,n} \tag{12.3.3}$$

$$X^*_{n,n-1} = X_n + N^*_{n,n-1} \tag{12.3.4}$$

and

$$Y_n = M_n X_n + N_n \tag{12.3.5}$$

we are able to rewrite (12.3.2) as (see Ex. 12.5)

$$N^*_{n,n} = (I - H_n M_n) N^*_{n,n-1} + H_n N_n \tag{12.3.6}$$

We now form the covariance matrix of $N^*_{n,n}$ and since N_n was assumed to be uncorrelated with its predecessors we obtain (see Ex. 12.5)

$$S^*(H_n) = (I - H_n M_n) S^*_{n,n-1} (I - H_n M_n)^T + H_n R_n H_n^T \tag{12.3.7}$$

Observe that S^* is shown as a matrix function of H_n.

Now we know that the Bayes and Kalman algorithms yield an estimate whose error covariance matrix, as given by (12.3.7), is minimum-variance. Thus the diagonal elements of the covariance matrices $S^*(\mathring{H}_n)$ and $S^*(H_n)$ satisfy

$$\left[S^*(\mathring{H}_n)\right]_{i,i} \le \left[S^*(H_n)\right]_{i,i} \tag{12.3.8}$$

where $\mathring{H}_n$ is given by (12.2.17) and H_n is *any* other matrix of the same order. Since the diagonal elements of $S^*(\mathring{H}_n)$ are *individually* least, then clearly their *sum* is also minimal, i.e.

$$\sum_{i=0}^{m} \left[S^*(\mathring{H}_n)\right]_{i,i} \le \sum_{i=0}^{m} \left[S^*(H_n)\right]_{i,i} \tag{12.3.9}$$

But the diagonal elements of the covariance matrix are precisely the *variances of the estimation errors,* and so the above is equivalent to the inequality

$$E\left\{\mathring{N}^{*T}_{n,n} \mathring{N}^*_{n,n}\right\} \le E\left\{N^{*T}_{n,n} N^*_{n,n}\right\} \tag{12.3.10}$$

where $\overset{\circ}{N}{}^*_{n,n}$ and $N^*_{n,n}$ are the error vectors obtained respectively when $\overset{\circ}{H}_n$ of (12.2.17) and any other H_n are used. Thus the fact that the Bayes and Kalman Filters are minimum-variance, means that they also minimize the expected value of $N^{*T}_{n,n}N^*_{n,n}$ i.e. *the sum of squares of the error in each element of the estimate.*†

The path that we followed in obtaining the Kalman Filter started from an application of Lagrange's method of undetermined multipliers using the exactness constraint in Chapter 6. This led to the minimum-variance rule which was later reformulated into the recursive Bayes Filter. Finally, by the use of the inversion lemma we obtained the Kalman Filter. A much more direct method would be to start from (12.3.7) *and then to choose* H_n *so that each of the diagonal elements of* $S^*(H_n)$ *is least.* Not surprisingly the H_n which accomplishes this is precisely $\overset{\circ}{H}_n$ of (12.2.17). The reader is referred to Ex. 12.6.‡

A further generalization of the foregoing is also possible. Let Γ be *positive definite* and define the quadratic form

$$e \equiv N^{*T}_{n,n}\Gamma N^*_{n,n} \tag{12.3.11}$$

Then the Bayes and Kalman Filters minimize $E\{e\}$ *for any such* Γ. This is also a direct consequence of the fact that their estimates are minimum-variance (see Ex. 12.7).

The above result shows us that a linear estimator whose weight-matrix H_n causes $E\left\{N^{*T}_{n,n}N^*_{n,n}\right\}$ to be minimal, would also minimize $E\left\{N^{*T}_{n,n}\Gamma N^*_{n,n}\right\}$. Hence, for example, if Γ were diagonal then

$$E\left\{N^{*T}_{n,n}\Gamma N^*_{n,n}\right\} = \sum_{i=0}^{m} \gamma_{ii} E\left\{\left(\nu^*_{n,n}\right)^2_{i,i}\right\} \tag{12.3.12}$$

which is seen to be a *weighted* sum of the error-variances. (The weights are of course all positive since Γ is positive definite.) Thus if the filter minimizes the sum of the error-variances then it also minimizes *any positively weighted* sum of those error-variances. (See also Ex. 12.8.)

†The scalar $E\left\{N^{*T}_{n,n}N^*_{n,n}\right\}$ is frequently termed the *expected quadratic loss.* It is the minimization of such a loss-function which is the *starting-point* of the derivation in [12.1]. This will be discussed further in Chapter 15.

‡The method used in Ex. 12.6 is purely algebraic and does not make any use of the differential calculus. It is accordingly a useful extension of our existing techniques developed earlier in this book.

We recall next, from Chapter 6 (see Section 6.9), that the minimum-variance estimate also satisfies the *weighted least-squares* criterion, i.e. it minimizes

$$e(X^*_{n,n}) \equiv \left[E(X^*_{n,n})\right]^T R^{-1}_{(n)} E(X^*_{n,n}) \tag{12.3.13}$$

where $E(X^*_{n,n})$ is the residual vector based on $X^*_{n,n}$. Suppose that we now set up the residual vector

$$E(X^*_{n,n}) = \begin{pmatrix} \mathring{X}^*_{n,n-1} - X^*_{n,n} \\ \hline Y_n - M_n X^*_{n,n} \end{pmatrix} \tag{12.3.14}$$

in which $\mathring{X}^*_{n,n-1}$ is a prediction of X_n, Y_n is an observation of $M_n X_n$ and $X^*_{n,n}$ is an estimate of X_n. As our covariance matrix we use

$$R_{(n)} \equiv \left(\begin{array}{c|c} \mathring{S}^*_{n,n-1} & \emptyset \\ \hline \emptyset & R_n \end{array}\right) \tag{12.3.15}$$

Then it is readily shown (see Ex. 12.9) that (12.3.13) leads to

$$e(X^*_{n,n}) = \left(\mathring{X}^*_{n,n-1} - X^*_{n,n}\right)^T \mathring{S}^{*-1}_{n,n-1}\left(\mathring{X}^*_{n,n-1} - X^*_{n,n}\right) \tag{12.3.16}$$
$$+ (Y_n - M_n X^*_{n,n})^T R_n^{-1} (Y_n - M_n X^*_{n,n})$$

and if we now apply the differential calculus to the minimization of the above $e(X^*_{n,n})$ over $X^*_{n,n}$, then the Bayes and Kalman filters are the direct outcomes (see Ex. 12.9). They are, in fact, *simply a recursive formulation of the minimum-variance rule for combining an a priori estimate with an observation vector.*

The error function of (12.3.16) is seen to be a sum of quadratic forms on the inverses of the covariance matrices of the observation and *a priori* state-vectors. These inverse matrices serve two purposes.

First we know from Section 6.8 that whereas the least-squares estimate depends on the units in which the observed quantities are expressed, the minimum-variance estimate is unique. The observation vectors can be of mixed dimensions, and regardless of the units used, by virtue of the presence of $\overset{\circ}{S}{}^{*-1}_{n,n-1}$ and R_n^{-1} in (12.3.16), the resultant estimate $\overset{\circ}{X}{}^{*}_{n,n}$ is unchanged. Those matrices thus serve as *normalizers* and permit us to mix any types of observations which we may obtain, and to express them in any convenient units.

In addition to serving the above function, the matrices $\overset{\circ}{S}{}^{*}_{n,n-1}$ and R_n^{-1} also serve as *weight matrices.* We desire, quite naturally, that the more error-free an observation is, the more heavily it should influence the estimate. This is precisely what will occur in the minimization of (12.3.16).

Finally we consider the following. If the errors in the vectors $\overset{\circ}{X}{}^{*}_{n,n-1}$ and Y_n of (12.3.14) are Gaussian, then the joint probability density function of the elements of those vectors is (c/f (5.6.3))

$$p\left(\overset{\circ}{X}{}^{*}_{n,n-1}, Y_n\right)$$

$$= K \exp\left\{-\frac{1}{2}\begin{pmatrix}\overset{\circ}{X}{}^{*}_{n,n-1} - X_n \\ \hline Y_n - M_n X_n\end{pmatrix}^T \left(\begin{array}{c|c}\overset{\circ}{S}{}^{*}_{n,n-1} & \emptyset \\ \hline \emptyset & R_n\end{array}\right)^{-1} \begin{pmatrix}\overset{\circ}{X}{}^{*}_{n,n-1} - X_n \\ \hline Y_n - M_n X_n\end{pmatrix}\right\} \tag{12.3.17}$$

where K is a normalizing constant. The Maximum-Likelihood Principle (see Ex. 6.19) states that X_n shall be estimated by the vector $X^*_{n,n}$ which maximizes the scalar p of (12.3.17). To do this we must minimize the exponent and this is the same as minimizing (12.3.16). Thus the Bayes and Kalman Filters are *also* the maximum-likelihood estimators based on an *a priori* estimate and an observation vector, when the errors are stage-wise uncorrelated and Gaussian.

We now turn our attention to some of the computational aspects of the Kalman Filter without driving-noise.

12.4 PROPERTIES OF THE KALMAN FILTER WITHOUT DRIVING-NOISE

In the normal course of events the Kalman Filter provides us with a very powerful way of obtaining the linear unbiased minimum-variance estimate on a steadily expanding data-base. Computationally it has certain advantages over the Bayes formulation, but these do not come without a price. We now turn our attention to a comparison of the two approaches and show that each has certain advantages over the other.

In the Kalman formulation *only one matrix inversion is called for,* namely the one which appears in the computation of (12.2.17). Clearly that matrix is of the same order as the observation vector. The Bayes Filter, on the other hand, requires the inversion of R_n as well as at least one matrix of the order of the state-vector (i.e. the inversion of $\left(\overset{\circ}{S}{}^*_{n,n}\right)^{-1}$). Thus it is clear that the Kalman Filter has a distinct computational advantage in this regard, particularly when the observation vector is of smaller order than the state-vector, as is very often the case. However there are other very important factors that enter into the choice, and as we now show, in some respects the Bayes formulation possesses certain definite advantages over the Kalman. It will have to be left to the user to weigh these aspects against each other in making a selection for each given application.

Consider first the problem of initialization. In the Bayes case we can, if we so choose,† select the initial vector $\overset{\circ}{X}{}^*_{1,0}$ completely arbitrarily, and then by setting

$$\left(\overset{\circ}{S}{}^*_{1,0}\right)^{-1} = \emptyset \tag{12.4.1}$$

in (12.2.7) we make the subsequent estimates *totally* independent of $\overset{\circ}{X}{}^*_{1,0}$. Thus the Bayes filter can be started with no *a priori* information whatever.‡ *The same thing cannot be done in the Kalman case.*

Thus suppose we wish to apply the counterpart of (12.4.1) to the Kalman Filter. Examination of (12.2.17) shows that only $\overset{\circ}{S}{}^*_{1,0}$ and *not* its inverse occurs. We might try using

$$\overset{\circ}{S}{}^*_{1,0} = \infty I \tag{12.4.2}$$

where by ∞ we mean a very large scalar, perhaps the largest that can be carried by the machine in which the computations are being carried out. Alternatively we could artificially set $R_1 = \emptyset$ and $\overset{\circ}{S}{}^*_{1,0} = I$, say. In either case, the matrix R disappears from (12.2.17), either by round-off or else by choice, and we obtain

$$\overset{\circ}{H}_1 = M_1^T\left(M_1 M_1^T\right)^{-1} \tag{12.4.3}$$

In order for $\overset{\circ}{H}_1$ to exist, the matrix $M_1 M_1^T$ must be nonsingular.

†See Section 10.3.
‡Assuming, of course, sufficient rank in the observation matrix.

Suppose first that M_1 has *more* rows than columns. Then $M_1 M_1^T$ is singular and so (12.4.3) is undefined. The filter cannot be initialized in this way.

Suppose next that M_1 has *fewer* rows than columns, and has full row-rank. Then $M_1 M_1^T$ is nonsingular. However we are now trying to estimate a state-vector of given order solely from an observation vector of smaller order. Although the filter will now be operable, *it will be found that the initializing vector has entered into the first estimate.* Again we have been unable to start up without dependence on the *a priori* vector. (See Ex. 12.11 and 12.12.)

Finally, consider the situation where M is square and nonsingular. Only in this case can the filter be started successfully so that the *a priori* vector exerts no influence. However as we now see, a further complication arises regardless of the properties of M, and so even this third case is invalidated. *In all cases then, the Kalman Filter cannot be started without the initializing vector appearing in the subsequent estimates.* This is in marked contrast to the Bayes case.

The complication mentioned above is as follows. By virtue of the fact that R_1 does not appear in (12.4.3) *we have, as we now show, inadvertently made the covariance matrix* $\mathring{S}^*_{1,1}$ *singular.* This is a serious difficulty and will thereafter make the filter virtually unworkable.

To see what effect the loss of R_1 has had, we examine the properties of the matrix (c/f (12.2.20))

$$C \equiv S - SM^T(R + MSM^T)^{-1}MS \tag{12.4.4}$$

when R is set to a *null matrix.* Thus we consider

$$C \equiv S - SM^T(MSM^T)^{-1}MS \tag{12.4.5}$$

If we post-multiply by M^T we obtain immediately

$$CM^T = \emptyset \tag{12.4.6}$$

This is true without either C or M being null, and so *when R is null C is singular.* That being the case $\mathring{S}^*_{1,1}$ of (12.2.20) will be singular. On the next cycle of the filter we use (12.2.16) to compute

$$\mathring{S}^*_{2,1} = \Phi(2,1)\mathring{S}^*_{1,1}\Phi(2,1)^T \tag{12.4.7}$$

and so $\mathring{S}^*_{2,1}$ *will be singular.* Then by (12.2.20) we obtain

$$\overset{\circ}{S}{}^*_{2,2} = \overset{\circ}{S}{}^*_{2,1} - \overset{\circ}{S}{}^*_{2,1} M_2{}^T \left(R_2 + M_2 \overset{\circ}{S}{}^*_{2,1} M_2{}^T\right)^{-1} M_2 \overset{\circ}{S}{}^*_{2,1} \tag{12.4.8}$$

But $\overset{\circ}{S}{}^*_{2,1}$ being singular implies the existence of one or more independent nonzero vectors, $\xi_1, \xi_2, \ldots, \xi_k$, such that

$$\overset{\circ}{S}{}^*_{2,1} \xi_i = \emptyset \qquad 1 \leq i \leq k \tag{12.4.9}$$

Hence by (12.4.8) $\overset{\circ}{S}{}^*_{2,2}$ will be singular and by induction, so will $\overset{\circ}{S}{}^*_{3,3} \ldots \overset{\circ}{S}{}^*_{n,n} \ldots$. In fact if once the covariance matrix of the Kalman filter becomes singular *it will thereafter remain permanently singular.*†

The fact that $\overset{\circ}{S}{}^*_{n,n}$ becomes singular and remains singular thereafter would, in itself, be no serious drawback. We do not require its inverse in the Kalman case. However in practice, owing to the fact that we are working with finite precision arithmetic, we must recognize that the matrix can now become *indefinite* (i.e. some eigenvalues positive and some negative). This could, in turn, lead to negative diagonal elements, (i.e. estimates with "negative variances") and to "correlation coefficients" in absolute excess of unity. Either of these is sufficient to make the filter algorithm untenable, and we are thus forced to conclude that it is not possible to make the subsequent behavior of the filter independent of the *a priori* estimate as we could in the Bayes case. *Thus the Bayes Filter can be started with no a priori data but the Kalman Filter cannot.* Much greater care must then be taken, in the Kalman case, in choosing the *a priori* estimate, and the elements of its covariance matrix should never be so large that R_1 is lost by roundoff in the formation of (12.2.17) when the filter is first cycled. (In the absence of reliable *a priori* information, the initializing estimate should be obtained by a fixed-memory estimator operating on the first few observations. If tracking assistance is essential, an Expanding-Memory Polynomial Filter of the type discussed in Chapter 9 can be used, and if the data-rate is kept at a sufficiently high level, the systematic errors can be kept manageably small until an initializing estimate for the Kalman Filter has been formed.)

The troubles encountered above arose because we supposed that R_1 had disappeared entirely in (12.2.17). This made the covariance matrix $\overset{\circ}{S}{}^*_{1,1}$ singular. We should also recognize, both here and in the discussion to follow, that even if the covariance matrix is made *almost* singular due to a partial rather than a total effect taking place, that the filter is still placed in

†This situation is true only when there is no driving-noise present and, as we shall show in the final chapter, the assumption of driving-noise can constantly renew the positive definiteness of the covariance matrix $\overset{\circ}{S}{}^*_{n,n}$.

jeopardy. An *almost* singular covariance matrix could, sooner or later, degenerate into an indefinite one and begin showing negative diagonal elements or off-diagonal terms which imply correlation coefficients in excess of unity.

We consider next the effects of incorporating a vector of *highly precise* data. Thus assume that for some n,

$$R_n = \epsilon Q_n \tag{12.4.10}$$

where ϵ is a small scalar and Q_n a positive definite matrix. Again by (12.2.17) we compute

$$\mathring{H}_n = \mathring{S}^*_{n,n-1} M_n^T \left(\epsilon Q_n + M_n \mathring{S}^*_{n,n-1} M_n^T \right)^{-1} \tag{12.4.11}$$

but then for ϵ sufficiently small we see that the matrix ϵQ_n could be lost by round-off, giving us

$$\mathring{H}_n \approx \mathring{S}^*_{n,n-1} M_n^T \left(M_n \mathring{S}^*_{n,n-1} M_n^T \right)^{-1} \tag{12.4.12}$$

It is possible that the matrix to be inverted on the right of this equation is nonsingular and that $\mathring{H}_n$ does exist. However we now see that, just as in the case of the initialization problems discussed above, *we again run into trouble with the properties of* $\mathring{S}^*_{n,n}$. For, by (12.2.20), the latter will be computed as

$$\mathring{S}^*_{n,n} = \mathring{S}^*_{n,n-1} - \mathring{S}^*_{n,n-1} M_n^T \left(\epsilon Q_n + M_n \mathring{S}^*_{n,n-1} M_n^T \right)^{-1} M_n \mathring{S}^*_{n,n-1} \tag{12.4.13}$$

and then if the matrix ϵQ_n is lost by round-off, this again means that $\mathring{S}^*_{n,n}$ is singular, *remaining singular thereafter for all subsequent recursion cycles.* If ϵQ_n is not wiped out entirely by roundoff, the matrix $\mathring{S}^*_{n,n}$ will be positive definite but only marginally so. Small subsequent roundoff errors could then serve to make it indefinite, and this could again give rise to "negative variances" and to "correlation coefficients" in absolute excess of unity. Thus the Kalman Filter runs into trouble in trying to accept observations which are too precise.

In the Bayes case on the other hand, we recall from Section 10.6 that troubles can be encountered when extremely precise data are incorporated, but only if the observation matrix (M or T) is of insufficient rank. When the highly precise observation vector is related to the true state-vector by an observation matrix *with full column-rank,* then no problems arise and the Bayes Filter simply forms a new estimate based almost exclusively on the new data.

One possible cure to the above problem would be to prevent R_n from being lost in the round-off process by multiplying it by a sufficiently large positive scalar. This artificial increase will, of course, result in a nonminimum-variance estimate, but that is a small price to pay for the ability to keep $\mathring{S}^*_{n,n}$ from becoming singular.

Next we consider the case of *perfectly correlated* observation errors, i.e. R_n positive semi-definite.† *Specifically let* R_n *be* 3×3 *and assume that it has a rank-defect of* 2, i.e. *rank unity. Let S be* 6×6. *Then show that*

$$C \equiv S - SM^T(R + MSM^T)^{-1} MS \tag{12.4.14}$$

has the same rank-defect as R.

Post-multiplying by M^T, which by assumption has full column-rank, we get

$$\begin{aligned} CM^T &= SM^T - SM^T(R + MSM^T)^{-1} MSM^T \\ &= SM^T - SM^T(R + MSM^T)^{-1}(MSM^T + R - R) \end{aligned} \tag{12.4.15}$$

which reduces to

$$CM^T = SM^T(R + MSM^T)^{-1} R \tag{12.4.16}$$

Now R has a rank-defect of 2 and so there are two linearly independent nonzero 3-vectors, ξ_1 and ξ_2, for which

$$R\xi_1 = \emptyset \qquad R\xi_2 = \emptyset \tag{12.4.17}$$

That being the case (12.4.16) gives us

$$CM^T\xi_1 = \emptyset \quad \text{and} \quad CM^T\xi_2 = \emptyset \tag{12.4.18}$$

But $M^T\xi_1$ and $M^T\xi_2$ cannot be null vectors since by assumption M^T has rank 3, and so its rows span the 3-space in which ξ_1 and ξ_2 are defined. *Thus C must have a rank-defect of* 2. The proof is easily extended to show that C has the same rank-defect as R in all cases. This completes the proof. ◆◆

†See p. 137.

The above argument thus generalizes our previous conclusion that when R_n is null (i.e. defect equal to its entire order), then $\overset{\circ}{S}{}^*_{n,n}$ of (12.2.20) is singular, and that when R_n is positive definite (defect zero), then $\overset{\circ}{S}{}^*_{n,n}$ is also positive definite. We see now that if a vector of observations is obtained with perfectly correlated errors, i.e. R_n singular with a certain rank-defect, *then that perfect correlation will immediately appear in the random output errors of the filter* in the form of a singular covariance matrix $\overset{\circ}{S}{}^*_{n,n}$, with corresponding rank-defect. Likewise if one of the observations in Y_n is assumed to be perfect, i.e. R_n has one zero row and column, then $\overset{\circ}{S}{}^*_{n,n}$ will be singular and will have a rank-defect of one.† *Moreover, once this rank-defect has been introduced into* $\overset{\circ}{S}{}^*_{n,n}$ *it will remain there permanently,* a fact which we have already proved earlier in this section. (See Ex. 12.13 and 12.14.) It is thus clear that if R_n is precisely singular, then $\overset{\circ}{S}{}^*_{n,n}$ will be likewise. This means that the algorithm is thereafter of marginal value and potentially in serious trouble, even though no overt computational problems arise.

The Bayes Filter again has the advantage here, since if R_n is precisely singular we simply cannot execute the Bayes algorithm because R_n^{-1} is required in (12.2.7). We are thus *immediately* confronted with this fact and steps can be taken then and there to correct the situation. This is almost certainly preferable to the Kalman case in which the cycling of the filter would proceed, but with the user being unaware of the existence of trouble until it is too late to correct matters.

In the event that R_n were *almost* singular, i.e. badly conditioned for inversion, a similar argument to the above demonstrates that the Bayes Filter would again be preferable to the Kalman Filter.

In all cases considered thus far, we saw that an abrupt degeneration could occur in $\overset{\circ}{S}{}^*_{n,n}$ which thereafter makes the value of the Kalman Filter questionable. We now point out a further serious problem which can and very frequently does arise with the Kalman Filter, and this consists of a *loss of precision* in the elements of $\overset{\circ}{S}{}^*_{n,n}$ in many cases of practical interest. The mechanism is as follows.

In Section 10.5 we pointed out that with a well-configured filtering scheme, the elements of $\overset{\circ}{S}{}^*_{n,n}$ will, in general, be shrinking monotonically as time progresses. In the Bayẽs case that shrinkage comes about as a result of executing (12.2.7) which amounts, essentially, to a shrinkage *by addition followed by inversion.* The shrinkage mechanism of (12.2.7) is analogous to the following: If a_n and b_n are positive and if

† See p. 137.

$$a_{n+1} \equiv \left(\frac{1}{a_n} + \frac{1}{b_n} \right)^{-1} \tag{12.4.19}$$

then a_{n+1} is less than a_n. Repeated cycling of this algorithm, using floating-point finite-precision arithmetic, *will not lead to a loss of precision in the answer.* The Bayes Filter thus encounters no problems in this respect.

However in the Kalman case, $\mathring{S}^*_{n,n}$ is shrunk by *repeated subtractions* as we see from (12.2.18), and as is well known, *this invariably does lead to a loss of precision.* Thus, consider the following trivial example: Let

$$a_{n+1} = a_n - b_n \tag{12.4.20}$$

where

$$a_n = 1.2165 \times 10^4 \qquad b_n = 1.2163 \times 10^4 \tag{12.4.21}$$

Then $a_{n+1} = 2 \times 10^0$. The data a_n and b_n were of *five*-digit accuracy whereas the answer of course has only *single*-digit accuracy, *a loss of four orders of precision.* (See Ex. 12.16.)

One way of circumventing this problem in the Kalman Filter would naturally be to use extended-precision arithmetic. Computationally however this increases the burden of cycling the equations. A compromise alternative could be to reconcile ourselves to a loss of precision in $\mathring{S}^*_{n,n}$, but to insist that it at least remain a covariance matrix, i.e. positive definite. This can be accomplished if we write $\mathring{S}^*_{n,n}$ as the product

$$\mathring{S}^*_{n,n} = K_n K_n^T \tag{12.4.22}$$

and thereafter reorganize the Kalman Filter so that K_n rather than $\mathring{S}^*_{n,n}$ is computed. If the latter is ever required, we obtain it from (12.4.22), and then, provided that K_n is nonsingular, $\mathring{S}^*_{n,n}$ will be positive definite. Such a procedure is discussed in [12.4].

An even better method is to insist that K_n be the *square-root* of $\mathring{S}^*_{n,n}$, i.e.

$$\mathring{S}^*_{n,n} = K_n K_n \tag{12.4.23}$$

and thereafter again to work with K_n rather than $\mathring{S}^*_{n,n}$. Then, as is shown in [12.5], both the objective of keeping $\mathring{S}^*_{n,n}$ positive definite as well as that of avoiding a loss of precision can be accomplished. (In fact, as it appears from experimental data in [12.5], the same precision in $\mathring{S}^*_{n,n}$ as that which would be accomplished on a 16-digit machine can, by this approach, be effectively maintained on an 8-digit one.) However additional computation is now required which may, in fact, make a straightforward execution of the Kalman algorithm on p. 465, using double precision arithmetic, to be preferred.

The trouble of course arises because we are eventually reducing the elements of the covariance matrix $\mathring{S}^*_{n,n}$ to values which are too small in relation to their initial values and to the precision of the arithmetic used. Perhaps the best way to overcome the problem is to avoid its occurrence.

Specifically, suppose that we were using 10-digit arithmetic and that we desire, ultimately, 3-digit precision in the elements of $\mathring{S}^*_{n,n}$. Then this could be accomplished by ensuring that the *a priori* covariance matrix $\mathring{S}^*_{0,0}$ has elements which are no greater than 10^7 times as large as those values which we hope ultimately to achieve. By appropriate prefiltering, using any of the non-Kalman schemes which we have discussed earlier, such an *a priori* covariance matrix can perhaps be obtained, and the problem of excessive loss of precision thereby entirely circumvented.

The reader may well be wondering why we bother with the Kalman Filter at all, if it is fraught with so many difficulties which the Bayes Filter does not seem to possess. The answer is that the Kalman Filter offers the definite computational advantage of requiring only a single matrix inversion of the order of the observation vector. Under a tightly constrained real-time situation this is an extremely significant factor, one which is important enough to make us consider very seriously using the Kalman Filter in preference to the Bayes, despite its possible drawbacks. Moreover, as we show in Chapter 15, if we assume driving-noise to be present, and there are many situations when we should, *then most of the drawbacks enumerated above tend to be strongly mitigated.*

Finally, consider the problem of systematic errors. We have seen in the chapter on the Bayes Filter, that if the scheme is appropriately organized, then our knowledge of the process should be steadily improving. This will manifest itself as a steadily shrinking covariance matrix and, as time passes, the new estimate formed from (12.2.19) will depend more and more heavily on $\mathring{X}^*_{n,n-1}$ as $\mathring{H}_n$ shrinks with $\mathring{S}^*_{n,n}$. If the model on which the filter is based is, in fact, very close to the true equations governing the process, then this is as it should be. $\mathring{X}^*_{n,n}$ should appear more and more like $\mathring{X}^*_{n,n-1}$. However if the assumed model and the true dynamics differ, either due to ignorance on the user's part or because of an admitted compromise, then bias errors will begin to manifest themselves in $\mathring{X}^*_{n,n}$.

In the final three chapters of this book, we consider the concept of a *fading memory* and as we shall see, the techniques discussed there provide us with methods for offsetting the possibility of systematic errors when the assumed model does not properly match the equations of motion. In [12.6] the question of divergence between the true trajectory and the estimated one is discussed, and a variety of possible modifications to the filter equations are considered. The problem of biases in the observations and their effects on the estimate is examined in [11.6] of Chapter 11.

The stability properties of the Kalman Filter without driving-noise will be investigated in Section 7 of Chapter 14.

12.5 ITERATIVE DIFFERENTIAL-CORRECTION

The Kalman Filter can be used as the basis of an iterative differential-correction scheme for the case where either the process differential equations or the observation relation is nonlinear. The method is in very close analogy to the Bayes Filter iterative differential-correction scheme developed in Sections 11.1 through 11.3.

Computationally the only difference lies in the way in which $\overset{\circ}{H}_n$ and $\overset{\circ}{S}^*_{n,n}$ are computed. Thus assuming that $\overline{X}_n$ is the state-vector of a trajectory close to the true one, the nonlinear differential equations

$$\frac{d}{dt} X(t) = F[X(t)] \tag{12.5.1}$$

which are assumed to be governing the process can be replaced, to first order accuracy, by the linear set

$$\frac{d}{dt} \delta X(t) = A\left[\overline{X}(t)\right] \delta X(t) \tag{12.5.2}$$

where

$$\left[A\left[\overline{X}(t)\right]\right]_{ij} = \left.\frac{\partial f_i}{\partial x_j}\right|_{X = \overline{X}(t)} \tag{12.5.3}$$

This gives rise to the transition relation

$$\delta X(t_n + \zeta) = \Phi(t_n + \zeta, t_n; \overline{X}) \delta X(t_n) \tag{12.5.4}$$

The nonlinear observation relation

$$Y_n = G(X_n) + N_n \tag{12.5.5}$$

is likewise replaced by the linear system

$$\delta Y_n = M(\overline{X}_n)\delta X_n + N_n \tag{12.5.6}$$

which neglects second and higher order terms, and where

$$\left[M(\overline{X}_n)\right]_{ij} = \left.\frac{\partial g_i}{\partial x_j}\right|_{X = \overline{X}_n} \tag{12.5.7}$$

The differentials δX_n and δY_n are, of course, defined by

$$\delta X_n \equiv X_n - \overline{X}_n \tag{12.5.8}$$

$$\delta Y_n = Y_n - G(\overline{X}_n) \tag{12.5.9}$$

This being the case we can write, as the Kalman estimator of δX_n, the equations

$$X^*_{n,n-1} \text{ obtained from } X^*_{n-1,n-1} \text{ by integration of (12.5.1)} \tag{12.5.10}$$

$$S^*_{n,n-1} = \Phi(t_n, t_{n-1}; \overline{X})S^*_{n-1,n-1}\,\Phi(t_n, t_{n-1}; \overline{X})^T \tag{12.5.11}$$

$$H_n = S^*_{n,n-1}\left[M(\overline{X}_n)\right]^T\left\{R_n + M(\overline{X}_n)S^*_{n,n-1}\left[M(\overline{X}_n)\right]^T\right\}^{-1} \tag{12.5.12}$$

$$S^*_{n,n} = \left[I - H_n M(\overline{X}_n)\right]S^*_{n,n-1} \tag{12.5.13}$$

$$\delta X^*_{n,n} = \delta X^*_{n,n-1} + H_n\left[\delta Y_n - M(\overline{X}_n)\delta X^*_{n,n-1}\right] \tag{12.5.14}$$

$$X^*_{n,n} = \delta X^*_{n,n} + \overline{X}_n \tag{12.5.15}$$

Using (12.5.8) and (12.5.9), the final two equations give us

$$X^*_{n,n} = X^*_{n,n-1} + H_n\left[Y_n - G(\overline{X}_n) - M(\overline{X}_n)\left(X^*_{n,n-1} - \overline{X}_n\right)\right] \tag{12.5.16}$$

and, as with the Bayes scheme, we use $X^*_{n,n-1}$ as the first vector-value of $\overline{X}_n$. Thereafter we iterate (12.5.16) using the algorithm

$$\left(X^*_{n,n}\right)_{r+1} \tag{12.5.17}$$

$$= X^*_{n,n-1} + H(\overline{X}_n)\left[Y_n - G(\overline{X}_n) - M(\overline{X}_n)\left(X^*_{n,n-1} - \overline{X}_n\right)\right]\Bigg|_{\overline{X}=\left(X^*_{n,n}\right)_r}$$

When this iteration converges† a new data-vector Y_{n+1} and covariance matrix R_{n+1} are read in, and a new iteration cycle initiated. If convergence does take place, then it is to the correct vector in the same sense as discussed in Section 11.4. Proof of this fact follows directly from the proof given in that section for the Bayes Filter.

We note that if $X^*_{n,n-1}$ is used as the first value of $\overline{X}_n$ in (12.5.17), then that equation becomes

$$\left(X^*_{n,n}\right)_1 = X^*_{n,n-1} + H\left(X^*_{n,n-1}\right)\left[Y_n - G\left(X^*_{n,n-1}\right)\right] \tag{12.5.18}$$

It is quite common in practice to iterate this equation only a *single* time, since it is found that subsequent iterations make very little difference when n increases. The Kalman Filter under these conditions then assumes the following form:

Kalman Filter with Differential-Correction

$$X^*_{n,n-1} \text{ obtained from } X^*_{n-1,n-1} \text{ by integration of (12.5.1)} \tag{12.5.19}$$

$$S^*_{n,n-1} = \Phi\left(t_n, t_{n-1}; X^*_{n,n-1}\right) S^*_{n-1,n-1} \Phi\left(t_n, t_{n-1}; X^*_{n,n-1}\right)^T \tag{12.5.20}$$

$$H_n = S^*_{n,n-1}\left[M\left(X^*_{n,n-1}\right)\right]^T \left\{R_n + M\left(X^*_{n,n-1}\right)S^*_{n,n-1}\left[M\left(X^*_{n,n-1}\right)\right]^T\right\}^{-1} \tag{12.5.21}$$

$$S^*_{n,n} = \left[I - H_n M\left(X^*_{n,n-1}\right)\right] S^*_{n,n-1} \tag{12.5.22}$$

$$X^*_{n,n} = X^*_{n,n-1} + H_n\left[Y_n - G\left(X^*_{n,n-1}\right)\right] \tag{12.5.23}$$

While the above algorithm is conceptually well-behaved, in practice many unexpected problems can arise. It is, after all, based on a linearized approach to the true problem, and the nonlinearities can (and will) cause difficulties if the discrepancy between the nominal and true trajectories is not sufficiently

†As the criterion of convergence we use (11.2.33) or an equivalent test on a norm of the difference $\left(X^*_{n,n}\right)_{r+1} - \left(X^*_{n,n}\right)_r$.

small. The best approach is to initialize the algorithm with a very carefully obtained *a priori* estimate and covariance matrix (derived e.g. by the use of polynomial least-squares), and to avoid excessive prediction times over which the abovementioned discrepancy might become excessive. Furthermore, in order to verify that the algorithm is working correctly, recourse should always be made (during the program debugging phase) to the use of the Chi-squared tests outlined at the end of Chapter 5 and discussed again in the closing paragraphs of Section 10.2.

NOTES

The techniques discussed in this chapter, in the preceding two chapters and in Chapter 15, are finding a rapidly expanding application in many important fields. For a discussion of the applications to *optimal control theory,* see e.g. Lee [12.2] and Sorenson [12.4]. Both of the above references are tutorial and cover the optimal filtering problem very effectively.

For a discussion on applications to *atmospheric reentry,* see e.g. Mowery [12.7]. The problem of *satellite orbit determination* is tutorially reviewed by Blackman [12.8]. Claus who worked on the Telstar tracking and prediction system has some excellent papers (see under References for Chapter 11).

The techniques have also been applied to the *space-navigation problem* by Schmidt et al. (see [12.3] and further references cited therein), and also independently of [12.1] by Battin in [12.9]. Almost all of the above are concerned with nonlinear processes.

Of course the list of references which we cite comprises only a very few of the already large and rapidly growing number in these areas. They were selected because of their tutorial value. Each of them includes further lists of valuable references. (See also [10.4] for a very extensive list of references.)

EXERCISES

12.1 a) Verify by direct expansion that

$$(S^{-1} + M^T R^{-1} M)\left[S - SM^T(R + MSM^T)^{-1} MS\right] = I$$

Hint:

$$M^T R^{-1} MSM^T(R + MSM^T)^{-1}MS$$

$$= M^T R^{-1}(R + MSM^T - R)(R + MSM^T)^{-1}MS$$

b) Let $\overset{\circ}{S}{}^*_{n,n-1} = \begin{pmatrix} 1 & 1 \\ 1 & 2 \end{pmatrix}$, $M_n = (1, \quad 0)$ and $R_n = 1$. Show that (12.2.7) and (12.2.20) both give

$$\overset{\circ}{S}{}^*_{n,n} = \frac{1}{2}\begin{pmatrix} 1 & 1 \\ 1 & 3 \end{pmatrix}$$

12.2 Verify that (12.2.12) follows from (12.2.11).

12.3† a) For the state-vector

$$Z_n \equiv \left(x, \quad \tau\dot{x}, \quad \frac{\tau^2}{2!}\ddot{x}, \quad \ldots, \quad \frac{\tau^m}{m!} D^m x \right)_n^T \tag{I}$$

the transition matrix is $\Phi(t_n + \zeta, t_n)$ whose i, jth element is (c/f (4.2.15)) $\binom{j}{i}\zeta^{j-i}$. Starting from (12.2.9), and assuming that Z^* estimates the vector Z in (I), verify that

$$\Phi(t_n, t_{n+1}) Z^*_{n+1,n} = Z^*_{n,n-1} + H_n\left(Y_n - M_n Z^*_{n,n-1}\right) \tag{II}$$

Letting M be the matrix $(1, \quad 0, \quad 0, \quad \ldots, \quad 0)$ (i.e. we are observing only $x(t)$) verify that (II) is of the same form as (9.4.8). Given that all observations are uncorrelated, have equal variance, and are equally spaced in time, infer that the weight matrix H_n in (9.4.8) is the same as H_n in (12.2.8) under the above conditions.

b) Verify that (12.2.8) becomes

$$H_n = \frac{1}{\sigma_\nu^2} S^*_{n,n} M^T \tag{III}$$

Starting from $\left(S^*_{0,-1}\right)^{-1} = \emptyset$ in (12.2.7), cycle it together with (12.2.6) and obtain

$$S^*_{n,n} = \sigma_\nu^2 \left\{ \sum_{k=0}^{n} [M\Phi(k,n)]^T M\Phi(k,n) \right\}^{-1} \tag{IV}$$

†This example makes use of the Bayes Filter to prove the general form given in (9.4.30).

(The reader should compare this result and its method of derivation to its counterpart in Ex. 6.18, part b).)

c) Let $P(n)$ be the matrix defined in (9.4.28). Verify that

$$P(k)\,P(n)^{-1} = \Phi(k,n) \tag{V}$$

and hence infer that (IV) above can be written

$$S^*_{n,n} = \sigma_\nu^2\, P(n)\left\{\sum_{k=0}^{n}[MP(k)]^T MP(k)\right\}^{-1} P(n)^T \tag{VI}$$

Combine this with (III) to give

$$H_n = P(n)\left\{\sum_{k=0}^{n}[MP(k)]^T MP(k)\right\}^{-1}[MP(n)]^T \tag{VII}$$

d) Applying the definition of $P(k)$, first reduce the inverse term in (VII) above to

$$\begin{pmatrix} \dfrac{1}{[c(0,n)]^2} & & & \\ & \dfrac{1}{[c(1,n)]^2} & & \\ & & \ddots & \\ & & & \dfrac{1}{[c(m,n)]^2} \end{pmatrix}$$

where $[c(j,n)]^2$ is defined in (9.2.8), and then post-multiply this by $[MP(n)]^T$ to obtain the vector $K(n)$ of (9.4.29).

e) Finally infer that (VII) above reduces to (c/f (9.4.30))

$$H_n = P(n)\,K(n)$$

12.4 By (12.3.1), $X^*_{n,n-1}$ is an unbiased estimate of X_n. Verify that (12.3.5) then implies that

$$E\{M_n X^*_{n,n-1}\} = E\{Y_n\}$$

and so, infer that (12.3.2) gives

$$E\{X^*_{n,n}\} = X_n$$

regardless of the choice for H_n.

12.5 Derive (12.3.6) and (12.3.7).

12.6 a) Verify, by direct expansion, that (c/f (12.3.7))

$$(I - HM)S(I - HM)^T + HRH^T = S - HMS - SM^TH^T + H(MSM^T + R)H^T$$

Now verify that the right-hand side of the above can be written as $A + B$ where

$$A \equiv \left[S - SM^T(R + MSM^T)^{-1} MS\right] \tag{I}$$

and

$$B \equiv \left[H - SM^T(R + MSM^T)^{-1}\right](MSM^T + R)\left[H - SM^T(R + MSM^T)^{-1}\right]^T \tag{II}$$

Note that only B contains terms involving H.

b) Prove that if S and R are positive definite then so is A of (I) above. Verify also that B in (II) above is nonnegative definite for any H. Hence infer that the diagonal elements of $A + B$ are individually least when B is a null matrix and verify that this takes place when

$$H = SM^T(R + MSM^T)^{-1}$$

We have thus proved that every diagonal element of $S^*(H_n)$ of (12.3.7) is least when H is given by (12.2.17). Note that this also means that the sum of the diagonal elements of S^*, namely $E\{N^*_{n,n}N^{*T}_{n,n}\}$ is least when $H = \mathring{H}_n$. Verify that under these conditions (12.3.7) is now equal to (12.2.20). Thus this approach constitutes a *completely independent method* of deriving the Kalman Filter and hence, by the inversion lemma, the Bayes Filter as well.

12.7 We wish to prove that $E\{N^{*T}_{n,n}\Gamma N^*_{n,n}\}$ is least when H in (12.3.2) is given by (12.2.17) and Γ is *any* positive definite matrix.

a) Verify that (c/f Ex. 6.21)

$$E\{N^T \Gamma N\} = \operatorname{Tr} E\{\Gamma N N^T\}$$

Hence prove that

$$\operatorname{Tr}\left(E\{\Gamma N N^T\}\right) = \operatorname{Tr}\left(\Gamma^{1/2} E\{N N^T\}\Gamma^{1/2}\right)$$

and so infer that

$$E\left\{N^{*T}_{n,n}\Gamma N^*_{n,n}\right\} = \operatorname{Tr}\left(\Gamma^{1/2} S^*_{n,n}\Gamma^{1/2}\right)$$

b) Let the covariance matrix which results when *any* H is used in (12.3.2) be $S^*_{n,n}$ and let it be $\mathring{S}^*_{n,n}$ when $\mathring{H}_n$ is used. Show that

$$\operatorname{Tr}\left(\Gamma^{1/2} S^*_{n,n}\Gamma^{1/2}\right) \geq \operatorname{Tr}\left(\Gamma^{1/2}\mathring{S}^*_{n,n}\Gamma^{1/2}\right)$$

Hint: See Corollary 6.6.1 on p. 194.

Hence infer finally that $E\left\{N^{*T}_{n,n}\Gamma N^*_{n,n}\right\}$ is least when H is given by (12.2.17).

12.8 Let Λ_n be a linear transformation of X_n, i.e.

$$\Lambda_n = CX_n$$

A vector of observations $Y(\Lambda)$ is made on Λ_n, with the corresponding set on X_n namely $Y(X)$ also being recorded. Let $Z^*_{n,n}$ be an unbiased linear estimate of Λ_n based on $Y(\Lambda)$. Prove that the scalar

$$E\left\{\left(Z^*_{n,n} - \Lambda_n\right)^T\left(Z^*_{n,n} - \Lambda_n\right)\right\}$$

is minimized by $Z^*_{n,n} \equiv C\mathring{X}^*_{n,n}$ where $\mathring{X}^*_{n,n}$ is the linear unbiased minimum-variance estimate of X_n, based on the observations $Y(X)$. (Hint: See Ex. 12.7.) Thus if $X^*_{n,n}$ is the minimum-variance estimate of X_n based on $Y(X)$, then $CX^*_{n,n}$ is the minimum-variance estimate of CX_n based on $Y(CX_n)$, an assertion which is subject to proof.

12.9 a) Show that (12.3.16) follows from (12.3.14) and (12.3.15).

b) Minimize $e\left(X^*_{n,n}\right)$ of (12.3.16) over $X^*_{n,n}$ by the differential calculus and show that we obtain

$$X^*_{n,n} = \left(\mathring{S}^{*-1}_{n,n-1} + M_n^T R_n^{-1} M_n\right)^{-1}\left(M_n^T R_n^{-1} Y_n + \mathring{S}^*_{n,n-1} \mathring{X}^*_{n,n-1}\right)$$

Now derive (12.2.8) and (12.2.9) of the Bayes Filter and infer, by the use of the inversion lemma, that we could also obtain the Kalman Filter.

12.10 a) Assume that Y_n is a two-vector and that

$$S^*_{n,n-1} = \begin{pmatrix} 1 & 2 \\ 2 & 5 \end{pmatrix} \qquad R_n = \begin{pmatrix} 1 & 0 \\ 0 & 2 \end{pmatrix} \qquad M = \begin{pmatrix} 1 & 0 \\ 0 & 1 \end{pmatrix}$$

Verify that (12.2.17) gives

$$\mathring{H}_n = \frac{1}{10}\begin{pmatrix} 3 & 2 \\ 4 & 6 \end{pmatrix}$$

and that (12.2.18) gives

$$S^*_{n,n} = \frac{1}{10}\begin{pmatrix} 3 & 4 \\ 4 & 12 \end{pmatrix} \tag{I}$$

b) Now process the elements of Y_n *one at a time.* Verify that (12.2.17) gives first

$$\left(\mathring{H}_n\right)_1 = \frac{1}{2}\begin{pmatrix} 1 \\ 2 \end{pmatrix}$$

and that (12.2.18) gives

$$\left(\mathring{S}^*_{n,n}\right)_1 = \frac{1}{2}\begin{pmatrix} 1 & 2 \\ 2 & 6 \end{pmatrix} \tag{II}$$

Now the second element of Y_n is processed. Verify that

$$\left(\overset{\circ}{H}_n\right)_2 = \frac{1}{5}\begin{pmatrix}1\\3\end{pmatrix}$$

and finally that

$$\left(\overset{\circ}{S}{}^*_{n,n}\right)_2 = \frac{1}{10}\begin{pmatrix}3 & 4\\4 & 12\end{pmatrix}$$

c) We infer then that the two methods of processing Y_n give the same result. Can we process the elements of Y_n separately if R_n is *not* diagonal? Are there any advantages to processing Y_n element by element rather than in a single pass?

d) Note from (I) in a) above that the variances of both elements of the state-vector are reduced by incorporating both of the observations. In (II) of part b) we see that they are *both* reduced, even though we only incorporated *one* of the observations. Explain.

12.11 a) Assume that M_1 is nonsingular, i.e. M_1 is square and M_1^{-1} exists. Show that the Bayes Filter can be made to ignore the *a priori* estimate by setting $\overset{\circ}{S}{}^{*-1}_{1,0} = \emptyset$ and that the Kalman Filter ignores its *a priori* estimate if we make $R_1 = \emptyset$.† Verify that in either case

$$\overset{\circ}{X}{}^*_{1,1} = M_1^{-1} Y_1$$

b) Assume that M_1 is now *rectangular.* Show that the Bayes Filter will ignore its *a priori* estimate if we set $\overset{\circ}{S}{}^*_{1,0} = \emptyset$, but that this approach only works if M_1 has full *column*-rank.

c) Show that when M_1 is rectangular, the Kalman Filter *cannot* be made to ignore its *a priori* estimate. Verify that if M_1 has full *column*-rank and we set $\overset{\circ}{S}{}^*_{1,0} = \infty I$, we cannot cycle (12.2.17). Verify, on the other hand, that if M_1 has full *row*-rank and we set $\overset{\circ}{S}{}^*_{1,0} = \infty I$, that the Kalman Filter *can* be cycled but that parts of the *a priori* estimate still appear in $\overset{\circ}{X}{}^*_{1,1}$.

†Note that setting $R_1 = \emptyset$ accomplishes the same result as setting $\overset{\circ}{S}{}^{*-1}_{1,0} = \infty I$ in (12.2.17), by virtue of (12.4.3).

12.12 Assume that

$$\mathring{S}^*_{n,n-1} = \begin{pmatrix} \alpha & \beta \\ \beta & \gamma \end{pmatrix}$$

and that we make a perfect observation on the first element of the state-vector, i.e. $M_n = (1, \quad 0)$ and $R_n = 0$.

a) Show that

$$H_n = \frac{1}{\alpha}\begin{pmatrix} \alpha \\ \beta \end{pmatrix}$$

and hence that

$$\mathring{X}^*_{n,n} = \begin{pmatrix} y_n \\ \left(\mathring{x}^*_1\right)_{n,n-1} + \frac{\beta}{\alpha}\left[y_n - \left(x^*_0\right)_{n,n-1}\right] \end{pmatrix} \tag{I}$$

Note that the perfect observation has completely replaced the estimate $\left(\mathring{x}^*_0\right)_{n,n-1}$ in (I) above. Is this what we would desire?

b) From (I) above, obtain the result

$$\mathring{S}^*_{n,n} = \begin{pmatrix} 0 & 0 \\ 0 & \gamma - \beta^2/\alpha \end{pmatrix} \tag{II}$$

c) Obtain (II) above by means of (12.2.18).

12.13 a) Let

$$\mathring{S}^*_{n,n-1} = \begin{pmatrix} 1 & 1 & 1 \\ 1 & 1 & 1 \\ 1 & 1 & 1 \end{pmatrix}$$

and assume that we make two perfect observations at t_n, one on $(x_0)_n$ and the other on $(x_1)_n$ where

$$X_n \equiv \left(x_0, \quad x_1, \quad x_2\right)^T_n$$

Show that after the incorporation of the first, we get $\overset{\circ}{S}^*_{n,n} = \emptyset$, i.e. we now have perfect knowledge of X_n. Explain how a *single* observation can give us perfect knowledge of *three* state-variables.

b) Verify next that we *cannot* now incorporate the second perfect observation even if we try to.

c) Assume instead that

$$\overset{\circ}{S}^*_{n,n-1} = \begin{pmatrix} 1 & 1 & 1 \\ 1 & 2 & 2 \\ 1 & 2 & 2 \end{pmatrix}$$

Show that incorporating the first perfect observation gives us perfect knowledge of only $(x_0)_n$, and that the second perfect observation can now be incorporated, giving $\overset{\circ}{S}^*_{n,n} = \emptyset$.

d) Explain why it is that in case c) two perfect observations can be incorporated as against only one in case a).

e) If

$$\overset{\circ}{S}^*_{n,n-1} = \begin{pmatrix} 3 & 1 & 1 \\ 1 & 3 & 1 \\ 1 & 1 & 4 \end{pmatrix}$$

how many perfect observations could we possibly incorporate?

12.14 Let

$$\overset{\circ}{S}^*_{n,n-1} = \begin{pmatrix} 2 & 1 & 1 \\ 1 & 2 & 1 \\ 1 & 1 & 3 \end{pmatrix}$$

$$M_n = \begin{pmatrix} 1 & 0 & 0 \\ 0 & 1 & 0 \end{pmatrix} \qquad R_n = \begin{pmatrix} 1 & 1 \\ 1 & 1 \end{pmatrix}$$

a) Verify that $\overset{\circ}{S}^*_{n,n-1}$ is positive definite, i.e. has rank 3, and that R_n has a rank-defect of 1.

b) Compute $\overset{\circ}{S}^*_{n,n}$ using (12.2.20) and verify that it has a rank-defect of 1.

c) Assuming that

$$\Phi(n+1,n) = \begin{pmatrix} 1 & -1 & 1 \\ & 1 & -2 \\ & & 1 \end{pmatrix}$$

compute $\mathring{S}^*_{n+1,n}$ using $\mathring{S}^*_{n,n}$ obtained in b) above. Show that $\mathring{S}^*_{n+1,n}$ also has a rank-defect of 1.

d) Now let

$$R_{n+1} = \begin{pmatrix} 1 & 1 \\ 1 & 2 \end{pmatrix}$$

with M_{n+1} the same as M_n given earlier. Obtain the resulting matrix $\mathring{S}^*_{n+1,n+1}$ and verify that it has a rank-defect of 1.

e) The above is a demonstration of how a rank-defect, *once generated, persists indefinitely.* Suppose that R_{n+1} in d) above were

$$R_{n+1} = \begin{pmatrix} 1 & 1 \\ 1 & 1 \end{pmatrix}$$

Verify that $\mathring{S}^*_{n+1,n+1}$ now acquires a rank-defect of 2.

f) If we were now to incorporate a further R matrix of the form given in e) above, infer that the result would be $\mathring{S}^*_{n+2,n+2} = \emptyset$. Verify that this is so.

12.15 In Chapters 10 and 11, problems were suggested for the application of the Bayes recursive algorithm. Rework those examples using the Kalman algorithm rather than the Bayes and contrast the two approaches.

12.16 a) Assume that a Kalman Filter *is being cycled on a machine which has 4-digit precision.* We are executing (12.2.20).† Let

$$\mathring{S}^*_{n,n-1} = 10^4 \begin{pmatrix} .1111 & .1111 \\ .1111 & .1112 \end{pmatrix} \qquad M = (1, \ 0) \qquad \text{(I)}$$

†i.e. (12.2.18) in which $\mathring{H}$ is given by (12.2.17).

and R_n (a scalar) be equal to unity. Verify that

$$\left(R_n + M\mathring{S}^*_{n,n-1}M^T\right) = .1112 \times 10^4$$

and that

$$\mathring{S}^*_{n,n-1}M^T\left(R_n + M\mathring{S}^*_{n,n-1}M^T\right)^{-1}M\mathring{S}^*_{n,n-1} = 10^4\begin{pmatrix} .1109 & .1109 \\ .1109 & .1109 \end{pmatrix}$$

Hence show that (12.2.10) gives us

$$\mathring{S}^*_{n,n} = 10\begin{pmatrix} .2 & .2 \\ .2 & .3 \end{pmatrix} \qquad \text{(II)}$$

b) We now compute $\mathring{S}^*_{n,n}$ using (12.2.7) of the Bayes Filter. Verify that (using 4-digit precision)

$$\mathring{S}^{*-1}_{n,n-1} = 10\begin{pmatrix} .1112 & -.1111 \\ -.1111 & .1111 \end{pmatrix}$$

and hence that

$$\mathring{S}^{*-1}_{n,n-1} + M^T R_n^{-1} M_n = 10\begin{pmatrix} .2112 & -.1111 \\ -.1111 & .1111 \end{pmatrix}$$

Now show finally that (12.2.7) gives

$$\mathring{S}^*_{n,n} = \begin{pmatrix} .9991 & .9991 \\ .9991 & 1.899 \end{pmatrix}$$

The Bayes algorithm is thus seen to retain *four* digits of precision, whereas the Kalman algorithm *loses three digits and retains only a single digit of precision.*

c) Using 8-digit precision, show that the "correct" answer is

$$\overset{\circ}{S}{}^{*}_{n,n} = \begin{pmatrix} .99910080 & .99910080 \\ .99910080 & 1.9991008 \end{pmatrix}$$

and so we see that the Bayes result is *both more precise as well as far more accurate* than the Kalman.

REFERENCES

1. Kalman, R. E., "A New Approach to Linear Filtering and Prediction Problems," Transactions of the ASME, Journal of Basic Engineering, Vol. 82, pp. 35-45, March 1960.
2. Lee, R. C. K., "Optimal Estimation, Identification and Control," Research Mono. No. 28, MIT Press, Cambridge, Mass., 1964, p. 35 et seq.
3. Schmidt, S. F., "Application of State-Space Methods to Navigation Problems," in "Advances in Control Systems," C. T. Leondes (editor), Academic Press, New York, Vol. 3, 1966, pp. 293-340.
4. Sorenson, H. W., "Kalman Filtering Techniques," in "Advances in Control Systems," C. T. Leondes (editor), Academic Press, New York, Vol. 3, 1966, pp. 219-292.
5. Bellantoni, J. F., and Dodge, K. W., "A Square Root Formulation of the Kalman-Schmidt Filter," American Inst. of Aeronautics and Astronautics (AIAA) Journal, Vol. 5, No. 7, pp. 1309-1314, July 1967.
6. Schlee, F. H., Standish, C. J., and Toda, N. F., "Divergence in the Kalman Filter," American Inst. of Aeronautics and Astronautics (AIAA) Journal, Vol. 5, No. 6, pp. 1114-1120, June 1967.
7. Mowery, V. O., "Least-Squares Recursive Differential-Correction Estimation in Nonlinear Problems," IEEE Trans. on Automatic Control, Vol. AC-10, No. 4, October 1965, pp. 399-407.
8. Blackman, R. B., "Methods of Orbit Refinement," Bell Systems Tech. Journ., Vol. 43, May 1964, pp. 885-909.
9. Battin, R. H., "A Statistical Optimizing Navigation Procedure for Space Flight," Journal of the Amer. Rocket Soc. (ARS), Vol. 32, November 1962, pp. 1681-1969. See also "Astronautical Guidance," R. H. Battin, McGraw-Hill Book Company, New York, 1964.

PART 4

FADING–MEMORY FILTERING

We have considered two approaches to the filtering problem up to now, namely the fixed-memory and expanding-memory techniques. In the former case we saw that, by an appropriate choice of the memory length, possible systematic errors can be balanced-off against the random output errors. In the latter case this was not possible since the memory length was growing steadily. In the event that a mismatch occurs, either by intention or by ignorance, between the model on which the filter is based and the true process, then the expanding-memory techniques will give rise to systematic errors which could become very large. However the expanding-memory schemes are definitely to be preferred to the fixed-memory ones because they are all recursive and require only that the most recent data be temporarily retained. They are also very compact in terms of the amount of computation required.

We now examine an approach which attempts to take the best out of each of the above two techniques. The algorithms we derive will be *recursive* and hence very compact. Moreover the memory-shape will be seen to be equivalent, in a sense, to a *fixed-memory,* so that systematic errors can be controlled at the expense of random errors. As will be seen, this is accomplished by the use of what we call a *fading* memory.

13

THE FADING-MEMORY POLYNOMIAL FILTER

13.1 INTRODUCTION

In this chapter we examine the filters which result when a polynomial is fitted to a sequence of equally-spaced observations by the use of *weighted* least-squares. The weight-factor which we propose to use is one which *decays exponentially* as time recedes into the past, and so this means that the emphasis being exerted by any given datum in the choice of the polynomial will taper off or fade out as the staleness of the datum increases. The filters which result are accordingly endowed with what we call a *fading-memory*.

The least-squares error-functional on which this chapter is based† has been studied independently by a number of other investigators, a partial list of references being [13.1, 13.2, 13.3, 13.4, 13.5]. Despite the fact that they all started from essentially the same point, their results are, at first sight, different from each other's and from ours. In most cases however, they can be reconciled without too much difficulty.

†See (13.2.1) on p. 498.

In [13.1], Bruckner and Ford based their approach on a set of orthogonal polynomials, but did not appear to recognize that these were related to the discrete Laguerre polynomials.† On the other hand, in [13.2] and [13.3] Duffin and Schmidt performed all of their analysis without reference to any orthogonal polynomials, but they do show, in the final section of [13.2], that their results are directly related to the discrete Laguerre polynomials.

The present discussion approaches the problem entirely on the basis of the discrete Laguerre polynomials. Our results can be readily reconciled with those of Duffin and Schmidt, and in fact the present chapter was strongly motivated by their work.

13.2 DISCOUNTED LEAST-SQUARES

Assume that the *semi-infinite* sequence of scalar observations

$$\ldots,\ y_{n-5},\ \ldots,\ y_{n-2},\ y_{n-1},\ y_n$$

has been obtained by taking measurements on a process at *equally-spaced instants,* τ seconds apart. In Figure 13.1 we depict the observations located on the t-axis. We also show the r-axis, r being a continuous variable with its origin located at $t = n\tau$, the integer n being assumed frozen for the present at some fixed value. Note that an *increase* in r corresponds to a *decrease* in n or t.

It is decided that the process shall be modelled by a polynomial in r of degree m, based on the above sequence of observations. We call this polynomial $[p^*(r)]_n$, the subscript n being included to show that this polynomial is founded on data up to y_n. When y_{n+1} becomes available a new polynomial $[p^*(r)]_{n+1}$ will be considered, based on data up to y_{n+1}, *and at that time the r-origin will be located at* $t = (n+1)\tau$.

Returning to $[p^*(r)]_n$, the difference between it and the observation made at $t = (n-r)\tau$ is $y_{n-r} - [p^*(r)]_n$. This is then the residual at that time, and so the scalar

$$e_n \equiv \sum_{r=0}^{\infty} \left\{ y_{n-r} - [p^*(r)]_n \right\}^2 \theta^r \qquad (13.2.1)$$

(See Note.)

†These were discussed in Section 3.3 where their general form was established.

Note: The criterion in (13.2.1) is based on *discrete* values of r. A complete dual, using a *continuous* variable, and *integration* in place of *summation,* has been developed in [13.7]. The filters which result are readily realized using resistors and capacitors.

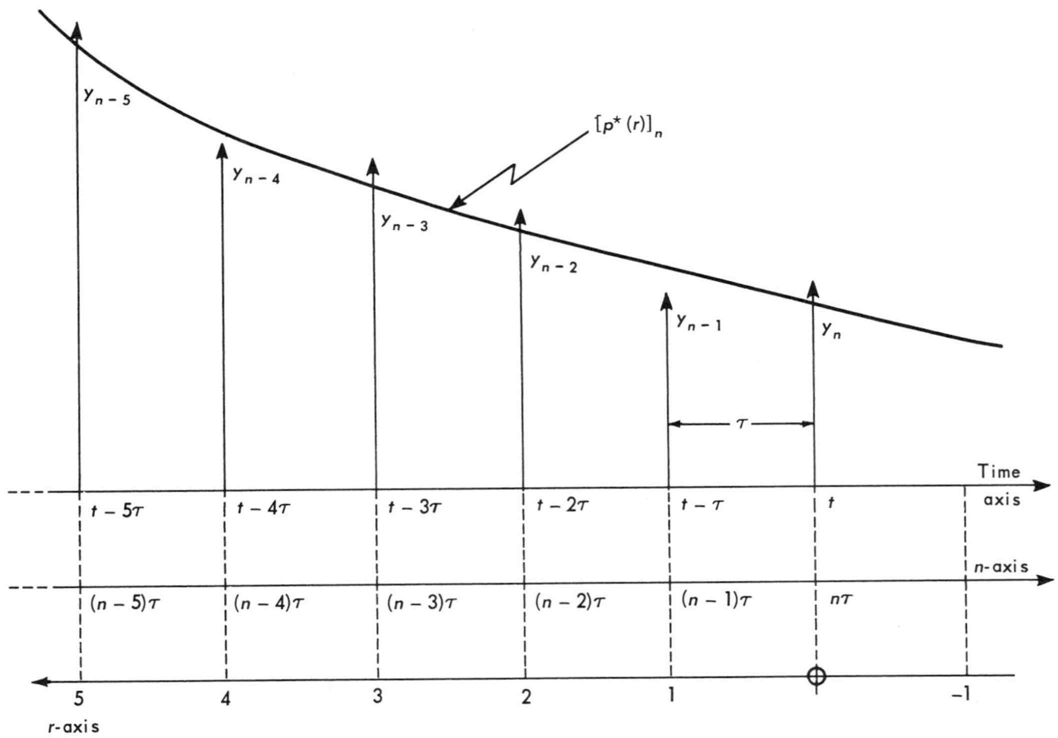

Fig. 13.1 *The approximating polynomial.*

is seen to be the sum of the squared residuals, weighted by the exponential function θ^r. The latter dies out as r increases if $|\theta| < 1$, which we assume to be the case. Hence if we take, as our criterion for the choice of $\left[p^*(r)\right]_n$, that the above e_n be minimized, it is clear that the *older* or *staler* a datum is, the *less* it will influence the choice. From (13.2.1) we see that the multiplier θ^r acts as a *discount* factor, hence the title for this section "discounted least-squares," a term coined by Duffin and Schmidt in [13.2].

The error-function (13.2.1) is very readily minimized over all polynomials of given degree if we first express $\left[p^*(r)\right]_n$ as a linear combination of the discrete Laguerre polynomials of Chapter 3. We recall from that discussion (see Section 3.3), that if we define a polynomial in r, of degree j, with parameter θ, by

$$p(r; j, \theta) \equiv \theta^j \sum_{\nu=0}^{j} (-1)^\nu \binom{j}{\nu} \left(\frac{1-\theta}{\theta}\right)^\nu \binom{r}{\nu} \tag{13.2.2}$$

then the resulting set of polynomials satisfies the discrete orthogonality condition

$$\sum_{r=0}^{\infty} p(r; i, \theta)\, p(r; j, \theta) = \begin{cases} 0 & j \neq i \\ [c(j,\theta)]^2 & j = i \end{cases} \tag{13.2.3}$$

where

$$[c(j,\theta)]^2 \equiv \frac{\theta^j}{1-\theta} \tag{13.2.4}$$

For the remainder of this chapter we abbreviate as follows:

$$p_j(r) \equiv p(r; j, \theta) \tag{13.2.5}$$

and

$$c_j \equiv c(j, \theta) \tag{13.2.6}$$

the parameter θ being implicit in both cases. Moreover we now let

$$K_j \equiv \frac{1}{c_j} = \left(\frac{1-\theta}{\theta^j}\right)^{1/2} \tag{13.2.7}$$

and then it is clear that the polynomial set defined by

$$\varphi_j(r) \equiv K_j p_j(r) \tag{13.2.8}$$

will satisfy the *orthonormal* condition

$$\sum_{r=0}^{\infty} \varphi_i(r)\,\varphi_j(r)\,\theta^r = \delta_{ij} \tag{13.2.9}$$

where δ_{ij} is the Kronecker delta.

For the moment let $[p^*(r)]_n$ be *any* polynomial of degree m in r. Then it is evident that there exist constants $(\beta_0)_n$, $(\beta_1)_n$, ... $(\beta_m)_n$, such that we can write

$$[p^*(r)]_n = \sum_{j=0}^{m} (\beta_j)_n \varphi_j(r) \tag{13.2.10}$$

i.e. $[p^*(r)]_n$ can be synthesized as a linear combination of the $\varphi_j(r)$'s. Note that the β's have been subscripted with an n to show their association with $[p^*(r)]_n$. We now set (13.2.10) into (13.2.1), thereby obtaining

$$e_n = \sum_{k=0}^{\infty} \left[y_{n-k} - \sum_{j=0}^{m} (\beta_j)_n \varphi_j(k) \right]^2 \theta^k \tag{13.2.11}$$

Observe that r has been replaced by k for convenience in what follows.

The above error-function e_n is seen to depend only on the β's, and so we can minimize it by setting

$$\frac{\partial e_n}{\partial (\beta_i)_n} = 0 \qquad 0 \le i \le m \tag{13.2.12}$$

Performing this operation yields the $m + 1$ equations

$$\sum_{k=0}^{\infty} \sum_{j=0}^{m} (\beta_j)_n \varphi_j(k) \varphi_i(k) \theta^k = \sum_{k=0}^{\infty} y_{n-k} \varphi_i(k) \theta^k \qquad 0 \le i \le m \tag{13.2.13}$$

and then, by an interchange of the order of summation on the left,

$$\sum_{j=0}^{m} (\beta_j)_n \sum_{k=0}^{\infty} \varphi_i(k) \varphi_j(k) \theta^k = \sum_{k=0}^{\infty} y_{n-k} \varphi_i(k) \theta^k \tag{13.2.14}$$

We now invoke the orthonormal property of the set of $\varphi_j(k)$ (see (13.2.9)) and obtain

$$(\beta_j)_n = \sum_{k=0}^{\infty} y_{n-k} \varphi_j(k) \theta^k, \qquad 0 \le j \le m \tag{13.2.15}$$

and so this is the expression for the jth constant to be used in (13.2.10) so that e_n of (13.2.1) be least.† We accordingly obtain, as the polynomial which minimizes e_n,

$$[p^*(r)]_n = \sum_{j=0}^{m} \left[\sum_{k=0}^{\infty} y_{n-k} \varphi_j(k) \theta^k \right] \varphi_j(r) \tag{13.2.16}$$

It has been decided that $[p^*(r)]_n$ shall, for better or for worse, serve as an estimate of the process, based on observations up to y_n. Thus we write

$$x^*_{n-r,n} \equiv [p^*(r)]_n \tag{13.2.17}$$

Setting $r = 0$, for example (see Figure 13.1), gives us the updated estimate $x^*_{n,n}$ or setting $r = -1$ gives us the 1-step prediction

$$x^*_{n+1,n} \equiv [p^*(-1)]_n \tag{13.2.18}$$

†These constants are termed the *expansion coefficients* or sometimes they are said to constitute the *Laguerre spectrum,* in analogy to the coefficients of a Fourier expansion which constitute the Fourier spectrum.

and so forth. The ability to move the validity instant of the estimate is thus clearly contained in the choice of the value assigned to the variable r in (13.2.16).

Reference to Figure 13.1 shows next that, to within a constant,

$$t = -r\tau \tag{13.2.19}$$

and so

$$\frac{d}{dt} = -\frac{1}{\tau}\frac{d}{dr} \tag{13.2.20}$$

and in general

$$\frac{d^i}{dt^i} = \left(-\frac{1}{\tau}\right)^i \frac{d^i}{dr^i} \tag{13.2.21}$$

Then, just as the process will be estimated by $\left[p^*(r)\right]_n$, so we decide that its *time-derivatives* will be estimated by the time-derivatives of $\left[p^*(r)\right]_n$. Thus by virtue of (13.2.17) and (13.2.21), we have

$$(D^i x^*)_{n-r,n} = \left(-\frac{1}{\tau}\right)^i \frac{d^i}{dr^i}\left[p^*(r)\right]_n \tag{13.2.22}$$

and so, from (13.2.10) it follows that

$$(D^i x^*)_{n-r,n} = \left(-\frac{1}{\tau}\right)^i \sum_{j=0}^{m} (\beta_j)_n \frac{d^i}{dr^i}\varphi_j(r) \tag{13.2.23}$$

where of course $D \equiv d/dt$. Finally if we define what we call the *unscaled-derivative estimate state-vector by*

$$X^*_{n-r,n} \equiv \begin{pmatrix} x^* \\ Dx^* \\ \vdots \\ D^m x^* \end{pmatrix}_{n-r,n} \tag{13.2.24}$$

then we see that its ith element is given by (13.2.23).

Once values are assigned to r and τ it is clear that (13.2.23) is just a linear sum of the β_j's. These in turn are seen from (13.2.15) to be a linear combination of the observations, and so it follows that each of the elements of $X^*_{n-r,n}$ above can be obtained by taking an appropriate *linear combination* of the data. At the same time, by varying r, prediction or retrodiction can be performed and, as we shall show presently, smoothing of the observation errors will also be accomplished by an amount which depends on the choice of the parameter θ.

13.3 RECURSIVE FORMULATION

The expression obtained in (13.2.15) for the jth expansion coefficient, $(\beta_j)_n$, is obviously unsatisfactory from a computational standpoint. However it can be recast into a very convenient *recursive* form as we now show.

Equation (13.2.15) gives us $(\beta_j)_n$. Suppose that we had previously used it to compute $(\beta_j)_{n-1}$. Then clearly we would have obtained the latter from

$$(\beta_j)_{n-1} = \sum_{k=0}^{\infty} y_{n-1-k}\,\varphi_j(k)\,\theta^k \tag{13.3.1}$$

which is readily seen to be equivalent to

$$\theta(\beta_j)_{n-1} = \sum_{k=1}^{\infty} y_{n-k}\,\varphi_j(k-1)\,\theta^k \tag{13.3.2}$$

We now extend this argument, and it becomes clear (see Ex. 13.1) that for the case of $(\beta_j)_{n-i}$, (13.3.2) generalizes to

$$\theta^i(\beta_j)_{n-i} = \sum_{k=i}^{\infty} y_{n-k}\,\varphi_j(k-i)\,\theta^k \tag{13.3.3}$$

Now consider the case $j = 0$. By (13.2.8) and (13.2.2) we know that $\varphi_0(r) = K_0$ (which is a constant independent of r) and so it follows quite easily from (13.2.15) and (13.3.2) (see Ex. 13.2), that

$$(\beta_0)_n = \theta(\beta_0)_{n-1} + K_0 y_n \tag{13.3.4}$$

This is seen to be a *recursion* for $(\beta_0)_n$ in terms $(\beta_0)_{n-1}$ and y_n, and is obviously a far more practical way of computing $(\beta_0)_n$ than (13.2.15) was. (It remains to be shown, however, that (13.3.4) can, in fact, be used; being a recursion it possesses potential stability problems. This question will be satisfactorily resolved in the next section.) Using the backward-shifting operator q, we observe further that (13.3.4) can be written in the *operational form*

$$(1 - q\theta)(\beta_0)_n = K_0 y_n \tag{13.3.5}$$

By repeating the above approach (see Ex. 13.3) the reader can readily verify that the recursion for $(\beta_1)_n$ assumes the form

$$(\beta_1)_n = 2\theta(\beta_1)_{n-1} - \theta^2(\beta_1)_{n-2} + K_1\theta(y_n - y_{n-1}) \tag{13.3.6}$$

with the associated operational form

$$(1 - q\theta)^2(\beta_1)_n = K_1\theta(1 - q)y_n \tag{13.3.7}$$

and we thus begin to suspect that, *in general,* the recursion for $(\beta_j)_n$ has the operational form

$$(1 - q\theta)^{j+1}(\beta_j)_n = K_j\theta^j(1 - q)^j y_n \tag{13.3.8}$$

This is in fact the case. The proof, while straightforward, is somewhat lengthy and so it would tend to distract us from our train of thought. We have accordingly located it separately in Appendix II.† *Equation* (13.3.8) *is a very fundamental result, and the entire remainder of this chapter will be based on it.* (See Ex. 13.5.) We now apply it to the results of preceding section.

By virtue of the discussion given in Section 2.2 (see (2.2.27) and (2.2.28)), we know that the operator $(1 - q\theta)^{j+1}$ on the left of (13.3.8) can be moved

†In addition to the proof given in that appendix, a proof of the validity of (13.3.8) is also given in [13.6]. The two proofs differ in their approach in that the one in Appendix II is very direct, making use of the Z-transform (the discrete counterpart of the Laplace transform). The one in [13.6] on the other hand follows a somewhat lengthier but purely algebraic approach. (See also Ex. 13.4.)

into the denominator on the right of that equation, i.e.

$$(\beta_j)_n = K_j \theta^j \frac{(1 - q)^j}{(1 - q\theta)^{j+1}} y_n \tag{13.3.9}$$

Then by (13.2.10) it follows that we can write

$$[p^*(r)]_n = \sum_{j=0}^{m} \left[K_j \theta^j \frac{(1 - q)^j}{(1 - q\theta)^{j+1}} y_n \right] \varphi_j(r) \tag{13.3.10}$$

and so finally, by (13.2.23), we obtain

$$(D^i x^*)_{n-r,n} = \left\{ \left(-\frac{1}{\tau} \right)^i \sum_{j=0}^{m} \left[\frac{d^i}{dr^i} \varphi_j(r) \right] \frac{K_j \theta^j (1 - q)^j}{(1 - q\theta)^{j+1}} \right\} y_n \tag{13.3.11}$$

This is the general operational form of a recursion for estimates of any of the derivatives of the process under observation, based on a polynomial of degree m and a prediction span determined by the choice of r, using discounted least-squares.

We pause for an example. Let $i = 0$ and $m = 1$ (i.e. we are estimating the zeroth derivative of the process using a first-degree polynomial approximation). Then (see Ex. 13.6), (13.3.11) reduces to

$$(1 - q\theta)^2 x^*_{n-r,n} = (1 - \theta) \left\{ (1 - q\theta) + [\theta - r(1 - \theta)](1 - q) \right\} y_n \tag{13.3.12}$$

which gives us the recursion algorithm

$$\begin{aligned} x^*_{n-r,n} = {} & 2\theta x^*_{n-1-r,n-1} - \theta^2 x^*_{n-2-r,n-2} \\ & + (1 - \theta) \left\{ y_n - \theta y_{n-1} + [\theta - r(1 - \theta)](y_n - y_{n-1}) \right\} \end{aligned} \tag{13.3.13}$$

Suppose moreover, that we desire a 1-step prediction. Then this is accomplished by setting $r = -1$ in the above equation which accordingly becomes

$$x^*_{n+1,n} = 2\theta x^*_{n,n-1} - \theta^2 x^*_{n-1,n-2} + 2(1-\theta)y_n - (1-\theta^2)y_{n-1} \tag{13.3.14}$$

This algorithm can now be readily implemented in a computer. ♦♦

In this way, given i, m and r, (13.3.11) can be reduced from its very general operational form to a directly programmable linear recursive form, and is thus seen to be a very basic formula for providing any of the algorithms which we might desire, in recursive form.

13.4 STABILITY

We now turn our attention to the question of the stability of the recursion algorithms which are generated by (13.3.11). As a first step, we reduce the right-hand side of that equation to its common denominator, obtaining

$$(D^i x^*)_{n-r,n} = \frac{f(q;r,\theta,\tau)}{(1-q\theta)^{m+1}} y_n \tag{13.4.1}$$

where $f(q;r,\theta,\tau)$ is a polynomial in q with parameters r, θ and τ. We now write this as

$$(1-q\theta)^{m+1}(D^i x^*)_{n-r,n} = f(q;r,\theta,\tau)\,y_n \tag{13.4.2}$$

and so it is clear that the homogeneous part of *any* recursion obtained from (13.3.11) will be, in operational form,

$$(1-q\theta)^{m+1}(D^i x^*)_{n-r,n} = 0 \tag{13.4.3}$$

From the results of Section 2.8 (see e.g. (2.8.12) and (2.8.13)) we know that (13.4.3) has solutions

$$(D^i x^*)_{n-r,n} = p_m(n)\,\theta^n \tag{13.4.4}$$

where $p_m(n)$ is a polynomial in n of degree m, whose coefficients depend on initial conditions. Thus the natural modes† of every algorithm obtained from (13.3.11) will be of the form of (13.4.4).

Now, in Section 9.5 we introduced the concept of *stability* as related to a recursion formula, and we stated there that an algorithm is said to be stable

†See p. 45.

if its natural modes go to zero as $n \to \infty$. From (13.4.4) we see *that this is in fact the case for the algorithms under consideration* if $|\theta| < 1$, since then

$$\lim_{n \to \infty} p_m(n)\,\theta^n = 0 \tag{13.4.5}$$

Hence we will restrict ourselves to this condition on θ throughout our discussion.† In order to avoid the occurrence of complex numbers in the outputs of the algorithms we further restrict θ to being *real,*‡ and finally, to avoid natural modes which alternate in sign on successive values of n, we restrict θ to being *positive.* We accordingly take θ to be a real number in the range

$$0 < \theta < 1 \tag{13.4.6}$$

When n is large, (13.4.4) shows that the natural modes of all of our algorithms will behave like $n^m\theta^n$.§ This means that they will persist longer for higher values of the degree of the estimating polynomial $[p^*(r)]_n$. It is almost inevitable that the natural modes will be excited sooner or later in practice, giving rise to what are termed *transient errors,* and so we see that it will be desirable to keep the degree of the estimator *as small as possible.* On the other hand it is intuitively clear that the outputs of the algorithms will contain systematic errors when the degree of the estimating polynomial is inadequate, and so for this reason we would want to *raise* the degree of the estimator. These two conflicting conditions mean that a compromise will have to be reached in which possible systematic errors are traded off against the persistence-time of the transient errors. (See Ex. 13.7.)

A second compromise is called for as the following shows. By making θ small (i.e. close to zero) we see from (13.4.4) that the natural modes die out more rapidly – a desirable situation. However when θ is small we see from (13.2.1) that the weight-function θ^r also dies out more rapidly. This in turn means that the estimate $[p^*(r)]_n$ will be based, effectively, on very recent data only, and that the older data will play very little part in the

†It was already assumed to hold when the discrete Laguerre polynomials were derived. See (3.3.1).

‡It has been pointed out to the author by A. J. Claus of Bell Laboratories that some very interesting possibilities for reducing bias errors are opened up by using complex values of θ, at a slight increase in the amount of computation. However, space does not permit us to dwell further on this point.

§The function $n^m\theta^n$ increases until $n = 1/\ln(1/\theta)^{1/m}$, thereafter tending to zero.

formation of the estimate. The filter can now be likened to a Fixed-Memory Polynomial Filter with a *short* observation time (i.e. L small). On the other hand when θ is close to unity then θ^r dies out slowly, and from (13.2.1) the filter is seen to be similar to a Fixed-Memory Filter with a *long* observation time (L large). It thus begins to appear as though a direct analogy might exist between θ and L, i.e. θ close to zero corresponding to L small, and θ close to unity corresponding to L large.

We recall from Chapter 7 that when L is large then smoothing is good but systematic errors are large, and that decreasing L reduces the smoothing but improves the systematic errors. Thus, on the assumption that the above-mentioned analogy does exist, making θ close to unity should result in good smoothing but in large systematic errors, and making θ close to zero, the reverse. *This is in fact the case* and we shall prove rigorously that as $\theta \to 1$:

a) The variance reduction factors of the Fading-Memory Polynomial Filter go to zero,

b) The systematic errors become unbounded.

Thus the effects of letting $\theta \to 1$ *will be seen to compare precisely with the effects of letting* $L \to \infty$ *in the filters of* Chapter 7.

We are thus not free to choose θ solely on the basis of transient error considerations. Clearly from (13.4.4) we see that $\theta \approx 0$ is most desirable if we are only concerned with rapidly expiring transients. This choice for θ is also most desirable from a systematic error standpoint. But $\theta \approx 1$ will be seen to be the most desirable from a smoothing standpoint, and so a compromise will have to be reached in which the above considerations are traded-off against each other.

One final fact should be pointed out. We will see that the analogy between the parameter θ, for these filters, and L, for those of Chapter 7, is a very strong one, and we will actually be able to compare the two filters quantitatively by setting up relationships between θ and L for equal variance reduction. However the filters of Chapter 7 are *nonrecursive* and so their transient errors, caused for example by an abrupt change in the signal, will persist for a *finite time,* expiring *completely* thereafter. On the other hand, the present filters are *recursive* and have transient errors which die out exponentially (see (13.4.4)). Thus, in theory, these transients *do not ever expire completely* – they merely decay until they drop below a perceptible value. The presence of these exponentially decaying transients is a distinct disadvantage which these filters possess relative to those of Chapter 7. However, as we show in the next section, they have the very strong counter-advantage of giving rise to algorithms which are extremely compact, and are thus easier to compute and require much less memory space than those of Chapter 7.

13.5 COUPLING PROCEDURE

We now show that an extremely compact set of algorithms can be obtained which recursively compute the state-vectors of the 1-step predictions. These are of particular interest in tracking systems where such predictions must constantly be provided in order to ensure that further observations will continue to be made. The 1-step predictions can of course be obtained from (13.3.11) by setting $r = -1$, but as will be seen, the following approach results in a far more economical procedure. Once the 1-step prediction state-vector has been obtained, we know that, by the use of the appropriate polynomial transition matrix, the state-vector can be obtained for any other validity instant. Thus the fact that the algorithms which we are about to derive are all 1-step predictors, does not constitute a limitation, and because of their computational convenience they will be readily seen to be the best way in which the Fading-Memory Filters should actually be implemented.

As a start it can easily be verified (see Ex. 13.8) that the polynomials $p_j(r)$ of (13.2.2) satisfy

$$p_j(-1) = 1 \tag{13.5.1}$$

for *all* j. That being the case, (13.2.8) gives us

$$\varphi_j(-1) = K_j \tag{13.5.2}$$

and so, setting $r = -1$ and $i = 0$ in (13.3.11), it follows that

$$x^*_{n+1,n} = \left[\sum_{j=0}^{m} \frac{K_j^2 \theta^j (1 - q)^j}{(1 - q\theta)^{j+1}}\right] y_n \tag{13.5.3}$$

Inserting (13.2.7) for K_j and carrying out the indicated summation (see Ex. 13.8), this reduces to

$$x^*_{n+1,n} = \frac{1}{q}\left[1 - \left(\frac{1 - q}{1 - q\theta}\right)^{m+1}\right] y_n \tag{13.5.4}$$

which is a recursion for $x^*_{n+1,n}$ in terms of its predecessors and the observations. Thus, assuming a first-degree estimating polynomial ($m = 1$), we obtain, after clearing fractions,

$$(1 - q\theta)^2 x^*_{n+1,n} = \left[2(1-\theta) - q(1-\theta^2)\right] y_n \tag{13.5.5}$$

which is easily seen to be in precise agreement with (13.3.14).

We now introduce the *scaled*-derivative estimate state-vector

$$Z^*_{n+1,n} \equiv \begin{pmatrix} z^*_0 \\ z^*_1 \\ \vdots \\ z^*_m \end{pmatrix}_{n+1,n} \equiv \begin{pmatrix} x^* \\ \tau \dot{x}^* \\ \vdots \\ \dfrac{\tau^m}{m!} D^m x^* \end{pmatrix}_{n+1,n} \tag{13.5.6}$$

(c/f (13.2.24) where we defined the *unscaled*-derivative estimate state-vector $X^*_{n-r,n}$). Then, (13.5.4) becomes (after multiplying both sides by q),

$$\left(z^*_0\right)_{n,n-1} = \left[1 - \left(\frac{1-q}{1-q\theta}\right)^{m+1}\right] y_n \tag{13.5.7}$$

Define the prediction error (c/f (9.3.21))

$$\epsilon_n \equiv y_n - \left(z^*_0\right)_{n,n-1} \tag{13.5.8}$$

i.e. the difference between the observation, y_n, and the prediction based on data up to the preceding instant, namely $\left(z^*_0\right)_{n,n-1}$. We combine (13.5.7) and (13.5.8) and obtain the result (see Ex. 13.9)

$$\epsilon_n = \left(\frac{1-q}{1-q\theta}\right)^{m+1} y_n \tag{13.5.9}$$

which shows that the prediction error, for these algorithms, can be readily computed by a recursion on the data. As an example, for $m = 1$, (13.5.9) gives us

$$\epsilon_n = 2\theta\epsilon_{n-1} - \theta^2\epsilon_{n-2} + y_n - 2y_{n-1} + y_{n-2} \tag{13.5.10}$$

We now eliminate y_n between (13.5.9) and (13.5.7), thereby obtaining (see Ex. 13.9)

$$\left(z_0^*\right)_{n+1,n} = \frac{1}{q}\left[\left(\frac{1-q\theta}{1-q}\right)^{m+1} - 1\right]\epsilon_n \tag{13.5.11}$$

which gives the prediction as a recursion on the prediction errors. Note that the q in the denominator on the right of (13.5.11) can always be divided into the numerator since the latter also possesses q as a factor (see Ex. 13.9).

We now make the same heuristic assumption as we did on p. 356, *namely that the updated estimate* $Z^*_{n,n}$ *is equal to the sum of the previous prediction* $Z^*_{n,n-1}$ *and a multiple of the prediction error,* i.e. that

$$Z^*_{n,n} = Z^*_{n,n-1} + H_n\epsilon_n \tag{13.5.12}$$

where H_n is a vector of weights, possibly dependent on n. The above is equivalent to (c/f (9.4.8))

$$\Phi(-1)\,Z^*_{n+1,n} = Z^*_{n,n-1} + H_n\epsilon_n \tag{13.5.13}$$

where $\Phi(-1)$ is defined by (4.2.15). The vector H_n will be chosen so that (13.5.13) gives an algorithm which is consistent with (13.5.11).

For simplicity consider the first-degree case ($m = 1$). Then (13.5.13) is

$$\begin{pmatrix}1 & -1\\ 0 & 1\end{pmatrix}\begin{pmatrix}z_0^*\\ z_1^*\end{pmatrix}_{n+1,n} = \begin{pmatrix}z_0^*\\ z_1^*\end{pmatrix}_{n,n-1} + \begin{pmatrix}h_0\\ h_1\end{pmatrix}_n\epsilon_n \tag{13.5.14}$$

which we write, using the backward-shifting operator q, as

$$\begin{pmatrix}1-q & -1\\ 0 & 1-q\end{pmatrix}\begin{pmatrix}z_0^*\\ z_1^*\end{pmatrix}_{n+1,n} = \begin{pmatrix}h_0\\ h_1\end{pmatrix}_n\epsilon_n \tag{13.5.15}$$

Solving for the vector on the left now gives us

$$\begin{pmatrix} z_0^* \\ z_1^* \end{pmatrix}_{n+1,n} = \frac{1}{(1-q)^2}\begin{pmatrix} 1-q & 1 \\ 0 & 1-q \end{pmatrix}\begin{pmatrix} h_0 \\ h_1 \end{pmatrix}_n \epsilon_n \tag{13.5.16}$$

from which

$$\left(z_0^*\right)_{n+1,n} = \frac{(1-q)\left(h_0\right)_n \epsilon_n + \left(h_1\right)_n \epsilon_n}{(1-q)^2} \tag{13.5.17}$$

i.e.

$$\left(z_0^*\right)_{n+1,n} = \frac{\left(h_0 + h_1\right)_n \epsilon_n - \left(h_0\right)_{n-1} \epsilon_{n-1}}{(1-q)^2} \tag{13.5.18}$$

We now return to (13.5.11) and setting $m = 1$ we obtain

$$\left(z_0^*\right)_{n+1,n} = \frac{1}{q}\left[\left(\frac{1-q\theta}{1-q}\right)^2 - 1\right]\epsilon_n \tag{13.5.19}$$

i.e.

$$\left(z_0^*\right)_{n+1,n} = \left[\frac{2(1-\theta) - q(1-\theta^2)}{(1-q)^2}\right]\epsilon_n \tag{13.5.20}$$

so that

$$\left(z_0^*\right)_{n+1,n} = \frac{2(1-\theta)\epsilon_n - (1-\theta^2)\epsilon_{n-1}}{(1-q)^2} \tag{13.5.21}$$

Equating coefficients of ϵ_n and ϵ_{n-1} in (13.5.18) and (13.5.21) then yields

$$\left(h_0 + h_1\right)_n = 2(1-\theta) \tag{13.5.22}$$

$$\left(h_0\right)_{n-1} = 1 - \theta^2 \tag{13.5.23}$$

and so finally

$$h_0 = 1 - \theta^2 \tag{13.5.24}$$

$$h_1 = (1 - \theta)^2 \tag{13.5.25}$$

We now insert these expressions (note that they in fact do *not* depend on n) in (13.5.14) and, by a slight rearrangement, we obtain the form best suited to actual computation, i.e.

$$\left(z_1^*\right)_{n+1,n} = \left(z_1^*\right)_{n,n-1} + (1 - \theta)^2 \epsilon_n \tag{13.5.26}$$

$$\left(z_0^*\right)_{n+1,n} = \left(z_0^*\right)_{n,n-1} + \left(z_1^*\right)_{n+1,n} + (1 - \theta^2)\epsilon_n \tag{13.5.27}$$

This is the algorithm which gives the 1-step predictions of position and scaled-velocity, based on the first-degree polynomial $\left[p^*(r)\right]_n$ which minimizes e_n of (13.2.1). We note that if we were to set up the algorithms to estimate the same two quantities starting from (13.3.11), we would obtain two *uncoupled* expressions of the same form as (13.3.14). However their execution would then be about twice as costly as the *coupled* pair (13.5.26) and (13.5.27). (See Ex. 13.10.)

Observe that only *three* memory locations are required above – two permanent and one temporary. In the temporary location we store ϵ_n, computed from (13.5.8) upon the receipt of y_n. Then $\left(z_1^*\right)_{n+1,n}$ is computed from (13.5.26) and is immediately reinserted into the permanent storage location containing $\left(z_1^*\right)_{n,n-1}$. (The latter is of course now no longer needed.) Finally $\left(z_0^*\right)_{n+1,n}$ is computed from (13.5.27), the result being stored in place of $\left(z_0^*\right)_{n,n-1}$. The weights $(1 - \theta)^2$ and $(1 - \theta^2)$ can be precomputed, *once and for all*, when θ is selected. It is thus clear that the above pair in fact constitutes a very compact algorithm for estimating the 1-step predictions, based on first-degree fading-memory polynomial approximation. The question of initialization will be discussed later.

In Table 13.1 (see p. 516) we display the algorithms up to degree 4, obtained in a manner analogous to the above. We are also able to give the general form of the weight-vector H, appearing in (13.5.13). Thus, define the $(m + 1) \times (m \times 1)$ matrix $P(r)$ whose i, j^{th} element is

$$\left[P(r)\right]_{ij} \equiv \frac{(-1)^i}{i!} \frac{d^i}{dr^i} \varphi_j(r) \qquad 0 \le i, j \le m \tag{13.5.28}$$

where $\varphi_j(r)$ is given by (13.2.8). Also define the $(m+1)$-vector L by

$$[L]_j \equiv K_j \theta^j \qquad 0 \le j \le m \tag{13.5.29}$$

where K_j is given by (13.2.7). Then the weight-vector H of (13.5.13) is given by

$$H = P(0)L \tag{13.5.30}$$

As an example, let $m = 1$. Then by (13.5.28),

$$P(r) = \begin{pmatrix} K_0 & K_1[\theta - r(1-\theta)] \\ 0 & K_1(1-\theta) \end{pmatrix} \tag{13.5.31}$$

and by (13.5.29),

$$L = \begin{pmatrix} K_0 \\ K_1\theta \end{pmatrix} \tag{13.5.32}$$

Hence, by (13.5.30),

$$P(0)L = \begin{pmatrix} K_0 & K_1\theta \\ 0 & K_1(1-\theta) \end{pmatrix} \begin{pmatrix} K_0 \\ K_1\theta \end{pmatrix} = \begin{pmatrix} 1-\theta^2 \\ (1-\theta)^2 \end{pmatrix} \tag{13.5.33}$$

This is in precise agreement with (13.5.24) and (13.5.25). ◆◆

The proof that (13.5.30) is true, is given as an exercise at the end of Chapter 14. (See Ex. 14.11.)

The algorithms of Table 13.1 and the expression (13.5.30) are in very close analogy to the coupled-form algorithms given in Table 9.1 on p. 360 and to (9.4.30) respectively. In fact the two sets of algorithms are seen to be *structurally identical* and differ *only in the multipliers* applied to the prediction-error, ϵ_n.

It is clear then, from an inspection of Table 13.1, that the Fading-Memory Polynomial Filters can be formualted as a set of extremely compact recursive algorithms. By comparison with the filters of Chapter 7, the present ones are indeed to be preferred from a computational and memory standpoint. In the present case, both the number of machine operations and the number of storage locations depend solely on the degree m, whereas in the

Table 13.1 The Fading-Memory Polynomial Filter

Define:

$$\begin{pmatrix} z_0^* \\ z_1^* \\ z_2^* \\ z_3^* \\ z_4^* \end{pmatrix}_{n+1,n} \equiv \begin{pmatrix} x^* \\ \tau D x^* \\ \dfrac{\tau^2}{2!} D^2 x^* \\ \dfrac{\tau^3}{3!} D^3 x^* \\ \dfrac{\tau^4}{4!} D^4 x^* \end{pmatrix}_{n+1,n}$$

$$\epsilon_n = y_n - \left(z_0^*\right)_{n,n-1}$$

Degree 0

$$\left(z_0^*\right)_{n+1,n} = \left(z_0^*\right)_{n,n-1} + (1-\theta)\,\epsilon_n$$

Degree 1

$$\left(z_1^*\right)_{n+1,n} = \left(z_1^*\right)_{n,n-1} + (1-\theta)^2\,\epsilon_n$$

$$\left(z_0^*\right)_{n+1,n} = \left(z_0^*\right)_{n,n-1} + \left(z_1^*\right)_{n+1,n} + (1-\theta^2)\,\epsilon_n$$

Degree 2

$$\left(z_2^*\right)_{n+1,n} = \left(z_2^*\right)_{n,n-1} + \frac{1}{2}(1-\theta)^3\,\epsilon_n$$

$$\left(z_1^*\right)_{n+1,n} = \left(z_1^*\right)_{n,n-1} + 2\left(z_2^*\right)_{n+1,n} + \frac{3}{2}(1-\theta)^2(1+\theta)\,\epsilon_n$$

$$\left(z_0^*\right)_{n+1,n} = \left(z_0^*\right)_{n,n-1} + \left(z_1^*\right)_{n+1,n} - \left(z_2^*\right)_{n+1,n} + (1-\theta^3)\,\epsilon_n$$

Table 13.1 The Fading-Memory Polynomial Filter *(Continued)*

Degree 3

$$\left(z_3^*\right)_{n+1,n} = \left(z_3^*\right)_{n,n-1} + \frac{1}{6}(1-\theta)^4 \epsilon_n$$

$$\left(z_2^*\right)_{n+1,n} = \left(z_2^*\right)_{n,n-1} + 3\left(z_3^*\right)_{n+1,n} + (1-\theta)^3(1+\theta)\epsilon_n$$

$$\left(z_1^*\right)_{n+1,n} = \left(z_1^*\right)_{n,n-1} + 2\left(z_2^*\right)_{n+1,n} - 3\left(z_3^*\right)_{n+1,n} + \frac{1}{6}(1-\theta)^2(11+14\theta+11\theta^2)\epsilon_n$$

$$\left(z_0^*\right)_{n+1,n} = \left(z_0^*\right)_{n,n-1} + \left(z_1^*\right)_{n+1,n} - \left(z_2^*\right)_{n+1,n} + \left(z_3^*\right)_{n+1,n} + (1-\theta^4)\epsilon_n$$

Degree 4

$$\left(z_4^*\right)_{n+1,n} = \left(z_4^*\right)_{n,n-1} + \frac{1}{24}(1-\theta)^5 \epsilon_n$$

$$\left(z_3^*\right)_{n+1,n} = \left(z_3^*\right)_{n,n-1} + 4\left(z_4^*\right)_{n+1,n} + \frac{5}{12}(1-\theta)^4(1+\theta)\epsilon_n$$

$$\left(z_2^*\right)_{n+1,n} = \left(z_2^*\right)_{n,n-1} + 3\left(z_3^*\right)_{n+1,n} - 6\left(z_4^*\right)_{n+1,n} + \frac{5}{24}(1-\theta)^3(7+10\theta+7\theta^2)\epsilon_n$$

$$\left(z_1^*\right)_{n+1,n} = \left(z_1^*\right)_{n,n-1} + 2\left(z_2^*\right)_{n+1,n} - 3\left(z_3^*\right)_{n+1,n} + 4\left(z_4^*\right)_{n+1,n} + \frac{5}{12}(1-\theta)^2(5+7\theta+7\theta^2+5\theta^3)\epsilon_n$$

$$\left(z_0^*\right)_{n+1,n} = \left(z_0^*\right)_{n,n-1} + \left(z_1^*\right)_{n+1,n} - \left(z_2^*\right)_{n+1,n} + \left(z_3^*\right)_{n+1,n} - \left(z_4^*\right)_{n+1,n} + (1-\theta^5)\epsilon_n$$

case of the Fixed-Memory Polynomial Filters dependence was on the *product* mL. Thus the drawback that the present filters have of possessing exponentially decaying transient errors (vs. transients which disappear completely in finite time for the filters of Chapter 7), is strongly counterbalanced by a decided computational advantage.

In conclusion we make the following very useful observation. By (13.5.13) we have that

$$Z^*_{n+1,n} = \Phi(1) Z^*_{n,n-1} + \Phi(1) H \left[y_n - \left(z^*_0\right)_{n,n-1} \right] \tag{13.5.34}$$

Define the row-vector with $m + 1$ elements

$$M \equiv (1, \ 0, \ 0, \ \ldots \ 0) \tag{13.5.35}$$

Then (13.5.34) can be written

$$Z^*_{n+1,n} = \Phi(1) Z^*_{n,n-1} + \Phi(1) H \left(y_n - M Z^*_{n,n-1} \right) \tag{13.5.36}$$

i.e.

$$Z^*_{n+1,n} = \Phi(1)(I - HM) Z^*_{n,n-1} + \Phi(1) H y_n \tag{13.5.37}$$

This is a set of coupled linear difference equations with forcing-vector $\Phi(1) H y_n$, and so removing the latter gives us as the homogeneous part,

$$Z^*_{n+1,n} = \Phi(1)(I - HM) Z^*_{n,n-1} \tag{13.5.38}$$

Now every one of the elements of $Z^*_{n+1,n}$ is known, by (13.4.3) to satisfy a homogeneous equation of the form

$$(1 - q\theta)^{m+1} \left(z^*_i\right)_{n+1,n} = 0 \tag{13.5.39}$$

This gives us the *eigenvalues* of the matrix $\Phi(1)(I - HM)$, as we now show.

Using the q-operator, (13.5.38) can be written

$$\left[\Phi(1)(I - HM) - q^{-1} I \right] Z^*_{n,n-1} = \emptyset \tag{13.5.40}$$

which is a set of coupled difference equations in operational form. In (2.8.31) we pointed out that by setting $q^{-1} = \lambda$, an operator can be converted into its associated *characteristic equation*. In the above case, we obtain

$$[\Phi(1)(I - HM) - \lambda I] Z^*_{n,n-1} = \emptyset \tag{13.5.41}$$

and so, excluding the trivial result $Z^*_{n,n-1} = \emptyset$, we must then have that the matrix $\Phi(1)(I - HM) - \lambda I$ is singular, i.e. that

$$\det[\Phi(1)(I - HM) - \lambda I] = 0 \tag{13.5.42}$$

This last result is a polynomial in λ of degree $m + 1$ and is the characteristic equation for each of the elements of $Z^*_{n,n-1}$ in (13.5.40). But then by (13.5.39)

$$\det[\Phi(1)(I - HM) - \lambda I] = (\lambda - \theta)^{m+1} \tag{13.5.43}$$

We thus see that (13.5.42) has the *single root* $\lambda = \theta$ with multiplicity $m + 1$.

Now (13.5.41) is also seen to be a statement of the *eigenvalue problem* for the matrix $\Phi(1)(I - HM)$. We have thus proved that this matrix has the value θ for each of its $m + 1$ eigenvalues, i.e. that it possesses a single eigenvalue with multiplicity $m + 1$. This result will be applied to the development of a computational procedure in the next section. The reader is also referred to Ex. 13.11.

13.6 VARIANCE REDUCTION

We now turn our attention to the question of variance reduction and to the general problem of obtaining an expression for the covariance matrix of the output errors given the statistics of the input errors.

It was shown in Section 13.2 that each element of the estimate state-vector is obtained by an appropriate linear transformation of the semi-infinite data-vector

$$Y_{(n)} \equiv \left(y_n, y_{n-1}, y_{n-2}, \ldots\right)^T \tag{13.6.1}$$

Thus there exists a matrix $W(r)$ so that the vector Z^* of (13.5.6) is given by

$$Z^*_{n-r,n} = W(r) Y_{(n)} \tag{13.6.2}$$

The matrix $W(r)$ has $m + 1$ rows and an infinite number of columns.

Given that the covariance matrix of the errors in $Y_{(n)}$ is the (infinite order) square matrix $R_{(n)}$, it follows immediately that the covariance matrix of the errors in $Z^*_{n-r,n}$ will be

$$S^*_{n-r,n} = W(r)\,R_{(n)}\,W(r)^T \tag{13.6.3}$$

In practice this matrix may be very difficult to evaluate since it involves infinite sums. However it can be evaluated in closed form in one case of particular interest, namely the situation where the observation errors are uncorrelated, stationary and of equal variance. Thus we shall assume that

$$R_{(n)} = \sigma_\nu^2 I \tag{13.6.4}$$

where σ_ν^2 does not depend on n, and where I is the infinite-order identity matrix. Under these circumstances (13.6.3) becomes

$$S^*_{n-r,n} = \sigma_\nu^2\, W(r)\, W(r)^T \tag{13.6.5}$$

and, as we now show, closed form expressions can be obtained for the elements of $S^*_{n-r,n}$. (The reader will thus note that we are restricting ourselves to the same situations as were considered in (7.5.5) and (9.7.2) of Chapters 7 and 9 respectively.)

Our first task is to obtain the form of the i, j^{th} element of the matrix $W(r)$ of (13.6.2). Returning to (13.2.22) and (13.2.16) we have that

$$\left(D^i x^*\right)_{n-r,n} = \left(-\frac{1}{\tau}\right)^i \sum_{\lambda=0}^{m}\left[\sum_{j=0}^{\infty} \frac{d^i}{dr^i}\,\varphi_\lambda(r)\,\varphi_\lambda(j)\,\theta^j\right] y_{n-j} \tag{13.6.6}$$

and by (13.5.6) we have

$$\left(z_i^*\right)_{n-r,n} \equiv \left(\frac{\tau^i}{i!}\,D^i x^*\right)_{n-r,n} \tag{13.6.7}$$

We thus see that by combining the above two equations

$$\left(z_i^*\right)_{n-r,n} = \sum_{j=0}^{\infty}\left[\sum_{\lambda=0}^{m} \frac{(-1)^i}{i!}\,\frac{d^i}{dr^i}\,\varphi_\lambda(r)\,\varphi_\lambda(j)\,\theta^j\right] y_{n-j} \tag{13.6.8}$$

and so it follows that the i, j^{th} element of $W(r)$ is

$$\left[W(r)\right]_{ij} = \sum_{\lambda=0}^{m} \frac{(-1)^i}{i!} \frac{d^i}{dr^i} \varphi_\lambda(r)\, \varphi_\lambda(j)\, \theta^j \tag{13.6.9}$$

for $0 \leq i \leq m$ and $0 \leq j \leq \infty$. This is now applied to (13.6.5).

By the definition of matrix multiplication, if

$$S = WW^T \tag{13.6.10}$$

then

$$[S]_{ij} = \sum_k [W]_{ik} [W]_{jk} \tag{13.6.11}$$

Thus, by (13.6.5) and (13.6.9),

$$\left[S^*_{n-r,n}\right]_{ij}$$

$$= \sigma_\nu^2 \sum_{k=0}^{\infty} \left[\sum_{\lambda=0}^{m} \frac{(-1)^i}{i!} \frac{d^i}{dr^i} \varphi_\lambda(r) \varphi_\lambda(k) \theta^k\right]\left[\sum_{\mu=0}^{m} \frac{(-1)^j}{j!} \frac{d^j}{dr^j} \varphi_\mu(r)\, \varphi_\mu(k) \theta^k\right]$$

$$= \sigma_\nu^2 \sum_{\lambda=0}^{m} \sum_{\mu=0}^{m} \left[\sum_{k=0}^{\infty} \varphi_\lambda(k)\, \varphi_\mu(k) \theta^{2k}\right] \frac{(-1)^i}{i!} \frac{d^i}{dr^i} \varphi_\lambda(r) \frac{(-1)^j}{j!} \frac{d^j}{dr^j} \varphi_\mu(r) \tag{13.6.12}$$

which we write, with the aid of (13.2.8) as

$$\left[S^*_{n-r,n}\right]_{i,j}$$

$$= \sigma_\nu^2 \sum_{\lambda=0}^{m} \sum_{\mu=0}^{m} \left[\sum_{k=0}^{\infty} K_\lambda \varphi_\lambda(k) K_\mu \varphi_\mu(k) \theta^{2k}\right] \frac{(-1)^i}{i!} \frac{d^i}{dr^i} p_\lambda(r) \frac{(-1)^j}{j!} \frac{d^j}{dr^j} p_\mu(r) \tag{13.6.13}$$

Now the term in parentheses is studied in [13.6], and it is shown in Appendix II of that reference that it can be evaluated *once and for all* for the given polynomials. In fact, defining the matrix A whose λ,μ^{th} element is

$$a_{\lambda,\mu} \equiv K_\lambda K_\mu \sum_{k=0}^{\infty} \varphi_\lambda(k)\, \varphi_\mu(k)\, \theta^{2k} \tag{13.6.14}$$

it is shown there that

$$a_{\lambda,\mu} = \binom{\lambda + \mu}{\lambda} \frac{1 - \theta}{(1 + \theta)^{\lambda + \mu + 1}} \tag{13.6.15}$$

We accordingly rewrite (13.6.13) as

$$\left[S^*_{n-r,n}\right]_{i,j} = \sigma_\nu^2 \sum_{\lambda=0}^{m} \sum_{\mu=0}^{m} a_{\lambda,\mu} \frac{(-1)^i}{i!} \frac{d^i}{dr^i} p_\lambda(r) \frac{(-1)^j}{j!} \frac{d^j}{dr^j} p_\mu(r) \tag{13.6.16}$$

Next we define the matrix $F(r)$ whose i, j^{th} element is (c/f (13.5.28))

$$\left[F(r)\right]_{i,j} \equiv \frac{(-1)^i}{i!} \frac{d^i}{dr^i} p_j(r) \tag{13.6.17}$$

Then recalling that $F(r)^T$ and $F(r)$ satisfy

$$\left[F(r)^T\right]_{i,j} = \left[F(r)\right]_{j,i} \tag{13.6.18}$$

we see that (13.6.16) can be written

$$\left[S^*_{n-r,n}\right]_{i,j} = \sigma_\nu^2 \sum_{\lambda=0}^{m} \left[F(r)\right]_{i,\lambda} \sum_{\mu=0}^{m} a_{\lambda,\mu} \left[F(r)^T\right]_{\mu,j} \tag{13.6.19}$$

which is equivalent to the matrix triple-product

$$S^*_{n-r,n} = \sigma_\nu^2 F(r)\, A F(r)^T \tag{13.6.20}$$

This then is the required closed-form expression for the covariance matrix of the errors in the scaled-derivative vector $Z^*_{n-r,n}$ *for the case where* (13.6.4) *applies.*

Since the unscaled-derivative vector $X^*_{n-r,n}$ is related to $Z^*_{n-r,n}$ by (c/f (4.2.18))

$$X^*_{n-r,n} = D(\tau) Z^*_{n-r,n} \tag{13.6.21}$$

it follows immediately that the covariance matrix of the errors in $X^*_{n-r,n}$ is given by

$$S^*_{n-r,n} = \sigma_\nu^2 D(\tau) F(r) A F(r)^T D(\tau) \tag{13.6.22}$$

where $D(\tau)$ was defined in (4.2.19).

Consider an example. Let $r = -1$ and $m = 1$, i.e. we are considering a 1-step prediction based on a first-degree estimator. Then (c/f (13.5.31))

$$F(r) = \begin{pmatrix} 1 & \theta - r(1-\theta) \\ 0 & 1-\theta \end{pmatrix} \tag{13.6.23}$$

and so

$$F(-1) = \begin{pmatrix} 1 & 1 \\ 0 & 1-\theta \end{pmatrix} \tag{13.6.24}$$

Moreover, by (13.6.15)

$$A = \begin{pmatrix} \dfrac{1-\theta}{1+\theta} & \dfrac{1-\theta}{(1+\theta)^2} \\[2ex] \dfrac{1-\theta}{(1+\theta)^2} & 2\,\dfrac{1-\theta}{(1+\theta)^3} \end{pmatrix} \tag{13.6.25}$$

and so (13.6.20) gives us

$$S^*_{n+1,n} = \sigma_\nu^2 \frac{1-\theta}{(1+\theta)^3} \begin{pmatrix} 5 + 4\theta + \theta^2 & (1-\theta)(3+\theta) \\ (1-\theta)(3+\theta) & 2(1-\theta)^2 \end{pmatrix} \tag{13.6.26}$$

This is the covariance matrix of the errors in the vector

$$Z^*_{n+1,n} \equiv (x^*,\ \tau Dx^*)^T_{n+1,n} \tag{13.6.27}$$

Then by (13.6.22) the covariance matrix of the errors in

$$X^*_{n+1,n} \equiv (x^*,\ Dx^*)^T_{n+1,n} \tag{13.6.28}$$

must be

$$S^*_{n+1,n} = \sigma_\nu^2 \frac{1-\theta}{(1+\theta)^3} \begin{pmatrix} 5 + 4\theta + \theta^2 & \frac{1}{\tau}(1-\theta)(3+\theta) \\ \frac{1}{\tau}(1-\theta)(3+\theta) & \frac{2}{\tau^2}(1-\theta)^2 \end{pmatrix} \tag{13.6.29}$$

◆◆

By (4.2.16) we have that

$$Z^*_{n+h-r,n} = \Phi(h)\, Z^*_{n-r,n} \tag{13.6.30}$$

where $\Phi(h)$ is defined in (4.2.15), and so by (5.5.22)

$$S^*_{n+h-r,n} = \Phi(h)\, S^*_{n-r,n}\, \Phi(h)^T \tag{13.6.31}$$

This result can be used to shift the validity-instant in the covariance matrix as computed from (13.6.20).

As an example, we have obtained $S^*_{n+1,n}$ in (13.6.26) for $m = 1$. Then $S^*_{n,n}$ can be obtained by setting both h and r equal to -1 in (13.6.31). Now by (4.2.15)

$$\Phi(-1) = \begin{pmatrix} 1 & -1 \\ 0 & 1 \end{pmatrix} \tag{13.6.32}$$

and so, forming

$$S^*_{n,n} \equiv \Phi(-1)\, S^*_{n+1,n}\, \Phi(-1)^T \tag{13.6.33}$$

we obtain

$$S^*_{n,n} = \sigma_\nu^2 \frac{1-\theta}{(1+\theta)^3} \begin{pmatrix} 1 + 4\theta + 5\theta^2 & (1-\theta)(1+3\theta) \\ (1-\theta)(1+3\theta) & 2(1-\theta)^2 \end{pmatrix} \tag{13.6.34}$$

Of course this last result could also have been obtained directly from (13.6.20) by setting $r = 0$. (See Ex. 13.13.) ◆◆

We now make the following fundamental observation. The covariance matrices for the filters of this chapter and for the class of errors considered in (13.6.4) are seen from (13.6.20) to be a congruence transformation on the matrix A. But by (13.6.15) we see that every element of A goes to zero as $\theta \to 1$, i.e.,

$$\lim_{\theta \to 1} A = \emptyset \tag{13.6.35}$$

It follows then that we can make the matrix $S^*_{n-r,n}$ *as close to a null matrix as we please by taking* θ *sufficient close to unity.* This in turn means that the diagonal elements of $S^*_{n-r,n}$ go to zero as $\theta \to 1$, and so the variances of the estimates can be made as small as we please by choosing θ close enough to unity.

The parameter θ thus clearly provides us with control over the smoothing properties of the algorithms derived in this chapter, and by selection of its value we can make the estimates as smooth as we wish. Of course making θ too close to unity will be undesirable from a transient-error standpoint, as we demonstrated in Section 13.4. Indeed, by (13.4.4) we see that for $\theta \approx 1$ the natural modes (and hence the transient errors) will be extremely persistent and very troublesome. From a transient-error standpoint we desire $\theta \approx 0$, but this conflicts with the choice we would make for θ based solely on smoothing considerations. The user must thus make a compromise, selecting θ as close to unity as he dare, taking into account the penalty he would have to pay for overly persistent transients. As we shall demonstrate, the systematic errors are also affected by the choice of θ and so they, too, must enter into the considerations when θ is being selected.

Of course, regardless of the value we choose for θ (other than $\theta = 1$), we would expect that $S^*_{n-r,n}$ of (13.6.20) be positive definite. This is heuristically obvious, but can also be proved analytically. The reader is referred to Ex. 13.14.

In Table 13.2 we give the exact form of the *diagonal* elements of $S^*_{n+1,n}$ (the 1-step predictors) as a function of θ for $m = 0, 1, 2, 3$. We see of

Table 13.2 VRF† for 1-step Predictor

Degree (m)	Output	VRF $(0 < \theta < 1)$
0	$x^*_{n+1,n}$	$\dfrac{1-\theta}{1+\theta}$
1	$Dx^*_{n+1,n}$	$\dfrac{2}{\tau^2}\dfrac{(1-\theta)^3}{(1+\theta)^3}$
	$x^*_{n+1,n}$	$\dfrac{1-\theta}{(1+\theta)^3}(5+4\theta+\theta^2)$
2	$D^2x^*_{n+1,n}$	$\dfrac{6}{\tau^4}\dfrac{(1-\theta)^5}{(1+\theta)^5}$
	$Dx^*_{n+1,n}$	$\dfrac{1}{\tau^2}\dfrac{(1-\theta)^3}{(1+\theta)^5}\left(\dfrac{49+50\theta+13\theta^2}{2}\right)$
	$x^*_{n+1,n}$	$\dfrac{1-\theta}{(1+\theta)^5}(19+24\theta+16\theta^2+6\theta^3+\theta^4)$

3	$D^3 x^*_{n+1,n}$	$\frac{20}{\tau^6} \frac{(1-\theta)^7}{(1+\theta)^7}$
	$D^2 x^*_{n+1,n}$	$\frac{1}{\tau^4} \frac{(1-\theta)^5}{(1+\theta)^7} (126 + 152\theta + 46\theta^2)$
	$Dx^*_{n+1,n}$	$\frac{1}{\tau^2} \frac{(1-\theta)^3}{(1+\theta)^7} \left(\frac{2797 + 4634\theta + 3810\theta^2 + 1706\theta^3 + 373\theta^4}{18}\right)$
	$x^*_{n+1,n}$	$\frac{1-\theta}{(1+\theta)^7} (69 + 104\theta + 97\theta^2 + 64\theta^3 + 29\theta^4 + 8\theta^5 + \theta^6)$

†The Variance Reduction Factor of $D^i x^*$ is defined as $\frac{E\left\{\left(D^i x^*_{n+1,n}\right)^2\right\}}{\sigma^2(\nu)}$ and is thus the diagonal element of the estimate covariance matrix when the variance of the input errors is unity.

course that each entry has a zero at $\theta = 1$. In fact, the multiplicity of the zero in the term pertaining to $D'x^*_{n-r,n}$ is $2i + 1$. Although we have not proved it, it is easy to show that, in general, the i, j^{th} element of $S^*_{n-r,n}$ has the form

$$\left[S^*_{n-r,n}\right]_{i,j} = \sigma_\nu^2 \frac{(1-\theta)^{i+j+1} \lambda_{i,j}(\theta, r; m)}{\tau^{i+j}} \tag{13.6.36}$$

where $\lambda_{i,j}(\theta, r; m)$ is a function of θ and r, (for each value of m) that is *nonzero* at $\theta = 1$. (This result can be derived from (13.6.20) and should be compared to (7.6.12) with which a strong similarity will be observed.)

When θ is close to unity, then (13.6.36) is close to

$$\left[S^*_{n-r,n}\right]_{i,j} = \sigma_\nu^2 \frac{(1-\theta)^{i+j+1}}{\tau^{i+j}} \lambda_{i,j}(1, r; m) \tag{13.6.37}$$

and the constants $\lambda_{i,j}(1, r; m)$ can be readily evaluated (see Ex. 13.15). In Table 13.3 we display those constants for the *diagonal* elements of $S^*_{n+1,n}$ (i.e., $r = -1$) up to i and m equal to 10. This table is extremely useful, and permits us to approximate the variance reduction factors† for these filters when θ is close to unity.‡

Moving across one of the rows of Table 13.3, say the top one, shows that as the degree increases so the VRF of the corresponding estimate also increases, i.e.

Degree	0	1	2	3
VRF of Position Estimate	$0.5(1-\theta)$	$1.25(1-\theta)$	$2.06(1-\theta)$	$2.9(1-\theta)$

The above fact is true for every row of that table and can be proved analytically without too much difficulty. We thus conclude that, from a variance reduction standpoint, it is desirable that we keep the degree of the estimator *as small as possible.* From the standpoint of systematic errors however, the

†See p. 257 for a definition of variance reduction factor.

‡The Table is asymptotically precise at $\theta = 1$ and becomes progressively worse as θ leaves unity. These approximations can be used with reasonable accuracy down to $\theta \approx 0.7$.

Table 13.3 Numerical Constants in the VRF Formula (λ_{ii})
1-step Prediction (θ close to 1)

i \ m	0	1	2	3	4	5	6	7	8	9	10
0	5.0(−1)	1.25(0)	2.063(0)	2.906(0)	3.770(0)	4.647(0)	5.534(0)	6.429(0)	7.331(0)	8.238(0)	9.150(0)
1		2.5 (−1)	1.75 (0)	5.781(0)	1.375(0)	2.714(1)	4.748(1)	7.636(1)	1.154(2)	1.661(2)	2.303(2)
2			1.875(−1)	2.531(0)	1.378(1)	4.906(1)	1.358(2)	3.177(2)	6.594(2)	1.251(3)	2.211(3)
3				1.563(−1)	3.438(0)	2.777(1)	1.377(2)	5.070(2)	1.525(3)	3.958(3)	9.184(3)
4					1.367(−1)	4.443(0)	4.982(1)	3.276(2)	1.546(3)	5.800(3)	1.839(4)
5						1.231(−1)	5.537(0)	8.218(1)	6.914(2)	4.065(3)	1.860(4)
6							1.128(−1)	6.711(0)	1.273(2)	1.333(3)	9.552(3)
7								1.047(−1)	7.960(0)	1.878(2)	2.396(3)
8									9.819(−2)	9.279(0)	2.666(2)
9										9.274(−2)	1.067(1)
10											8.810(−2)

Note 1: 5.070(2) means 5.070×10^2.

Note 2: $\text{VRF} = \dfrac{(1-\theta)^{2i+1}}{\tau^{2i}} \lambda_{ii}(m)$ (see (13.6.37)). Ex: $i = 2, m = 3$, $\text{VRF} = \dfrac{(1-\theta)^5}{\tau^4} \times 2.531 \times 10^0$.

degree should be as high as possible – a fact which is intuitively obvious and to which we shall return at a later stage. Thus, as with the selection of θ, so the degree must be chosen to balance the variance reduction properties against the systematic errors. The transient errors were shown to be least troublesome for the lowest degree, and so in this case transient errors and variance reduction are in concert, both calling for the smallest possible value of m.

In conclusion we observe the following. We have shown in Section 13.5 that the 1-step predictors can be arranged into the compact form

$$\Phi(-1) Z^*_{n+1,n} = Z^*_{n,n-1} + H\epsilon_n \tag{13.6.38}$$

and as was seen in (13.5.37), this can be reorganized into

$$Z^*_{n+1,n} = \Phi(1)(I - HM) Z^*_{n,n-1} + \Phi(1) H y_n \tag{13.6.39}$$

Then the error equation will be

$$N^*_{n+1,n} = \Phi(1)(I - HM) N^*_{n,n-1} + \Phi(1) H \nu_n \tag{13.6.40}$$

Assuming that the observation errors satisfy (13.6.4), we now form the covariance matrix of $N^*_{n+1.n}$ obtaining (see Ex. 12.5)

$$S^*_{n+1,n} = BS^*_{n,n-1} B^T + \sigma_\nu^2 GG^T \tag{13.6.41}$$

where

$$B \equiv \Phi(1)(I - HM) \tag{13.6.42}$$

and

$$G \equiv \Phi(1) H \tag{13.6.43}$$

This has the following very useful practical application.

Consider the recursion

$$J_{k+1} = BJ_k B^T + \sigma_\nu^2 GG^T \tag{13.6.44}$$

where J is a matrix, and B and G are as defined above. Its homogeneous part is

$$J_{k+1} = BJ_k B^T \tag{13.6.45}$$

which means that

$$J_k = B^k J_0 (B^T)^k \tag{13.6.46}$$

But from p. 519 we know that B of (13.6.42) has all of its eigenvalues equal to θ. Then, since $|\theta| < 1$, this means that (see e.g. [13.8])

$$\lim_{k \to \infty} B^k = \emptyset \tag{13.6.47}$$

and so, regardless of the choice of the initial matrix J_0, (13.6.46) shows that

$$\lim_{k \to \infty} J_k = \emptyset \tag{13.6.48}$$

Thus the natural modes of (13.6.44) will, in the limit, go to a null matrix as $k \to \infty$, *and so that equation will give us a unique value for* J *regardless of starting value* J_0. i.e. after repeated cycling of (13.6.44), we will have obtained the unique matrix J satisfying

$$J = BJB^T + \sigma_\nu^2 GG^T \tag{13.6.49}$$

Now, reference to (13.6.41) shows us that $S^*_{n+1,n}$ also satisfies (13.6.49). Indeed, as we know from (13.6.20) for example, $S^*_{n+1,n}$ is independent of n, and so

$$S^*_{n+1,n} = S^*_{n,n-1} \tag{13.6.50}$$

Thus (13.6.41) can be written (c/f (13.6.49))

$$S^*_{n+1,n} = BS^*_{n+1,n} B^T + \sigma_\nu^2 GG^T \tag{13.6.51}$$

and so the matrix J *obtained in* (13.6.49) *must be precisely* $S^*_{n+1,n}$.

Specifically then, once θ has been selected, we can, with the aid of a computer, cycle (13.6.44) repeatedly, starting from $J_0 = \emptyset$ say. If $0 < \theta < 1$, *we are guaranteed that this iteration will converge to a unique limiting matrix.* Moreover, we also know that that limit is precisely $S^*_{n+1,n}$ as given by (13.6.20) for $r = -1$. This is a very convenient way in which to obtain an exact numerical evaluation of (13.6.20) for $r = -1$, once m and θ are chosen. (See Ex. 13.17, 13.18 and 13.19.)

13.7 COMPARISON WITH THE FIXED-MEMORY POLYNOMIAL FILTERS

It has become clear that the filters of this chapter and those of Chapter 7 possess some very strong similarities. Chief among them is the fact that, for these two classes of filters, the VRF's do not depend on n as they do in the case of the Expanding-Memory Filters of Chapter 9. We accordingly think of the Fading-Memory approach *as one which bases its estimates on the data over a fixed observation time.*

In the case of Chapter 7 the observation time was $L\tau$ seconds (a well-defined quantity, both numerically as well as conceptually), and the estimate was formed over $L + 1$ observations, each entering with equal weight. However the Fading-Memory estimates are based on a semi-infinite data-vector together with an exponentially decaying weight-function. Strictly speaking then, the observation interval is infinite, but it is clear that, because of the exponential weighting, *beyond a certain point a datum's effect on the estimate is, to all intents and purposes, negligible.* The observation interval for the present filters can thus be regarded as being of *finite* length.

In order to be able to compare quantitatively the Fading and Fixed-Memory filters, it would be desirable if we could relate a chosen value of θ to a certain value of L, i.e. if we could say, for example, that a Fading-Memory Filter with some value of θ is effectively equivalent to a Fixed-Memory Filter with some corresponding value of L. *This can in fact be done quite easily by equating variance reduction factors.*

We proceed as follows. For a specific value of θ we compute the numerical value of a Fading-Memory Filter's covariance matrix. We then equate a term of this to the corresponding term for the Fixed-Memory Filter, and solve for L. This gives us what we call the *effective smoothing time* of that Fading-Memory Filter. We demonstrate the idea by a simple example.

From p. 523, we see that for the Fading-Memory 1-step predictor of first-degree:

$$S^*_{n+1,n} = \sigma_\nu^2 \begin{pmatrix} \dfrac{(1-\theta)(5+4\theta+\theta^2)}{(1+\theta)^3} & \dfrac{(1-\theta)^2(3+\theta)}{(1+\theta)^3} \\ \dfrac{(1-\theta)^2(3+\theta)}{(1+\theta)^3} & \dfrac{2(1-\theta)^3}{(1+\theta)^3} \end{pmatrix} \tag{13.7.1}$$

On the other hand, on p. 245 we have, for the Fixed-Memory 1-step predictor of first-degree:

$$S^*_{n+1,n} = \sigma_\nu^2 \begin{pmatrix} \dfrac{2(2L+3)}{(L+1)L} & \dfrac{6}{(L+1)L} \\ \dfrac{6}{(L+1)L} & \dfrac{12}{(L+2)(L+1)L} \end{pmatrix} \tag{13.7.2}$$

Equating the 0,0 elements of the above two matrices gives us

$$\frac{2(2L+3)}{(L+1)L} = \frac{(1-\theta)(5+4\theta+\theta^2)}{(1+\theta)^3} \tag{13.7.3}$$

Now assume that $\theta = 0.9$ say. Then the above becomes

$$\frac{2(2L+3)}{(L+1)L} = 0.137 \tag{13.7.4}$$

from which we obtain

$$L \approx 30 \tag{13.7.5}$$

Hence on this basis the Fading-Memory Filter has an effective smoothing-time of about 30τ *seconds.* On the other hand, equating the 1,1 elements, using the same value of θ, gives us

$$\frac{12}{(L+2)(L+1)L} = \frac{2(1-\theta)^3}{(1+\theta)^3} \approx 0.29 \times 10^{-3} \tag{13.7.6}$$

from which

$$L \approx 34 \tag{13.7.7}$$

which is in close, but not precise, agreement with (13.7.5). Finally, equating the off-diagonal elements gives us

$$L \approx 32 \tag{13.7.8}$$

and so we see that the two covariance matrices match each other (term by term) fairly well, even if not exactly. ◆◆

If θ is close to unity, and if we restrict ourselves to 1-step prediction, we know that the formulae of Table 13.3 provide us with fairly good estimates of the Fading-Memory VRF's. These can be equated to the corresponding formulae for the VRF's of the Fixed-Memory Filter *when L is large,* obtained from Table 7.2 on p. 258. The result is the array of constants given in Table 13.4 on p. 535.

To see what those constants mean we consider a few simple examples. Let $m = 1$. Then the Fading-Memory VRF of the *position* estimate is obtained from Table 13.3 as $1.25(1-\theta)$ and for the Fixed-Memory Filter the corresponding entry in Table 7.2 is $4/L$. We equate these, i.e. we set

$$\frac{4}{L} = 1.25(1 - \theta) \tag{13.7.9}$$

from which we obtain

$$L = \frac{3.2}{1 - \theta} \tag{13.7.10}$$

Then the constant 3.2 is displayed in the appropriate position in Table 13.4. On the other hand, if we equate *velocity* VRF's of the same two filters, we obtain the relationship

$$\frac{12}{\tau^2 L^3} = \frac{0.25(1 - \theta)^3}{\tau^2} \tag{13.7.11}$$

from which

$$L = \frac{3.63}{1 - \theta} \tag{13.7.12}$$

The constant 3.63 is accordingly displayed in Table 13.4 in its appropriate position.

Suppose that $\theta = 0.9$. Then (13.7.10) gives us

$$L = 32 \tag{13.7.13}$$

whereas (13.7.12) gives us

$$L \approx 36 \tag{13.7.14}$$

Table 13.4 Constants in the L, θ Relation

i \ m	0	1	2	3	4	5	6	7	8	9	10
0	2.00	3.20	4.36	5.51	6.63	7.75	8.85	9.96	11.0	12.1	13.2
1		3.63	4.79	5.92	7.04	8.15	9.25	10.4	11.4	12.5	13.6
2			5.21	6.34	7.46	8.56	9.66	10.8	11.8	12.9	14.0
3				6.76	7.87	8.98	10.1	11.2	12.3	13.3	14.4
4					8.29	9.40	10.5	11.6	12.7	13.7	14.8
5						9.82	10.9	12.0	13.1	14.2	15.2
6							11.3	12.4	13.5	14.6	15.7
7								12.8	13.9	15.0	16.1
8									14.4	15.4	16.5
9										15.8	16.9
10											17.4

Note: $L = \dfrac{\text{constant}}{1-\theta}$ Ex: Let $m = 1$. Equating position VRF's gives $L = \dfrac{3.2}{1-\theta}$. Equating velocity VRF's gives $L = \dfrac{3.63}{1-\theta}$. Let $\theta = 0.9$. Then $L = 32$ or $L = 36$. See p. 534.

These compare well with each other and with (13.7.5), (13.7.7) and (13.7.8). ♦♦

In this manner then, we obtain a quantitative method of comparison between a given Fading-Memory and a Fixed-Memory Filter. We note, of course, that the relationship between θ and L varies slightly, depending on which of the elements of the covariance matrices that we choose to equate. This divergence exists because the two procedures are, after all, not identical, but at the same time it is small because they are not completely dissimilar. However the relationship that we are establishing is only an approximate one, and we can therefore afford to overlook these small discrepancies. (See Ex. 13.20.)

13.8 INITIALIZATION

In order to be able to cycle the algorithms of Table 13.1 it is necessary that we initialize the vector Z^*. The initializing values should, of necessity, be as close as possible to their correct values in order to avoid exciting objectionable transients, particularly when θ exceeds about 0.9 or m exceeds 2.

The word "correct" that we have used above bears further scrutiny. If the data were error-free then we know that the correct value with which to initialize is very close to the value obtainable by performing a polynomial fit to the first $m + 1$ observations (see e.g. Section 4.5). However, when the data contain errors then the correct estimate ceases to be the above one. Indeed, if we were to perform such a polynomial fit to data containing errors and we were to use the resultant state-vector for the initialization of a Fading-Memory Filter, we would almost invariably find that objectionable transients are excited. It is now clearly no longer the "correct" approach.

Consider instead the following argument. Suppose that a Fading-Memory Filter has been operating satisfactorily for some time and that there are no transients present. Then the property which distinguishes its state-vector $Z^*_{n,n-1}$ from any other vector of the same order, is that this vector possesses first, a particular set of *systematic errors* and second, *a set of random errors which are drawn from an ensemble whose covariance matrix is* $S^*_{n,n-1}$ *of* (13.6.20). In order to initialize the filter when cycling is first commenced, what we would thus really like to use is just such a vector.

While it is hard to produce a vector whose errors are *precisely* as required, we do have a relatively simple way of producing one whose random errors match the required covariance matrix quite closely and whose systematic errors also match fairly well. *This is accomplished by the use of the Expanding-Memory Polynomial Filter of the same degree as the Fading-Memory Filter which we wish to initialize.* The procedure is as follows.

Suppose, for definiteness, that we wish to initialize the Fading-Memory Filter of *first* degree. We start out by using the Expanding-Memory algorithm (see Table 9.1, p. 360)

$$\left(z_1^*\right)_{n+1,n} = \left(z_1^*\right)_{n,n-1} + h_1\left[y_n - \left(z_0^*\right)_{n,n-1}\right] \tag{13.8.1}$$

$$\left(z_0^*\right)_{n+1,n} = \left(z_0^*\right)_{n,n-1} + \left(z_1^*\right)_{n+1,n} + h_0\left[y_n - \left(z_0^*\right)_{n,n-1}\right] \tag{13.8.2}$$

where

$$h_1 = \frac{6}{(n+2)(n+1)} \qquad h_0 = \frac{2(2n+1)}{(n+2)(n+1)} \tag{13.8.3}$$

This algorithm is *completely self-starting* (see Section 9.6) and so, to initialize it when $n = 0$, we can use for example

$$\left(z_1^*\right)_{0,-1} = 0 = \left(z_0^*\right)_{0,-1} \tag{13.8.4}$$

It then starts up without any difficulties and, on successive values of n, produces the least-squares estimates based on the expanding data-base $y_0, y_1, \ldots, y_n$. At each cycling of the above filter its error covariance matrix is given by (9.7.3) (assuming that (9.7.2) holds).

We now designate the covariance matrices of the Fading, Fixed and Expanding-Memory Filters by $S^*_{\mathrm{fa}}(\theta)$, $S^*_{\mathrm{fi}}(L)$ and $S^*_{\mathrm{ex}}(n)$ respectively. Then from the preceding section we know that when $\theta = \theta_o$ say, there is a value of L, say L_o, such that S^*_{fa} and S^*_{fi} are approximately equal, i.e.

$$S^*_{\mathrm{fa}}(\theta_o) \approx S^*_{\mathrm{fi}}(L_o) \tag{13.8.5}$$

Moreover when $L = n$, we recall from Chapter 9 (see p. 369) that

$$S^*_{\mathrm{fi}}(L) = S^*_{\mathrm{ex}}(n) \tag{13.8.6}$$

Hence at the time when $n = L_o$, we must then have

$$S^*_{\mathrm{fa}}(\theta_o) \approx S^*_{\mathrm{ex}}(n_o) \tag{13.8.7}$$

which means that at this time the random errors in the Expanding Filter's state-vector have a covariance matrix which is approximately as required in order to initialize the Fading Filter. Moreover the systematic errors are also approximately as required, since both filters are based on polynomials of the same degree. Thus if we were to switch from the Expanding to the Fading Filter when $n = n_o$, then initialization should take place satisfactorily.

We consider a simple example. From Table 13.4, we see that for $m = 1$, by equating the VRF's of the position estimates,

$$L_o = \frac{3.2}{1 - \theta_o} = n_o \tag{13.8.8}$$

(c/f (13.7.10)). Thus for $\theta_o = 0.9$ say, we get $n_o = 32$, *which means that we can switch to the Fading Filter after the* 32nd *cycle of the Expanding Filter.*

The switch is readily implemented when required, if we continue to operate (13.8.1) and (13.8.2), and merely change (13.8.3) to

$$h_1 = (1 - \theta)^2 \qquad h_0 = 1 - \theta^2 \tag{13.8.9}$$

The filter in operation will thereafter be the Fading-Memory Filter – satisfactorily initialized. Table 13.4 thus serves the useful function of providing the switching-times required for the implementation of the above initialization procedure. ◆◆

In Figures 13.2 and 13.4 we show two examples of starting transients encountered when a Fading-Memory Filter was initialized using *polynomial interpolation* on the first $m + 1$ observations. They are clearly quite objectionable. After each case, we also show the corresponding runs *when initialization was carried out by the method described in this section.* Starting transients are now seen to be essentially absent.†

The reader is referred to Ex. 13.21 where we consider values of the switching instant n_0 for typical values of θ and m.

13.9 SYSTEMATIC ERRORS

In Chapter 7 a fair amount of effort was expended in analyzing both the variance reduction and the systematic errors of the Fixed-Memory Polynomial Filters, and much of the methodology developed there is directly applicable to this chapter. With regard to the systematic errors, perhaps the most

†In Figures 13.2 through 13.5, the signal was in all cases zero and the input consisted solely of random numbers whose covariance matrices were of the form $R_{(n)} = \sigma_\nu^2 I$.

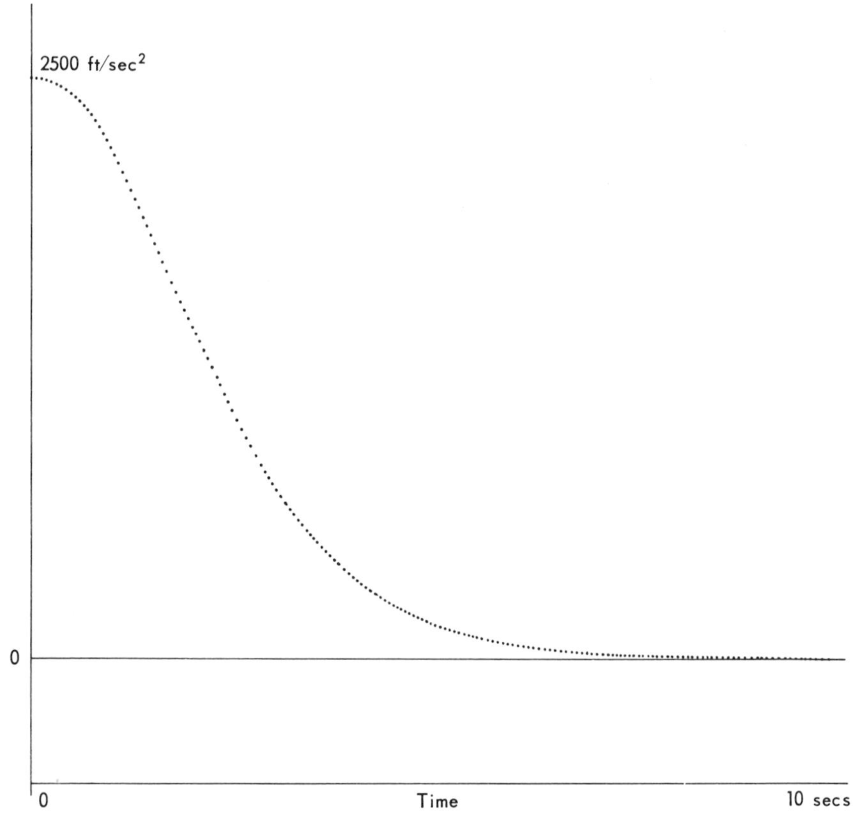

Fig. 13.2 *Starting error in* $D^2 x^*_{n+1,n}$ *Fading-Memory 1-step predictor initialized by interpolation. Degree* 2, $\theta = 0.942$, $\tau = 0.05$. *White noise,* $\sigma = 5$ *ft. (c/f Figure 13.3)*

important result that we developed was that for the filters of Chapter 7, they can, under many practical circumstances, be balanced off against the random errors by appropriate choice of the parameters L and m. In the case of the Fading-Memory Filters the same result applies, the *balance being accomplished by appropriate choice of* θ *and* m. The remaining aspects of the systematic errors for the present filters will be touched on only very briefly, but the reader can readily verify that almost every one of the comments made on systematic errors in Chapter 7 also applies here.

Assume now that the true process, which we call $\pi(t)$, is a polynomial of degree d,† and that we are able to observe it without errors. Call these

†If $\pi(t)$ is not a polynomial we assume that it has a convergent power-series, and so it can be approximated (by a truncation of that series) to within arbitrarily small errors by a polynomial of degree d.

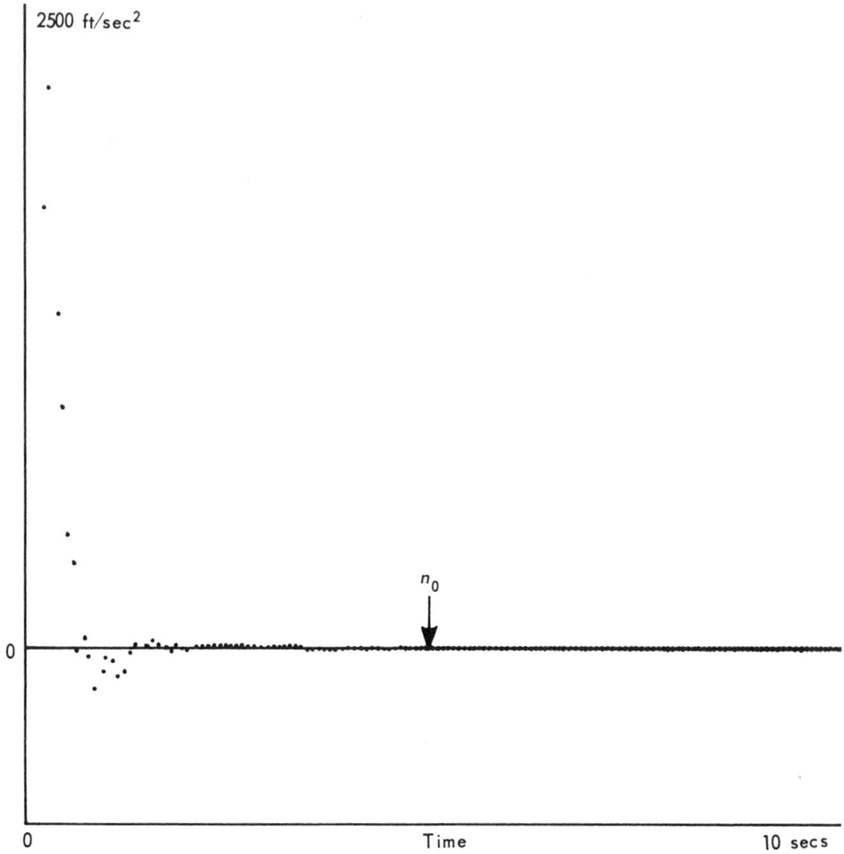

Fig. 13.3 *Starting error in* $D^2 x^*_{n+1,n}$ *Fading-Memory 1-step predictor initialized by Expanding-Memory* ($n_0 = 75$). *Degree* 2, $\theta = 0.942$, $\tau = 0.05$. *White noise,* $\sigma = 5$ *ft. (c/f Figure 13.2)*

observations $\tilde{y}_n$, i.e.

$$\tilde{y}_{n-r} = \pi[(n - r)\tau] \qquad r = 0, 1, 2, \ldots \tag{13.9.1}$$

Then, for n fixed, $\tilde{y}_{n-r}$ is a polynomial in r of degree d and so we can write

$$\tilde{y}_{n-r} = \sum_{j=0}^{d} \left(\tilde{y}_j\right)_n p_j(r) \tag{13.9.2}$$

where the $\tilde{y}$'s are constant (for fixed n), and $p_j(r)$ is given in (13.2.2). The r-origin is located at $t = n\tau$. In direct analogy with (7.10.9) and (7.10.11) it

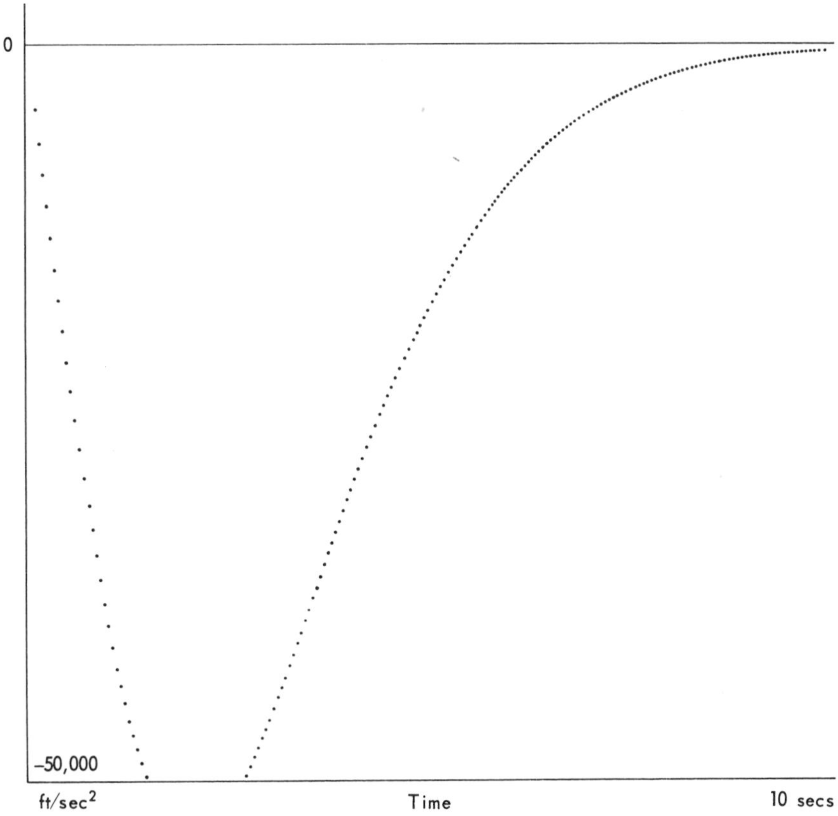

Fig. 13.4 *Starting error in* $D^2 x^*_{n+1,n}$ *Fading-Memory 1-step predictor initialized by interpolation. Degree* 3, $\theta = 0.945$, $\tau = 0.05$. *White noise,* $\sigma = 5$ *ft. (c/f Figure 13.5)*

then follows that

$$\left(\tilde{\gamma}_j\right)_n = \frac{1}{c_j^2} \sum_{k=0}^{\infty} \tilde{y}_{n-k} p_j(k)\, \theta^k \qquad 0 \le j \le d \tag{13.9.3}$$

the constant c_j^2 being defined in (13.2.4).

On the other hand, if $\pi(t)$ were observed without errors and the resultant observations fed to a Fading-Memory Polynomial Filter, we would have as the estimating polynomial (c/f (13.2.10))

$$\left[p^*(r)\right]_n = \sum_{j=0}^{m} (\gamma_j)_n p_j(r) \tag{13.9.4}$$

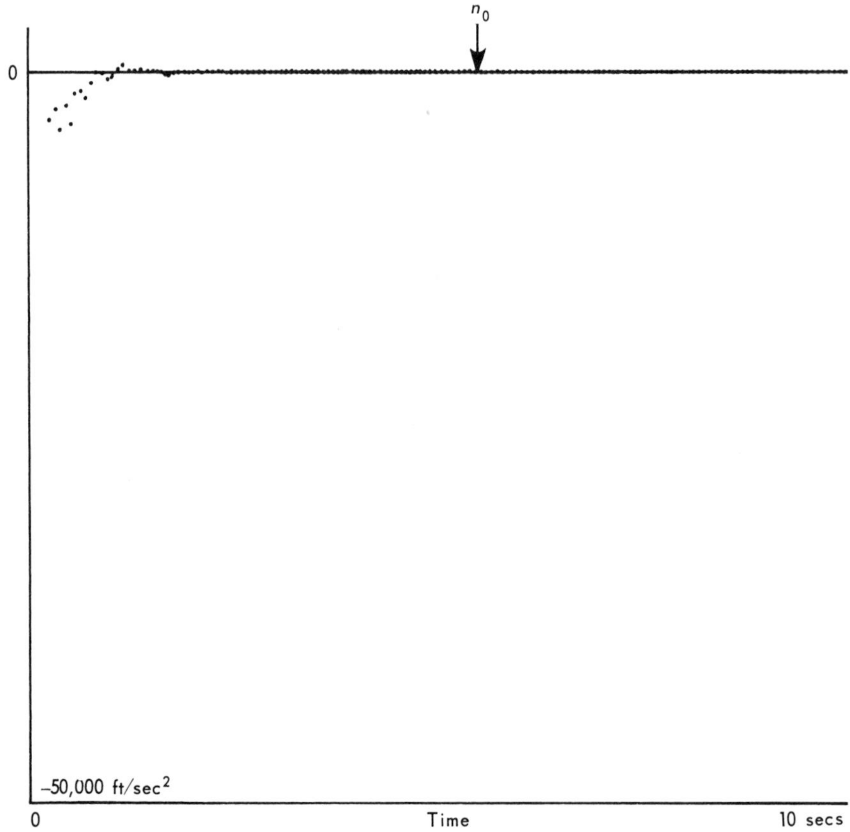

Fig. 13.5 *Starting error in* $D^2 x^*_{n+1,n}$ *Fading-Memory 1-step predictor initialized by Expanding-Memory* (n_0 = 100). *Degree* 3, θ = 0.945, τ = 0.05. *White noise,* σ = 5 *ft. (c/f Figure 13.4)*

where, in analogy with (13.2.15),

$$(\gamma_j)_n = \frac{1}{c_j^2} \sum_{k=0}^{\infty} \tilde{y}_{n-k} p_j(k)\, \theta^j \qquad 0 \le j \le m \tag{13.9.5}$$

We define the systematic error as

$$\left[b^*(r)\right]_n \equiv \pi_{n-r} - \left[p^*(r)\right]_n \tag{13.9.6}$$

and so we have that

$$[b^*(r)]_n = \sum_{j=0}^{d} (\tilde{\gamma}_j)_n p_j(n) - \sum_{j=0}^{m} (\gamma_j)_n p_j(n) \tag{13.9.7}$$

But by (13.9.3) and (13.9.5) we see that

$$(\gamma_j)_n = (\tilde{\gamma}_j)_n \qquad 0 \leq j \leq m \tag{13.9.8}$$

and of course

$$(\gamma_j)_n = 0 \qquad j > m \tag{13.9.9}$$

Hence if $d \leq m$, (13.9.7) shows us that $b^*(r)$ is precisely zero, i.e. there will be no systematic error, and if $d > m$ then the systematic error is given by the expression

$$[b^*(r)]_n = \sum_{j=m+1}^{d} (\tilde{\gamma}_j)_n p_j(r) \tag{13.9.10}$$

The above result is analogous to (7.10.17) of Chapter 7, and shows that the systematic error, if it exists, is a linear combination of the orthogonal polynomials. We comment on it briefly.

In (13.2.1) we set up the error criterion for the selection of $[p^*(r)]_n$, and an exponentially decaying stress was placed on the data in this respect. Thus the error-functional e_n forces $[p^*(r)]_n$ to fit closely when r is small and positive (i.e. θ^r large), *but places no conditions whatever on the goodness of fit when r is negative* (i.e. *in the prediction region*) *or when r is large and positive.* It thus follows that the systematic errors† are forced to be small in the first of these regions and become rapidly larger as we proceed into either of the other two. We accordingly should not expect too good a result if we were to attempt either large-interval prediction or else a large-interval retrodiction, and the Fading-Memory Filters are accordingly not recommended for such applications. *However for updated estimates or 1-step prediction they do work surprisingly well.*

†Assuming, of course, that these errors exist. This will be the case whenever $\pi(t)$ is not a polynomial of degree m or less.

We now examine, in greater detail, how the systematic errors depend on θ and we show, specifically, *that they become unbounded at* $r = 0$ *as* $\theta \to 1$. Of course this result is heuristically obvious from (13.2.1), since the filter is now based on a polynomial which is fitted to the data by equally weighted least-squares over the semi-infinite interval $0 \le r < \infty$. Clearly there must be infinite divergence between that polynomial and the data at $r = 0$, for if not then there certainly will be at $r = \infty$, in which case we simply reverse the interval and the argument is proved. However we now also prove this assertion by more formal means, since certain valuable intermediate results are thereby obtained.

Specifically, assume again that $\tilde{y}_n$ is an error-free observation on $\pi(t)$, assumed to be a polynomial of degree d. By Taylor's expansion theorem

$$\pi(t - r\tau) = \sum_{k=0}^{d} \frac{(-1)^k}{k!} \tau^k r^k \frac{d^k}{dt^k} \pi(t) \tag{13.9.11}$$

and so

$$\tilde{y}_{n-r} = \sum_{k=0}^{d} (a_k)_n \tau^k r^k \tag{13.9.12}$$

where we define

$$(a_k)_n \equiv \frac{(-1)^k}{k!} \frac{d^k}{dt^k} \pi(t) \Bigg|_{t = n\tau} \tag{13.9.13}$$

Using the Stirling numbers of the second kind (c/f (2.4.13) on p. 24), the equation (13.9.12) can be written

$$\tilde{y}_{n-r} = \sum_{k=0}^{d} (a_k)_n \tau^k \sum_{i=0}^{k} [S^{-1}]_{ki} r^{(i)} \tag{13.9.14}$$

and so, by interchanging the order of summation,

$$\begin{aligned} \tilde{y}_{n-r} &= \sum_{i=0}^{d} r^{(i)} \sum_{k=i}^{d} (a_k)_n \tau^k [S^{-1}]_{ki} \\ &= \sum_{i=0}^{d} [\lambda_i(\tau)]_n r^{(i)} \end{aligned} \tag{13.9.15}$$

where

$$[\lambda_i(\tau)]_n \equiv \sum_{k=i}^{d} (a_k)_n \tau^k [S^{-1}]_{ki} \tag{13.9.16}$$

Now, in Ex. 13.22 it is shown that $r^{(i)}$ can be expressed as a linear combination of the discrete Laguerre polynomials, i.e. that

$$r^{(i)} = i!\left(\frac{\theta}{1-\theta}\right)^i \sum_{j=0}^{i} (-1)^j \binom{i}{j} \frac{p_j(r)}{\theta^j} \tag{13.9.17}$$

Define

$$g_{ij} = i!(-1)^j \binom{i}{j} \frac{1}{\theta^j} \tag{13.9.18}$$

Then (13.9.17) can be written

$$r^{(i)} = \left(\frac{\theta}{1-\theta}\right)^i \sum_{j=0}^{i} g_{ij} p_j(r) \tag{13.9.19}$$

and so (13.9.15) gives us

$$\tilde{y}_{n-r} = \sum_{i=0}^{d} [\lambda_i(\tau)]_n \left(\frac{\theta}{1-\theta}\right)^i \sum_{j=0}^{i} g_{ij} p_j(r) \tag{13.9.20}$$

Finally, by interchanging the order of summation once more, we get

$$\tilde{y}_{n-r} = \sum_{j=0}^{d} \left\{ \sum_{i=j}^{d} [\lambda_i(\tau)]_n \left(\frac{\theta}{1-\theta}\right)^i g_{ij} \right\} p_j(r) \tag{13.9.21}$$

Comparison of this result with (13.9.2) shows that we have obtained an expression for $(\tilde{\gamma}_j)_n$ of (13.9.3), i.e. that we can write

$$(\tilde{\gamma}_j)_n = \sum_{i=j}^{d} [\lambda_i(\tau)]_n \left(\frac{\theta}{1-\theta}\right)^i g_{ij} \tag{13.9.22}$$

Then by (13.9.10) the systematic error is

$$\left[b^*(r)\right]_n = \sum_{j=m+1}^{d} \left\{ \sum_{i=j}^{d} \left[\lambda_i(\tau)\right]_n \left(\frac{\theta}{1-\theta}\right)^i g_{ij} \right\} p_j(r) \tag{13.9.23}$$

Now suppose we set $r = 0$. Then by (13.2.2),

$$\lim_{\theta \to 1} p_j(0) = 1 \tag{13.9.24}$$

Moreover, by (13.9.18)

$$\lim_{\theta \to 1} g_{ij} = i!\,(-1)^j \binom{i}{j} \tag{13.9.25}$$

and so we see that (13.9.23) gives us

$$\lim_{\theta \to 1} \left[b^*(0)\right]_n = \sum_{j=m+1}^{d} \sum_{i=j}^{d} \left[\lambda_i(\tau)\right]_n \frac{i!\,(-1)^j \binom{i}{j}}{\lim\limits_{\theta \to 1} (1-\theta)^i} \tag{13.9.26}$$

which is clearly infinite. Thus as $\theta \to 1$ *the systematic errors become unbounded in the neighborhood of* $r = 0$.

In conclusion then, we now know that θ close to unity is best from a variance reduction standpoint, and worst from a systematic error and transient error standpoint. We have also seen that making m small is best from a variance reduction and transient error standpoint and, as is evident either intuitively or else from (13.9.10), making m small is worst from systematic error considerations. In Chapter 7 we discussed a method by which the parameters L and m can be chosen (see Section 7.13). The reader can now readily repeat that analysis for the filters of the present chapter and develop an analogous scheme for the selection of θ and m which balances systematic against random errors, *omitting the transient errors.* When a value of θ and m has been arrived at, he can then numerically test the *impulse response* of the filter (i.e. the response to the input sequence

$$y_n = \delta_{n,0} \tag{13.9.27}$$

where $\delta_{n,0}$ is the Kronecker delta and where the vector $Z^*_{n,n-1}$ is set to a null vector initially). The form and duration of the impulse response provide good insight into how the filter will behave when its natural modes are excited. In the event that this impulse response is unsatisfactory, i.e. if it dies out too slowly, then he can adopt one of two approaches:

a. Reduce θ. This improves the transient response but worsens the smoothing properties. However it reduces the systematic errors.

b. Reduce τ. This causes the transients to expire more rapidly at the expense of more computation per unit time, but worsens the smoothing properties for the derivatives from first on up (see (13.6.29)). However it improves the systematic errors which can then be again traded off against the worsened smoothing properties by varying θ.

In following the method of Section 7.13 a central result is the counterpart to (7.13.5). It is easily shown (see Ex. 13.23) that for the present filters, the equivalent result is (to within sign)

$$\left[b^*(r)\right]_n \approx \frac{\left(a_{m+1}\right)_n \tau^{m+1}(m+1)!}{(1-\theta)^{m+1}} p_{m+1}(r) \tag{13.9.28}$$

with an analogous result for (7.13.6).

We now turn our attention to the possibility of generalizing these filters so that they can be used with models other than polynomials and with unequally spaced data.

EXERCISES

13.1 Show that (13.3.2) generalizes into (13.3.3).

13.2 Obtain (13.3.4) from (13.2.15) and (13.3.2).

13.3 Obtain (13.3.6).

13.4 Contrast the two proofs for (13.3.8) given in Appendix II and in [13.6].

13.5 a) Expand (13.3.8) for the case $j = 2$ and obtain the recursion

$$\begin{aligned}(\beta_2)_n &= 3\theta(\beta_2)_{n-1} - 3\theta^2(\beta_2)_{n-2} + \theta^3(\beta_2)_{n-3} \\ &\quad + K_2\,\theta^2(y_n - 2y_{n-1} + y_{n-2})\end{aligned}$$

b) Repeat for $j = 3$ and obtain the corresponding recursion for $(\beta_3)_n$.

13.6 Obtain (13.3.12), (13.3.13) and (13.3.14) from (13.3.11).

13.7 a) Implement (13.3.14) on a computer. For y_n use both a step-function (i.e. $y_n = 0$ $(n < 0)$ and $y_n = 1$ $(n \geq 0)$) as well as a Kronecker delta, initializing the recursents at zero. Study the behavior of the algorithm for various values of θ (0.1, 0.5, 0.9, 0.99, 0.999), and note how the stability weakens as $\theta \to 1$.

b) Repeat a) above for (13.3.8) for the cases $j = 0,1,2$ and compare the stability properties for various values of θ. Verify that for given θ, the stability worsens as the degree increases.

13.8 a) Verify (13.5.1).

b) Show that (13.5.4) follows from (13.5.3).

13.9 a) Obtain (13.5.9).

b) Obtain (13.5.11) and show that a factor q can be canceled from both numerator and denominator.

13.10 Starting from (13.3.11) set up the two algorithms for the 1-step predictions of position and velocity respectively. Contrast them with the coupled system in (13.5.26) and (13.5.27), and compare the relative amounts of computation and memory space required for their execution.

13.11 Starting with the degree-1 algorithm of Table 13.1 show that the homogeneous part is

$$\left(z_1^*\right)_{n+1,n} = \left(z_1^*\right)_{n,n-1} - (1-\theta)^2\left(z_0^*\right)_{n,n-1}$$

$$\left(z_0^*\right)_{n+1,n} = \theta^2\left(z_0^*\right)_{n,n-1} + \left(z_1^*\right)_{n+1,n}$$

Express this, using the q-operator, in operational form and verify that its characteristic equation is

$$\det\begin{pmatrix} (1-\theta)^2\lambda^{-1} & 1-\lambda^{-1} \\ 1-\theta^2\lambda^{-1} & -1 \end{pmatrix} = 0$$

Show that this gives $\lambda = \theta$ with multiplicity 2.

13.12 Perform the indicated summation in (13.6.14) for the three cases a) $\lambda = 0, \mu = 0$, b) $\lambda = 0, \mu = 1$, c) $\lambda = 1, \mu = 1$ and verify that the results agree with (13.6.15) in each case.

13.13 Obtain (13.6.34) from (13.6.20) by setting $r = 0$.

13.14 Starting from (13.6.9) verify that $W(r)$ has full row-rank. Hence verify that S of (13.6.10) is positive definite. Now show that $F(r)$ of (13.6.17) is nonsingular for all r and so finally infer that A in (13.6.20) must be positive definite.

13.15 a) Define the matrix M whose i, j^{th} element is

$$[M]_{ij} \equiv \frac{1}{i!} \Delta^i p_j(r) \Big|_{r=0}$$

Verify, using (3.3.10) that

$$[M]_{ij} = \frac{(-1)^i}{i!} \theta^j \binom{j}{i} \left(\frac{1-\theta}{\theta} \right)^i$$

and hence that for $\theta \sim 1$,

$$[M]_{ij} = \frac{(-1)^i}{i!} \binom{j}{i} (1-\theta)^i$$

b) Using (4.4.14) verify that F of (13.6.17) is related to M above by

$$F(0) = J S^T M$$

where J is the identity matrix with alternate rows negated. Show that for $\theta \sim 1$, this reduces to

$$[F]_{ij} = \frac{1}{i!} \binom{j}{i} (1-\theta)^i$$

c) Show that for $\theta \sim 1$, the matrix A of (13.6.15) reduces to

$$[A]_{ij} = \binom{i+j}{i} \frac{(1-\theta)}{2^{i+j+1}}$$

d) Infer finally that the elements of the mth column of Table 13.3 can be obtained from the diagonal elements of the matrix GBG^T where

$$[G]_{ij} = \binom{j}{i} \qquad 0 \leq i, j \leq m$$

and where

$$[B]_{ij} = \binom{i+j}{i} \frac{1}{2^{i+j+1}} \qquad 0 \leq i, j \leq m$$

e) Verify that for $m = 2$, G and B above are

$$G = \begin{pmatrix} 1 & 1 & 1 \\ 0 & 1 & 2 \\ 0 & 0 & 1 \end{pmatrix}$$

and

$$B = \begin{pmatrix} \frac{1}{2} & \frac{1}{4} & \frac{1}{8} \\ \frac{1}{4} & \frac{2}{8} & \frac{3}{16} \\ \frac{1}{8} & \frac{3}{16} & \frac{6}{32} \end{pmatrix}$$

Now show that the diagonal of GBG^T is $\frac{33}{16}, \frac{7}{4}, \frac{3}{16}$, which are the numbers in column 2 of Table 13.3.

13.16 Show, using the approach of Ex. 13.15 that for $\theta \approx 1$, the covariance matrix for the 1-step predictor of degree 2 is

$$[S]_{ij} = \frac{(1 - \theta)^{i + j + 1}}{\tau^{i + j}} [K]_{ij}$$

where the matrix K is given by

$$K = \begin{pmatrix} 2.0625 & 1.6875 & 0.5 \\ 1.6875 & 1.75 & 0.5625 \\ 0.5 & 0.5625 & 0.1875 \end{pmatrix}$$

13.17 a) Let the input to a Fading-Memory Polynomial Filter be a Kronecker delta. Show that if we form the sum of the squares of the output of a given channel (e.g. position, velocity, etc.), that the sum tends, in the limit, to the VRF for that channel.

b) For degree zero, the algorithm is $x_n = \theta x_{n-1} + (1 - \theta) y_n$. Letting y_n be the Kronecker delta $\delta_{n,0}$, verify that when $x_o = 0$, we obtain

$$x_n = \theta^{n-1}(1 - \theta) \qquad n \geq 1$$

and hence that

$$\sum_{n=0}^{\infty} x_n^2 = \frac{1-\theta}{1+\theta}$$

Reconcile this with Table 13.2.

13.18 a) Implement the degree-1 algorithm on a computer and *compute* the VRF's (using the method of Ex. 13.17) for various values of θ. Reconcile the results with the entries in Table 13.2.

b) Prove that if we multiply the outputs of the position and velocity channels together and add the results (assuming a Kronecker delta input as in Ex. 13.7), that the sum tends to the off-diagonal element of the covariance matrix.

c) Use this method to compute the entire covariance matrix and reconcile the result with (13.6.26).

13.19 Cycle (13.6.44) on a computer for the 2×2 case, and reconcile the results with (13.6.26). Compare this method of computing a covariance matrix to the method of Ex. 13.18.

13.20 Show that for $\theta = 0.99$ the Fading-Memory Polynomial Filter of degree 2 has a smoothing time of approximately 500τ seconds. (Hint: Use Table 13.4.)

13.21 a) We wish to employ the initialization scheme outlined in Section 13.8. Verify that for degree 2 and $\theta = 0.93$, the switch from the Expanding to the Fading Filter should take place when $n \approx 68$.

b) For degree 3 and $\theta = 0.81$, verify that $n \approx 32$.

c) For degree 1 and $\theta = 0.98$, verify that $n \approx 170$.

13.22 a) Define the matrix C whose i, j^{th} term is

$$[C]_{ij} = (-1)^j \binom{i}{j}$$

Verify that C is its own inverse, i.e. that $C^2 = I$.

b) Using this result verify that if

$$b_i = \sum_{j=0}^{i} (-1)^j \binom{i}{j} a_j$$

then

$$a_i = \sum_{j=0}^{i} (-1)^j \binom{i}{j} b_j$$

c) Infer from the above that (13.2.2) gives

$$r^{(i)} = i! \left(\frac{\theta}{1-\theta}\right)^i \sum_{j=0}^{i} (-1)^j \binom{i}{j} \frac{p_j(r)}{\theta^j}$$

thus proving (13.9.17).

13.23 a) Assume that the power-series expansion for $\pi(t)$ converges so that if truncated, the first neglected term dominates the truncation error. Verify that in this case (13.9.10) gives

$$\left[b^*(r)\right]_n \approx \left(\tilde{\gamma}_{m+1}\right)_n p_{m+1}(r) \qquad \text{(I)}$$

b) Show that under the above circumstances (13.9.22) gives

$$\left(\tilde{\gamma}_{m+1}\right)_n \approx \left[\lambda_{m+1}(\tau)\right]_n \left(\frac{\theta}{1-\theta}\right)^{m+1} g_{m+1,\, m+1}$$

Now make use of (13.9.16) and (13.9.18) to reduce (I) above to (13.9.28), i.e.

$$\left[b^*(r)\right]_n \approx (-1)^{m+1} \frac{\left(a_{m+1}\right)_n \tau^{m+1} (m+1)!}{(1-\theta)^{m+1}} p_{m+1}(r)$$

13.24 We consider the systematic errors of a zeroth-degree Fading-Memory Polynomial Filter when the input is a first-degree polynomial.

a) Verify that the zeroth-degree algorithm can be written as

$$x_n^* = \theta x_{n-1}^* + (1-\theta) y_n$$

Letting the input be $y_n = n$, show that the steady-state solution is

$$x_n^* = n - \frac{\theta}{1-\theta} \qquad \text{(I)}$$

b) Verify that since the output of the filter is a zeroth-degree polynomial in r, that (I) above gives us, as the estimate of y_{n-r},

$$x^*_{n-r} = n - \frac{\theta}{1-\theta}$$

The correct estimate should be $x^*_{n-r} = n - r$. Verify then that the bias error is

$$\left[b^*(r)\right]_n = \frac{\theta}{1-\theta} - r \tag{II}$$

c) Show that (13.9.2) gives

$$\tilde{y}_{n-r} = \left(n - \frac{\theta}{1-\theta}\right)p_0(r) + \frac{1}{1-\theta}p_1(r)$$

Hence infer from (13.9.10) that

$$\left[b^*(r)\right]_n = \frac{1}{1-\theta}p_1(r) \tag{III}$$

and reconcile this with (II) above.

d) Show that (13.9.28) gives the same result as (II) and (III) above.

REFERENCES

1. Bruckner, J., and Ford, L. R., Jr., "Recursive Digital Smoothing," Releasable Internal Memo, Defense Research Corporation, Santa Barbara, California, February 1964.
2. Duffin, R. J., and Schmidt, Th. W., "An Extrapolator and Scrutator," Jour. Math. Anal. and Applications, Vol. 1, No. 2, September 1960, pp. 215-227.
3. Duffin, R. J., and Schmidt, Th. W., "A Simple Formula for Prediction and Automatic Scrutation," Am. Rocket Soc. Flight Testing Conference, Daytona Beach, Florida, March 1959.
4. Helms, H. D., "Maximally Reliable Exponential Prediction Equations for Data-Rate-Limited Tracking Servomechanisms," B.S.T.J., Vol. XLIV, December 1965, pp. 2337-2362.

5. Larson, R. E., "Analysis of the Exponentially Weighted Polynomial Filter," Tech. Memo 7, Contract DA-01-021-AMC-90006(Y), Stanford Research Institute, Menlo Park, California, 1964.
6. Morrison, N., "Smoothing and Extrapolation of Time Series by Means of Discrete Laguerre Polynomials," SIAM (Soc. of Ind. and App. Math) Journal, Vol. 15, No. 3, May 1967, pp. 516-538.
7. Morrison, N., "Smoothing and Extrapolation of Continuous Time-Series Using Laguerre Polynomials," SIAM Journal, Vol. 16, No. 6, Nov. 1968, pp. 1280-1304.
8. Varga, R. S., "Matrix Iterative Analysis," Prentice-Hall, New Jersey, 1962, Theorem 1.4, p. 13.
9. Ragazzini, J. R., and Franklin, G. F., "Sampled-Data Control Systems," McGraw-Hill, New York, 1958.

14

GENERALIZED FADING-MEMORY FILTERS†

14.1 INTRODUCTION

In Chapter 13 we obtained the algorithms which recursively derive the best fitting polynomial in the least-squares sense, using an exponentially decaying weight-factor. We now generalize that procedure, thereby obtaining a scheme which makes it possible to incorporate all of the following features.

a. The model selected can be an arbitrary differential equation rather than just a polynomial.
b. Vectors of observations can be accepted, rather than just scalar observations.
c. The data can be obtained at unequally spaced instants along the time-axis.
d. The quality of the data, as exemplified by their covariance matrices, can be taken into account. High quality observations will be able to influence the estimate more strongly than low quality ones.

As with the filters of Chapter 13, the present ones will be seen to have a *fading memory.* The algorithms of this chapter will contain within them a scalar parameter whose value the user can select in the range from zero to

† The author is deeply indebted to P.J. Buxbaum of Bell Telephone Laboratories who provided the stimulus for this chapter.

unity. The effect of altering that parameter will be to vary the rate at which the filter's memory fades out.

At one extreme, if the parameter is set to unity (this will be possible in this chapter even though it was not so in Chapter 13), the filters which result will be identical to the Bayes or Kalman algorithms of Chapters 10 and 12. The memory is in this case *steadily expanding,* i.e. there is no fading at all, and so the influence which an observation has on the estimate is not affected in any way by its age or staleness.

When the parameter is set to less than unity then the influence which each observation has on the current estimate will be seen to fall off exponentially as the staleness of that observation increases. This implies that under these conditions the filter has a *fading* memory.

Finally if the parameter is set to zero then the estimate will be based solely on the most recent data-vector. None of the preceding vectors have any influence whatever on the estimate, and so at this extreme the fading of the memory will be instantaneous.

The ability to modify the shape of the filter's memory in this way will be seen to offer us some very interesting possibilities, particularly in regard to the control of systematic errors by trading them off against the random errors. In this respect we will consider an algorithm whereby systematic errors might be detected *on-line,* and if their presence is established, then both the fade-rate and the data-rate of the filter can be varied so as to keep those errors down to a desired level.

14.2 EXPONENTIAL STRESS

The minimum-variance unbiased linear estimator satisfies a number of criteria, among them being what we have called the *weighted least-squares criterion.* This was discussed in detail in Section 6.9, and we now review it very briefly here.

Assume that the total vector of observations presented to a filter from time t_1 up to time t_n is

$$\mathsf{Y}_{(n)} \equiv \begin{pmatrix} Y_n \\ \hline Y_{n-1} \\ \hline \vdots \\ \hline Y_1 \end{pmatrix} \tag{14.2.1}$$

where each of the subvectors in $Y_{(n)}$ is related to the model's state-vector by a linear observation relation of the form

$$Y_k = M_k X_k + N_k \tag{14.2.2}$$

Let $R_{(n)}$ be the covariance matrix of the errors in $Y_{(n)}$. Assume moreover that the model state-vector $X(t)$ satisfies a linear differential equation, i.e. that

$$\frac{d}{dt} X(t) = A(t)X(t) \tag{14.2.3}$$

This in turn implies the existence of the transition relation

$$X(t_n + \zeta) = \Phi(t_n + \zeta, t_n)X(t_n) \tag{14.2.4}$$

Suppose now that $X^*_{n,n}$ is *any* estimate of X_n. Then (see p. 180) the *simulated observation vector* based on $X^*_{n,n}$ would be

$$Y_S \equiv \begin{pmatrix} M_n X^*_{n,n} \\ \hline M_{n-1}\Phi(n-1,n)X^*_{n,n} \\ \hline \vdots \\ \hline M_1 \Phi(1,n)X^*_{n,n} \end{pmatrix} \tag{14.2.5}$$

and so the *total residual vector* based on $X^*_{n,n}$ is the difference between $Y_{(n)}$ and Y_S, namely

$$E(X^*_{n,n}) \equiv \begin{pmatrix} Y_n - M_n X^*_{n,n} \\ \hline Y_{n-1} - M_{n-1}\Phi(n-1,n)X^*_{n,n} \\ \hline \vdots \\ \hline Y_1 - M_1 \Phi(1,n)X^*_{n,n} \end{pmatrix} \tag{14.2.6}$$

Then the weighted least-squares criterion requires that $X^*_{n,n}$ be selected so as to minimize the scalar quantity

$$e(X^*_{n,n}) \equiv \left[E(X^*_{n,n})\right]^T R_{(n)}^{-1} E(X^*_{n,n}) \tag{14.2.7}$$

As we showed in Section 6.9, this leads directly to the minimum-variance rule (see (6.9.6)) which formed the cornerstone of Chapters 8, 10, 11 and 12.

We now assume, as we have done frequently before, that the errors in $Y_{(n)}$ are *stage-wise uncorrelated*, i.e. that $R_{(n)}$ has the block-diagonal form

$$R_{(n)} \equiv \begin{pmatrix} R_n & & & \\ & R_{n-1} & & \\ & & \ddots & \\ & & & R_1 \end{pmatrix} \tag{14.2.8}$$

where R_k is the covariance matrix of N_k in (14.2.2). This being the case, it is readily verified (see Ex. 14.1) that (14.2.7) reduces to

$$e(X^*_{n,n}) = \sum_{k=1}^{n} \left[Y_k - M_k \Phi(k,n) X^*_{n,n}\right]^T R_k^{-1} \left[Y_k - M_k \Phi(k,n) X^*_{n,n}\right] \tag{14.2.9}$$

from which we see that Y_k enters into $e(X^*_{n,n})$ through an associated quadratic form on R_k^{-1}. As we know from Sections 6.8 and 6.9, the matrices R_k^{-1} in (14.2.9) accomplish two very useful results. In the first place they act as normalizers, permitting us to mix the dimensions of the quantities which go into the formation of each observation vector. Secondly they serve to stress the more precise observations more heavily. The net result is that an estimate $\mathring{X}^*_{n,n}$ is obtained which is unaffected by the choice of units in which the observations are expressed, and whose error covariance matrix $\mathring{S}^*_{n,n}$ has, what we have termed, the minimum-variance property (see Section 6.7).

We now propose to examine the effects of modifying (14.2.9) slightly. Thus we replace that equation by the error-function

$$e(X^*_{n,n}) = \sum_{k=1}^{n} \left[Y_k - M_k \Phi(k,n) X^*_{n,n}\right]^T R_k^{-1} \left[Y_k - M_k \Phi(k,n) X^*_{n,n}\right] \beta^{t_n - t_k} \tag{14.2.10}$$

which is a sum of the same quadratic forms as before, *except that now each is multiplied by a scalar of the form* $\beta^{t_n - t_k}$ *where*

$$0 < \beta \leq 1 \tag{14.2.11}$$

We can see qualitatively what effect this modification will have. Thus suppose that t_n is fixed. Then as k goes from n to 1, the scalar $\beta^{t_n - t_k}$ in (14.2.10) *falls off exponentially* as $t_n - t_k$ increases. This in turn means that each of the observation vectors Y_k is being made to enter into the selection of $X^*_{n,n}$ *with an importance depending on its staleness* (i.e. the elapsed time $t_n - t_k$ since Y_k was obtained). On the other hand if we think of t_n as moving forward and t_k as being fixed, then again Y_k is seen to enter into each of the successive estimates $X^*_{n,n}, X^*_{n+1,n+1}, \ldots$ *with exponentially diminishing importance.* The result of introducing $\beta^{t_n - t_k}$ into (14.2.10) is thus seen to give the filters we are about to derive an *exponentially fading memory*. It is important to note that we can, if we so desire, let β be unity. This will result in the present filters becoming identical to the Bayes and Kalman Filters previously developed. Under slight restrictions those also included the filters of Chapter 9. We note further that (14.2.10) is a generalization of the error criterion on which the filters of Chapter 13 were based, namely (13.2.1), *and so the present discussion can be seen to be a generalization of all of the schemes derived in* Chapters 9 *through* 13.

An alternate approach to (14.2.10) is also possible. Thus we might choose to consider the error function

$$e(X^*_{n,n}) = \sum_{k=1}^{n} \left[Y_k - M_k \Phi(k,n) X^*_{n,n}\right]^T R_k^{-1} \left[Y_k - M_k \Phi(k,n) X^*_{n,n}\right] \theta^{n-k} \tag{14.2.12}$$

where

$$0 < \theta \leq 1 \tag{14.2.13}$$

In this case we see that the exponential stress-function θ^{n-k} depends solely on the integer $n-k$, i.e. on the *counting-number staleness* of Y_k (whereas in (14.2.10) the stress-function $\beta^{t_n - t_k}$ depended on the actual *elapsed-time* since the observation vector Y_k was taken). While the notion of using the counting-number staleness may at first sight seem to be rather arbitrary, we will see that it does in fact lead to a useful result. Of course if the time between observations is fixed, then the two approaches as exemplified by (14.2.10) and (14.2.12) become identical. We shall pursue our analysis based on (14.2.10) and will return to (14.2.12) at a later stage.

14.3 MINIMIZATION PROCEDURE

We now obtain an expression for the vector $X^*_{n,n}$ which minimizes $e(X^*_{n,n})$ of (14.2.10). This can best be done by reorganizing that equation into a form which we have already studied and from which we will be able to write down the required results by inspection.

First we define the matrix (c/f (6.3.9))

$$T_n \equiv \begin{pmatrix} M_n \\ \hline M_{n-1}\Phi(n-1, n) \\ \hline \vdots \\ \hline M_1\Phi(1, n) \end{pmatrix} \tag{14.3.1}$$

and so (14.2.5) can be written

$$Y_s = T_n X^*_{n,n} \tag{14.3.2}$$

Then the residual vector of (14.2.6) becomes

$$E(X^*_{n,n}) = Y_{(n)} - T_n X^*_{n,n} \tag{14.3.3}$$

a form with which we are already very familiar (see e.g. (6.5.4)).

Consider next the matrix $R_{(n)}$ of (14.2.8). If we multiply its kth diagonal block by the scalar $\beta^{-(t_n - t_k)}$, we obtain the matrix

$$Q_{(n)}(\beta) \equiv \begin{pmatrix} R_n & & & \\ & R_{n-1}\beta^{-(t_n - t_{n-1})} & & \\ & & \ddots & \\ & & & R_1\beta^{-(t_n - t_1)} \end{pmatrix} \tag{14.3.4}$$

Note the functional dependence of this matrix on β, a fact which we display by showing β as an argument on the left. Thus for example by setting $\beta = 1$, it is immediately obvious that

$$Q_{(n)}(1) = R_{(n)} \tag{14.3.5}$$

Next we form the *weighted least-squares error criterion*

$$e(X^*_{n,n}) = \left[E(X^*_{n,n})\right]^T \left[Q_{(n)}(\beta)\right]^{-1} E(X^*_{n,n}) \tag{14.3.6}$$

which is of the same general form as (14.2.7). Making use of (14.3.3) above this now becomes

$$e(X^*_{n,n}) = (Y_{(n)} - T_n X^*_{n,n})^T \left[Q_{(n)}(\beta)\right]^{-1} (Y_{(n)} - T_n X^*_{n,n}) \tag{14.3.7}$$

and it is then easily shown (see Ex. 14.2) that (14.3.7) is identical to (14.2.10). Thus minimizing the former will give the same result as minimizing the latter.

Now it is clear that (14.3.7) has the same form as (6.9.5), and so it follows immediately from Ex. 6.17 that (14.3.7) will be least if we use

$$X^*_{n,n} = \left\{T_n^T \left[Q_{(n)}(\beta)\right]^{-1} T_n\right\}^{-1} T_n^T \left[Q_{(n)}(\beta)\right]^{-1} Y_{(n)} \tag{14.3.8}$$

This is a linear transformation on $Y_{(n)}$, i.e. (14.3.8) is of the form

$$X^*_{n,n} = W_n Y_{(n)} \tag{14.3.9}$$

where

$$W_n \equiv \left\{T_n^T \left[Q_{(n)}(\beta)\right]^{-1} T_n\right\}^{-1} T_n^T \left[Q_{(n)}(\beta)\right]^{-1} \tag{14.3.10}$$

This then is the algorithm for computing the vector $X^*_{n,n}$ which minimizes (14.2.10).

We note first that if $\beta = 1$, then by virtue of (14.3.5) the above algorithm for $X^*_{n,n}$ is simply the minimum-variance rule. Thus for $\beta = 1$, by (6.6.29) and (6.6.30), $X^*_{n,n}$ and W_n of the above two equations satisfy

$$X^*_{n,n} = \mathring{X}^*_{n,n} \tag{14.3.11}$$

and

$$W_n = \overset{\circ}{W}_n \tag{14.3.12}$$

Moreover if we define

$$P_{n,n}(\beta) \equiv \left\{T_n^T \left[Q_{(n)}(\beta)\right]^{-1} T_n\right\}^{-1} \tag{14.3.13}$$

then when β is unity, $P_{n,n}(\beta)$ is seen from (6.6.32) to be the covariance matrix of $\overset{\circ}{X}{}^*_{n,n}$, i.e.

$$P_{n,n}(1) = \overset{\circ}{S}{}^*_{n,n} \tag{14.3.14}$$

However for $\beta < 1$, $P_{n,n}(\beta)$ above is no longer the covariance matrix of $X^*_{n,n}$, and to obtain that matrix we must proceed instead as follows.

From (14.3.9) we see that

$$S^*_{n,n} = W_n R_{(n)} W_n^T \tag{14.3.15}$$

and so by (14.3.10)

$$S^*_{n,n}(\beta) = \left\{\left\{T_n^T \left[Q_{(n)}(\beta)\right]^{-1} T_n\right\}^{-1} T_n^T \left[Q_{(n)}(\beta)\right]^{-1}\right\} \cdot R_{(n)} \left\{\left[Q_{(n)}(\beta)\right]^{-1} T_n \left\{T_n^T \left[Q_{(n)}(\beta)\right]^{-1} T_n\right\}^{-1}\right\} \tag{14.3.16}$$

Considering first the triple product $\left[Q_{(n)}(\beta)\right]^{-1} R_{(n)} \left[Q_{(n)}(\beta)\right]^{-1}$ appearing in the center of the above equation, it can be shown quite readily (see Ex. 14.3) that

$$\left[Q_{(n)}(\beta)\right]^{-1} R_{(n)} \left[Q_{(n)}(\beta)\right]^{-1} = \left[Q_n(\beta^2)\right]^{-1} \tag{14.3.17}$$

This means that (14.3.16) can be written as

$$S^*_{n,n}(\beta) = \left\{T_n^T\left[Q_{(n)}(\beta)\right]^{-1}T_n\right\}^{-1}\left\{T_n^T\left[Q_{(n)}(\beta^2)\right]^{-1}T_n\right\}$$
$$\left\{T_n^T\left[Q_{(n)}(\beta)\right]^{-1}T_n\right\}^{-1} \tag{14.3.18}$$

and if we now make use of (14.3.13), then the above is equivalent to

$$S^*_{n,n}(\beta) = P_{n,n}(\beta)\left[P_{n,n}(\beta^2)\right]^{-1}P_{n,n}(\beta) \tag{14.3.19}$$

This then is an expression for the covariance matrix of $X^*_{n,n}$. Clearly when $\beta = 1$, it follows from (14.3.14) that

$$S^*_{n,n}(1) = P_{n,n}(1) = \mathring{S}^*_{n,n} \tag{14.3.20}$$

Thus setting $\beta = 1$, makes $S^*_{n,n}(\beta)$ of (14.3.19) the minimum-variance error covariance matrix.

As with (14.3.8), we see that (14.3.19) above is not a convenient form in which to carry out numerical computations. We now show that a recursive procedure can be set up which gives us computational methods for $X^*_{n,n}$ and $S^*_{n,n}$ which are far more preferable.

14.4 RECURSIVE FORMULATION

From equations (14.3.9), (14.3.10) and (14.3.13), of the preceding section, we see that the vector $X^*_{n,n}$ which minimizes the chosen error function is given by

$$P_{n,n} = \left(T_n^T Q_{(n)}^{-1} T_n\right)^{-1} \tag{14.4.1}$$

$$W_n = P_{n,n} T_n^T Q_{(n)}^{-1} \tag{14.4.2}$$

$$X^*_{n,n} = W_n Y_{(n)} \tag{14.4.3}$$

(Note that, for convenience, we are now no longer displaying explicitly that $P_{n,n}$ and $Q_{(n)}$ depend on β, as we did before.) In this section we show how the above set of equations can be recast into a recursive form.

When $\beta = 1$ we know that we are considering the minimum-variance estimator, and in Chapter 10 (see Section 10.2) we showed how the Bayes Filter can be obtained as a recursive restatement of the minimum-variance rule. The discussion to follow parallels that of Section 10.2 very closely, and the only differences which arise are because of the presence of the exponentially fading stress-function $\beta^{t_n - t_k}$ which we have introduced.

Consider first, that by (14.3.1)

$$T_n \equiv \begin{pmatrix} M_n \\ \hline M_{n-1}\Phi(n-1,n) \\ \hline M_{n-2}\Phi(n-2,n) \\ \hline \vdots \\ \hline M_1\Phi(1,n) \end{pmatrix}$$

$$= \begin{pmatrix} M_n \\ \hline M_{n-1}\Phi(n-1,n-1)\Phi(n-1,n) \\ \hline M_{n-2}\Phi(n-2,n-1)\Phi(n-1,n) \\ \hline \vdots \\ \hline M_1\Phi(1,n-1)\Phi(n-1,n) \end{pmatrix} \tag{14.4.4}$$

and so we see that

$$T_n = \begin{pmatrix} M_n \\ \hline T_{n-1}\Phi(n-1,n) \end{pmatrix} \tag{14.4.5}$$

Next, by (14.3.4),

$$
\mathbf{Q}_{(n)} = \begin{pmatrix} R_n & & & & \\ & R_{n-1}\beta^{-(t_n - t_{n-1})} & & & \\ & & R_{n-2}\beta^{-(t_n - t_{n-2})} & & \\ & & & \ddots & \\ & & & & R_1\beta^{-(t_n - t_1)} \end{pmatrix}
$$

$$
= \begin{pmatrix} R_n & & & & \\ & R_{n-1}\beta^{-(t_{n-1} - t_{n-1})}\beta^{-(t_n - t_{n-1})} & & & \\ & & R_{n-2}\beta^{-(t_{n-1} - t_{n-2})}\beta^{-(t_n - t_{n-1})} & & \\ & & & \ddots & \\ & & & & R_1\beta^{-(t_{n-1} - t_1)}\beta^{-(t_n - t_{n-1})} \end{pmatrix} \tag{14.4.6}
$$

and so we can write $Q_{(n)}$ in the form

$$Q_{(n)} = \left(\begin{array}{c|c} R_n & \\ \hline & Q_{(n-1)}\beta^{-(t_n - t_{n-1})} \end{array}\right) \tag{14.4.7}$$

We now apply (14.4.5) and (14.4.7) to (14.4.1). Thus

$$\begin{aligned} P_{n,n}^{-1} &= T_n^T Q_{(n)}^{-1} T_n \\ &= \left(M_n^T \;\middle|\; \Phi(n-1,n)^T T_{n-1}^T\right) \left(\begin{array}{c|c} R_n^{-1} & \\ \hline & Q_{(n-1)}^{-1}\beta^{t_n - t_{n-1}} \end{array}\right) \left(\begin{array}{c} M_n \\ \hline T_{n-1}\Phi(n-1,n) \end{array}\right) \\ &= M_n^T R_n^{-1} M_n + \beta^{t_n - t_{n-1}} \Phi(n-1,n)^T T_{n-1}^T Q_{(n-1)}^{-1} T_{n-1} \Phi(n-1,n) \end{aligned} \tag{14.4.8}$$

Next we define

$$P_{n,n-1} \equiv \Phi(n, n-1) P_{n-1,n-1} \Phi(n, n-1)^T \tag{14.4.9}$$

and so it follows that

$$\begin{aligned} P_{n,n-1}^{-1} &= \Phi(n-1,n)^T P_{n-1,n-1}^{-1} \Phi(n-1,n) \\ &= \Phi(n-1,n)^T \left(T_{n-1}^T Q_{(n-1)}^{-1} T_{n-1}\right) \Phi(n-1,n) \end{aligned} \tag{14.4.10}$$

This now means that (14.4.8) can be written

$$P_{n,n}^{-1} = M_n^T R_n^{-1} M_n + \beta^{t_n - t_{n-1}} P_{n,n-1}^{-1} \tag{14.4.11}$$

and so we have that

$$P_{n,n} = \left(\beta^{t_n - t_{n-1}} P_{n,n-1}^{-1} + M_n^T R_n^{-1} M_n\right)^{-1} \tag{14.4.12}$$

This last result together with (14.4.9) constitutes a recursive algorithm for obtaining $P_{n,n}$ given $P_{n-1,n-1}$ and R_n.

Our next task is to consider the product $T_n^T Q_{(n)}^{-1}$ on the right of (14.4.2). By (14.4.5) and (14.4.7),

$$T_n^T Q_{(n)}^{-1} = \left(M_n^T \,\middle|\, \Phi(n-1,n)^T T_{n-1}^T\right)\left(\begin{array}{c|c} R_n^{-1} & \\ \hline & \beta^{t_n - t_{n-1}} Q_{(n-1)}^{-1}\end{array}\right) \tag{14.4.13}$$

$$= \left(M_n^T R_n^{-1} \,\middle|\, \beta^{t_n - t_{n-1}} \Phi(n-1,n)^T T_{n-1}^T Q_{(n-1)}^{-1}\right)$$

Then if we partition $Y_{(n)}$ into the form

$$Y_{(n)} = \left(\begin{array}{c} Y_n \\ \hline Y_{(n-1)}\end{array}\right) \tag{14.4.14}$$

it follows from (14.4.13) that

$$T_n^T Q_{(n)}^{-1} Y_{(n)} = M_n^T R_n^{-1} Y_n + \beta^{t_n - t_{n-1}} \Phi(n-1,n)^T T_{n-1}^T Q_{(n-1)}^{-1} Y_{(n-1)} \tag{14.4.15}$$

Consider the final term of this equation.

By (14.4.2) and (14.4.3) we have that

$$X_{n-1,n-1}^* = P_{n-1,n-1} T_{n-1}^T Q_{(n-1)}^{-1} Y_{(n-1)} \tag{14.4.16}$$

and so, by multiplying both sides by $P_{n-1,n-1}^{-1}$ we obtain

$$T_{n-1}^T Q_{(n-1)}^{-1} Y_{(n-1)} = P_{n-1,n-1}^{-1} X_{n-1,n-1}^* \tag{14.4.17}$$

This now means that

$$\begin{aligned}
&\Phi(n-1,n)^T T_{n-1}^T Q_{(n-1)}^{-1} Y_{(n-1)} \\
&\quad = \Phi(n-1,n)^T P_{n-1,n-1}^{-1} X_{n-1,n-1}^* \\
&\quad = \Phi(n-1,n)^T P_{n-1,n-1}^{-1} \Phi(n-1,n)\Phi(n,n-1) X_{n-1,n-1}^* \\
&\quad = \left[\Phi(n,n-1) P_{n-1,n-1} \Phi(n,n-1)^T\right]^{-1} \Phi(n,n-1) X_{n-1,n-1}^*
\end{aligned} \tag{14.4.18}$$

Thus if we write

$$X^*_{n,n-1} = \Phi(n, n-1) X^*_{n-1,n-1} \tag{14.4.19}$$

and if we make use of (14.4.9), then it follows that the final line of (14.4.18) can be written as $P^{-1}_{n,n-1} X^*_{n,n-1}$ Hence (14.4.15) becomes

$$T_n^T Q_{(n)}^{-1} Y_{(n)} = M_n^T R_n^{-1} Y_n + \beta^{t_n - t_{n-1}} P^{-1}_{n,n-1} X^*_{n,n-1} \tag{14.4.20}$$

We now consider (14.4.3). By (14.4.20) we see that it can be written as

$$X^*_{n,n} = P_{n,n}\left(M_n^T R_n^{-1} Y_n + \beta^{t_n - t_{n-1}} P^{-1}_{n,n-1} X^*_{n,n-1}\right) \tag{14.4.21}$$

and it is then a simple matter (see Ex. 14.4) to show that the above is equivalent to

$$X^*_{n,n} = X^*_{n,n-1} + H_n (Y_n - M_n X^*_{n,n-1}) \tag{14.4.22}$$

where

$$H_n \equiv P_{n,n} M_n^T R_n^{-1} \tag{14.4.23}$$

This pair of equations together with (14.4.19) constitutes a recursion algorithm for obtaining $X^*_{n,n}$, given $X^*_{n-1,n-1}$ and Y_n.

Finally we must derive a recursive method for obtaining $S^*_{n,n}$. But by comparing (14.4.22) to (12.3.2) we see immediately from (12.3.7) that

$$S^*_{n,n} = (I - H_n M_n) S^*_{n,n-1} (I - H_n M_n)^T + H_n R_n H_n^T \tag{14.4.24}$$

where of course

$$S^*_{n,n-1} \equiv \Phi(n, n-1) S^*_{n-1,n-1} \Phi(n, n-1)^T \tag{14.4.25}$$

We are now in a position to assemble the complete algorithm, and as its name indicates, its form is very suggestive of the Bayes Filter on p. 389. Thus by the arguments given above we can now write the following:

Fading-Memory Bayes Filter

$$X^*_{n,n-1} = \Phi(n, n-1)X^*_{n-1,n-1} \tag{14.4.26}$$

$$P_{n,n-1} = \Phi(n, n-1)P_{n-1,n-1}\Phi(n, n-1)^T \tag{14.4.27}$$

$$P_{n,n} = \left(\beta^{t_n - t_{n-1}}P^{-1}_{n,n-1} + M_n{}^T R_n{}^{-1} M_n\right)^{-1} \tag{14.4.28}$$

$$H_n = P_{n,n}M_n{}^T R_n{}^{-1} \tag{14.4.29}$$

$$X^*_{n,n} = X^*_{n,n-1} + H_n(Y_n - M_n X^*_{n,n-1}) \tag{14.4.30}$$

$$S^*_{n,n-1} = \Phi(n, n-1)S^*_{n-1,n-1}\Phi(n, n-1)^T \tag{14.4.31}$$

$$S^*_{n,n} = (I - H_n M_n)S^*_{n,n-1}(I - H_n M_n)^T + H_n R_n H_n{}^T \tag{14.4.32}$$

The inputs to this algorithm are Y_n and R_n and its outputs are $X^*_{n,n}$ and $S^*_{n,n}$. Storage must be provided, between cycles, for $X^*_{n,n}$, $S^*_{n,n}$ and $P_{n,n}$ for use on the next execution.

Just as the inversion lemma permitted us to convert the Bayes Filter into the Kalman Filter on p. 465, so it now permits us to convert the above algorithm into one closely resembling the Kalman Filter. To accomplish this we first write (14.4.28) in the form

$$P_{n,n} = \left[\left(\frac{P_{n,n-1}}{\beta^{t_n - t_{n-1}}}\right)^{-1} + M_n{}^T R_n{}^{-1} M_n\right]^{-1} \tag{14.4.33}$$

and we then obtain quite readily (see Ex. 14.5) the following extension of the Kalman Filter:

Fading-Memory Kalman Filter

$$X^*_{n,n-1} = \Phi(n, n-1)X^*_{n-1,n-1} \tag{14.4.34}$$

$$P_{n,n-1} = \Phi(n, n-1)P_{n-1,n-1}\Phi(n, n-1)^T \tag{14.4.35}$$

$$H_n = P_{n,n-1}M_n{}^T\left(\beta^{t_n - t_{n-1}}R_n + M_n P_{n,n-1}M_n{}^T\right)^{-1} \tag{14.4.36}$$

(continued)

$$X^*_{n,n} = X^*_{n,n-1} + H_n(Y_n - M_n X^*_{n,n-1}) \tag{14.4.37}$$

$$P_{n,n} = \frac{1}{\beta^{t_n - t_{n-1}}}(I - H_n M_n) P_{n,n-1} \tag{14.4.38}$$

$$S^*_{n,n-1} = \Phi(n, n-1) S^*_{n-1,n-1} \Phi(n, n-1)^T \tag{14.4.39}$$

$$S^*_{n,n} = (I - H_n M_n) S^*_{n,n-1} (I - H_n M_n)^T + H_n R_n H_n^T \tag{14.4.40}$$

As with the Fading Bayes Filter on p. 569, the inputs to the above algorithm are Y_n and R_n, its outputs are $X^*_{n,n}$ and $S^*_{n,n}$, and intermediate storage must be provided for $X^*_{n,n}$, $S^*_{n,n}$ and $P_{n,n}$ for use on the next cycle. In the next section we examine some of the more important properties of the Fading Bayes and Kalman Filters.

14.5 PROPERTIES OF THE GENERALIZED FADING-MEMORY FILTERS

The recursive algorithms which were derived in the preceding section provide us with two very powerful ways in which to obtain the estimate vector which minimizes (14.2.10). They are computationally extremely efficient and from a memory standpoint they require only the retention of the most recent estimate, its covariance matrix, and the matrix $P_{n,n}$. By comparison with their nonrecursive counterparts obtained in Section 14.3, they are thus clearly seen to be the more desirable ways in which to implement the filters of this chapter.

Structurally they bear very strong resemblances to their respective expanding-memory counterparts, the Bayes and Kalman algorithms of Chapters 10 and 12, and as we already know *when* $\beta = 1$ *they become identical to those filters in all respects.*

In the course of studying the Expanding Bayes and Kalman Filters† we examined the following topics.

a. The number of matrix inversions required.
b. Initialization techniques.
c. Problems arising out of the introduction of highly precise data.
d. Problems caused by the introduction of highly correlated data.

†By "Expanding" Bayes or "Expanding" Kalman Filter we refer to the versions derived in Chapters 10 and 12 respectively. The term "Fading" on the other hand refers to the versions derived in this chapter when $\beta < 1$.

e. Loss of precision in the covariance matrix.

We saw that, by virtue of differences in their basic structures, the two filters possess differing and sometimes markedly contrasting attributes in each case. Now it follows immediately that because they are *structrually indistinguishable,* the Fading Bayes will have the same properties as the Expanding Bayes in each of the above categories, and similarly for the Fading and Expanding Kalman algorithms. *Thus the same arguments which might be applied in choosing between an Expanding Bayes and an Expanding Kalman Filter, would also be applied in choosing between a Fading Bayes and Fading Kalman Filter.* We do not intend to restate here the results of our analysis on the topics itemized above, and we content ourselves simply with referring the reader back to Sections 10.6 and 12.4. Almost every one of the comments made there on the attributes of the Expanding Bayes and Kalman algorithms apply virtually unchanged to their fading-memory counterparts that we are now examining.

In itself the preceding paragraph has already said a considerable amount about the filters of this chapter. However there are additional properties which we have yet to uncover that arise because of the added flexibility provided by the parameter β. We now turn our attention to these.

As a first step we note that because of the incorporation of the exponential stress-function $\beta^{t_n - t_k}$, the matrix $P_{n,n}$ is not the estimate covariance matrix, unless $\beta = 1$. When $\beta < 1$ we compute the covariance matrix by the extra pair of equations involving $S^*_{n,n-1}$ and $S^*_{n,n}$ which appears at the end of each of the algorithms on pp. 569 and 570. One question which immediately presents itself is, how does one initialize those algorithms? As an answer we suggest that at the time of initialization β be set to unity, in which case P^* and S^* are identical. Then the same approach as was used to initialize the Expanding Bayes and Kalman Filters can be used to initialize these. At some convenient time after initialization, β can be reduced to its chosen value and the filters will then be operating as required.

We have already heuristically examined some of the effects which the introduction of β can be expected to have on our filters, and as we pointed out in discussing the choice of (14.2.10) as our error criterion, when β is less than unity, then the resultant filters can be expected to have a fading memory. We now examine this aspect further.

Thus suppose first that $\beta = 1$. Then the Fading Filters are identical to their Expanding counterparts and so the influence which each observation vector has on the estimate is determined once and for all by its quality, as exemplified by its covariance matrix. This influence is completely unrelated to the age or staleness of the observation, and so at this extreme value of β there is no fading whatever.

At the other extreme suppose we set $\beta = 0$. While this value was not contemplated when we stated that β lies in the range $0 < \beta \leq 1$, we use it for discussion purposes rather than using some number infinitesimally close to zero. Note that in the Kalman case, as we see from (14.4.38), it is not possible to let $\beta = 0$ precisely. However in the Bayes case we see that we can, and under these conditions (14.4.28) becomes

$$P_{n,n} = \left(M_n{}^T R_n{}^{-1} M_n\right)^{-1} \tag{14.5.1}$$

from which it follows readily that (14.4.30) becomes

$$X^*_{n,n} = \left(M_n{}^T R_n{}^{-1} M_n\right)^{-1} M_n{}^T R_n{}^{-1} Y_n \tag{14.5.2}$$

Thus, assuming that $\left(M_n{}^T R_n{}^{-1} M_n\right)^{-1}$ exists, *we see that at* $\beta = 0$ *the Fading Bayes becomes precisely the minimum-variance Fixed-Memory Filter based on only the most recent observation vector.* The memory shape is in this case seen to be a Kronecker delta, incorporating only the most recent observation and excluding all of its predecessors, which is equivalent to a fading memory with *instantaneous* fading. Thus we see that, at the extremes at any rate, our heuristic arguments on what to expect as β is varied are now confirmed, namely that β serves to control the fading-rate of the filter's memory, from no fading at all when $\beta = 1$, to instantaneous fading when $\beta = 0$.

The Kalman formulation is theoretically equivalent to the Bayes by virtue of the inversion lemma, and so it follows that the above arguments also apply in the Kalman case when β is very close to zero, even though $\beta = 0$ itself is an unacceptable value. We note that this inability to set $\beta = 0$ in the Kalman case is another manifestation of the fact, which we pointed out on p. 472, that the Kalman Filter *cannot* form an estimate based on only a single observation vector. It *must* also have an *a priori* estimate which it then combines with the given observation vector to form an estimate. The Bayes Filter on the other hand *can* form an estimate solely on Y_n if the matrix $\left(M_n{}^T R_n{}^{-1} M_n\right)^{-1}$ exists, (see (14.5.1) and (14.5.2)). This will be the case if M_n has full column-rank.

We have seen that at the two extremes, $\beta = 1$ and $\beta = 0$, the fading of the filter's memory is either entirely absent or else instantaneous, and so it follows, by the continuity of the algorithms with respect to the parameter β, that a continuous transition must exist between those two extremes. Heuristically it is easy to see what that transition implies, and in Figure 14.1 we have sketched out a set of exponentially decaying curves for various values of the parameter β. Evidently when $\beta = 1$ we have the flat non-decaying curve, signifying equal stress on all of the data with respect to age, and at $\beta = 0$ we obtain, in the limit, a Kronecker delta since

$$\lim_{\beta \to 0} \beta^{t_n - t_k} = \delta_{t_n, t_k} \tag{14.5.3}$$

The intermediate cases show a decaying stress on the data, which increases more or less rapidly with age as β is made close to zero or closer to unity.

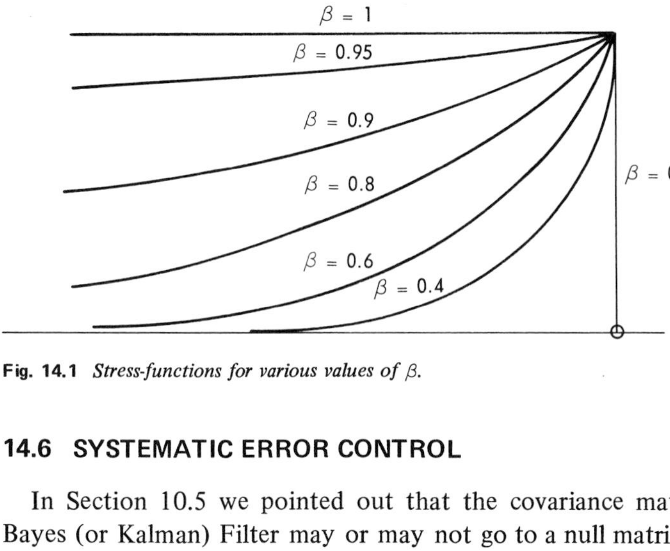

Fig. 14.1 *Stress-functions for various values of β.*

14.6 SYSTEMATIC ERROR CONTROL

In Section 10.5 we pointed out that the covariance matrix $S^*_{n,n}$ of the Bayes (or Kalman) Filter may or may not go to a null matrix as n increases. In the event that the observation scheme is properly configured and if the observations are of good enough quality and are made often enough, then as time passes $S^*_{n,n}$ will go to a null matrix.

Assume that we have such a situation, *but that the model on which the filter is based differs from the true process equations.* (This might be either because of ignorance of the true process equations or else out of a desire to simplify the model in order to reduce the amount of computation.) Now as $S^*_{n,n}$ shrinks we see from (12.2.8) that H_n also shrinks, and eventually the time will come when the estimate, obtained from

$$X^*_{n,n} = X^*_{n,n-1} + H_n(Y_n - M_n X^*_{n,n-1}) \tag{14.6.1}$$

is virtually unaffected by the vector $Y_n - M_n X^*_{n,n-1}$. The estimate $X^*_{n,n}$ is then essentially $X^*_{n,n-1}$ and the filter simply follows the particular trajectory on which it has settled, without being amenable to modification by the observations. Under these conditions the estimated and true trajectories will begin to diverge by a deterministic amount and so the errors in $X^*_{n,n}$ are now no longer zero-mean random variables, but develop a bias.

In addition to the above mechanism, systematic errors can also arise even when there is a supposedly perfect match between the true process and the model. All of our considerations up to now have presupposed that the true process is a homogeneous, i.e. an *unforced,* linear differential equation of the form

$$\frac{d}{dt}\Pi(t) = A(t)\Pi(t) \tag{14.6.2}$$

We assume that $A(t)$ is known and so the filter model is also (14.6.2). However suppose that at some instant a disturbance takes place and that the true process is subjected to an impulsive forcing function for a very brief period at some time. Thereafter it returns to (14.6.2) *but the effect of this forcing function will have been to cause a change in trajectory to a new one still governed by* (14.6.2). The estimate given by the filter will thereafter contain systematic errors, since the estimate is now based on data obtained partly from the old trajectory (i.e. before the impulse) and partly from the new one.† It is the purpose of this and later sections to consider how such systematic errors can be controlled.

The key to systematic error control lies in the recognition of the fact that our filters are essentially *curve-fitters.* If the true curve and the approximating one are chosen from the same ensemble, i.e. solutions to the same differential equation, then the fit can be made arbitrarily good, in spite of the observation errors, by extending the fitting interval over a sufficiently long time. However if the given curve and the one we use for approximating it are not intrinsically identical, either due to a mismatch in their (homogeneous) differential equations or because of an impulsive forcing function not accounted for, then a good fit can only be made over a sufficiently short period. Increasing the length over which the fit is made will lead to errors which could exceed the errors caused by corruption in the observations. In the parlance of our filter theory, *we must keep the filter's*

†Strictly speaking, of course, this second mechanism also constitutes a mismatch between the true process and the model. However it differs from the preceding one in which even the homogeneous portions of the differential equations were assumed to differ. Note that systematic errors could arise *simultaneously* from both sources.

memory-length from becoming too large if we wish to control the systematic errors. As a consequence we must also be prepared to accept random errors in the estimate which have a covariance matrix that now no longer goes to a null matrix as time increases.

The Bayes and Kalman Filters of Chapters 10 and 12 have, as we know, a steadily expanding memory. Thus unless there is a very precise match between the true process and the model, we must expect systematic errors to develop if we decide to estimate by their use. Of course there are many situations where a very precise match is indeed possible. However if a mismatch exists, or if an occasional forcing impulse can be expected in the true process, then their use is not recommended over too long a period. (In exactly the same way the use of the Expanding-Memory Polynomial Filter is not recommended for too long a period, unless the true process is precisely a polynomial of the same degree (or less) as the one on which the filter is based.)

We have already verified that the filters of the present chapter possess a fading memory for $\beta < 1$. This means that the memory-length does not expand steadily *but that the observations are forgotten or discounted at a rate exponentially proportional to their age.* As a result the curve-fitting process takes place over an essentially *shortened and nonexpanding* time interval and so, making $\beta < 1$, will serve to keep the systematic errors within prescribed bounds.

At the extreme when $\beta = 0$, we have seen that if $M_n{}^T R_n{}^{-1} M_n$ is nonsingular then the Bayes estimate is based purely on the most recent observation vector, and so clearly it will be entirely free of systematic errors. This then suggests that by making β sufficiently small we can keep the effective memory-length sufficiently short, thereby keeping the systematic errors down to any desired level. Of course the price we pay is that with a shortened effective memory-length the smoothness of the estimate is reduced, i.e. the random output errors will be larger. But a balance can be struck in which the systematic and random errors are made roughly comparable. This is certainly preferable to an estimate which becomes steadily smoother with time but also more and more biased.

In the case of impulsive forcing functions in the true process, we can keep β close to unity for most of the time. If at some time we are able to detect that an impulse has acted (in a later section we will consider this problem more closely), then we can *temporarily* reduce β, thereby essentially freeing our filter from data pertaining to the now obsolete trajectory. Then we bring β back up to unity and so permit the memory to fill up with data on the new trajectory. The covariance matrix of the estimate will be temporarily enlarged while β is small, but after β is set back to a value near unity that covariance matrix will begin to shrink again, and we obtain once more a smooth estimate which is also relatively free of systematic errors.

We now devote some effort to an analytical study of the effects that β has on the covariance matrix. What we propose to do is to analyze the behavior of the covariance matrix of the Generalized Fading-Memory Filter assuming (1) a constant-coefficient linear differential equation as the model, (2) an invariant linear observation scheme, (3) a constant observation-error covariance matrix R, and (4) equal spacing between the observations. The last three assumptions are not very restrictive and the first, namely that the model is a constant-coefficient linear differential equation, while being somewhat restrictive has the advantage of keeping the algebra manageable. Moreover, in spite of its seemingly restrictive nature, the assumption of such a model nevertheless gives us a considerable amount of insight into the mechanisms underlying the Bayes and Kalman algorithms *both with and without* fading.

We shall consider only the Bayes Filter, since the Kalman Filter without driving-noise is algebraically equivalent to it.

14.7 STABILITY ANALYSIS

Up to now we have not made any attempt to discuss the stability properties of the Bayes or Kalman algorithms, and being recursive in structure, such properties, if they exist, are certainly subject to proof. In Section 9.5 (see p. 365) we did point out the need for such a discussion, and what we now propose to do is to consider the somewhat restricted case of the constant-coefficient linear model and constant-coefficient linear observation scheme, and a nonchanging input-error covariance matrix. We will show, specifically, that in the Expanding Bayes and Kalman Filters *a relationship exists*† *between* M *and* Φ *whose presence is both necessary and sufficient* for $S^*_{n,n}$ *to tend to a null matrix regardless of the choice of initial conditions* $S^*_{1,0}$. That being the case, if there is a perfect match between the true process and the model, the estimate will ultimately be precisely equal to the true state-vector with probability one, regardless of the choice of initial conditions $X^*_{1,0}$. Thus $S^*_{n,n}$ tending to a null matrix as $n \to \infty$ constitutes proof of stability.

Our approach will be to study the recursion for the matrix $P_{n,n}$ of the Fading Bayes Filter. We shall show first that $P_{n,n}(1)$ goes to a null matrix as $n \to \infty$,‡ and since by (14.3.14) we have

†Identical to the conditions on M and Φ given in Theorem 8.2 on p. 325.
‡Subject to the above-mentioned relationship between M and Φ being satisfied.

$$P_{n,n}(1) = \overset{\circ}{S}{}^{*}_{n,n} \tag{14.7.1}$$

will thus also have proved that for the Expanding Bayes and Kalman Filters

$$\lim_{n \to \infty} \overset{\circ}{S}{}^{*}_{n,n} = \emptyset \tag{14.7.2}$$

When β is less than one we will see that the above limits are not null matrices. Thus in the Fading Bayes and Kalman cases *the descent of both* $S^{*}_{n,n}$ *and the weight matrix* H_n *of* (14.4.29) *or* (14.4.36) *to null matrices is forestalled.* This indicates that if a mismatch exists between the true process and the filter's model, then the concomitant systematic errors can be kept within bounds, at the expense of larger random errors, by an appropriate choice of β.

Assuming a linear constant-coefficient differential equation as the model, we know from Chapter 4 that the transition matrix has the form

$$\Phi(\tau) = \exp(\tau A) \tag{14.7.3}$$

where τ is the intersample time, assumed constant. We now write (14.4.27) as

$$P_{n,n-1} = \Phi P_{n-1,n-1} \Phi^{T} \tag{14.7.4}$$

where we are suppressing the argument of Φ for brevity. The inverse of (14.7.4) is

$$P^{-1}_{n,n-1} = \Phi_{T}^{-1} P^{-1}_{n-1,n-1} \Phi^{-1} \qquad \text{(See Note)} \tag{14.7.5}$$

and making use of this, (14.4.28) can be written as

$$P^{-1}_{n,n} = \beta \Phi_{T}^{-1} P^{-1}_{n-1,n-1} \Phi^{-1} + M^{T} R^{-1} M \tag{14.7.6}$$

in which the exponent of β, namely τ, has been taken without loss of generality to be unity. Note that neither M nor R is subscripted with an n, a consequence of the assumptions which we made earlier that the observation matrix and error covariance matrix were both constant.

Equation (14.7.6) is a linear recursion for $P^{-1}_{n,n}$ in terms of its predecessor, with the constant matrix $M^{T} R^{-1} M$ serving as a forcing function. For further

Note: By the symbol Φ_{T}^{-1} we naturally mean the transpose of Φ^{-1}.

brevity we define the two matrices Z and G by

$$Z_n \equiv P_{n,n}^{-1} \tag{14.7.7}$$

and

$$G \equiv M^T R^{-1} M \tag{14.7.8}$$

and so (14.7.6) becomes

$$Z_n(\beta) = \beta \Phi_T^{-1} Z_{n-1}(\beta) \Phi^{-1} + G \tag{14.7.9}$$

in which form our study will be conducted.

First we note that G is a nonnegative definite matrix. Moreover if Z_0 (namely $P_{0,0}^{-1}$) is positive definite, then so is $\beta \Phi_T^{-1} Z_0 \Phi^{-1}$ and hence so is Z_1, the sum of a positive definite and a nonnegative definite matrix being positive definite. We thus have the following.

Lemma 14.1

Let Z_n, Φ, G and β be related as in (14.7.9) where Z_0 is positive definite. Then for any n, Z_n is positive definite. ◆◆

We note next that all of our matrices including Φ are real. However Φ may or may not be symmetric and so it can have complex eigenvalues and eigenvectors. If the eigenvalues of Φ happen to be complex then they will always occur in conjugate pairs.

Consider now a constant-coefficient, linear model characterized by the transition relation

$$X_n = \Phi X_{n-1} \tag{14.7.10}$$

This can be viewed as a linear homogeneous vector difference equation, and its solutions constitute the form which the vector X_n assumes. Thus given Φ there is a class of X_n's generated by (14.7.10). The elements of these X_n's are related to the eigenvalues of Φ in the following manner, which we state without proof.†

Lemma 14.2

Let λ be an eigenvalue of Φ with multiplicity m, and let the space formed by the eigenvectors associated with λ have dimension ρ. Then in general

†Proof can be supplied quite readily by the reader, making use of the material in Examples 8.19 through 8.21.

there will be terms in the elements of X_n of the form $p_n \lambda^n$, where p_n is a polynomial of degree $m - \rho$ whose coefficients depend on the initial conditions of X. ◆◆

Example: The matrix

$$\Phi = \begin{pmatrix} 3 & 0 \\ 0 & 2 \end{pmatrix} \tag{14.7.11}$$

has eigenvalues 3 and 2, m being unity in both cases. Associated with these eigenvalues are the two eigenvectors

$$V_1 \equiv \begin{pmatrix} 1 \\ 0 \end{pmatrix} \qquad V_2 \equiv \begin{pmatrix} 0 \\ 1 \end{pmatrix} \tag{14.7.12}$$

respectively,† and so in each case $\rho = 1$. Hence in both cases $m - \rho = 0$ and so the multiplicative polynomial p_n is a constant, i.e. a typical element in X_n has the form

$$(x_i)_n = a3^n + b2^n \tag{14.7.13}$$

where a and b are constants that depend on the initial conditions. On the other hand the matrix

$$\Phi \equiv \begin{pmatrix} \frac{1}{2} & 1 & 0 \\ 0 & \frac{1}{2} & 0 \\ 0 & 0 & \frac{1}{2} \end{pmatrix} \tag{14.7.14}$$

has the single eigenvalue 1/2 of multiplicity $m = 3$, and has two eigenvectors, namely

$$V_1 \equiv \begin{pmatrix} 1 \\ 0 \\ 0 \end{pmatrix} \qquad V_2 \equiv \begin{pmatrix} 0 \\ 0 \\ 1 \end{pmatrix} \qquad \text{(See Note)} \tag{14.7.15}$$

†In general the *length* of an eigenvector is arbitrary. For definiteness we shall always use the eigenvector with *unit length.*

Note: Because the eigenvalues which gave rise to these two eigenvectors are equal, it follows that *any linear combination* of these two vectors are also eigenvectors.

which means that $\rho = 2$. Hence in this case $m - \rho = 1$ and so we can expect to find an element in X_n of the form

$$(x_i)_n = (a + nb)\left(\frac{1}{2}\right)^n \qquad (14.7.16)$$

◆◆

It is clear then that the location of Φ's eigenvalues, their multiplicities and the number of associated linearly independent eigenvectors are all that we need to know in order to determine the behavior of the elements of X_n as $n \to \infty$. We thus have the following corollaries to Lemma 14.2.

Corollary

If Φ has all of its eigenvalues strictly within the unit circle, then for each element of X_n,

$$\lim_{n\to\infty} (x_i)_n = 0 \qquad (14.7.17)$$

We call such a system an *unconditionally stable system.* ◆◆

Corollary

If Φ has any of its eigenvalues outside the unit circle then† *for at least one element of X_n*

$$\lim_{n\to\infty} \left|(x_i)_n\right| = \infty \qquad (14.7.18)$$

We call such a system *unconditionally unstable.* ◆◆

Neither of the above two cases is of really great interest from a practical standpoint, since as we shall see, the covariance matrix of the former goes to a null matrix *without even making observations,* and the covariance matrix of the latter *can only be made a null matrix if* R is *also going to a null matrix,* i.e. if the observation errors have variances which approach zero as $n \to \infty$.

The only case of real interest then is the situation where some of Φ's eigenvalues lie *on* the unit circle and none lie outside of it. Moreover, as time passes, the terms in X_n associated with any eigenvalues which lie *inside* the unit circle die out exponentially, and so they do not require observing in order to improve our knowledge of them. We need merely

†Barring very exceptional initial conditions.

declare, with perfect certainty, that if we wait long enough, the state of those terms is precisely zero. Hence we shall devote our energy to the case where *all of* Φ*'s eigenvalues lie on the unit circle.* This includes the polynomial models which have all of Φ's eigenvalues at unity (see Ex. 8.29) as well as the "Fourier series" models, and in practice these are the only really useful sets of constant-coefficient linear models.

Example: The matrix

$$\Phi = \begin{pmatrix} 1 & 1 & 1 \\ & 1 & 2 \\ & & 1 \end{pmatrix} \tag{14.7.19}$$

is a well-known polynomial transition matrix (quadratic). Clearly Φ has one eigenvalue, $\lambda = 1$, of multiplicity $m = 3$. It has only the single eigenvector

$$V = \begin{pmatrix} 1 \\ 0 \\ 0 \end{pmatrix} \tag{14.7.20}$$

and so the rank of the space formed by Φ's eigenvectors is $\rho = 1$. Thus Lemma 14.2 shows that there will be an element in X_n of the form

$$\left(x_i\right)_n = p_n \tag{14.7.21}$$

where p_n is a polynomial in n of degree $m - \rho = 2$, i.e. there will be elements in X_n which become unbounded like n^2 as $n \to \infty$. ◆◆

We now begin our proof of the following proposition, *which is the first of the two main results* of this section.

Let Z_n *be given by the recursion* (14.7.9) *where* Φ *has all of its eigenvalues on the unit circle. Then*

$$\lim_{n \to \infty} \left[Z_n(1)\right]^{-1} = \emptyset \tag{14.7.22}$$

if and only if none of the eigenvectors of Φ *is annihilated by* M. (See Note)

Note: Comparison with Theorem 8.2 on p 325 shows that the constraint on M given Φ is identical in both cases.

Thus once Φ is chosen, we can find its eigenvectors, V_1, V_2, etc. In general *these may not span the entire space* and in most cases M is *singular.* Hence our theorem tells us that if we desire $\left[Z_n(1)\right]^{-1} \to \emptyset$ as $n \to \infty$, then we must be sure that for any of the eigenvectors of Φ,

$$MV_k \neq \emptyset \tag{14.7.23}$$

In the event that V_1 and V_2 are eigenvectors associated with the *same* eigenvalue, then this also means that M shall not annihilate any linear combination of these vectors, *since such a linear combination is again an eigenvector of* Φ. (If V_1 and V_2 stem from different eigenvalues, then M is permitted to annihilate a linear combination of them.)

We note from (14.7.7) that

$$\left[Z_n(\beta)\right]^{-1} = P_{n,n}(\beta) \tag{14.7.24}$$

and for $\beta = 1$, $P_{n,n}$ and $S^*_{n,n}$ are equal. Thus the above theorem, which actually concerns $Z_n(1)$, shows how the matrix $S^*_{n,n}$ of the Expanding Bayes and Kalman Filters behaves, as $n \to \infty$. If we keep β in $0 < \beta < 1$, then *the second of our main results* will tell us precisely how $S^*_{n,n}(\beta)$ behaves for the Fading Bayes and Kalman Filters as $n \to \infty$.

Prior to proving the first theorem we present two examples showing its use.

Example 1: Let

$$\Phi(\tau) = \begin{pmatrix} 1 - \tau & \tau \\ -\tau & 1 + \tau \end{pmatrix} \tag{14.7.25}$$

This is a transition matrix (see p. 98) which has one eigenvalue, $\lambda = 1$, of multiplicity $m = 2$ It has the single eigenvector

$$V = \frac{1}{\sqrt{2}} \begin{pmatrix} 1 \\ 1 \end{pmatrix} \tag{14.7.26}$$

and so, according to the theorem, if we wish that $S^*_{n,n} \to \emptyset$ for $\beta = 1$, we must avoid for example, using

$$M = (-1, 1) \tag{14.7.27}$$

since then $MV = \emptyset$.

Now the state-vector for Φ of (14.7.25) is

$$X_n = \begin{pmatrix} x_n + \dot{x}_n \\ x_n + 2\dot{x}_n \end{pmatrix} \tag{14.7.28}$$

We see then that if M is as in (14.7.27), then the observation relation would be

$$\begin{aligned} y_n &= MX_n + \nu_n \\ &= \dot{x}_n + \nu_n \end{aligned} \tag{14.7.29}$$

showing that our observation scheme is such that only $\dot{x}_n$ and not x_n is being observed. Intuitively we know that our knowledge of a first-degree polynomial will never become perfect unless our observations contain information on the zeroth derivative.

On the other hand, by our theorem we could, for example, use

$$M = (1, \ 1) \tag{14.7.30}$$

for in this case $MV \neq \emptyset$, and so we are assured that $S^*_{n,n} \to \emptyset$. Our observations now contain data on the zeroth derivative. ◆◆

Example 2: Let

$$\Phi(\tau) = \begin{pmatrix} \cos\tau & \sin\tau & 0 & 0 \\ -\sin\tau & \cos\tau & 0 & 0 \\ 0 & 0 & 1 & 0 \\ 0 & 0 & 0 & 1 \end{pmatrix} \tag{14.7.31}$$

Assuming that τ is not an integral multiple of π radians, this matrix has *three distinct eigenvalues* $e^{j\tau}$, $e^{-j\tau}$ *and* 1 which give rise to four eigenvectors, two real and two complex. The eigenvectors associated with the eigenvalues $e^{j\tau}$ and $e^{-j\tau}$ are, respectively,

$$V_1 \equiv \frac{1}{\sqrt{2}}\begin{pmatrix} 1 \\ j \\ 0 \\ 0 \end{pmatrix} \qquad V_2 \equiv \frac{1}{\sqrt{2}}\begin{pmatrix} 1 \\ -j \\ 0 \\ 0 \end{pmatrix} \tag{14.7.32}$$

and the eigenvectors associated with the eigenvalue 1 are

$$V_3 \equiv \begin{pmatrix} 0 \\ 0 \\ 1 \\ 0 \end{pmatrix} \qquad V_4 = \begin{pmatrix} 0 \\ 0 \\ 0 \\ 1 \end{pmatrix} \tag{14.7.33}$$

as well as any linear combinations of these vectors. We thus see that M's first two columns must be such that both $MV_1 \neq \emptyset$ and $MV_2 \neq \emptyset$. Moreover the third and fourth columns must have rank two in order that $M(\alpha V_3 + \alpha' V_4) \neq \emptyset$, for any constants α and α'.

The state-vector for Φ of (14.7.31) is

$$X_n = \begin{pmatrix} a \sin t + b \cos t \\ a \cos t - b \sin t \\ c \\ d \end{pmatrix} \tag{14.7.34}$$

where a, b, c and d are constants. The first two elements of X_n are of the form $x(t)$ and $\dot{x}(t)$, and constitute the state-vector of a simple harmonic oscillator. Observing *either* $x(t)$ or $\dot{x}(t)$ will be sufficient to give us an ultimately perfect knowledge of the state of that oscillator. In order to determine c and d we must observe them in two linearly independent ways. In Examples 14.6 and 14.7 we consider the cases where τ is a multiple of π or 2π radians. ♦♦

We now turn our attention to the proof of the first main result, and as a start we prove a lemma concerning the eigenvectors of Φ. Note that these are also the eigenvectors of Φ^{-1} and that the eigenvalues of Φ are the reciprocals of the eigenvalues of Φ^{-1} (see Ex. 14.8).

Lemma 14.3

Let $Z_n(\beta)$ be defined by (14.7.9) where Φ has all of its eigenvalues on the unit circle, and let V be any one of Φ's eigenvectors. Then

$$\lim_{n\to\infty} \overline{V}^T Z_n(1) V = \infty \qquad \text{(See Note)} \tag{14.7.35}$$

if and only if $MV \neq \emptyset$.

Note: By $\overline{V}$ we mean the complex conjugate of V.

Proof

By (14.7.9)

$$\begin{aligned}\overline{V}^T Z_n(1) V &= \overline{V}^T \Phi_T^{-1} Z_{n-1}(1) \Phi^{-1} V + \overline{V}^T M^T R^{-1} M V \\ &= \overline{V}^T Z_{n-1}(1) V + \overline{V}^T M^T R^{-1} M V\end{aligned} \tag{14.7.36}$$

and so it follows that

$$\overline{V}^T Z_n(1) V = \overline{V}^T Z_0(1) V + n\overline{V}^T M^T R^{-1} M V \tag{14.7.37}$$

Hence

$$\lim_{n\to\infty} \overline{V}^T Z_n(1) V = \infty \tag{14.7.38}$$

if and only if

$$\overline{V}^T M^T R^{-1} M V \neq 0 \tag{14.7.39}$$

But this last condition will be true if and only if

$$MV \neq \emptyset \tag{14.7.40}$$

since by assumption R^{-1} is positive definite, and so the lemma is proved. ◆◆

We now apply this lemma to the proof of the following result concerning the eigenvectors of $Z_n(1)$. Note that since $Z_n(1)$ is a real symmetric matrix, it has a set of eigenvectors which span the space completely. Calling them $(U_0)_n$, $(U_1)_n$, ... we assume that they have been orthonormalized so that

$$(U_i)_n^T (U_j)_n = \delta_{ij} \tag{14.7.41}$$

Lemma 14.4

Let $Z_n(\beta)$ be defined by (14.7.9) where Φ has all of its eigenvalues on the unit circle. Then for each of its eigenvectors $(U)_n$,

$$\lim_{n\to\infty} (U)_n^T Z_n(1) (U)_n = \infty \tag{14.7.42}$$

if and only if $MV \neq \emptyset$, where V is any eigenvector of Φ.

Proof

Divide the vectors $(U)_n$ into two mutually exclusive classes C_1 and C_2. Into C_1 we put all those for which (14.7.42) is satisfied, and into C_2 we put all the remainder. Then C_1 and C_2 can be used as bases for two mutually orthogonal spaces which we call S_1 and S_2 respectively.

Consider a vector N which is a *linear combination* of the U's, e.g.

$$N \equiv \alpha_0(U_0)_n + \alpha_1(U_1)_n \tag{14.7.43}$$

Then

$$\begin{aligned}\bar{N}^T Z_n(1) N &= \left[\bar{\alpha}_0(U_0)_n^T + \bar{\alpha}_1(U_1)_n^T\right] Z_n(1)\left[\alpha_0(U_0)_n + \alpha_1(U_1)_n\right] \\ &= \left[\bar{\alpha}_0(U_0)_n^T + \bar{\alpha}_1(U_1)_n^T\right]\left[\alpha_0(\lambda_0 U_0)_n + \alpha_1(\lambda_1 U_1)_n\right]\end{aligned} \tag{14.7.44}$$

where $(\lambda_0)_n$ and $(\lambda_1)_n$ are the eigenvalues of $Z_n(1)$ which gave rise to the eigenvectors $(U_0)_n$ and $(U_1)_n$ respectively. But now by virtue of (14.7.41) the last result in (14.7.44) becomes $|\alpha_0|^2(\lambda_0)_n + |\alpha_1|^2(\lambda_1)_n$ which then enables us to write

$$\bar{N}^T Z_n(1) N = |\alpha_0|^2 (U_0)_n^T Z_n(1)(U_0)_n + |\alpha_1|^2 (U_1)_n^T Z_n(1)(U_1)_n \tag{14.7.45}$$

From this it follows that *any vector N which is in S_1 also satisfies a relation of the form of* (14.7.42) *and if in S_2 then it does not.*

Assume now that $MV \neq \emptyset$. Then we prove that (14.7.42) must hold.

Let N be a vector in S_2. By (14.7.9)

$$\bar{N}^T Z_n(1) N = \bar{N}^T \Phi_T^{-1} Z_{n-1}(1)\Phi^{-1}N + \bar{N}^T GN \tag{14.7.46}$$

and so the first term on the right is equal to the one on the left to within the constant term $\bar{N}^T GN$. But this then means that if N is in S_2 then so is $\Phi^{-1}N$, showing that Φ^{-1} maps the space S_2 onto itself. Then Φ^{-1} must have an eigenvector in S_2 (which is also an eigenvector of Φ).† Call that eigenvector V_0 say. But from Lemma 14.3, if $MV_0 \neq \emptyset$, which we have assumed to be the case, then

†See Ex. 14.10.

$$\lim_{n\to\infty} \bar{V}_0^T Z_n(1) V_0 = \infty \tag{14.7.47}$$

and so V_0 must be in S_1. This is clearly a contradiction, and so S_2 must be a null-space, proving the sufficiency portion of the statement of the lemma.

We now prove the necessity. Thus assume that $MV = \emptyset$, for V an eigenvector of Φ. Let V be expressed in terms of the U's, e.g.

$$V = \alpha_0 (U_0)_n + \alpha_1 (U_1)_n \tag{14.7.48}$$

Then by Lemma 14.3, (14.7.35) does not hold. Now, by (14.7.45) we have

$$\bar{V}^T Z_n(1) V = \left[|\alpha_0|^2 U_0^T Z(1) U_0 + |\alpha_1|^2 U_1^T Z(1) U_1 \right]_n \tag{14.7.49}$$

and since the right-hand side of the sum of two positive terms, if the left does not tend to ∞ as $n \to \infty$, then neither can either term on the right. This completes the proof of the lemma. ◆◆

Corollary

Each of the eigenvalues of $Z_n(1)$ satisfies

$$\lim_{n\to\infty} (\lambda)_n = \infty \tag{14.7.50}$$

if and only if $MV \neq \emptyset$.

Proof follows immediately from (14.7.42). ◆◆

We are now able to prove the following.

Theorem 14.1

For the Expanding Bayes and Kalman Filters, assuming

i) *a constant-coefficient linear model with a transition matrix that has all of its eigenvalues on the unit circle,*
ii) *a constant-coefficient linear observation scheme,*
iii) *a constant input-error covariance matrix,*
iv) *constant inter-sample time,*

$$\lim_{n\to\infty} \overset{\circ}{S}{}^*_{n,n} \to \emptyset \tag{14.7.51}$$

if and only if $MV \neq \emptyset$ where V is any eigenvector of Φ.

Proof

By (14.3.14) we know that $\overset{\circ}{S}^*_{n,n}$ for these filters is equal to $P_{n,n}(\beta)$ of their fading counterparts when $\beta = 1$.

Let Q_n be the orthogonal matrix made up of $P_{n,n}(1)$'s orthonormalized eigenvectors. Then we have

$$P_{n,n}(1) = Q_n^T \Lambda_n Q_n \tag{14.7.52}$$

where Λ_n is the diagonal matrix made up of P's eigenvalues. These are the inverses of the eigenvalues of $Z_n(1)$, and so by (14.7.50)

$$\lim_{n \to \infty} \Lambda_n = \emptyset \tag{14.7.53}$$

if and only if $MV \neq \emptyset$. This completes the proof. ◆◆

We have thus shown that under the assumed conditions, the Bayes and Kalman Filters of Chapters 10 and 12 are stable. The reader can repeat the above analysis for the case where *all of* Φ*'s eigenvalues lie inside the unit circle* and he will see quite readily that $\overset{\circ}{S}^*_{n,n}$ goes to a null-matrix as $n \to \infty$ *without any constraints on* M *whatever.* Similarly he can verify that *if any of* Φ*'s eigenvalues are outside the unit circle,* then for the associated eigenvectors (14.7.35) is no longer true, the right-hand side of that equation being replaced by a *finite number* instead. This then means that (14.7.42) no longer holds, and so $P_{n,n}(1)$ no longer tends to a null matrix. However, if in (14.7.6) the matrix R is replaced by a matrix R' of the form

$$R'_n = \frac{1}{n} R \tag{14.7.54}$$

then the term $\overline{V}^T M^T R^{-1} MV$ in (14.7.36) goes to ∞ as $n \to \infty$. In this case (14.7.35) still holds and so does (14.7.50). We thus require that the observation errors shall go to zero as $n \to \infty$ in order that the covariance matrix $S^*_{n,n}$ go to a null-matrix. (See Examples 14.12 through 14.15.)

We now turn our attention to the case where $0 < \beta < 1$, i.e. the *fading-memory* filters, and we show that $S^*_{n,n}(\beta)$ tends to a well-defined positive definite limit (i.e. *not* a null matrix) whose value is independent of the initial choice used to initialize the procedure. We start by proving the following result.

Lemma 14.5

Let

$$Z_n = \beta \Phi_T^{-1} Z_{n-1} \Phi^{-1} + G \tag{14.7.55}$$

where $0 < \beta < 1$ *and where* Φ *has all of its eigenvalues on the unit circle. Then* $\lim_{n \to \infty} Z_n$ *exists and is equal to the matrix* $Z_\infty(\beta)$ *satisfying*

$$Z_\infty(\beta) = \beta \Phi_T^{-1} Z_\infty(\beta) \Phi^{-1} + G \tag{14.7.56}$$

Proof

The homogeneous part of (14.7.55) is

$$Z_n^H = \beta \Phi_T^{-1} Z_{n-1}^H \Phi^{-1} = \beta^{1/2} \Phi_T^{-1} Z_{n-1}^H \Phi^{-1} \beta^{1/2} \tag{14.7.57}$$

which means that

$$Z_n^H = \left(\beta^{1/2} \Phi_T^{-1}\right)^n Z_0^H \left(\beta^{1/2} \Phi^{-1}\right)^n \tag{14.7.58}$$

Now the eigenvalues of Φ are all on the unit circle, hence so are the eigenvalues of Φ^{-1} and Φ_T^{-1} (see Ex. 14.8a)). Thus the eigenvalues of $\beta^{1/2}\Phi^{-1}$ and $\beta^{1/2}\Phi_T^{-1}$ all have modulus $\beta^{1/2}$ (see Ex. 14.8b)). In that case, for $0 < \beta < 1$,

$$\lim_{n \to \infty} \left(\beta^{1/2} \Phi^{-1}\right)^n = \emptyset = \lim_{n \to \infty} \left(\beta^{1/2} \Phi_T^{-1}\right)^n \tag{14.7.59}$$

(see e.g. [14.1]). By (14.7.57) this now means that

$$\lim_{n \to \infty} Z_n^H = \emptyset \tag{14.7.60}$$

and so the natural modes of (14.7.55) all die out. Its solution then depends only on the forcing function and is independent of initial conditions. But the forcing function is the constant matrix G, and so Z_n tends to a constant matrix. The latter can only be $Z_\infty(\beta)$, if it exists, satisfying (14.7.56).

To show that such a $Z_\infty(\beta)$ does in fact exist for any G, we proceed as follows. First we write (14.7.56) as

$$\beta\Phi_T^{-1}Z_\infty - Z_\infty\Phi = -G\Phi \tag{14.7.61}$$

which is of the form

$$AZ_\infty + Z_\infty B = C \tag{14.7.62}$$

A necessary and sufficient condition that this equation have a solution for any C (see [14.2]) is that

$$\lambda_A + \lambda_B \neq 0 \tag{14.7.63}$$

where λ_A and λ_B are respectively *any* of the eigenvalues of A and B. But in this case, by virtue of our assumption that all of Φ's eigenvalues lie on the unit circle, it follows that all the eigenvalues of A have modulus β (which is less than unity) and all those of B have modulus unity. Thus (14.7.63) is satisfied and so $Z_\infty(\beta)$ exists. This proves the lemma. ◆◆

We have thus established that regardless of the choice of M, the recursion (14.7.9) tends to a unique, well-defined, limiting matrix for any β in $0 < \beta < 1$. We now place a constraint on M, as the next lemma shows.

Lemma 14.6

Let $Z_\infty(\beta)$, Φ and G be defined as before. Then $Z_\infty(\beta)$ is nonsingular for $0 < \beta < 1$ if and only if $MV \neq \emptyset$ where V is any eigenvector of Φ.

Proof

Assume first that $MV \neq \emptyset$ and suppose that $Z_\infty(\beta)$ is singular. Let S_0 be its null-space and let N be a vector in S_0. Then by (14.7.56)

$$\overline{N}^T Z_\infty(\beta) N = \beta\overline{N}^T\Phi_T^{-1}Z_\infty(\beta)\Phi^{-1}N + \overline{N}^T GN \tag{14.7.64}$$

where, by assumption, the left-hand side is zero. Now $Z_\infty(\beta)$ and G are both nonnegative definite, and so both

$$\beta\overline{N}^T\Phi_T^{-1}Z_\infty(\beta)\Phi^{-1}N \geq 0 \tag{14.7.65}$$

and

$$\overline{N}^T GN \geq 0 \tag{14.7.66}$$

Then since the left of (14.7.64) is zero, it follows that

$$\bar{N}^T \Phi_T^{-1} Z_\infty(\beta) \Phi^{-1} N = 0 \tag{14.7.67}$$

$$\bar{N}^T G N = 0 \tag{14.7.68}$$

By (14.7.67) we thus see that if N is in S_0 then $\Phi^{-1}N$ is also in S_0. Hence Φ^{-1} maps S_0 onto itself. But then there is at least one eigenvector of Φ in S_0. Let that vector be V_0. Then using the above arguments, (14.7.68) shows that we must have

$$\bar{V}_0^T M^T R^{-1} M V_0 = 0 \tag{14.7.69}$$

But this contradicts our assumption, namely $MV_0 \neq \emptyset$. Hence $Z_\infty(\beta)$ must be nonsingular.

On the other hand, suppose that $MV = \emptyset$ for one of Φ's eigenvectors. Then (14.7.9) gives us

$$\bar{V}^T Z_n V = \beta \bar{V}^T \Phi_T^{-1} Z_{n-1} \Phi^{-1} V = \beta \bar{V}^T Z_{n-1} V \tag{14.7.70}$$

Hence, since $0 < \beta < 1$, this means that

$$\lim_{n \to \infty} \bar{V}^T Z_n V = \left(\lim_{n \to \infty} \beta^n\right) \bar{V}^T Z_0 V = 0 \tag{14.7.71}$$

which means that $Z_\infty(\beta)$ is singular. This proves the lemma. ♦♦

Corollary

For $0 < \beta < 1$, the matrices $P_{n,n}(\beta)$ and $S^*_{n,n}(\beta)$ tend to well-defined positive definite limiting matrices, independent of their initial values, if and only if $MV \neq \emptyset$.

Proof for $P_{n,n}(\beta)$ follows directly from the above two lemmas, by virtue of (14.7.7). Proof for $S^*_{n,n}(\beta)$ then follows from (14.3.19). ♦♦

Our task is thus completed and we have shown that in the expanding Bayes and Kalman algorithms, the matrix $\mathring{S}^*_{n,n}$ shrinks to a null-matrix if and only if $MV \neq \emptyset$ for all of Φ's eigenvectors. However if the memory is fading ($\beta < 1$) then the shrinkage of both $S^*_{n,n}(\beta)$ and $H_n(\beta)$ in (14.4.29) is halted at some level.

In Figures 14.2 and 14.3, we show the time-histories of the two diagonal elements of $P_{n,n}(\beta)$ for various values of β, for the case where

$$\Phi(\tau) = \begin{pmatrix} 1 & \tau \\ 0 & 1 \end{pmatrix} \tag{14.7.72}$$

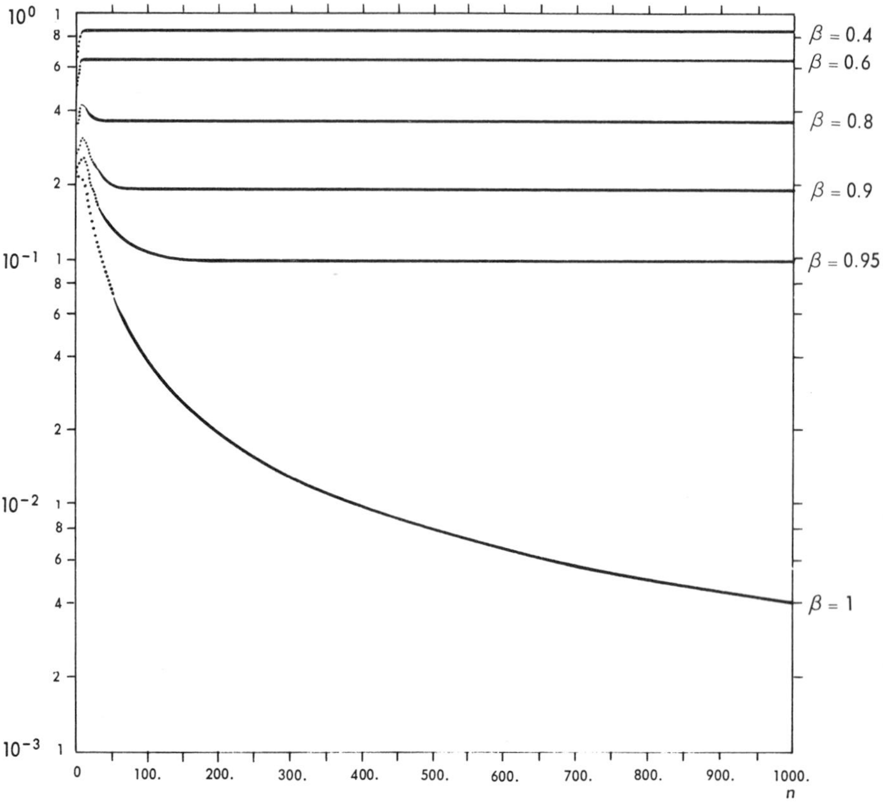

Fig. 14.2 p_{00} *family of curves.*

i.e. a first-degree polynomial, and where

$$M = (1,\ 0) \tag{14.7.73}$$

The case

$$M = (0,\ 1) \tag{14.7.74}$$

was run at the same time as these plots were made, and the element $(p_{0,0})_{n,n}$ became unbounded for $0 < \beta \leq 1$, whereas $(p_{1,1})_{n,n}$ remained bounded for $\beta < 1$, and tended to zero for $\beta = 1$.

It is easily inferred that if Φ is the transition matrix of a polynomial state-vector and if M is such that we are observing only the zeroth derivative of that polynomial, then if $\beta = 1$ and $R_{(n)} = \sigma^2 I$, the filters of this chapter

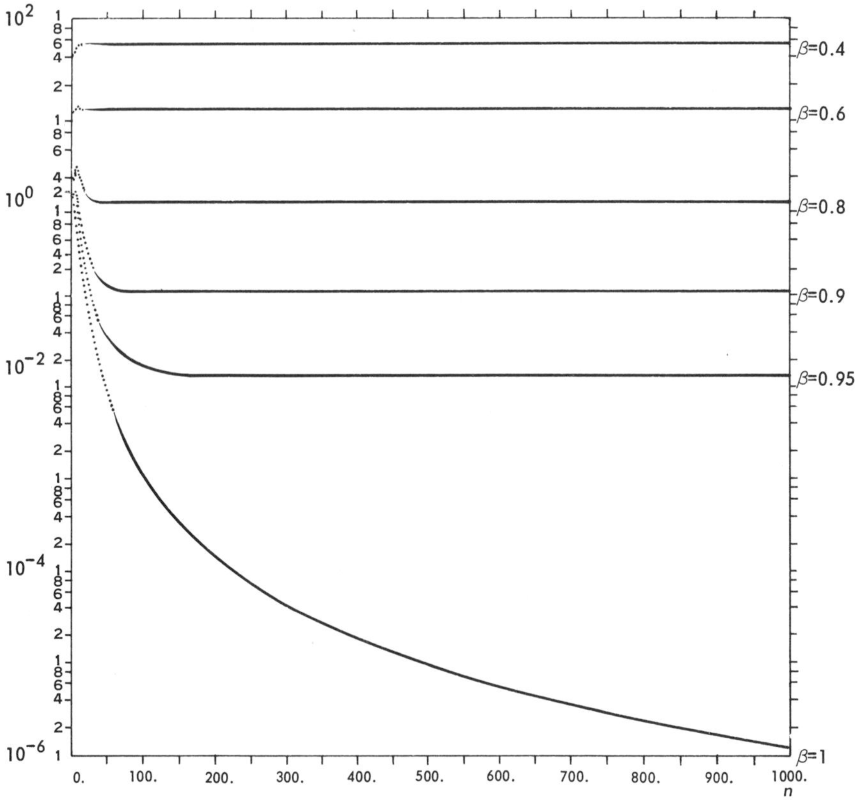

Fig. 14.3 p_{11} *family of curves.*

are essentially the same as the Expanding-Memory Polynomial Filters of Chapter 9. On the other hand when $\beta < 1$, then as $n \to \infty$ the filters of this chapter become the same as the Fading-Memory Polynomial Filters of Chapter 13. Thus the filters of the present chapter are very general and cover, in effect, all of the work of Chapters 9 through 13. In Ex.14.11 we use the present filters to prove (13.5.30).

In conclusion then, we note that precisely the same conditions which, in Chapter 8, were shown to be necessary in order that the Fixed-Memory Filters be operable, are also required to ensure the good behavior of the Expanding-Memory Filters. This is not really surprising, since we would hardly expect an estimate to be obtainable on a given data-base by one method and not by another. We note that the condition between M and Φ of this section is identical to the condition for Theorem 8.2, and the latter was shown to be fully equivalent to Theorem 8.1. *It thus follows that the conditions for* Theorem 8.1 *can be used in place of the condition on* M *and* Φ *used in this section.* Thus, to ensure the good behavior of an Expanding-Memory Filter we must be sure that the matrix G of (8.9.5) has full column-rank.

We have also shown, by the analysis of this section, that if the errors are stage-wise uncorrelated then the covariance matrix of a *Fixed*-Memory Filter can be made arbitrarily close to a null matrix by taking the data-base large enough, if Theorem 8.1 or 8.2 is satisfied.

The time-varying or nonlinear cases are beyond the scope of this book, but it is hoped that the elementary analysis given on the constant-coefficient cases will provide some intuitive basis for an approach to the more general problem. We also refer the reader back to the comments made at the end of Section 10.5 relating to this question.

14.8 DETECTION OF SYSTEMATIC ERRORS

The analysis of the preceding section has verified that when $\beta < 1$, then the covariance matrix $S^*_{n,n}(\beta)$ and the weight matrix $H_n(\beta)$ of the generalized Fading-Memory Filters do not tend to null matrices, but flatten out at some level. The proof was for the constant-coefficient case, but we assert that this behavior is essentially also manifested by time-varying systems whose $S^*_{n,n} \to \emptyset$ when $\beta = 1$. In this section we make some practical suggestions concerning the detection and control of bias errors.

In the normal course of events, *assuming that bias errors are not present,* the term $Y_n - M_n X^*_{n,n-1}$ appearing in (14.4.30) is a vector of *zero-mean* random variables. This follows from the fact that

$$Y_n - M_n X^*_{n,n-1} = N_n - M_n N^*_{n,n-1} \tag{14.8.1}$$

Then, since both N_n and $N^*_{n,n-1}$ are zero-mean (assuming zero-mean observations and an unbiased estimate), it follows that

$$E\left\{Y_n - M_n X^*_{n,n-1}\right\} = \emptyset \tag{14.8.2}$$

It is also clear that if the observation errors are stage-wise uncorrelated, then

$$E\left\{N_n N^{*T}_{n,n-1}\right\} = \emptyset \tag{14.8.3}$$

and so the covariance matrix of $Y_n - M_n X^*_{n,n-1}$ is easily seen to be equal to

$$T_n \equiv R_n + M_n S^*_{n,n-1} M_n^T \tag{14.8.4}$$

Since R_n is given and since $S^*_{n,n-1}$ is being computed in the filter algorithms, we see that T_n can be computed quite simply, if required.†

Assume now that the input errors are multivariate Gaussian. Then, since the output errors are linear transformations on the input errors, it follows (see Section 5.6) that they too are Gaussian and hence so also is the vector $Y_n - M_n X^*_{n,n-1}$. Now, in Section 5.6 we discussed a method, based on the Chi-squared distribution function, whereby we can decide whether a given vector of numbers constitutes a likely or an unlikely draw from an ensemble of Gaussian random vectors whose covariance matrix and mean-vector are given. We can apply that method here.

On each cycle of the Fading Bayes or Kalman Filter, a numerical realization of $Y_n - M_n X^*_{n,n-1}$ is obtained and its covariance matrix T_n can easily be computed. Suppose that we now form the scalar

$$k^2 \equiv (Y_n - M_n X^*_{n,n-1})^T T_n^{-1} (Y_n - M_n X^*_{n,n-1}) \tag{14.8.5}$$

Then as was demonstrated in Ex. 5.16, if the vector $Y_n - M_n X^*_{n,n-1}$ is zero-mean Gaussian, the scalar k^2 has a Chi-squared distribution. From the appropriate Chi-squared tables‡ we can, on each cycle, look-up the cumulative distribution function of the given k^2, namely $P(k^2)$. In the normal course of events, $P(k^2)$ will be a relatively small number, dependent on the matrix R_n. Only on very rare occasions could it become large. Based on experience with the details of the situation, it should thus be possible to place a threshold on $P(k^2)$, or, what amounts to the same thing on k^2 itself, (the look-up operation can thus be dispensed with in practice) and when k^2 exceeds that threshold, *a rare event can be deemed to have occurred.*

Now assume that bias errors begin to develop. Then it is clear that the value of k^2 as computed by (14.8.5) will become inflated, and if the threshold is chosen appropriately then the latter will be exceeded *repeatedly* when the bias attains some value. The threshold being exceeded on isolated instants can perhaps reasonably be attributed to the occurrence of a rare event, *but it is clear that if it is exceeded repeatedly, then a bias error can be assumed to have developed.* Once this decision has been made, β can be temporarily reduced until the threshold crossings cease, signalling that the bias has been eliminated. Then the memory can again be lengthened by bringing β closer to unity.

†Note that (14.8.4) is normally computed in the course of cycling the Expanding Kalman Filter (see (12.2.17)). However in the Fading Kalman Filter it is not normally being computed.

‡i.e. the one with the number of "degrees of freedom" equal to the order of the vector $Y_n - M_n X^*_{n,n-1}$.

In one scheme† the following approach was adopted. First of all it was decided that fading should be on the basis of *count-number* rather than age, i.e. (14.2.12) was minimized rather than (14.2.10). This results in $\beta^{t_n - t_{n-1}}$ in (14.4.28), (14.4.36) and (14.4.38) being replaced by θ (see Ex. 14.16) where $0 < \theta \leq 1$. The value of θ was computed on-line using the algorithm

$$\theta_n = c_1 \exp\left\{-c_2 (Y_n - M_n X^*_{n,n-1})^T T_n^{-1} (Y_n - M_n X^*_{n,n-1})\right\} \tag{14.8.6}$$

where the constants c_1 and c_2 were adjusted experimentally. Note that θ_n depends essentially on the quantity k^2 of (14.8.5), and so this choice for θ results in a fade-rate which depends, in some sense, on the bias. In addition to varying θ by this *on-line* procedure, the time between observations could also be varied. Thus when the presence of bias errors was detected and θ was reduced, the sampling rate was simultaneously increased. This resulted in the memory of the filter being filled very rapidly with new data, which in turn resulted in the rapid disappearance of the bias errors.

14.9 ITERATIVE DIFFERENTIAL-CORRECTION

For completeness we point out that the techniques of this chapter can be applied to situations where either the model or the observation scheme, or both, are nonlinear. The details of the procedure are based on the application of differential-correction, and with obvious modifications to include the fading parameter, the algorithms which result for the Kalman case are given below. They are naturally very similar to those given on p. 481. Thus with *single iteration* differential-correction, the algorithm now becomes the following:

Fading Kalman Filter With Differential-Correction

$X^*_{n,n-1}$ obtained from $X^*_{n-1,n-1}$ by integration of the state-equations (14.9.1)

$$\overline{X}_n \equiv X^*_{n,n-1} \tag{14.9.2}$$

$$P_{n,n-1} = \Phi(t_n, t_{n-1}; \overline{X}) P_{n-1,n-1} \Phi(t_n, t_{n-1}; \overline{X})^T \tag{14.9.3}$$

$$H_n = P_{n,n-1}\left[M(\overline{X}_n)\right]^T \left\{\beta^{t_n - t_{n-1}} R_n + M(\overline{X}_n) P_{n,n-1} \left[M(\overline{X}_n)\right]^T\right\}^{-1} \tag{14.9.4}$$

(continued)

†Devised and tested by P. J. Buxbaum of Bell Telephone Laboratories.

$$X^*_{n,n} = X^*_{n,n-1} + H_n\left[Y_n - G(\overline{X}_n)\right] \quad (14.9.5)$$

$$P_{n,n} = \frac{1}{\beta^{t_n - t_{n-1}}}\left[I - H_n M(\overline{X}_n)\right] P_{n,n-1} \quad (14.9.6)$$

$$S^*_{n,n-1} = \Phi(t_n, t_{n-1}; \overline{X}) S^*_{n-1,n-1} \Phi(t_n, t_{n-1}; \overline{X})^T \quad (14.9.7)$$

$$S^*_{n,n} = \left[I - H_n M(\overline{X}_n)\right] S^*_{n,n-1}\left[I - H_n M(\overline{X}_n)\right]^T + H_n R_n H_n^T \quad (14.9.8)$$

An equivalent algorithm exists in the Bayes Case.

EXERCISES

14.1 Show that (14.2.9) follows from (14.2.7).

14.2 Verify that (14.3.7) is identical to (14.2.10).

14.3 Verify (14.3.17).

14.4 Show that (14.4.22) follows from (14.4.21). Hint: use (14.4.12).

14.5 Apply the inversion lemma† to obtain the Fading-Memory Kalman Filter on p. 569 from the Fading-Memory Bayes algorithm on p. 569.

14.6 a) Assume that in (14.7.31), $\tau = \pi$ seconds. Show that the first two columns of M must have rank two and so must the second two columns, in order that M does not annihilate any linear combination of eigenvectors from the same eigenvalue. Infer that M need only have rank equal to at least 2.

b) Let $\tau = 2\pi$ radians. Show that M must now have rank at least 4.

14.7 a) For the situation in Ex. 14.6 part a), show that (14.7.9) assumes the form

$$\left(\begin{array}{c|c} Z_{11} & Z_{12} \\ \hline Z_{12}^T & Z_{22} \end{array}\right)_n = \beta\left(\begin{array}{c|c} -I & \emptyset \\ \hline \emptyset & I \end{array}\right)\left(\begin{array}{c|c} Z_{11} & Z_{12} \\ \hline Z_{12}^T & Z_{22} \end{array}\right)_{n-1}\left(\begin{array}{c|c} -I & \emptyset \\ \hline \emptyset & I \end{array}\right) + M^T R^{-1} M$$

Let R be the identity matrix and let M be in the form $(M_1 \mid M_2)$. Show that in the limit the above gives

†See (12.2.10) and (12.2.13).

$$\left(\begin{array}{c|c} Z_{11} & Z_{12} \\ \hline Z_{12}^T & Z_{22} \end{array}\right)_\infty = \left(\begin{array}{c|c} \dfrac{1}{1-\beta} M_1{}^T M_1 & \dfrac{1}{1+\beta} M_1{}^T M_2 \\ \hline \dfrac{1}{1+\beta} M_2{}^T M_1 & \dfrac{1}{1-\beta} M_2{}^T M_2 \end{array}\right)$$

Infer that as $\beta \to 1$, for Z_∞ to remain positive definite we require only that M_1 and M_2 have rank 2, and that M need not have rank 4.

b) Show that for Ex. 14.6 part b),

$$Z_\infty = \frac{1}{1-\beta} M^T M$$

and hence verify that for Z_∞ to be positive definite, M must have rank 4.

14.8 a) Starting from the eigenvalue equation $\Phi V = \lambda V$, (Φ nonsingular) show that Φ^{-1} and $\Phi_T{}^{-1}$ have the same eigenvalues and that they are the reciprocals of Φ's eigenvalues. Infer that if all of Φ's eigenvalues are on the unit circle then so are the eigenvalues of Φ^{-1} and $\Phi_T{}^{-1}$.

b) Show that if Φ's eigenvalues lie on the unit circle then the eigenvalues of $\beta^{1/2}\Phi^{-1}$ and $\beta^{1/2}\Phi_T{}^{-1}$ all have modulus $\beta^{1/2}$.

c) Show that if V is an eigenvector of Φ then it is also an eigenvector of Φ^{-1}.

14.9 a) Show that the recursion (14.7.9), for the case where Z, Φ and G are 2×2 and where $\Phi = I$, gives

$$Z_\infty(\beta) = \frac{1}{1-\beta} G$$

Now infer that if $R = I$ and $M = (1,\ 1)$ then

$$Z_\infty(\beta) = \frac{1}{1-\beta}\begin{pmatrix} 1 & 1 \\ 1 & 1 \end{pmatrix}$$

and so $Z_\infty(\beta)$ is singular.

b) Assume instead that

$$\Phi = \begin{pmatrix} 1 & 0 \\ 0 & -1 \end{pmatrix}$$

Show that

$$Z_\infty(\beta) = \frac{1}{1-\beta}\begin{pmatrix} 1 & 0 \\ 0 & 1 \end{pmatrix} + \frac{1}{1+\beta}\begin{pmatrix} 0 & 1 \\ 1 & 0 \end{pmatrix}$$

and verify that $Z_\infty(\beta)$ is positive definite for $0 < \beta < 1$.

c) Repeat a) and b) above, starting from (14.7.9) using $\beta = 1$ before letting $n \to \infty$. Investigate $\lim_{n\to\infty} Z_n(1)$.

14.10 Let S_2 be the space defined in Lemma 14.4 on p. 586. Then if N is in S_2, so is $\Phi^{-1}N$. Thus Φ^{-1} maps S_2 onto itself. Prove that there is at least one eigenvector of Φ^{-1} in S_2.

14.11† a) Let $Z^*_{n,n}$ be the scaled derivative state-vector of (13.5.6) and $\Phi(k)$ the associated transition matrix. Let $M \equiv (1, 0, 0, \ldots, 0)$ and assume that all observations have equal variance, are uncorrelated, and that they are equally spaced in time. Verify that (14.2.12) gives (as $n \to \infty$)

$$e\left(Z^*_{n,n}\right) = \sum_{k=0}^{\infty}\left[y_{n-k} - M\Phi(-k)\,Z^*_{n,n}\right]^2 \theta^k \tag{I}$$

b) Minimize (I) above over $Z^*_{n,n}$ to give

$$Z^*_{n,n} = \left[Q(\theta)\right]^{-1}\sum_{k=0}^{\infty}\left[M\Phi(-k)\right]^T \theta^k y_{n-k} \tag{II}$$

where

$$Q(\theta) \equiv \sum_{k=0}^{\infty}\left[M\Phi(-k)\right]^T M\Phi(-k)\,\theta^k \tag{III}$$

†This example makes use of the general form of the Fading-Memory Filter to prove the result (13.5.30). Note: $Z^*_{n,n}$ in this example is a *vector*, completely unrelated to the matrix Z_n appearing throughout Section 14.7.

and verify that this is the same as would be obtained from (14.4.1), (14.4.2) and (14.4.3) under these circumstances.

c) Infer from (14.4.30) and the above, that (II) can be reformulated as

$$Z^*_{n,n} = Z^*_{n,n-1} + H\epsilon_n \tag{IV}$$

in which

$$\epsilon_n \equiv y_n - (z_0^*)_{n,n-1}$$

and

$$H \equiv [Q(\theta)]^{-1} M^T \tag{V}$$

Verify that (IV) is the same as (13.5.12) and hence infer that (V) is the same as the H-vector of Section 13.5.

d) Prove that (13.5.30) follows from (V) above.

Hint: Follow the technique of Ex. 12.3 but make use instead of the discrete Laguerre polynomials. Show that for P of (13.5.28),

$$\Phi(-k) = P(k)P(0)^{-1}$$

and apply this to (III) of part b) above.

14.12 Assume that Φ has all of its eigenvalues *inside* the unit circle.

a) Let $\beta = 1$ and show that Lemma 14.3 and hence Theorem 14.1 hold *with no constraints whatever* on M. Thus $\mathring{S}^*_{n,n} \to \emptyset$ even if $M = \emptyset$, i.e. without even observing the process.

b) Let λ be the eigenvalue with the largest modulus and V the associated eigenvector. Let $|\lambda|^2 < \beta < 1$. Show that (14.7.9) gives

$$\overline{V}^T Z_n(\beta) V = \frac{\beta}{|\lambda|^2} \overline{V}^T Z_{n-1}(\beta) V + \overline{V}^T G V$$

and hence that $\lim_{n\to\infty} \overline{V}^T Z_n(\beta) V = \infty$ regardless of the choice of M. Infer that $S^*_{n,n} \to \emptyset$ for *any* M.

c) Investigate the case where $\beta \le |\lambda|^2 < 1$

14.13 a) Assume that Φ has at least one of its eigenvalues outside the unit circle. Show that if λ is that eigenvalue and V is the associated eigenvector, then (14.7.9) gives

$$\bar{V}^T Z_n V = \frac{\beta}{|\lambda|^2} \bar{V}^T Z_{n-1} V + \bar{V}^T G_n V$$

where $\beta/|\lambda|^2 < 1$ for $0 < \beta \leq 1$. Note that we are subscripting G with n. Now show that $\lim_{n \to \infty} \bar{V}^T Z_n(1) V < \infty$ if G_n is a constant matrix.

b) Infer that the only way in which $\bar{V}^T Z_n(1) V$ can go to ∞ as $n \to \infty$ is if $\bar{V}^T G_n V \to \infty$ as $n \to \infty$. Investigate what this implies for R_n if M is a constant matrix.

c) Infer that if R_n is constant then $S^*_{n,n}$ cannot go to a null matrix as $n \to \infty$.

14.14 a) Consider the (scalar) differential equation $\dot{x}(t) = -x(t)$. Verify that its transition "matrix" has its (one and only) eigenvalue *inside* the unit circle, namely at e^{-1}.

b) Show that (14.7.6) gives

$$p^{-1}_{n,n} = \beta^\tau e^{2\tau} p^{-1}_{n-1,n-1} + \frac{m^2}{\sigma_\nu^2}$$

and show that if $e^{-2} < \beta \leq 1$ then $p_{n,n}(\beta) \to 0$ as $n \to \infty$; *whether observations are taken or not.* Infer that $s^*_{n,n} \to 0$ with or without observations. (This is a demonstration of the fact that when all of Φ's eigenvalues lie inside the unit circle, $S^*_{n,n} \to \emptyset$ as $n \to \infty$ *whether or not we observe the process.)*

14.15 a) Consider the (scalar) differential equation $\dot{x}(t) = x(t)$. Verify that its transition "matrix" has one eigenvalue lying *outside* the unit circle.

b) Show that (14.7.6) gives

$$p^{-1}_{n,n} = \beta^\tau e^{-2\tau} p^{-1}_{n-1,n-1} + \frac{m^2}{\sigma_\nu^2} \qquad \text{(I)}$$

and hence that (for $m = 1$)

$$\lim_{n \to \infty} p_{n,n} = \sigma_\nu^2 (1 - \beta^\tau e^{-2\tau})$$

Infer that the variance of the Expanding Bayes or Kalman estimate (i.e. $\beta = 1$) does not go to zero as $n \to \infty$.

c) Assume that the variance of the observations goes to zero with n. Specifically let $(\sigma_\nu^2)_n = \frac{\alpha}{n}$ so that (I) now becomes

$$P_{n,n}^{-1} = \beta^\tau e^{-2\tau} P_{n-1,n-1}^{-1} + \frac{n}{\alpha}$$

Show that in this case, for $\beta = 1$,

$$\lim_{n \to \infty} s^*_{n,n} = 0$$

Thus when an eigenvalue of Φ lies outside the unit circle, the covariance matrix of the estimate cannot go to a null matrix if the covariance matrix of the observations is constant.

14.16 Repeat the analysis of Sections 14.3 and 14.4 but starting from (14.2.12) rather than (14.2.10). Show that in the algorithms on pp. 569 and 570, the quantity $\beta^{t_n - t_{n-1}}$ is replaced by θ, with no other changes.

14.17 Rerun the filters programmed for Chapters 10, 11 and 12 but introduce fading, using the methods developed in this chapter. Compare time-histories of the covariance matrices with and without fading. Introduce biases by making the model slightly different from the true process and study the magnitude of the bias errors as the fading parameter is varied.

REFERENCES

1. Varga, R., "Matrix Iterative Analysis," Prentice Hall, New Jersey, 1962, Theorem 1.4, p. 13.
2. Bellman, R., "Introduction to Matrix Analysis," McGraw-Hill, New York, 1960, Theorem 4, p. 231.

15

THE KALMAN FILTER WITH DRIVING-NOISE

15.1 INTRODUCTION

The Fading-Memory algorithms derived in the preceding chapter provide us with recursive schemes which are structurally very similar to the Expanding Bayes and Kalman Filters obtained earlier. As we saw in Chapter 14, the Fading-Memory schemes contain within them a scalar parameter which permits the user to vary the memory fade-rate of those filters. This enables systematic errors, caused by possible model discrepancies, to be balanced against the random errors, and a set of very practical filters is thereby obtained for use under a wide range of circumstances.

However, by virtue of the structural similarities between the Expanding and Fading algorithms, most of the numerical drawbacks of the former are also inherent in the latter. We showed in Section 12.4 for example, that the covariance matrix of the Expanding Kalman estimate can become singular due to numerical roundoff when a vector of highly precise observations is processed. Once singular it will remain singular thereafter. This drawback, and others enumerated in that section, apply also to the Fading Kalman Filter.

We now consider an alternate method of controlling the systematic errors at the expense of increased random errors. While the scheme to be developed will not have the conceptual simplicity of an exponentially fading memory, it does have the very attractive feature that in addition to providing a measure of control over possible bias errors *it also constantly reinforces the positive definiteness of the estimate covariance matrix.* Should this matrix ever become singular it can immediately be made positive definite by the algorithm on the next cycle of the recursion.

15.2 RANDOM FORCING-FUNCTIONS

In Section 12.2 we displayed an algorithm called the Kalman Filter without driving-noise, which serves to estimate the state-vector of a model consisting of a homogeneous linear differential equation. We now extend our choice of models to include linear differential equations of the form

$$\frac{d}{dt} X(t) = A(t) X(t) + D(t) U(t) \tag{15.2.1}$$

where $D(t)$ is a time-varying matrix† and where $U(t)$ is a forcing-function, restricted to being a vector of random variables whose properties will be detailed below. The inclusion of $U(t)$, known as *driving-noise,* will be shown to have certain very beneficial effects on the properties of the resultant filter.

We examine first the form which the solution to (15.2.1) assumes. Consider initially its homogeneous portion

$$\frac{d}{dt} X(t) = A(t) X(t) \tag{15.2.2}$$

where we have temporarily set $U(t) = \emptyset$. As we already know, associated with (15.2.2) is a transition matrix $\Phi(t, t_{n-1})$ such that

$$X(t) = \Phi(t, t_{n-1}) X(t_{n-1}) \tag{15.2.3}$$

†$D(t)$ need not be a square matrix and so $U(t)$ need not necessarily be a vector of the same order as $X(t)$.

The matrix Φ satisfies the differential equation

$$\frac{\partial}{\partial t}\Phi(t, t_{n-1}) = A(t)\Phi(t, t_{n-1}) \tag{15.2.4}$$

with initial conditions

$$\Phi(t_{n-1}, t_{n-1}) = I \tag{15.2.5}$$

This was discussed in detail in Section 4.7 of Chapter 4.

Consider now the vector-function $\widetilde{X}(t)$ defined by

$$\widetilde{X}(t) \equiv \Phi(t, t_{n-1})\widetilde{X}(t_{n-1}) + \int_{t_{n-1}}^{t} \Phi(t,\lambda)D(\lambda)U(\lambda)\,d\lambda \tag{15.2.6}$$

where Φ is the same matrix as before, $D(t)$ is the matrix appearing in (15.2.1), $U(t)$ is a vector of *arbitrary* forcing-functions and λ is a dummy variable of integration. *We show that* $\widetilde{X}(t)$ *so defined, satisfies the forced differential equation* (15.2.1) *for initial conditions* $\widetilde{X}(t_{n-1})$.

By direct differentiation, (15.2.6) gives us

$$\frac{d}{dt}\widetilde{X}(t) = \frac{\partial}{\partial t}\Phi(t, t_{n-1})\widetilde{X}(t_{n-1}) + \frac{\partial}{\partial t}\int_{t_{n-1}}^{t} \Phi(t,\lambda)D(\lambda)U(\lambda)\,d\lambda \tag{15.2.7}$$

and it is shown in texts on advanced calculus that for any function $G(t,\lambda)$,

$$\frac{\partial}{\partial t}\int_{t_{n-1}}^{t} G(t,\lambda)\,d\lambda = G(t,t) + \int_{t_{n-1}}^{t} \frac{\partial}{\partial t}G(t,\lambda)\,d\lambda \tag{15.2.8}$$

provided that $\partial G(t,\lambda)/\partial t$ is continuous. Applying both this and (15.2.4) to (15.2.7) now gives us

$$\frac{d}{dt}\widetilde{X}(t) = A(t)\left[\Phi(t, t_{n-1})\widetilde{X}(t_{n-1}) + \int_{t_{n-1}}^{t} \Phi(t,\lambda)D(\lambda)U(\lambda)\,d\lambda\right] + D(t)U(t) \tag{15.2.9}$$

But by (15.2.6) the term post-multiplying $A(t)$ in the above equation is just $\tilde{X}(t)$. Hence finally

$$\frac{d}{dt}\tilde{X}(t) = A(t)\tilde{X}(t) + D(t)U(t) \tag{15.2.10}$$

showing that (15.2.6) *is, in fact, the solution of the linear forced system* (15.2.1) *under consideration.*

We now define

$$V(t, t_{n-1}) \equiv \int_{t_{n-1}}^{t} \Phi(t,\lambda)\, D(\lambda)\, U(\lambda)\, d\lambda \tag{15.2.11}$$

Then (15.2.6) can be written (after deleting the tildes),

$$X(t_n) = \Phi(t_n, t_{n-1})\, X(t_{n-1}) + V(t_n, t_{n-1}) \tag{15.2.12}$$

This difference equation is thus the discretized counterpart of (15.2.1) and will be regarded, equivalently, as defining the process-model when a forcing-function is present.

At this stage we specify the properties of the driving-vector $U(t)$. We have seen earlier that when the observational errors are uncorrelated from time t_n to t_m $(n \neq m)$, then the linear minimum-variance estimator, as derived in Chapter 6, could in a very natural way be reformulated to give the recursive form of the Expanding-Memory Kalman Filter. The fact that those errors were stage-wise uncorrelated gave rise to a block-diagonal total covariance matrix, and it was this property which permitted us to arrive at the recursive form of the algorithms.

Thus, letting N_n and N_m be the errors in the observation vectors Y_n and Y_m, assumed to be two inputs to a Bayes or Kalman filter, we required that they satisfy

$$E\{N_n\} = \emptyset \tag{15.2.13}$$

$$E\{N_n N_m^T\} = R_n \delta_{n,m} \tag{15.2.14}$$

where $\delta_{n,m}$ is the Kronecker delta defined by

$$\delta_{n,m} = \begin{cases} 0 & n \neq m \\ 1 & n = m \end{cases} \tag{15.2.15}$$

We now choose the driving-vector $U(t)$ to be a vector of random variables which satisfy the *continuous* counterpart of the discrete conditions on N_n above. Thus we define $U(t)$ to be a random vector such that

$$E\{U(t)\} = \emptyset \tag{15.2.16}$$

$$E\{U(t)U(t')^T\} = K(t)\delta(t - t') \tag{15.2.17}$$

where $K(t)$ is a given nonnegative definite matrix function of t, and where $\delta(t - t')$ is the Dirac delta function, characterized by

$$\delta(t - t') = 0 \qquad t' \neq t \tag{15.2.18}$$

$$\int_a^b \delta(t - t')\,dt = 1 \qquad a < t' < b \tag{15.2.19}$$

The above definition characterizes $U(t)$ as being a possibly nonstationary, white random process, i.e. completely uncorrelated with itself when observed at two discrete time instants, no matter how close. Of course such a process exists only in the ideal sense, but then so does the vector N_n of (15.2.13) and (15.2.14) if we let t_m approach arbitrarily close to t_n.

Note that $\delta(t - t')$ is undefined for $t' = t$, and it is generally conceived of as a limiting process, as for example the limit of

$$f(t - t') \equiv \frac{k}{\sqrt{\pi}} \exp\left[-k^2(t - t')^2\right] \tag{15.2.20}$$

as k becomes larger without bound. For this function, it can be verified that in the limit as $k \to \infty$, both (15.2.18) and (15.2.19) are satisfied. In particular, as $k \to \infty$, $\delta(t - t')$ so defined is unbounded at $t = t'$, and so it is not possible to talk about the covariance matrix of $U(t)$, as defined by (15.2.17), when $t' = t$. Thus despite the fact that, by assumption, $K(t)$ contains bounded elements, the matrix $E\{U(\lambda)U(\lambda)^T\}$ will not be defined. However, as we shall see, this idealized concept of a white-noise forcing-function, defined precisely as above, does in fact lead to well-defined results in the end.

We close this section by deriving the covariance matrix of $V(t, t_{n-1})$ of (15.2.11). Thus letting $Q(t, t_{n-1})$ be that matrix, we have by definition,

$$
\begin{aligned}
Q(t, t_{n-1}) &\equiv E\left\{V(t, t_{n-1})V(t, t_{n-1})^T\right\} \\
&= E\left\{\int_{t_{n-1}}^{t} \Phi(t, \lambda)\, D(\lambda)\, U(\lambda)\, d\lambda \int_{t_{n-1}}^{t} U(\gamma)^T D(\gamma)^T \Phi(t, \gamma)^T\, d\gamma\right\} \\
&= \int_{t_{n-1}}^{t} \int_{t_{n-1}}^{t} \Phi(t, \lambda)\, D(\lambda)\, E\left\{U(\lambda)\, U(\gamma)^T\right\} D(\gamma)^T\, \Phi(t, \gamma)^T\, d\gamma d\lambda \\
&= \int_{t_{n-1}}^{t} \int_{t_{n-1}}^{t} \Phi(t, \lambda)\, D(\lambda)\, K(\lambda)\, D(\gamma)^T\, \Phi(t, \gamma)^T\, \delta(\lambda - \gamma)\, d\gamma d\lambda
\end{aligned}
\tag{15.2.21}
$$

Now it is a consequence of (15.2.18) and (15.2.19) that if $g(t)$ is any continuous function of t and if t' is in the interval (a, b), then

$$
\int_{a}^{b} g(t)\, \delta(t - t')\, dt = g(t') \tag{15.2.22}
$$

If t' is not in the interval (a, b) then this integral is zero. Clearly in the final line of (15.2.21), since λ is somewhere in the interval of integration of γ, it follows that

$$
Q(t, t_{n-1}) = \int_{t_{n-1}}^{t} \Phi(t, \lambda)\, D(\lambda)\, K(\lambda)\, D(\lambda)^T\, \Phi(t, \lambda)^T\, d\lambda \tag{15.2.23}
$$

which is a well-defined matrix function of t given the matrices K, D and Φ. Then since $K(t)$ is nonnegative definite by assumption for all t, it follows quite readily that $Q(t, t_{n-1})$ is also a nonnegative definite matrix. In particular, if $K(t)$ is positive definite and $D(t)$ has full row-rank, then $Q(t, t_{n-1})$ is positive definite. We also note that by (15.2.11) and (15.2.16), $V(t, t_{n-1})$ is a vector of zero-mean random variables.

It is important to recognize that the state-vector $X(t_n)$ of (15.2.12) *is now no longer a deterministic function of time*, by virtue of the *presence* of the random vector $V(t_n, t_{n-1})$. Thus the solutions to (15.2.12) are of the nature of a random-walk in which the present state is a linear combination of its previous value and a random vector. Clearly if V is not too large, $X(t_n)$ will be strongly related to its immediate predecessors, but its dependence on those vectors becomes increasingly weaker as they recede into the past. *The assumption of* (15.2.12) *as our model thus means that we can expect the memory of our filters to fade out, and that the contribution of observations to the current estimate can be expected to diminish as the staleness of those observations increases.*

15.3 ESTIMATING IN THE PRESENCE OF DRIVING-NOISE

All of the estimation schemes which we have developed in the earlier chapters were based on the assumption that the quantities which we were estimating were *deterministic.* The situation is now changed, since with driving-noise present the quantities in question become *random variables.* As a result our techniques must also be modified appropriately.

From a historical standpoint, our methods thus far have been essentially applications and extensions of the work of Gauss and other Nineteenth century mathematicians. A notable contribution in the present century was made by Swerling who, in 1958, introduced the recursive approach (see [11.3]) on which Chapters 10, 11 and 12 were based.

The rapid expansion which took place in the present century in communications and electronics, motivated largely by the development programs of World War II, led to a new class of problems in which research workers were faced with a need to design *electrical filters* which could estimate signals in the presence of noise. Neither the classical methods of estimation theory nor existing filter-design techniques based on the Fourier integral were adequate, and so an entirely new approach was developed, independently by Kolmogorov in 1941 [15.1] and Wiener in 1942 [15.2],† *in which both signal and noise were treated as random variables.* The criterion on which their work was based is known as the *least mean-squared error criterion* which requires that the *ensemble expectation of the squared difference between the true value and the estimate be minimized.* This was a departure from either the classical method of least-squares or the weighted least-squares and minimum-variance techniques already in existence, as discussed in Chapter 6.

In 1960 and 1961 an extension was made to the Kolmogorov-Wiener theory by Kalman and Bucy ([15.7] and [15.8]) who addressed themselves

†See also [15.3] through [15.6].

to the *estimation of random variables which satisfy a linear differential equation driven by white-noise.* This is precisely the problem which we are now considering, and in what follows we present a very brief summary of [15.7] and merely state its results. We shall then show that *when driving-noise is absent,* the results of [15.7] can be precisely reconciled with the algorithms presented in Chapter 12 which were based on minimum-variance. Thus in spite of the use of a somewhat different criterion, namely that of the least mean-squared error, the results of [15.7]are completely consistent with those which we have obtained in the preceding chapters based on least-squares and minimum-variance. *However the presence of driving-noise invalidates our earlier techniques, and so the discussion which we are about to present constitutes a definite extension.* Only in the special case of no driving-noise do the algorithms become identical.

The estimation problem under consideration is stated in [15.7] as follows.†

a. Let $\mathsf{Y}_{(n)} \equiv (y_0, y_1, \ldots, y_n)^T$ be a vector of random variables and let x_n be a further random variable, all of which share the joint probability density function $p(x_n, \mathsf{Y}_{(n)})$.

b. Let $x^*_{n,n}$ be an estimate of x_n. Then the error in the estimate is $\epsilon \equiv x^*_{n,n} - x_n$.

c. The scalar function $L(\epsilon)$ is termed a *loss-function* if it satisfies

$$\begin{aligned} &\text{i)}\ L(0) = 0 \\ &\text{ii)}\ L(\epsilon') \geq L(\epsilon'') \geq 0 \quad \text{if} \quad \epsilon' \geq \epsilon'' \geq 0 \\ &\text{iii)}\ L(\epsilon) = L(-\epsilon)\,^{\ddagger} \end{aligned} \tag{15.3.1}$$

d. Then the *optimal estimator,* relative to a chosen loss-function, is the one (if it exists) that minimizes $E\{L(\epsilon)\}$, i.e. the expected loss.

Theorem 1 of [15.7] presents the following very powerful result concerning the optimal estimate:

Let the conditional density function for x_n given sample values of $\mathsf{Y}_{(n)}$ be $p(x_n \mid \mathsf{Y}_{(n)})$, and assume that it is

a) *unimodal*

b) *symmetric about its conditional expectation* $E\left\{x_n \mid \mathsf{Y}_{(n)}\right\}$.

Then the optimal estimate of x_n based on the vector of observations $\mathsf{Y}_{(n)}$ with respect to a chosen loss-function $L(\epsilon)$, is precisely

†We restrict our comments, for simplicity, to the case where the quantity being estimated is a *scalar.* The work in [15.7] considers *vectors.*

‡For example $L(\epsilon) \equiv \epsilon^2$ and $L(\epsilon) \equiv |\epsilon|$ are both loss-functions.

$$x^*_{n,n} \equiv E\{x_n \mid Y_{(n)}\} \tag{15.3.2}$$

Thus the conditional expectation of x_n given the observations is the optimal estimate. Since this conditional expectation depends only on the form of $p(x_n \mid Y_{(n)})$ and does not depend on the choice for $L(\epsilon)$, it follows that subject to a) and b) above, *every loss-function* $L(\epsilon)$ *is minimized by the conditional expectation of* x_n. The Gaussian conditional density function is again Gaussian and so it satisfies a) and b). We note that, in general $E\{x_n \mid Y_{(n)}\}$ is a nonlinear function of $Y_{(n)}$ which may be very difficult to compute. In the event that we are interested only in the *quadratic* loss-function,† then conditions a) and b) can be relaxed. It is now only necessary that $p(x_n \mid Y_{(n)})$ have a finite second moment in order that the optimal estimate with respect to this particular loss-function be given by $E\{x_n \mid Y_{(n)}\}$.

The concept of an *orthogonal projection* as related to random variables is now briefly examined. (This idea follows by virtue of a one-for-one analogy with the theory of linear vector spaces.)

Thus, let λ_i and λ_j be two random variables. If λ_i is not a constant multiple of λ_j then, in vector terminology, we say that they are *linearly independent.* Any further random variable

$$\lambda \equiv \alpha_i \lambda_i + \alpha_j \lambda_j \tag{15.3.3}$$

which is a linear combination of λ_i and λ_j is now said to *lie in the two-dimensional space defined by* λ_i *and* λ_j.

We form a basis for this space as follows (c/f Gram-Schmidt orthogonalization): Let

$$e_i \equiv \lambda_i \tag{15.3.4}$$

$$e_j \equiv \lambda_j - \frac{E\{\lambda_i \lambda_j\}}{E\{\lambda_i^2\}} \lambda_i \tag{15.3.5}$$

Then clearly

$$E\{e_i e_j\} = 0 \qquad i \neq j \tag{15.3.6}$$

This last equation constitutes an *orthogonality condition,* in which the familiar *inner-product* of linear algebra is now paralleled by the *expectation* of the given product.

†i.e., $L(\epsilon) \equiv \epsilon^2$.

Finally if we were to divide e_i and e_j above by their respective standard deviations then they would have "unit length," *and would form an orthonormal basis for the space defined by* λ_i *and* λ_j, i.e. the resulting e_i and e_j would satisfy

$$E\{e_i e_j\} = \delta_{ij} \tag{15.3.7}$$

where δ_{ij} is the Kronecker delta.

Suppose next that β is *any* random variable, not necessarily a linear combination of λ_i and λ_j. Then the *orthogonal projection of* β *on the* λ_i, λ_j *space* is defined as

$$\overline{\beta} \equiv e_i E\{\beta e_i\} + e_j E\{\beta e_j\} \tag{15.3.8}$$

We define

$$\widetilde{\beta} \equiv \beta - \overline{\beta} \tag{15.3.9}$$

Then as is readily verified

$$E\{\widetilde{\beta} e_i\} = 0 = E\{\widetilde{\beta} e_j\} \tag{15.3.10}$$

i.e. $\widetilde{\beta}$ is seen to be orthogonal to the λ_i, λ_j space. In this manner β can be partitioned into a sum of $\overline{\beta}$ and $\widetilde{\beta}$, where the former is entirely in the λ_i, λ_j space and the latter is orthogonal to that space.

The above ideas can be readily generalized to cover the orthogonal projection of vectors of random variables onto arbitrarily dimensioned spaces. Theorem 2 of [15.7] points out the following:

Assume that the random variables $x_n, y_0, y_1, \ldots, y_n$ *are all zero-mean and let a sample value be given for* $Y_{(n)} \equiv (y_0, y_1, \ldots, y_n)^T$. *If either*

i) *the random variables in* x_n *and* $Y_{(n)}$ *are jointly Gaussian, or else if*

ii) *the estimator is restricted to being linear and* $L(\epsilon) \equiv \epsilon^2$,

then the optimal estimate of x_n *given* $Y_{(n)}$ *is equal to the orthogonal projection of* x_n *on the space defined by* $Y_{(n)}$.

Since the orthogonal projection onto $Y_{(n)}$ is a linear combination of the elements of $Y_{(n)}$, it thus follows that in the Gaussian case the optimal estimator is a unique linear transformation on the observations. However as we have already noted, in general the optimal estimate, i.e. the conditional expectation of x_n given $Y_{(n)}$, does not result in a linear estimator. But in the class of *linear estimators* the orthogonal projection will always minimize the *expected quadratic loss* $E\{\epsilon^2\}$. (See Ex. 15.1)

Consider now the linear dynamic system defined by

$$\frac{d}{dt} X(t) = A(t)X(t) + D(t)U(t) \tag{15.3.11}$$

together with the observation relation

$$Y(t) = M(t)X(t) \tag{15.3.12}$$

Note that the above differential equation is basically an unforced linear time-varying system as considered throughout this book, but has here the additional forcing-vector $U(t)$ acting through the transformation matrix $D(t)$. The vector $U(t)$ is the white-noise vector which we discussed earlier in this chapter.

We have shown that the above differential equation can be discretized to give the difference equation

$$X(t_{n+1}) = \Phi(t_{n+1}, t_n)X(t_n) + V(t_{n+1}, t_n) \tag{15.3.13}$$

in which Φ is the transition matrix of the *unforced* system and the vector $V(t_{n+1}, t_n)$ arises out of $U(t)$ acting over the time-span $t_n \leq t \leq t_{n+1}$. *We note that the observation relation* (15.3.12) *does not contain an error term.* This results in the algorithm as derived in [15.7] being apparently different from what we have called the Kalman Filter in Chapter 12. The main estimation problem can now be stated as follows:

Given the above linear dynamic model with Gaussian statistics, and given sample values of $Y_0, Y_1, \ldots, Y_n$, *find the prediction estimate* $X^*_{n+1,n}$ *of* X_{n+1} *which minimizes* $E\{L(\epsilon)\}$.

The solution, by virtue of Theorem 2 of [15.7] is just the orthogonal projection† of X_{n+1} onto the space

$$Y_{(n)} \equiv \begin{pmatrix} Y_n \\ \hline Y_{n-1} \\ \hline \vdots \\ \hline Y_0 \end{pmatrix} \tag{15.3.14}$$

†As was pointed out in ii) of the statement Theorem 2 given earlier, the orthogonal projection is also the optimal estimate with respect to the quadratic loss-function for any type of noise statistics, provided only that the conditional expectation $E\{x \mid Y\}$ has a finite second moment.

and as is shown in that reference this is given *recursively* by the algorithm:

$$\Delta_n^* = \Phi(n+1,n)P_n^* M_n^T \left(M_n P_n^* M_n^T\right)^{-1} \tag{15.3.15}$$

$$\Phi^*(n+1,n) = \Phi(n+1,n) - \Delta_n^* M_n \tag{15.3.16}$$

$$X_{n+1,n}^* = \Phi_{n+1,n}^* X_{n,n-1}^* + \Delta_n^* Y_n \tag{15.3.17}$$

$$P_{n+1}^* = \Phi^*(n+1,n)P_n^* \Phi^*(n+1,n)^T + Q_{n+1,n} \tag{15.3.18}$$

(In the above four equations we have essentially retained the nomenclature of [15.7]. The matrix M_n is the observation matrix and $\Phi(n+1,n)$ is the transition matrix of the unforced system. The matrix $Q_{n+1,n}$ is the covariance matrix of the vector $V(t_{n+1}, t_n)$ and the matrix P_{n+1}^* is the covariance matrix of the errors in the estimate $X_{n+1,n}^*$.)

These then are the equations of the Kalman Filter with driving-noise. On the surface they appear to differ very strongly from those displayed on p. 465, but as we now show a very close similarity does exist.

15.4 THE KALMAN FILTER WITH AND WITHOUT DRIVING-NOISE

In the preceding section we displayed the recursive algorithm which estimates the state-vector of a linear differential equation driven by white-noise, based on the least mean-squared error criterion. We now show that by an appropriate rearrangement and redefinition of the quantities involved, that algorithm can be rewritten in a form very similar to the Kalman Filter without driving-noise, obtained in Chapter 12.

Thus consider the model

$$X_{n+1} = \Phi(n+1,n)X_n + V_{n+1,n} \tag{15.4.1}$$

and the observation relation

$$Y_n = M_n X_n + N_n \tag{15.4.2}$$

Note the presence of a driving-noise term in the former, *as well as* an error-vector in the latter. This error-vector is not present in (15.3.12). Define the augmented state-vector

$$X_n' \equiv \begin{pmatrix} X_n \\ -- \\ N_n \end{pmatrix} \tag{15.4.3}$$

and the augmented driving-vector

$$V'_{n+1,n} \equiv \begin{pmatrix} V_{n+1,n} \\ \hline N_{n+1} \end{pmatrix} \tag{15.4.4}$$

Also define the augmented transition matrix

$$\Phi'(n+1,n) \equiv \left(\begin{array}{c|c} \Phi(n+1,n) & \emptyset \\ \hline \emptyset & \emptyset \end{array}\right) \tag{15.4.5}$$

and the augmented observation matrix

$$M'_n \equiv \left(M_n \,\middle|\, I\right) \tag{15.4.6}$$

Then it is readily verified (see Ex. 15.2) that (15.4.1) can be written as

$$X'_{n+1} = \Phi'(n+1,n)\, X'_n + V'_{n+1,n} \tag{15.4.7}$$

and that (15.4.2) can be written as

$$Y_n = M'_n X'_n \tag{15.4.8}$$

which are seen to be in precisely the same form as (15.3.13) and (15.3.12) respectively.

Assuming next that the vectors $V_{n+1,n}$ and N_n of (15.4.1) and (15.4.2) are uncorrelated, it is immediately obvious that V' in (15.4.4) has the covariance matrix

$$Q'_{n+1,n} = \left(\begin{array}{c|c} Q_{n+1,n} & \\ \hline & R_{n+1} \end{array}\right) \tag{15.4.9}$$

where $Q_{n+1,n}$ is the covariance matrix of $V_{n+1,n}$ (see (15.2.23)) and of course R_{n+1} is the covariance matrix of N_{n+1}.

It is now a relatively simple matter, using the definitions given above, to replace the *unprimed* quantities appearing in (15.3.15) through (15.3.18) with their *primed* counterparts appearing in (15.4.3) through (15.4.6). This is carried out in a sequence of three exercises (see Ex. 15.3 through 15.5) and provides us with the following algorithm:

Kalman Filter with Driving-Noise

$$X^*_{n,n-1} = \Phi(n, n-1)X^*_{n-1,n-1} \tag{15.4.10}$$

$$S^*_{n,n-1} = \Phi(n, n-1)S^*_{n-1,n-1}\Phi(n, n-1)^T + Q_{n,n-1} \tag{15.4.11}$$

$$H_n = S^*_{n,n-1}M_n^T\left(R_n + M_nS^*_{n,n-1}M_n^T\right)^{-1} \tag{15.4.12}$$

$$S^*_{n,n} = (I - H_nM_n)S^*_{n,n-1} \tag{15.4.13}$$

$$X^*_{n,n} = X^*_{n,n-1} + H_n\left(Y_n - M_nX^*_{n,n-1}\right) \tag{15.4.14}$$

15.5 PRACTICAL ASPECTS

The Kalman Filter *with* driving-noise is strikingly similar to the Kalman Filter *without* driving-noise, as we see by comparing the algorithm given above to the one on page 465. The only difference lies in the presence of the matrix Q in (15.4.11).

This similarity is of course not coincidental, despite the radically different ways in which the two algorithms were obtained. On the contrary, we would have been concerned if they had not been so similar, because by multiplying the vector $V_{n+1,n}$ in (15.4.1) by a scalar, say α, *we see that the cases with and without driving-noise can be made to be as close to one another as we please simply by taking α sufficiently small.* In the limit if α is set to zero then the two cases are identical. Now it is a simple matter to see that if $V_{n+1,n}$ is (15.4.1) is multiplied by α, then $Q_{n+1,n}$ in (15.4.11) becomes multiplied by α^2 and so, as $\alpha \to 0$, (15.4.11) becomes identical to (12.2.16), which is precisely what we should expect.

When driving-noise *is* present, then the approach used to obtain the algorithm of Chapter 12 can no longer be applied. It is thus clear that the material in this chapter constitutes a definite extension of the earlier work on which Chapter 12 was based, and it is reassuring to see that by two such radically different paths, algorithms are obtained which are consistent with one another and which merge into each other precisely as required.

The presence of driving-noise gives rise, as we mentioned earlier, to a process with a fading memory. Thus if V were absent in (15.4.1), then the state X_n would be completely defined once X_{n-1} was given. The memory of the process is, in this case, perfect. However when V is present, then X_n is not determined solely by X_{n-1} *but depends on each particular realization of the random-vector V.* The vector X_n is now a random vector whose value, while related to X_{n-1}, is no longer completely known unless V is given, *and in practice we do not know V itself but only its statistics.* Hence if V is "small" then the dependence of X_n on X_{n-1} is strong (i.e. the memory is strong) whereas if V is "large" then the dependence of X_n on X_{n-1} is weak (which implies a rapidly fading memory).

The above heuristic argument is borne out very well by the presence of Q in (15.4.11). When the driving-noise is "small" then Q is a matrix with small elements and so $S^*_{n,n-1}$ is very close to the value given in (12.2.16). In this case the estimation errors propagate mainly because of the prediction along the trajectory. However when V is "large" then Q in (15.4.11) makes a sizeable contribution to $S^*_{n,n-1}$, showing that the basic uncertainties in the estimate have now been strongly increased.

We have seen in Chapter 14 that if we arbitrarily enlarge $S^*_{n,n-1}$ before combining $X^*_{n,n-1}$ with Y_n by the rule derived in Chapter 10 (see (10.4.13)), then the resultant estimate will be made dependent on Y_n to a greater extent and on $X^*_{n,n-1}$ to a lesser extent. This change of stress which accompanies the enlargement of $S^*_{n,n-1}$ is, after all, what led to the *fading* aspect of the memory which the filters of Chapter 14 possess. It is thus also clear then, that the enlargement of $S^*_{n,n-1}$ by Q in (15.4.11) will result in a reduced dependence of $X^*_{n,n}$ on $X^*_{n,n-1}$ in (15.4.14), and an increased dependence on Y_n in that equation.

In this way the filter which includes driving-noise in its model is seen to have a fading memory. Removing Q from the algorithm eliminates all fading entirely, and enlarging Q hastens the rate at which fading occurs.

It is not as easy to discuss the *rate* at which fading occurs in the present filters as it was in Chapter 14. We recall that in that case the stress-factor which led to the fade was a decaying exponential, and so the fade-rate of the filter could be specified very accurately. It is perhaps in this ability to control the fade-rate by the choice of a *single scalar,* where the greatest power of the filters of Chapter 14 lies. However in the present case the fading mechanism is much more subtle, and in practice what is usually done is to select Q to be a diagonal matrix whose entries are in some sense (if possible) related to the uncertainties in the model. Thus, recalling that Q is the covariance matrix of the driving vector V in (15.4.1), we see that the entries of Q would in this case be the variances of the random perturbations in the model.

The matrix Q is often also introduced for purely practical reasons *regardless of whether or not the presence of driving-noise can justifiably be presumed.* There are two reasons for this.

In the event that the selected model is known to differ from the true process, then the introduction of the matrix Q can be used to control bias-error build-up. This follows directly from the fact that its presence results in a fading memory. In practice the effect of introducing Q is to provide an inhibiting influence on the descent of $S^*_{n,n}$ to a null-matrix which in turn results in the bias errors being reduced at the expense of an increase in the random ones.

The second and perhaps greatest benefit which the introduction of the matrix Q brings about, lies in the following. In Chapter 12 we pointed out that the covariance matrix $S^*_{n,n}$ could, under certain circumstances, become precisely or nearly singular. This in turn might easily cause it to cease being positive definite.

However, as we see from (15.4.11), *the matrix Q can be made to reinforce the positive-definiteness of $S^*_{n,n-1}$ particularly if we take it to be a diagonal matrix.*† In this case, if at any time $S^*_{n,n}$ becomes singular, then on the following cycle of the filter $S^*_{n,n-1}$ will again become positive definite, thereby breaking the mechanism which we saw in Chapter 12 keeps $S^*_{n,n}$ singular should it ever become singular. It is because of these reasons that the matrix Q, with appropriately selected values, is usually introduced. Note also that Q can be *varied dynamically* if the user so desires. For example an on-line algorithm could easily be devised which varies Q's entries when bias errors are detected by the method outlined in Section 14.8.

In conclusion we point out that the techniques of this chapter can also be applied to nonlinear iterative differential-correction. This would be done by the inclusion of a matrix Q in (12.5.20).

Our treatment of the Kalman Filter with driving-noise has, of necessity, been extremely brief and superficial. Strictly speaking it is beyond the scope of this book and we merely touched on it for completeness. There is a rapidly growing literature related to this filter and it will have to be left to the reader to pursue the matter further by himself. (See references both at the end of this chapter and at the end of Chapter 12.) It is hoped however that the stated objectives of this book have been met and that the reader has been introduced, in an orderly way, to the field of sequential smoothing and prediction.

EXERCISES

15.1 Assume that we have two correlated zero-mean Gaussian random variables x and y.

a) Starting from the joint Gaussian density function for x and y, i.e.

†With positive entries of course.

$$p(x, y) = \frac{1}{2\pi \, |R|^{1/2}} \exp\left\{-\frac{1}{2} V^T R^{-1} V\right\}$$

where

$$V \equiv \begin{pmatrix} x \\ y \end{pmatrix} \quad \text{and} \quad R \equiv \begin{pmatrix} \sigma_x^{\,2} & E\{xy\} \\ E\{xy\} & \sigma_y^{\,2} \end{pmatrix}$$

verify that

$$p(x \mid y) = \frac{\sigma_y}{\sqrt{2\pi} \, |R|^{1/2}} \exp\left\{-\frac{1}{2}\left(V^T R^{-1} V - \frac{y^2}{\sigma_y^{\,2}}\right)\right\}$$

Now show that $E\{x \mid y\}$ is

$$\bar{x} = y \frac{E\{xy\}}{\sigma_y^{\,2}} \tag{I}$$

Thus by the theorem on p. 610, (I) above is the optimal estimate of x given the observation y.

b) By the theorem on p. 612, the optimal estimate of x given y is also equal to the orthogonal projection of x on y. Using (15.3.8) show that

$$\bar{x} = y \frac{E\{xy\}}{\sigma_y^{\,2}}$$

which is the same as the algorithm obtained in (I) above.

15.2 Verify that (15.4.7) and (15.4.8) are equivalent to (15.4.1) and (15.4.2).

15.3 a) Show that (15.3.15) and (15.3.16) in (15.3.18) gives us

$$P^*_{n+1} = \Phi(n+1, n)\left[P^*_n - P^*_n M_n^{\,T}\left(M_n P^*_n M_n^{\,T}\right)^{-1} M_n P^*_n\right]\Phi(n+1, n)^T + Q_{n+1,n} \quad \text{(I)}$$

b) The matrix P_n^* is the covariance matrix of the errors in the prediction $X_{n,n-1}^*$. Since the latter was based solely on observations up to t_{n-1}, its errors are uncorrelated with those in Y_n. Letting $P_n^{*\prime}$ be the covariance matrix of the vector

$$\begin{pmatrix} X_{n,n-1}^* \\ \hline N_n \end{pmatrix}$$

infer that

$$P_n^{*\prime} = \left(\begin{array}{c|c} S_{n,n-1}^* & \emptyset \\ \hline \emptyset & R_n \end{array}\right) \tag{II}$$

15.4 Replace all matrices in (I) of Ex. 15.3 by their primed counterparts, i.e. Φ' of (15.4.5), M' of (15.4.6), $P^{*\prime}$ of Ex. 25.3 and Q' of (15.4.9). Show that this results in the equation

$$\left(\begin{array}{c|c} S_{n+1,n}^* & \emptyset \\ \hline \emptyset & R_{n+1} \end{array}\right) = \left(\begin{array}{c|c} \Phi(n+1,n)S_{n,n}^*\Phi(n+1,n)^T + Q_{n+1,n} & \emptyset \\ \hline \emptyset & R_{n+1} \end{array}\right) \tag{I}$$

in which

$$S_{n,n}^* = S_{n,n-1}^* - S_{n,n-1}^* M_n^T \left(R_n + M_n S_{n,n-1}^* M_n^T\right)^{-1} M_n S_{n,n-1}^* \tag{II}$$

We have thus derived the rules for computing $S_{n+1,n}^*$ and $S_{n,n}^*$. Note the presence of $Q_{n+1,n}$ in the algorithm for $S_{n+1,n}^*$.

15.5 a) Using the primed matrices Φ', M', and $P^{*\prime}$ discussed in Ex. 15.4 show that (15.3.15) reduces to

$$\Delta_n^* = \left(\frac{\Phi(n+1,n) S_{n,n-1}^* M_n^T \left(R_n + M_n S_{n,n-1}^* M_n^T\right)^{-1}}{\emptyset} \right)$$

b) Combine (15.3.16) and (15.3.17) to give

$$X_{n+1,n}^* = \Phi(n+1,n) X_{n,n-1}^* + \Delta_n^* (Y_n - M_n X_{n,n-1}^*)$$

Now replace Φ and X by their primed equivalents (for X' see (15.4.3)), and for Δ_n^* use the result of part a) above. Show that this gives us

$$X_{n,n}^* = X_{n,n-1}^* + H_n (Y_n - M_n X_{n,n-1}^* - N_n^*) \qquad \text{(I)}$$

$$N_{n+1}^* = \emptyset \qquad \text{(II)}$$

in which H_n is defined in (12.2.17). The vector N_{n+1}^* is the prediction of what N_{n+1} will be, based on data up to t_n. The algorithm thus predicts a null vector for the observation errors, which is precisely as it should be. Note that (II) above also implies that N_n^* in (I) is a null vector. We have thus reduced the algorithm on p. 614 to the form given on p. 616.

REFERENCES

1. Kolmogorov, A. N., "Interpolation and Extrapolation of Stationary Random Sequences," (First published 1941 in the USSR Science Academy Bulletin). Translation by W. Doyle and J. Selin, RM-3090-PR, 1962, Rand Corp., Santa Monica, Calif.
2. Wiener, N., "The Extrapolation, Interpolation, and Smoothing of Stationary Time Series," OSRD 370, Report to the Services 19, Research Project DIC-6037, MIT, February 1942.
   See also: Book by the same title, John Wiley and Sons, 1949.
3. Levinson, N., "A Heuristic Exposition of Wiener's Mathematical Theory of Prediction and Filtering," J. Math. Phys., 26 (1947) 110-119.
4. Darlington, S., "Linear Least-Squares, Smoothing and Prediction, with Applications," Bell System Tech. Journal, 37 (1958) 1221-1294.
5. Lee, Y. W., "Statistical Theory of Communication," John Wiley and Sons, 1960.
6. Davenport, W. J., Jr., and Root, W. L., "Random Signals and Noise," McGraw-Hill, 1957.

7. Kalman, R. E., "New Methods and Results in Linear Prediction and Filtering Theory," Journal of Basic Engineering, Trans. ASME 82D (1960) 33-45.
8. Kalman, R. E., and Bucy, R. S., "New Results in Linear Filtering and Prediction Theory," Journal of Basic Engineering, Trans. ASME 83D (1961) 95-108.

   See also
9. Swerling, P., "Topics in Generalized Least-Squares Signal Estimation," SIAM Journal, 14-5 Sept. 1966, 998-1031.
10. Deutsch, R., "Estimation Theory," Prentice-Hall, 1965.

APPENDIX I

PROPERTIES OF THE DISCRETE LEGENDRE POLYNOMIALS

The discrete Legendre polynomials satisfy the recursion

$$\frac{d^i}{dx^i} p_j(x) = (A_j x + B_j) \frac{d^i}{dx^i} p_{j-1}(x) - C_j \frac{d^i}{dx^i} p_{j-2}(x) + iA_j \frac{d^{i-1}}{dx^{i-1}} p_{j-1}(x) \tag{A1.1}$$

where

$$p_j(x) \equiv p(x; j, L) \tag{A1.2}$$

and where

$$A_j = -\frac{2(2j-1)}{j(L-j+1)} \tag{A1.3}$$

$$B_j = \frac{L(2j-1)}{j(L-j+1)} \tag{A1.4}$$

$$C_j = \frac{(j-1)(L+j)}{j(L-j+1)} \tag{A1.5}$$

To get the recursion started we use (3.2.21), i.e.,

$$p_0(x) = 1 \qquad p_1(x) = 1 - 2\frac{x}{L} \tag{A1.6}$$

This recursion is extremely easy to implement on a computer, and can be used to tabulate $(d^i/dx^i)\,p_j(x)$ for any values of i and j, once x and L have been assigned numerical values. In this way the matrix P defined in (7.4.4) can be computed.

The fact that these polynomials satisfy the above recursion relation is not unique to them alone, but is a direct consequence of their orthogonality property. All sets of orthogonal polynomials satisfy such a recursion, with A_j, B_j and C_j defined appropriately.

We now prove the following relationship, from which (7.4.13) can be inferred.

Let ∇ be the backward-difference operator with respect to x. Then

$$\nabla^i p_j(x)\Big|_{x=L} = (-1)^j \binom{j+i}{j} \frac{j^{(i)}}{L^{(i)}} \tag{A1.7}$$

Proof

By definition

$$p_j(x) \equiv \sum_{\nu=0}^{j} (-1)^\nu \binom{j}{\nu}\binom{j+\nu}{\nu} \frac{x^{(\nu)}}{L^{(\nu)}} \tag{A1.8}$$

and so, by (2.4.17)

$$\nabla^i p_j(x) = \sum_{\nu=0}^{j} (-1)^\nu \binom{j}{\nu}\binom{j+\nu}{\nu} \nu^{(i)} \frac{(x-i)^{(\nu-i)}}{L^{(\nu)}} \tag{A1.9}$$

Now

$$\left.\frac{(x-i)^{(\nu-i)}}{L^{(\nu)}}\right|_{x=L} = \frac{(L-i)^{(\nu-i)}}{L^{(i)}(L-i)^{(\nu-i)}} = \frac{1}{L^{(i)}} \tag{A1.10}$$

Hence (A1.9) becomes

$$\left.\nabla^i p_j(x)\right|_{x=L} = \sum_{\nu=0}^{j} (-1)^{\nu}\binom{j}{\nu}\binom{j+\nu}{\nu}\frac{\nu^{(i)}}{L^{(i)}} \tag{A1.11}$$

Next, by (2.4.20)

$$(j+\nu)^{(j)} = (-1)^j(-\nu-1)^{(j)} \tag{A1.12}$$

and so

$$\binom{j+\nu}{\nu} = \binom{j+\nu}{j} = \frac{(j+\nu)^{(j)}}{j!} = (-1)^j\frac{(-\nu-1)^{(j)}}{j!} \tag{A1.13}$$

Similarly

$$\nu^{(i)} = (-1)^i(i-\nu-1)^{(i)} \tag{A1.14}$$

Thus

$$\binom{j+\nu}{\nu}\nu^{(i)} = (-1)^{j+i}\frac{(i-\nu-1)^{(j+i)}}{j!} \tag{A1.15}$$

Hence (A1.11) can be written

$$\left.\nabla^i p_j(x)\right|_{x=L} = \frac{(-1)^{j+i}}{j!\,L^{(i)}}\sum_{\nu=0}^{j}(-1)^{\nu}\binom{j}{\nu}(i-1-\nu)^{(j+i)} \tag{A1.16}$$

The summand is in the form of (2.5.22). Thus we obtain

$$\nabla^i p_j(x)\Big|_{x=L} = \frac{(-1)^{j+i}}{j!\,L^{(i)}} \nabla^j (n-1)^{(j+i)}\Big|_{n=i} \qquad \text{(See Note)} \tag{A1.17}$$

and so by (2.4.17)

$$\nabla^i p_j(x)\Big|_{x=L} = \frac{(-1)^{j+i}}{j!\,L^{(i)}} (j+i)^{(j)} (i-1-j)^{(i)} = (-1)^j \binom{j+i}{j} \frac{j^{(i)}}{L^{(i)}} \tag{A1.18}$$

This completes the proof. ◆◆

We now turn our attention to the proof of (7.12.8). First, we state the following: Let

$$a_j = \sum_{\nu=0}^{j} \binom{j+\nu}{j-\nu} b_\nu \tag{A1.19}$$

Then

$$b_j = \sum_{\nu=0}^{j} (-1)^{j+\nu} \left(\frac{2\nu+1}{2j+1}\right) \binom{2j+1}{j-\nu} a_\nu \tag{A1.20}$$

This theorem is one of many such inversion pairs due to J. Riordan of Bell Telephone Laboratories.† The proof is rather lengthy and we omit it.

Consider now

$$p_j(x) \equiv \sum_{\nu=0}^{j} (-1)^\nu \binom{j}{\nu} \binom{j+\nu}{\nu} \frac{x^{(\nu)}}{L^{(\nu)}} \tag{A1.21}$$

But by the use of the A and B transformations defined by (2.6.6) and (2.6.7) respectively,

$$\binom{j+\nu}{j}\binom{j}{\nu} = \binom{j+\nu}{j}\binom{j}{j-\nu} = \binom{j+\nu}{j-\nu}\binom{2\nu}{\nu} \tag{A1.22}$$

Note: The ∇^j operates on the n.

†Communicated to the author in an unpublished memorandum.

and so

$$p_j(x) = \sum_{\nu=0}^{j} \binom{j+\nu}{j-\nu} \left[(-1)^{\nu} \binom{2\nu}{\nu} \frac{x^{(\nu)}}{L^{(\nu)}} \right] \tag{A1.23}$$

This is of the same form as (A1.19). Thus it follows by (A1.20) that

$$(-1)^j \binom{2j}{j} \frac{x^{(j)}}{L^{(j)}} = \sum_{\nu=0}^{j} (-1)^{j+\nu} \left(\frac{2\nu + 1}{2j + 1} \right) \binom{2j+1}{j-\nu} p_{\nu}(x) \tag{A1.24}$$

We have thus proved that

$$x^{(j)} = \frac{L^{(j)}}{\binom{2j}{j}} \sum_{\nu=0}^{j} (-1)^{\nu} \left(\frac{2\nu + 1}{2j + 1} \right) \binom{2j+1}{j-\nu} p_{\nu}(x) \tag{A1.25}$$

This is the *general inverse relation* for the discrete Legendre polynomials.

◆◆

As an example, letting $j = 2$, (A1.25) gives us

$$\begin{aligned} x^{(2)} &= \frac{L^{(2)}}{\binom{4}{2}} \left[\frac{1}{5} \binom{5}{2} p_0(x) - \frac{3}{5} \binom{5}{1} p_1(x) + \frac{5}{5} \binom{5}{0} p_2(x) \right] \\ &= L^{(2)} \left[\frac{1}{3} p_0(x) - \frac{1}{2} p_1(x) + \frac{1}{6} p_2(x) \right] \end{aligned} \tag{A1.26}$$

◆◆

APPENDIX II

PROOF OF EQUATION (13.3.8)

Given that (c/f (13.2.15))

$$\left(\beta_j\right)_n \equiv \sum_{k=0}^{\infty} y_{n-k}\, \varphi_j(k)\, \theta^k \tag{A2.1}$$

we must show that (c/f (13.3.8))

$$(1 - q\theta)^{j+1} \left(\beta_j\right)_n = K_j \theta^j (1 - q)^j y_n \tag{A2.2}$$

Proof

Let f_n be a discrete function of n for which

$$f_n \equiv 0 \qquad n < 0 \tag{A2.3}$$

Then we define the Z-transform† of f_n by

$$Z(f_n) \equiv \sum_{n=0}^{\infty} f_n z^{-n} \tag{A2.4}$$

where z is a complex-variable. (This is the discrete counterpart of the Laplace transform of a function $f(t)$ which satisfies

$$f(t) \equiv 0 \qquad t < 0 \tag{A2.5}$$

namely,

$$L[f(t)] = \int_0^{\infty} f(t)\, e^{-st}\, dt \tag{A2.6}$$

Just as $L[f(t)]$ is a function of the complex-variable s, so $Z(f_n)$ is a function of the complex-variable z.)

We apply (A2.4) to both sides of (A2.1) i.e., we form

$$Z\left[(\beta_j)_n\right] \equiv Z\left[\sum_{k=0}^{\infty} y_{n-k}\, \varphi_j(k)\, \theta^k\right] \tag{A2.7}$$

The right-hand side of the above gives us

$$\begin{aligned}\sum_{n=0}^{\infty}\left[\sum_{k=0}^{\infty} y_{n-k}\, \varphi_j(k)\, \theta^k\right] z^{-n} &= \sum_{n=0}^{\infty}\sum_{k=0}^{\infty} y_{n-k}\, z^{-(n-k)}\, \varphi_j(k)\, \theta^k z^{-k} \\ &= \left[\sum_{k=0}^{\infty} \varphi_j(k)\, \theta^k z^{-k}\right]\left(\sum_{n=0}^{\infty} y_{n-k}\, z^{-(n-k)}\right)\end{aligned} \tag{A2.8}$$

Consider the two terms in parentheses in the final line of the above equation. By definition

$$\sum_{k=0}^{\infty} \varphi_j(k)\, \theta^k z^{-k} \equiv Z\left[\varphi_j(n)\, \theta^n\right] \tag{A2.9}$$

†See e.g. [13.9].

Moreover, assuming that

$$y_n \equiv 0 \qquad n < 0 \tag{A2.10}$$

we see that

$$\sum_{n=0}^{\infty} y_{n-k} z^{-(n-k)} \equiv Z(y_n) \tag{A2.11}$$

We have thus shown that (A2.1) is equivalent to

$$Z\left[\left(\beta_j\right)_n\right] = Z\left[\varphi_j(n)\,\theta^n\right] Z(y_n) \tag{A2.12}$$

(This is the counterpart of the fact that the Laplace transform of a convolution product equals the product of the Laplace transforms.)

We now analyze (A2.9) above, in further detail. By (13.2.8) and (13.2.2),

$$\begin{aligned} Z\left[\varphi_j(n)\,\theta^n\right] &\equiv \sum_{n=0}^{\infty} K_j\theta^j \sum_{\nu=0}^{j} (-1)^\nu \left(\frac{1-\theta}{\theta}\right)^\nu \binom{j}{\nu}\binom{n}{\nu} \theta^n z^{-n} \\ &= K_j\theta^j \sum_{\nu=0}^{j} (-1)^\nu \left(\frac{1-\theta}{\theta}\right)^\nu \binom{j}{\nu} \sum_{n=0}^{\infty} \binom{n}{\nu} \theta^n z^{-n} \end{aligned} \tag{A2.13}$$

But, as is readily verified

$$\sum_{n=0}^{\infty} \binom{n}{\nu} \theta^n z^{-n} = \frac{\theta^\nu z^{-\nu}}{(1-\theta z^{-1})^{\nu+1}} \qquad \text{(See Note)} \tag{A2.14}$$

and so (A2.13) gives us

†Note: We assume that $|\theta z^{-1}| < 1$, thereby formally assuring convergence.

$$Z\left[\varphi_j(n)\,\theta^n\right] = K_j\theta^j \sum_{\nu=0}^{j} (-1)^\nu \left(\frac{1-\theta}{\theta}\right)^\nu \binom{j}{\nu} \frac{\theta^\nu z^{-\nu}}{(1-\theta z^{-1})^{\nu+1}}$$

$$= \frac{K_j\theta^j}{1-\theta z^{-1}} \sum_{\nu=0}^{j} (-1)^\nu \binom{j}{\nu} \left[\frac{(1-\theta)\,z^{-1}}{1-\theta z^{-1}}\right]^\nu$$

$$= \frac{K_j\theta^j}{1-\theta z^{-1}} \left[1 - \frac{(1-\theta)\,z^{-1}}{1-\theta z^{-1}}\right]^j \tag{A2.15}$$

which reduces to

$$Z\left[\varphi_j(n)\,\theta^n\right] = K_j\theta^j \frac{(1-z^{-1})^j}{(1-\theta z^{-1})^{j+1}} \tag{A2.16}$$

Finally by inserting the above result into (A2.12) we obtain

$$(1-\theta z^{-1})^{j+1} Z\left[(\beta_j)_n\right] = K_j\theta^j (1-z^{-1})^j Z(y_n) \tag{A2.17}$$

Suppose next that we were to seek the Z-transform of $\left(\sum_{j=0}^{m} \alpha_j q^j\right) f_n$, i.e. of f_n operated on by a polynomial in q with constant coefficients.† Then we would obtain, by the definition of Z,

$$Z\left(\sum_{j=0}^{m} \alpha_j q^j f_n\right) \equiv \sum_{n=0}^{\infty}\left(\sum_{j=0}^{m} \alpha_j q^j f_n\right) z^{-n} = \sum_{n=0}^{\infty}\left(\sum_{j=0}^{m} \alpha_j f_{n-j}\right) z^{-n} \tag{A2.18}$$

$$= \sum_{j=0}^{m} \alpha_j z^{-j} \sum_{n=0}^{\infty} f_{n-j} z^{-(n-j)}$$

and so, by virtue of (A2.3),

$$Z\left(\sum_{j=0}^{m} \alpha_j q^j f_n\right) = \left(\sum_{j=0}^{m} \alpha_j z^{-j}\right) Z(f_n) \tag{A2.19}$$

†q is the backward-shifting operator.

This shows that the operators Z and $\sum_{j=0}^{m} \alpha_j q^j$ can be commuted if we change q^j into z^{-j}. In the reverse case, given $\sum_{j=0}^{m} \alpha_j z^{-j} Z(f_n)$, we see then that we can also write it as $Z\left(\sum_{j=0}^{m} \alpha_j q^j f_n\right)$.

Applying the above result to both sides of (A2.17) now gives us

$$Z\left[(1 - \theta q)^{j+1}\left(\beta_j\right)_n\right] = Z\left[K_j \theta^j (1 - q)^j y_n\right] \qquad \text{(A2.20)}$$

and so, finally, it must also be true that

$$(1 - q\theta)^{j+1}\left(\beta_j\right)_n = K_j \theta^j (1 - q)^j y_n \qquad \text{(See Note)} \qquad \text{(A2.21)}$$

This completes the proof. ◆◆

Note: In going from (A2.20) to (A2.21), we are assuming a *unique inverse relation* between $Z(f_n)$ and f_n. Just as with the Laplace transform in which $L[f(t)]$ and $f(t)$ are uniquely related by an inversion integral, subject to (A2.5), so are $Z(f_n)$ and f_n subject to (A2.3).

INDEX

INDEX